DEPARTMENT OF COMMERCE

BULLETIN

OF THE

UNITED STATES
BUREAU OF FISHERIES

VOL. XLIII
1927

IN TWO PARTS—PART I

HENRY O'MALLEY
COMMISSIONER

THE BLACKBURN PRESS

Reprint of First Edition, Published 1928 by the U.S. Department of Commerce

Fishes of Chesapeake Bay
Bulletin of the United Stated Bureau of Fisheries
Vol. XLIII 1927 Part I

ISBN-10: 1-930665-74-1
ISBN-13: 978-1-930665-74-3

Library of Congress Control Number: 2002107349

THE BLACKBURN PRESS
P. O. Box 287
Caldwell, New Jersey 07006 U.S.A.
973-228-7077
www.BlackburnPress.com

ERRATA

Page 1, eighth line from the bottom of the page: *Sphyrinidæ* should read *Sphyrnidæ*.

Page 47, first line: *Carcharpinus* should read *Carcharhinus*.
Page 49, Family VI: *Sphyrinidæ* should read *Sphyrnidæ*.
Page 169: Figure 87 is wrong side up.

FISHES OF CHESAPEAKE BAY

✧

By SAMUEL F. HILDEBRAND, *Director, U. S. Fisheries Biological Station, Beaufort, N. C.,*
and WILLIAM C. SCHROEDER, *Assistant Aquatic Biologist, U. S. Bureau of Fisheries*

✧

CONTENTS

NEW GENERA AND SPECIES

GENERA

SPECIES

49826—28——2

INTRODUCTION

The field work that resulted in the present report was begun in 1912 when Lewis Radcliffe and the late William W. Welsh undertook a study of the anadromous clupeoids principally on the Potomac River and at the head of Chesapeake Bay. These studies were continued more or less intermittently until the winter of 1914–15, when the Fisheries steamer *Fish Hawk* was assigned to this work and the scope was enlarged to include a general biological and physical examination of Chesapeake Bay. This work, which was then under the supervision of Lewis Radcliffe, was interrupted by the World War. It was resumed in 1920 under the immediate supervision of Dr. R. P. Cowles, of Johns Hopkins University. In 1921 the general survey was supplemented by a special investigation of the fishes of Chesapeake Bay by the authors of the present report and was continued at intervals until the fall of 1922, when all the field operations pertaining to the Chesapeake Bay investigations were brought to a close.

Collections of fishes were made during the general survey, and especially many young fish were taken. The operations of the general survey were almost wholly in offshore waters and particularly in the "deep holes." These collections were supplemented by the special survey, chiefly with collections made in the shallow inshore waters. Much attention was given to the spawning and feeding habits of fishes, also to migrations, seasonal abundance, etc. Special attention was directed to the methods employed in the fisheries, manner of handling and marketing the catches, prices received by the fishermen, wholesale dealers, and retailers, etc.

Scientific descriptions and keys, made as nontechnical as is consistent with the purpose of the work in hand, have been introduced, all based upon specimens, so far as available, collected in Chesapeake Bay. For the species of which no specimens were at hand, the source of the account given is stated. An attempt was made in the descriptions drawn up directly from specimens always to discuss the various characters commonly described in the same sequence. It is hoped that this arrangement will prove to be a convenience to those who may have occasion to use the descriptions.

Preceding each description, and following the scientific name of the species, are one or more common names. Those that are of more or less local use only are placed inside quotation marks. Next follow certain references to literature. The first of these gives the exact name used by the discoverer of the species and a sufficiently complete reference to the work in which the species was described and also the type locality for the species. Then follow references to the local fauna and to the general work by Jordan and Evermann—namely, Bulletin No. 47 of the U. S. National Museum. For all references except the first one only the date of publication and the page number or numbers on which the particular species is discussed are given Complete titles to the works referred to are found in the bibliography (pp. 358–366).

In the matter that follows the descriptions, the subheads mentioned below are discussed without naming them in the text in the sequence in which they are listed here.

(a) A brief statement of the number and range in size of the specimens upon which the description was based.

(*b*) A mention of the chief diagnostic characters, naming those, so far as possible, that are readily noticed in the field.

(*c*) Variations among individuals; variations with age; also sexual differences.

(*d*) Food and feeding habits.

(*e*) Spawning, embryology, larval development.

(*f*) Rate of growth.

(*g*) Relative and seasonal abundance in Chesapeake Bay; how taken.

(*h*) Commercial importance.

(*i*) Size attained.

(*j*) Habitat—i. e., general range of distribution.

(*k*) Previous Chesapeake Bay records.

(*l*) Specimens in collection; individuals observed in the field; where, when, and how taken.

It is understood, of course, that for many species nothing is known relative to some of the subheads, and in others they do not apply. In such cases the subject or subjects are not mentioned or are passed over with the remark that little or nothing is known about them.

The scope of the work was fixed arbitrarily to include all fishes taken in the salt water of the bay as well as those taken in the mouths of streams, where the water was brackish to only slightly brackish. This arbitrary division resulted in bringing several species of "fresh-water" fishes within the limits of this report. Species not taken during the present investigation, but previously recorded from the bay or reliably reported by fishermen, also have been included.

In the arrangement of the orders and families Dr. David Starr Jordan's recent work, "A Classification of Fishes," has been followed. Jordan's "Genera of Fishes," too, has been consulted freely.

The collection of the specimens and data and the preparation of the report have extended over a long time, and so many persons have helped at one time or another to further the work that it will be impossible to give a complete list of all who have made contributions of one kind or another. The authors are particularly grateful to the former officers of the Bureau of Fisheries—namely, Dr. Hugh M. Smith, former commissioner, Dr. H. F. Moore, former deputy commissioner, and Dr. R. E. Coker, formerly in charge of scientific research, as well as to the officers succeeding them in the same positions. These gentlemen, of course, made the undertaking possible, have rendered advice and encouragement, and have been patient with us, as the preparation of the report (the writers claim because of other duties) appeared to progress very slowly.

The work was undertaken originally by Messrs. Radcliffe and Welsh, as stated elsewhere. We have had the collections and the notes of these workers, of which we have made use freely. Mr. Radcliffe had already prepared an indexed card catalogue of the various species of fishes known from the vicinity of Maryland and Virginia when the work of preparing the report was assigned to us, and this catalogue has been of great convenience. During the later stages of the work we also received specimens and helpful data from Dr. R. P. Cowles, of Johns Hopkins University.

We are especially indebted to the Buchanan brothers—John, Roland, and Richard—of Norfolk, Va., who allowed us full freedom of their fishery at James Siding, as well as their unusually complete records of catches made since 1908. These records are of great value in indicating the trend of the fishery with respect to species commonly taken in pound nets. Tables and graphs have been prepared from these records and they appear elsewhere in this report. Thanks also are due to the Parkerson brothers, of Ocean View, for permission to take specimens from their 1,800-foot haul-seine catches and for records of the fish taken at their fishery during the autumn of 1922. We wish to acknowledge, too, the courtesy of Messrs. E. E. Bennett and H. W. Bennett, of Bennett's North Carolina Line, Norfolk, Va., in allowing us the use of their warehouse for storing equipment. Thanks are due the fishermen of Chesapeake Bay generally for their interest in this work and for their helpfulness in giving information and in securing specimens.

We wish to thank Dr. Edward Linton, of Augusta, Ga., for examining the contents of a large number of stomachs of various species of fishes. Much valuable assistance also was rendered by Thomas K. Chamberlain, now director of the United States Fisheries Biological Station at Fairport, Iowa, and by Isaac Ginsburg and Irving L. Towers, junior aquatic biologists with the Bureau of Fisheries. Mr. Chamberlain assisted us in arranging the collection and notes in order to make both readily accessible. Mr. Ginsburg made many of the preliminary identifications of specimens, as well as a large part of the measurements and scale and fin-ray counts, etc., used in the descriptions. Mr. Towers examined stomach contents, assisted in the preparation of many of the tables included in the report, and made the final drafts of nearly all of the graphs and several of the drawings of fishes appearing in the report.

LITERATURE ON FISHES OF CHESAPEAKE BAY

The most comprehensive work on the fishes of Chesapeake Bay is the List of Fish of Maryland, by P. R. Uhler and Otto Lugger, published by the Maryland Fish Commission in 1876 in the report of the commissioner of fisheries to the governor, on pages 83 to 208, and dated January 1, 1876. The second edition of the list appeared the same year, in a reprint, with few alterations, of the same report. The list in the reprint occurs on page 69 to 176. This work, however, is much more than a "list" of fishes of Maryland, for a description (often very inadequate) for every species is offered, together with a brief synonomy, common names, and notes on occurrence, abundance, habits, etc. Nor do the authors confine themselves merely to the fishes of Maryland. "A Catalogue of the Fishes of Maryland and Virginia" would have been a much more appropriate title for this work. This catalogue was supplemented in 1877 by Otto Lugger, through the addition of 29 species, and again in 1878 with 10 species.

Shorter lists, with notes on the fishes from various sections of Chesapeake Bay, were prepared by the following authors: Tarlton H. Bean, 1883; Barton A. Bean, 1891; Hugh M. Smith, 1892; and Barton W. Evermann and Samuel F. Hildebrand, 1910. Complete titles and references to the publications by these authors are given in the bibliography.

Several species of fishes from Chesapeake Bay also are mentioned in various lists by Henry W. Fowler. References to these lists will be found in the text under the particular species that this author mentions. Notes on the species propagated on

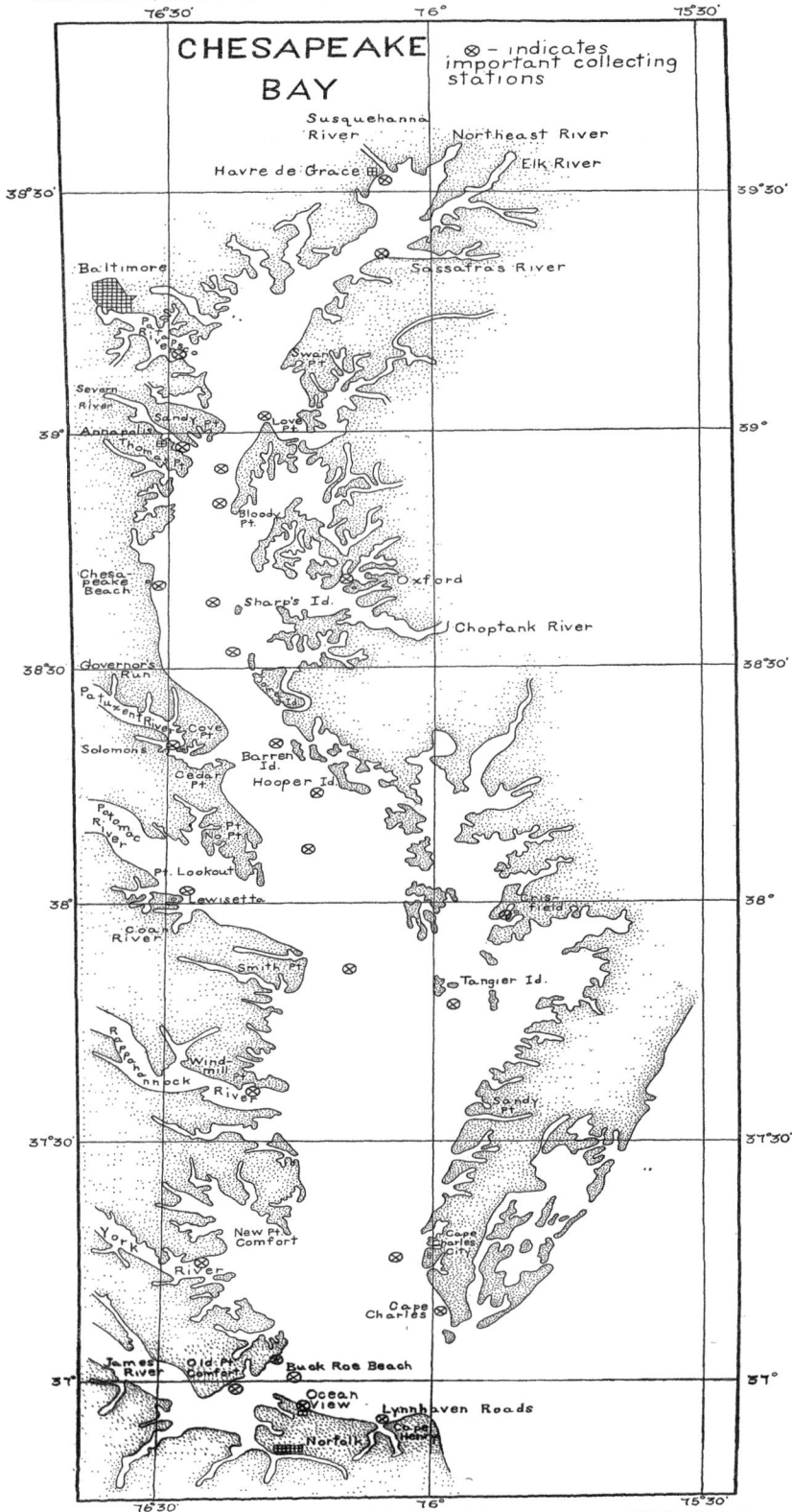

CHESAPEAKE
BAY

⊗ – indicates
important collecting
stations

Susquehanna
River

Northeast River

Elk River

Havre de Grace

Sassafras River

Baltimore

Swan
Pt.

Severn
River

Sandy

Love
Pt.

Annapolis
Thomas

Bloody
Pt.

Chesa-
peake
Beach

Oxford

Sharp's Id.

Choptank River

Governor's
Run

Patuxent River Cove

Solomon's

Barren
Id.

Cedar
Pt.

Hooper Id.

Potomac
River

Pt.
No Pt.

Pt. Lookout

Lewisette

Coan
River

Smith Pt.

Tangier Id.

Rappahannock
River

Wind-
mill

Sandy
Pt.

York

River

New Pt.
Comfort

Cape
Charles
City

Cape
Charles

James
River

Old Pt.
Comfort

Buck Roe Beach

Ocean
View

Lynnhaven Roads

Norfolk

Cape
Henry

Fig. 1.—Map of Chesapeake Bay 49826—28. (Face p. 10)

Chesapeake Bay and tributary streams are scattered through the numerous reports of the United States Fish Commission and those of the fish commissions of Maryland and Virginia. Finally, various fishes from Chesapeake Bay are mentioned in an array of miscellaneous papers. Some of these are short and deal with a single fish, others are of a general nature, and one or more Chesapeake fishes are mentioned more or less incidentally. References to such publications occur in appropriate places in the text, and the complete titles are included in the bibliography.

GENERAL CHARACTER OF THE FAUNA

The fishes of Chesapeake Bay are not of a peculiar or distinctive type. It will be seen from the following table that of the 202 species described the great majority range both north and south of Chesapeake Bay. Present information indicates that the bay is the stopping point for 27 species of southern distribution, whereas only 12 species of northern distribution reach their southernmost range in Chesapeake Bay. One species, recently described, and four new species described in the present work, so far as known to date, are the only ones peculiar to Chesapeake Bay. We have included 44 species that do not appear to have been recorded previously from Chesapeake Bay. Other species undoubtedly will be taken, probably as stragglers, from time to time, as not a few coastwise species range both north and south of the entrance to Chesapeake Bay but have not been observed to date within the bay by a naturalist. Such species, of course, may stray past the capes and into the bay at almost any time.

The anadromous species, chief among which are the shad, alewives, and the striped bass, are especially numerous, and they constitute a very important part of the products of the fisheries of Chesapeake Bay. They are particularly important in that section of the bay lying within the State of Maryland, as many of the more strictly salt-water species common in the southern sections of the bay do not reach the Maryland waters in large numbers.

Distribution of species

[An X in the first column indicates that the species ranges both north and south of Chesapeake Bay; an X in the second column shows that it is found in Chesapeake Bay and southward, only; an X in the third column shows that it is found in Chesapeake Bay and northward, only; and an X in the last column shows that, to date, it has been taken only in Chesapeake Bay]

Species	North and south	South only	North only	Chesapeake Bay only	Species	North and south	South only	North only	Chesapeake Bay only
Branchiostoma virginiæ				X	Paralichthys dentatus	X			
Petromyzon marinus	X				Limanda ferruginea			X	
Ginglymostoma cirratum		X			Pseudopleuronectes americanus	X			
Carcharodon carcharias	X				Lophopsetta maculata	X			
Mustelus mustelus	X				Etropus microstomus	X			
Carcharhinus milberti	X				Etropus crossotus		X		
Scoliodon terræ-novæ	X				Neoetropus macrops gen. et sp. nov.				X
Sphyrna zygæna	X				Achirus fasciatus	X			
Sphyrna tiburo	X				Symphurus plagiusa		X		
Squalus acanthias	X				Gasterosteus aculeatus			X	
Squatina dumeril	X				Apeltes quadracus			X	
Pristis pectinatus	X				Syngnathus fuscus	X			
Raja diaphanes			X		Syngnathus floridæ		X		
Raja eglanteria	X				Syngnathus louisianæ		X		
Raja stabuliforis	X				Hippocampus hudsonius	X			
Raja erinacea	X				Fistularia tabacaria	X			
Torpedo nobiliana	X				Menidia menidia	X			
Dasyatis centrura	X				Menidia beryllina	X			
Dasyatis americana sp. nov	X				Membras vagrans	X			
Dasyatis say		X			Mugil cephalus	X			
Dasyatis sabina	X				Mugil curema	X			
Pteroplatea micrura	X				Sphyræna guachancho	X			
Myliobatis freminvillii	X				Sphyræna borealis	X			
Aëtobatus narinari		X			Polynemus octonemus	X			
Rhinoptera quadriloba	X				Scomber scombrus	X			
Manta birostris	X				Pneumatophorus colias	X			
Acipenser oxyrhynchus	X				Scomberomorus maculatus	X			
Acipenser brevirostrum	X				Scomberomorus regalis	X			
Lepisosteus osseus	X				Sarda sarda	X			
Elops saurus	X				Thunnus thynnus	X			
Tarpon atlanticus	X				Trichiurus lepturus	X			
Clupea harengus	X				Xiphias gladius	X			
Pomolobus mediocris	X				Peprilus alepidotus	X			
Pomolobus æstivalis	X				Poronotus triacanthus	X			
Pomolobus pseudoharengus	X				Selar crumenophthalmus	X			
Alosa sapidissima	X				Seriola dumerili	X			
Opisthonema oglinum	X				Oligoplites saurus	X			
Brevoortia tyrannus	X				Chloroscombrus chrysurus	X			
Dorosoma cepedianum	X				Caranx hippos	X			
Anchoviella mitchilli	X				Caranx crysos	X			
Anchoviella epsetus	X				Caranx latus		X		
Anguilla rostrata	X				Alectis ciliaris	X			
Conger conger	X				Selene vomer	X			
Erimyzon sucetta	X				Vomer setipinnis	X			
Minytrema melanops	X				Trachinotus falcatus	X			
Catostomus commersonii	X				Trachinotus glaucus		X		
Cyprinus carpio	X				Trachinotus carolinus	X			
Notemigonus crysoleucas	X				Pomatomus saltatrix	X			
Hybognathus nuchalis	X				Rachycentron canadus	X			
Notropis hudsonius amarus			X		Perca flavescens	X			
Notropis bifrenatus			X		Boleosoma olmstedi	X			
Felichthys felis	X				Pomoxis annularis	X			
Ameiurus catus	X				Enneacanthus gloriosus	X			
Synodus foetens	X				Lepomis gibbosus	X			
Esox reticulatus	X				Micropterus dolomieu	X			
Esox americanus	X				Micropterus salmoides	X			
Cyprinodon variegatus	X				Morone americana	X			
Lucania parva	X				Roccus lineatus	X			
Fundulus heteroclitus	X				Mycteroperca microlepis		X		
Fundulus majalis	X				Centropristes striatus	X			
Fundulus ocellaris		X			Priacanthus arenatus	X			
Fundulus diaphanus	X				Pseudopriacanthus altus	X			
Fundulus luciæ	X				Lobotes surinamensis	X			
Gambusia holbrooki	X				Lutianus griseus	X			
Tylosurus marinus	X				Orthopristis chrysopterus	X			
Tylosurus acus	X				Hæmulon plumieri		X		
Ablennes hians	X				Bathystoma rimator		X		
Scomberesox saurus	X				Stenotomus chrysops	X			
Hyporhamphus unifasciatus	X				Stenotomus aculeatus		X		
Hemiramphus brasiliensis		X			Lagodon rhomboides	X			
Exocoetus heterurus	X				Archosargus probatocephalus	X			
Pollachius virens	X				Diplodus holbrookii		X		
Gadus callarias	X				Kyphosus sectatrix	X			
Urophycis chuss			X		Eucinostomus californiensis		X		
Urophycis regius	X				Eucinostomus gula	X			
Merluccius bilinearis	X				Leiostomus xanthurus	X			

Distribution of species—Continued

Species	North and south	South only	North only	Chesapeake Bay only	Species	North and south	South only	North only	Chesapeake Bay only
Sciænops ocellatus	X				Gobiosoma bosci	X			
Larimus fasciatus	X				Gobiosoma ginsburgi sp. nov				X
Bairdiella chrysura	X				Microgobius holmesi		X		
Stellifer lanceolatus		X			Microgobius eulepis		X		
Micropogon undulatus	X				Mugilostoma goblo gen. et sp. nov				X
Pogonias cromis	X				Echeneis naucrates	X			
Umbrina coroides		X			Astroscopus guttatus			X	
Menticirrhus saxatilis	X				Chasmodes bosquianus	X			
Menticirrhus americanus	X				Hypsoblennius hentz		X		
Menticirrhus littoralis			X		Blennius fucorum	X			
Cynoscion nebulosus	X				Rissola marginata	X			
Cynoscion nothus		X			Opsanus tau	X			
Cynoscion regalis	X				Gobiesox strumosus		X		
Lopholatilus chamæleonticeps			X		Balistes carolinensis	X			
Chætodipterus faber	X				Monacanthus hispidus	X			
Chætodon ocellatus	X				Ceratacanthus schœpfi	X			
Hemitripterus americanus			X		Lactophrys trigonus	X			
Cyclopterus lumpus			X		Lagocephalus lævigatus	X			
Prionotus evolans	X				Tetraodon maculatus	X			
Prionotus carolinus	X				Tetraodon testudineus	X			
Prionotus affinis sp. nov.				X	Diodon hystrix	X			
Cephalacanthus volitans	X				Chilomycterus schœpfi	X			
Tautoga onitis	X				Lophius piscatorius	X			
Tautogolabrus adspersus			X		Histrio histrio	X			
Scarus cæruleus		X			Ogcocephalus vespertilio		X		

GENERAL STATISTICS[1] AND REMARKS ON FISHERIES OF CHESAPEAKE BAY

Fishing in Chesapeake Bay is confined almost wholly to the period extending from about March 1 to November 1. Activities begin in the lower sections of the bay early in March, whereas the fishermen at the head of the bay usually do not set their nets until early in April. The first catches of the season consist of shad and herring, which arrive at the entrance about a month earlier than at the head of the bay. The first catches generally are small but remunerative, because they bring fancy prices, and therefore the nets are set early enough to intercept the earliest arrivals.

The biological fact that, exclusive of the rockfish, the white perch, the common eel, and a few other species of little importance, the commercial fishes leave the bay during the fall of each year and return the following spring is brought out in the discussions of the various species. This migration leaves the waters of the bay largely barren of fish during the winter months, and it is for that reason that nearly all fishing operations are discontinued by about the 1st of November and are resumed the following March or April, when the fish begin to return. The earliest to arrive, as already shown, are the shad and herrings, followed rather shortly by the croaker, kingfish, and several other species.

---

[1] The statistical data given here and elsewhere in this work, unless otherwise stated, are largely taken from the reports of the United States Commissioner of Fisheries. Since the statistics are given by counties in these reports, it was necessary to estimate the part taken within the bay proper for those counties not wholly on Chesapeake Bay. However, the original working sheets on which the data were compiled were available in the Bureau of Fisheries for our use for the statistics of 1920. These sheets contained the catches by localities, and for this year we were able to obtain fairly definite figures on the amount taken within the bay; and for those years where the amounts for certain counties had to be estimated, the relative proportion of 1920 was used in arriving at the estimated quantities taken within the bay itself. It is quite certain, however, that the figures are approximately correct. It will be noticed, also, that in some instances the figures given in the present report have been reduced to round numbers.

The quantity and value of each species of food fish taken in Chesapeake Bay during 1920 and apportioned between the States of Maryland and Virginia are given in the following table:

Value and weight of food fishes taken in Chesapeake Bay in 1920 [1]

Common name	Scientific name	Maryland				Virginia				Entire Bay			
		Pounds	Rank	Value	Rank	Pounds	Rank	Value	Rank	Pounds	Rank	Value	Rank
Alewives	{Pomolobus æstivalis / P. pseudoharengus	}6,604,891	1	$163,544	3	16,381,267	1	$253,424	4	22,986,158	1	$416,968	2
Croaker	Micropogon undulatus	1,130,590	3	31,683	6	13,039,795	2	361,479	2	14,170,385	2	393,162	3
Shad	Alosa sapidissima	1,816,346	2	344,110	1	7,257,987	3	1,138,184	1	9,074,333	3	1,482,294	1
Squeteague	Cynoscion regalis	678,470	5	44,143	4	7,240,243	4	345,958	3	7,918,713	4	390,101	4
Rockfish	Roccus lineatus	1,040,274	4	193,295	2	370,356	8	68,623	5	1,410,630	5	261,918	5
Butterfish	Poronotus triacanthus	15,062	14	603	15	1,263,566	5	42,000	7	1,278,628	6	42,603	9
Spot	Leiostomus xanthurus	51,692	8	3,138	9	786,153	6	60,000	6	837,845	7	63,138	6
White perch	Morone americana	316,915	6	32,026	5	218,165	12	19,888	9	535,080	8	51,914	7
Spotted squeteague	Cynoscion nebulosus	20,000	13	2,000	11	418,797	7	41,879	8	438,797	9	43,879	8
Starfish	Peprilus alepidotus	3,765	17	150	17	315,916	9	10,500	13	319,681	10	10,650	13
Eel	Anguilla rostrata	197,293	7	21,395	7	120,715	14	12,309	12	318,008	11	33,704	10
Flounder	Paralichthys dentatus	26,746	12	1,150	13	258,354	10	12,613	11	285,100	12	13,763	11
Mullet	Mugil cephalus and curema	35,337	10	1,861	12	246,683	11	8,346	14	282,020	13	10,207	14
Hickory shad	Pomolobus mediocris	2,100	18	95	18	216,520	13	8,150	15	218,620	14	8,245	15
Gizzard shad	Dorosoma cepedianum	30,067	11	913	14	42,785	15	1,100	21	72,852	15	2,013	20
Winter flounder	Pseudopleuronectes americanus	40,119	9	4,012	8	13,600	21	1,360	20	53,719	16	5,372	17
Bluefish	Pomatomus saltatrix	14,989	15	2,112	10	36,979	16	4,925	16	51,968	17	7,037	16
Hogfish	Orthopristes chrysopterus					31,725	17	2,348	17	31,725	18	2,348	18
Sturgeon (including caviar)	Acipenser oxyrhynchus	734	19	259	16	24,808	18	12,712	10	25,542	19	12,971	12
Black drum	Pogonias cromis	700	20	8	21	23,000	19	230	26	23,700	20	238	27
King whiting	Menticirrhus saxatilis, americanus, and littoralis					17,933	20	1,606	19	17,933	21	1,606	21
Red drum	Sciænops ocellatus	4,835	16	76	18	12,730	23	204	27	17,565	22	280	26
Spanish mackerel	Scomberomorus maculatus	337	21	62	20	13,429	22	2,052	18	13,766	23	2,114	19
Scup	Stenotomus chrysops and aculeatus					7,165	24	585	22	7,165	24	585	22
Sea bass	Centropristes striatus					5,100	25	492	23	5,100	25	492	23
Black bonito	Rachycentron canadus					²3,000	26	300	25	²3,000	26	300	25
Whiting	Merluccius bilinearis					3,000	27	60	34	3,000	27	60	34
Tautog	Tautoga onitis					³2,000	28	80	32	³2,000	28	80	32
Pompano	Trachinotus carolinus					1,650	29	330	24	1,650	29	330	24
Bonito	Sarda sarda					1,400	30	192	28	1,400	30	192	28
Crevalles	Caranx crysos and hippos					1,200	31	120	30	1,200	31	120	30
Spadefish	Chætodipterus faber					³1,000	32	80	31	³1,000	32	80	31
Tripletail	Lobotes surinamensis					³1,000	33	80	33	³1,000	33	80	33
Sheepshead	Archosargus probatocephalus					863	34	129	29	863	34	129	29
Total		12,031,262		846,635		48,378,884		2,412,338		60,410,146		3,258,973	

[1] Large quantities of white sand perch (*Bairdiella chrysura*) are caught in Chesapeake Bay, but only a small part of the catch is marketed. As the amount marketed can not be determined, this species is not included in the statistics for the bay.
² Estimated for 1921.
³ Estimated for 1922.

The total catch of fish taken in the salt and brackish waters of Chesapeake Bay in 1920, in round numbers, amounted to 60,000,000 pounds. Of this amount, 12,000,000 pounds, valued at $850,000, were caught in Maryland and 48,000,000 pounds, worth $2,400,000, were taken in Virginia. About 90 per cent of the entire catch consisted of alewives, croakers, shad, and squeteagues. According to the apparatus used, the catch may be divided as follows: Pound nets, 81½ per cent; gill nets, 7 per cent; seines, 6 per cent; fyke nets, 2 per cent; lines, 2 per cent; eel pots, one-half of 1 per cent; and miscellaneous, 1 per cent. The catch by States, expressed in per cent, according to apparatus used, may be divided as follows:

FIG. 2.—Haul-seining for spots and other fish at Ocean View, Va. The power boat towing the seine boat is about to leave the beach to pay out the 300-fathom seine

FIG. 3.—A winch, operated by a gasoline engine, is used for hauling in the seine in localities out of reach of electric power. Within Ocean View proper electric power is used. Note that only one person is required to manipulate the line as the seine is being drawn in. Later, as the seine approaches shore and man power supplants gasoline, 22 men are required

Fig. 4.—The bunt of the seine near shore. At this stage of a haul 2 or 3 men are required to foot the lead line and hold up the cork line of the bunt to prevent the fish from escaping

5.—The catch landed on the beach. In this instance the catch is small and can be drawn up on the beach in the seine. Frequently, however, when a large catch is made, the fish are bailed out with dip nets. Sometimes it requires an hour or more to remove the fish

Apparatus	Mary-land	Virginia	Apparatus	Mary-land	Virginia
	Per cent	*Per cent*		*Per cent*	*Per cent*
Pound nets	68.0	85.0	Lines	0.5	2.5
Seines	16.0	3.0	Eel pots	1.5	.2
Gill nets	10.0	6.5	Miscellaneous	1.5	1.0
Fyke nets	2.5	1.8			

The pound net, as shown by the data given in the preceding paragraph, is by far the most important apparatus employed in the fisheries of Chesapeake Bay. It is used throughout the bay, as well as in the lower parts of the larger tributary streams. The majority of the pound nets, particularly in the northern sections of the bay, are drawn up in midsummer, when fish, for a time, appear to be scarce, but are again operated during the autumn. Many nets are used only in the spring for catching striped bass, shad, and herrings. In the lower parts of the bay and in a few favorable localities elsewhere the nets are operated throughout the entire season—namely, from March to November. The principal species of fish taken in pound nets are indicated in tables and graphs that appear elsewhere in this report.

Seines rank next to pound nets in importance in the fisheries of Chesapeake Bay and are used almost everywhere. Seining, like pound-net fishing, is more profitable at certain seasons of the year than others. At Ocean View, Va., for example, where very large nets are used, operations do not begin until sometime in July, and large catches usually are not made until late in September or in October. Fair to large catches of spots, spotted and gray weakfish, striped bass, white perch, and occasionally bluefish and pompanoes, are taken. An unusually large catch of spots was obtained in an 1,800-foot seine at Ocean View, Va., in October, 1922, when 90,000 fish, weighing approximately 50,000 pounds, were taken in a single haul.

Gill nets appear to be somewhat less important than seines in the fisheries of the Chesapeake. They are used to a limited extent throughout the bay, however, and rather extensively in the lower Potomac, Rappahannock, and York Rivers; also in the vicinity of Love Point, Crisfield, and Cape Charles. The nets are used either as stationary nets or they are allowed to drift with the tide and current. Frequently fair to large catches of striped bass, croakers, weakfish, spots, kingfish, and bluefish are taken.

Fyke nets, too, are used almost everywhere in the bay. These nets are generally used in small coves and other places too small for pound nets and in places where pound nets are not permitted. Although the quantity of fish taken with fyke nets is comparatively small, many nets of this type are used. Nevertheless, the operation of the fyke net probably is quite remunerative, as the net itself is inexpensive and it can be fished by one man. Furthermore, the fyke net often is used far into the winter, when virtually all other methods of fishing have been abandoned. The fish caught at such times, of course, bring a fancy price. The species caught are chiefly winter flounders, white perch, yellow perch, croakers, and squeteagues.

Comparatively little hand-line fishing is done in Chesapeake Bay, because it does not appear to be as profitable as other methods. The only species that are taken almost exclusively with hand lines are the sea bass and the tautog, and of these fish only small quantities are caught. In May and June, particularly at the mouth of the York River, croakers are caught with hand lines. This happens to be at a time when few of these fish are taken in pound nets. A limited amount of hand-line fishing for large squeteagues is done in the lower parts of the York and Rappahannock Rivers in October. About the same time many hand-line fishermen in small boats are seen off Ocean View fishing for spots, which appear to collect there, presumably preparatory to leaving the bay.

Eel pots are used throughout the Chesapeake region, but chiefly in the vicinity of the lower Choptank River and at the head of the bay. Virtually nothing except eels is caught in these traps.

In 1920 about 40,000 persons were engaged in the fisheries of Maryland and Virginia, and the shore property, boats, and gear employed (exclusive of the menhaden industry) were valued at about $12,000,000. The property of the menhaden industry, including factories, boats, and gear, was valued at about $5,000,000, and about 350,000,000 pounds of menhaden, worth about $2,000,000, were caught in and near the mouth of Chesapeake Bay.

It may be of interest to make a comparison here of the catches of fish taken from Chesapeake Bay and Georges Bank, both intensively fished areas, the one protected by land and fed by numerous streams and the other in the open ocean. Chesapeake Bay and the brackish parts of its tributaries contain about 2,700 square miles and produced about 11 tons of fish per square mile in 1920, whereas Georges Bank, with an area of about 7,000 square miles, produced about 3 tons of fish to the square mile.

It is apparent from the statistics collected by the United States Bureau of Fisheries that, as a whole, no serious decline in the quantities of fish caught in Chesapeake Bay has taken place during recent years. The catch, however, probably is kept up to a certain extent through more intensive fishing and by the use of more efficient gear. It has been shown elsewhere that a much larger part ($81\frac{1}{2}$ per cent) of the total quantity of fish taken in Chesapeake Bay is caught with pound nets than with all other gear combined. Unfortunately, this apparatus is often very wasteful of young and undersized fish, especially if the operators are indifferent and careless. It may be said with great credit to some of the operators (as, for example, the Buchanan brothers, who run pound nets in Lynnhaven Roads, at James Siding, and others) that they are very careful to return to the water uninjured small and unmarketable fish. On the other hand, not a few pound-net operators empty the entire catch into their boats and later, at their leisure and after the fish are all dead, sort out the small fish and throw them overboard; it sometimes happens that only comparatively few fish of marketable size are contained in the catch. In fact, it is not unusual for some 5,000 young spots, croakers, or butterfish, all just slightly under marketable size, to be destroyed in one day at a set of two pound nets. Such a practice can not be condemned too strongly. Fishermen with forethought and with a sense of duty to the future will not do this, of course, but will cull their catch at the net (whenever weather conditions are not too unfavorable) and reduce the waste to a minimum.

BUCHANAN BROTHERS' FISHERY

LOCATION AND DESCRIPTION

For over 50 years a fishery has been in existence in Lynnhaven Roads, Va., at a place now known as James Siding. This place is only about 3 miles west of Cape Henry. The fishery (herein called the Buchanan Brothers' fishery, because it is owned and has been operated during recent years by three brothers of that name), therefore, is near the entrance of Chesapeake Bay.

Pound nets and seines only have been used in this fishery, and they have always been operated in the same immediate vicinity and no evident physical changes have taken place during the period (1908 to 1922) for which statistics are available.

RECORDS OF THE FISHERY AND THE GEAR EMPLOYED

Records of the quantities of fish caught at this fishery have been kept for many years in the form of duplicate bills of lading. The amounts listed, therefore, are quite accurate, as the fish are shipped by rail directly from the fishery at James Siding to Norfolk. In general, if 10 pounds or more of any one species were included in the shipment, the species was listed separately. The only discrepancy that occurs is in small catches consisting of only a few pounds, for these were listed as "mixed" fish.

Through the courtesy of the Buchanan brothers we have had free access to the records, which are complete for most of the species (exclusive of 1911) since 1908. Subsequent to the close of the field work in 1922, the records of the shad caught in 1923 also were obtained.

Unfortunately for our purpose, the statistics from the fishery, for all the species taken, are not directly comparable for the entire period covered, as the gear was not uniformly employed. From 1908 to 1911 a set of two pound nets was operated from early March until about July 20, and for the remainder of the season, or until about the 1st of November, an 1,800-foot haul seine alone was used. From 1912 to 1917 a set of two pound nets was operated throughout the season, and in addition an 1,800-foot seine was used after about July 20. Finally, from 1918 to 1922 a set of two pound nets alone was used throughout the fishing seasons. Since the pound nets alone were used during the spring—that is, during the shad and herring runs—throughout the period of years covered by the records, the changes in apparatus do not apply to these species, and for them the data are directly comparable. Similarly, the data for the months of March, April, May, and June, for all the species, are directly comparable.

VALUE OF THE RECORDS

Tables and graphs (in so far as they seemed useful) have been prepared from the statistics in order to show the yearly fluctuations and the trend of the various species caught at this fishery. Regardless of the change in the apparatus employed, it seems probable that the tables serve the purpose not only of showing the trend in the abundance of the species commonly caught in pound nets in Lynnhaven Roads, but that, in a measure, they may reflect the general rise and fall in the abundance of these species over a series of years for the entire bay. We are unable to produce definite proof for the last-mentioned hypothesis as no statistics (exclusive of those of 1909 and 1915 for the shad and herrings) covering the vicinity of the bay are available for comparison from 1908 to 1920. In comparing the Bureau of Fisheries statistics for 1908, 1909, 1915, 1920, and 1921, published in Appendix IX of the report of the United States Commissioner of Fisheries for 1922 (p. 85), for the shad and herrings, with those compiled from the records of the Buchanan brothers' fishery, it is seen that (disregarding changes in the gear used or in the number of men and boats employed in the fishery for the entire bay) the general downward trend for both shad and herrings is reflected in each group of statistics. For individual years, however, the statistics do not always agree; as, for example, the bureau's records show a larger catch for 1908 than for 1909. The records of the fishery under consideration, on the other hand, show that the larger catch there was made in 1909. Both sets of statistics, however, show that a very small catch was taken in 1915 and that better catches were made in 1920 and 1921. Nevertheless, the banner year (1921) at the Buchanan brothers' fishery is not reflected for the rest of the bay, as the bureau's report shows a larger catch for 1920 than for 1921.

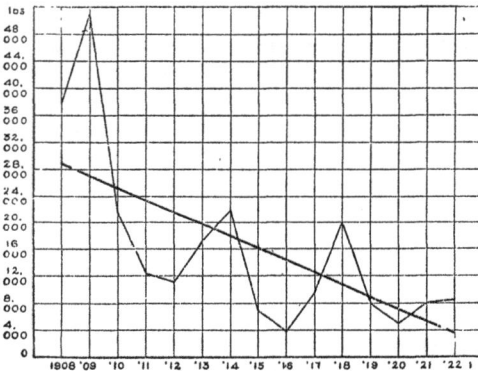

FIG. 6.—Graphic representation of the number of pounds of herrings (*Pomolobus pseudoharengus* and *P. æstivalis*) taken from 1908 to 1922 at the Buchanan Bros. fishery, arranged by years. The straight, heavy line shows the general trend in the quantities caught

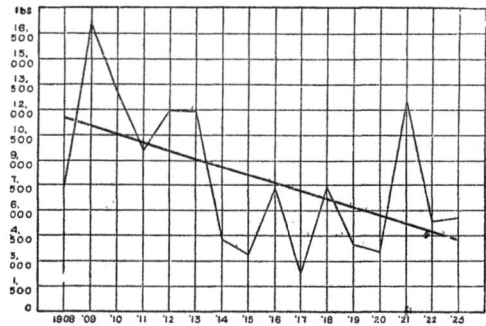

FIG. 7.—Graphic representation of the number of pounds of shad (*Alosa sapidissima*) taken from 1908 to 1923 at Buchanan Bros. fishery, arranged by years. The straight, heavy line shows the general trend in the quantities caught

For the herrings, as for the shad, when individual years are compared the banner years at the fishery do not always correspond with the better years for the bay generally; as, for example, the catch at the fishery in 1909 was larger than that for 1908. The bureau's statistics for those years, nevertheless, show a larger catch in 1908 than in 1909. A small catch in 1915 and a still smaller one in 1920 are indicated by both sets of statistics, and, similarly, both records show a larger catch for 1921 than for the preceding year. A further analysis of the records for the catches of shad and herrings at the fishery under discussion will be given in a succeeding paragraph.

It has been shown in the preceding paragraph that the general trend in the abundance of the shad and herrings for Chesapeake Bay appears to be reflected by the catches made at the Buchanan brothers' fishery, when statistics for a series of years are compared. No reason is evident to the writers why the same apparent fact should not hold for the other species, for which unfortunately insufficient records are available to afford similar comparisons. Furthermore, it has been shown on page 13, as well as in the discussion of the various species, that most commercial species, including nearly all the fish that commonly are caught in pound nets, leave the bay upon the approach of cold weather in the fall and that they return the following spring. Because of the especially strategic position of the present fishery—almost within the mouth of the bay—it seems probable that a somewhat equal percentage of the entire body of migrating fish may be caught from year to year. The only exception that has been found to this supposition in the study of the records

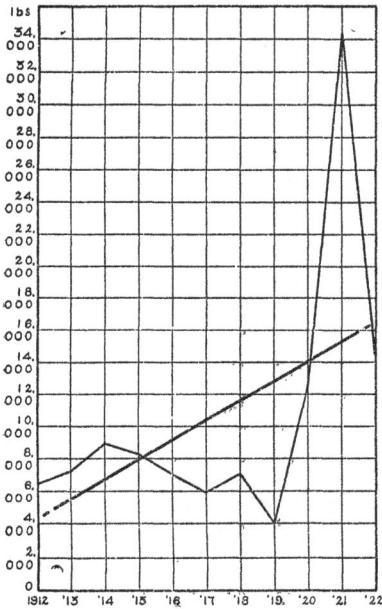

FIG. 8.—Graphic representation showing the number of pounds of summer flounders (*Paralichthys dentatus*) taken from 1912 to 1922 at the Buchanan Bros. fishery, arranged by years. Note the apparent increase in the abundance of this fish. The straight, heavy line shows the general trend in the quantities caught

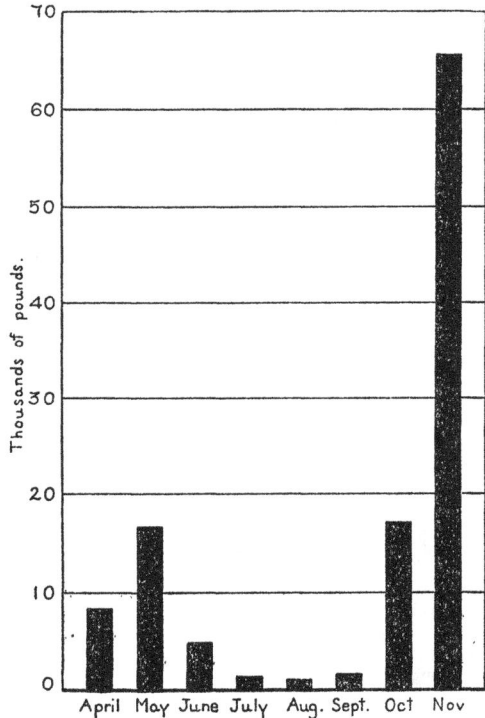

FIG. 9.—Graphic representation showing the number of pounds of summer flounders (*Paralichthys dentatus*) taken from 1912 to 1922 at the Buchanan Bros. fishery, arranged by months

is brought about by exceptionally large catches sometimes made within the course of a day or two, when apparently large schools of fish are intercepted by the nets.

In addition to such value as the tables may have in showing the trend of the fishery, they also show at what time the various species appeared in Lynnhaven Roads in commercial abundance from year to year over the period covered by the records, and also when they again became scarce in that vicinity. These dates, in each instance, because of the location of the fishery, may be interpreted to show, in general, the time of arrival in and time of departure from the bay of the species listed.

FLUCTUATIONS IN YEARLY CATCHES

It is evident from the table and graphs that comparatively large yearly fluctuations in the catch of the various species take place. It is shown also that a species may decline seriously for a year or two and then return to occupy its previous place of importance. The common shad, for example, although suffering a general

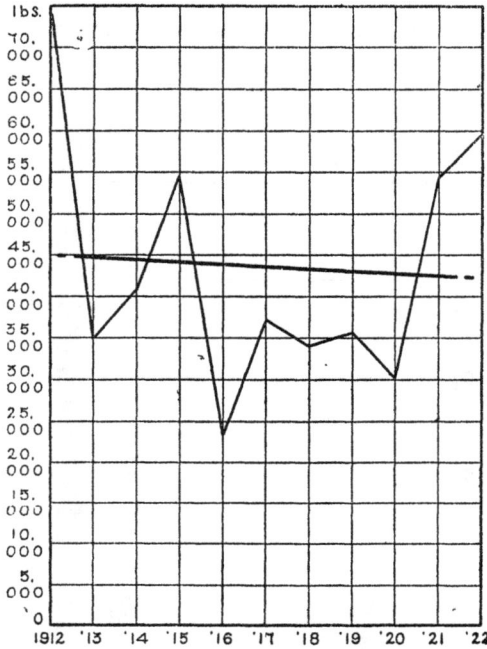

FIG. 10.—Graphic representation of the number of pounds of starfish (*Peprilus alepidotus*) taken from 1912 to 1922 at the Buchanan Bros. fishery, arranged by years. The straight, heavy line shows the general trend in the quantities caught

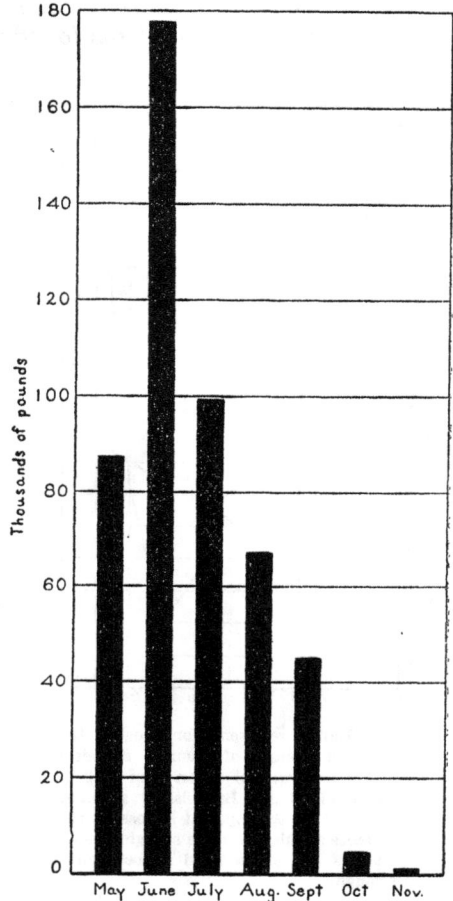

FIG. 11.—Graphic representation of the number of pounds of starfish (*Peprilus alepidotus*) taken from 1912 to 1922 at the Buchanan Bros. fishery, arranged by months. This species is rarely taken later than the last of October

decline over the series of years for which statistics are available, recovered from a new low mark (2,225 pounds) in 1917 to one of the largest catches (12,460 pounds) made in recent years in 1921. Similarly, the catch of branch and glut herring dropped to 3,800 pounds in 1916, but in 1918 it consisted of 20,020 pounds and it compared favorably with the catches made during the earlier years for which statistics are available. The next year a great decline (7,915 pounds) again took place. Somewhat similar fluctuations have taken place in the catch of nearly all the species commonly taken in pound nets in Lynnhaven Roads, and they are especially pronounced for the croaker and the kingfish.

ANALYSIS OF THE DATA

Attention already has been called to the fact that, due to a change in the apparatus used, only the statistics for the shad and the herrings are directly comparable for the entire period covered. The operation of the pound nets was discontinued about July 20 and an 1,800-foot seine was used for the remainder of

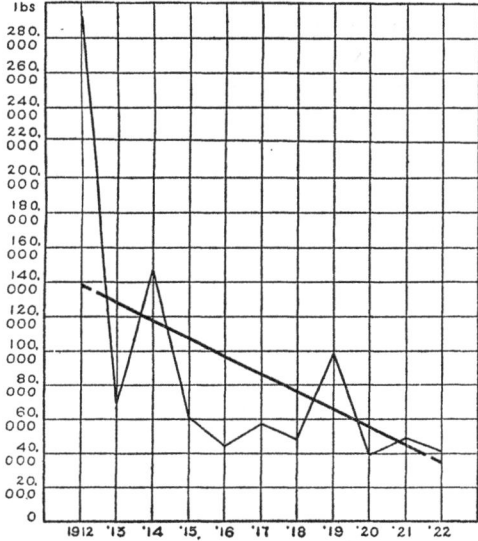

FIG. 12.—Graphic representation of the number of pounds of butterfish (*Poronotus triacanthus*) taken from 1912 to 1922 at the Buchanan Bros. fishery, arranged by years. Although a seine, in addition to a set of two pound nets, was used from about July 20 to October, from 1912 to 1917, few fish were caught by this method, and this does not affect the final results greatly. Note the great abundance of this fish in 1912. This species was taken in large quantities throughout May and June of 1912, the largest single catch consisting of 19,400 pounds and was taken on June 25. The straight, heavy line shows the general trend in the quantities caught

the season from 1908 to 1911, because this gear, during that part of the fishing season, was thought to yield more profitable results. Then followed the period (1912 to 1917) when the pound nets were operated throughout the fishing season,

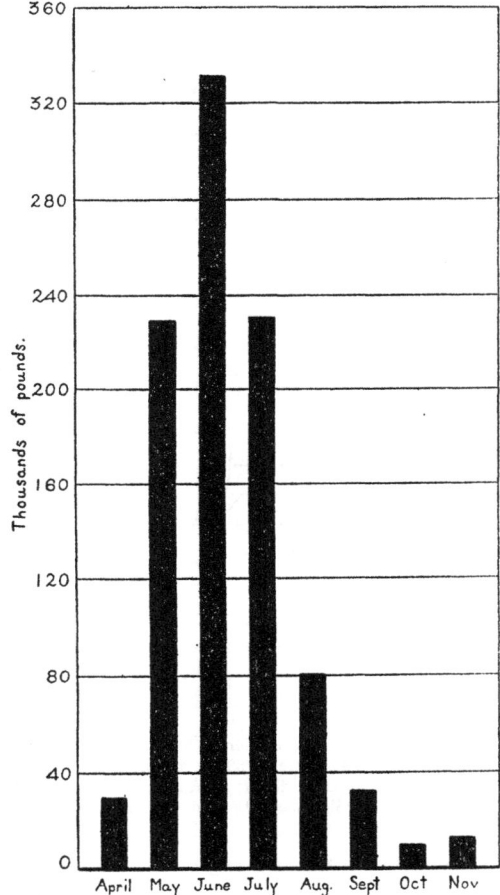

FIG. 13.—Graphic representation of the number of pounds of butterfish (*Poronotus triacanthus*) taken from 1912 to 1922 at the Buchanan Bros. fishery, arranged by months

and in addition an 1,800-foot seine was used after about July 20 until the close of the fishing season, and thereafter pound nets only were used. It is probable that a larger quantity of fish was caught with the seine than would have been taken with the pound nets during the same number of fishing seasons, and the annual catch undoubtedly was considerably increased for most of the species from 1912 to 1917 by the operation of both gears. The tables and graphs for all the species, exclusive of the shad and herrings, therefore, must not be interpreted too literally, as the decline shown for those species for which the catch has diminished quite certainly is not as pronounced as indicated. On the other hand, in those species where an upward trend is shown, regardless of the discontinuance of the use of the seine, the increase very probably is greater than shown.

In summing up the statistics it may be concluded that an unmistakable and definite decline has taken place for the shad and herrings for the period covered. The decline, based on the average yearly catch for the first and second halves of the period covered by the statistics is 39.4 per cent for the shad and 60.2 per cent for the herrings. A very pronounced decline in the catch of shad took place in 1914 and 1915. After that time a partial recovery is shown, as averages (arrived at as

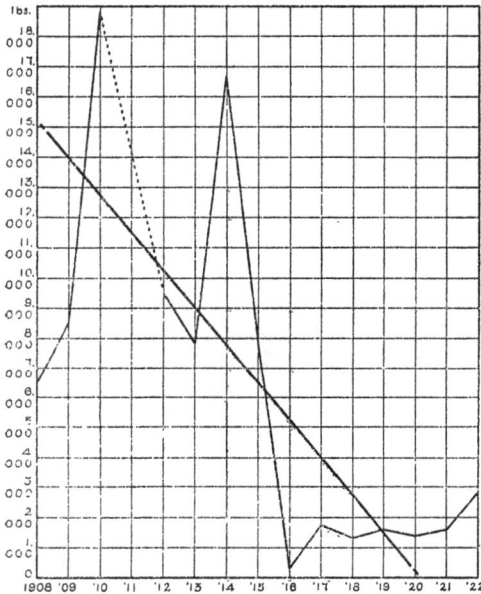

Fig. 14.—Graphic representation of the number of pounds of bluefish (*Pomatomus saltatrix*) taken from 1908 to 1922 at the Buchanan Bros. fishery, arranged by years. The straight, heavy line shows the general trend in the quantities caught

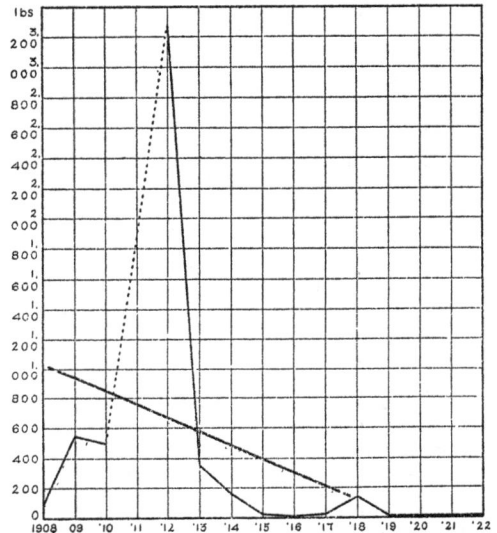

Fig. 15.—Graphic representation of the number of pounds of sheepshead (*Archosargus probatocephalus*) taken at the Buchanan Bros. fishery from 1908 to 1922, arranged by years. During those years for which no catch is listed, a few, no doubt, were taken; but the daily catch consisted of less than 10 pounds and no separate record was made. The straight, heavy line shows the general trend in the quantities caught

before) for 1914 to 1923 show an increase of 12.6 per cent. The first two years (1908 and 1909) for which data are available for the herrings appear to have been banner years and a large decline took place in 1910. The lowest mark, however, resulted in 1916. Excluding from consideration the large catches for 1908 and 1909, general averages show a decline of 34.5 per cent for the period 1910 to 1922, as compared with 60.2 per cent for the entire period. The species was rather stationary from 1915 to 1922, as only a slight increase is shown. It is at least somewhat encouraging that the shad has shown an upward trend and the herrings no further downward trend during recent years (that is, since 1915), as shown by the records of the fishery under discussion supported by the bureau's statistics for Maryland and Virginia for 1915, 1920, and 1921.

A really serious decline during recent years is shown by the records for the important commercial species known locally as the gray squeteague and the king-fishes. The squeteague was almost stationary from 1908 to 1918. Then occurred a sudden decline, which was not overcome during the next four years, or up to the end of the period for which statistics are available. The decline for the entire period (1908 to 1922) covered by the records, as shown by average yearly catches arrived at as in the preceding paragraph, was 35 per cent.

FIG. 16.—Graphic representation of the number of pounds of spots (*Leiostomus xanthurus*) taken from 1912 to 1922 at the Buchanan Bros. fishery, arranged by years. The spot is caught in large quantities in seines during the autumn. Therefore, the smaller catches since 1918 (the seine was used in 1917, which evidently was a very poor year) do not necessarily indicate a decline in the abundance of the species. The straight, heavy line shows the general trend in the quantities caught

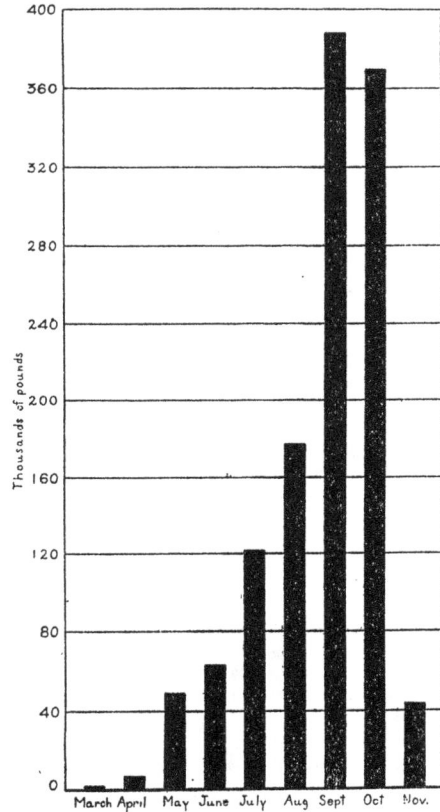

FIG. 17.—Graphic representation of the number of pounds of spots (*Leiostomus xanthurus*) taken from 1912 to 1922 at the Buchanan Bros. fishery, arranged by months. The first commercial catches of spots usually are made sometime in April. In 1920, however, the fish were caught in relatively large quantities in March

Large yearly fluctuations took place in the catch of kingfish from 1908 to 1917, the trend being upward until the banner year, 1912. Then followed a very greatly reduced catch in 1913 and another large catch in 1914. Thereafter the trend was strongly downward, the catch falling so low in 1918 that the species became of minor commercial importance in the fishery. The following year the catch was still smaller, and no recovery had taken place by the end of the period covered by the records (1922).

Very large catches of spot were made from 1912 to 1916, followed by much smaller catches, causing a decline of 55.8 per cent from 1912 to 1922, as shown by general averages. A recovery (amounting to an increase of 30 per cent) took place after the sharp decline of 1917, or from 1917 to 1922. Should these data be somewhat representative of the catches for the entire bay, some hope for the rehabilitation of the species remains.

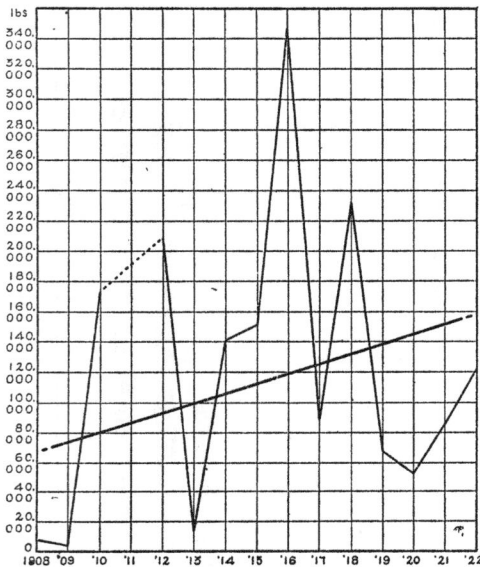

FIG. 18.—Graphic representation of the number of pounds of croakers (*Micropogon undulatus*) taken at the Buchanan Bros. fishery from 1908 to 1922, arranged by years. The straight, heavy line shows the general trend in the quantities caught

The decline for the butterfish (as shown by general averages, based on the total catch for each half of the period 1912 to 1922 for which data are at hand) is 51 per cent. This high percentage of decline is due in great measure to the enormously large catch of 1912. As this catch comes at the very beginning of the period for which we have records, it is impossible to know whether this was a much larger

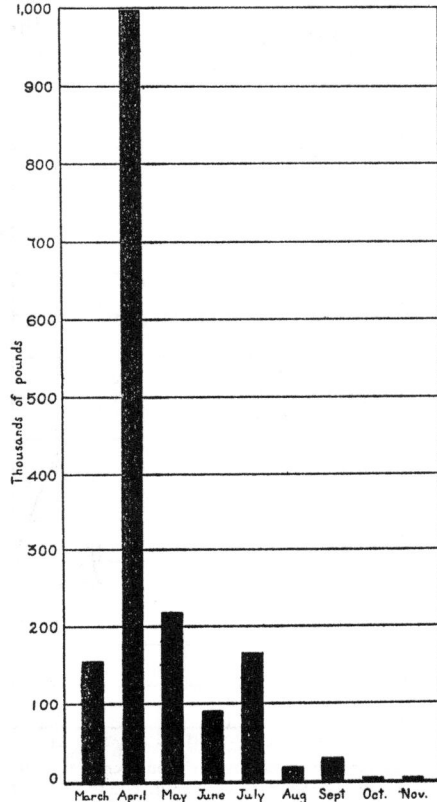

FIG. 19.—Graphic representation of the total number of pounds of croakers (*Micropogon undulatus*) taken from 1908 to 1922 at the Buchanan Bros. fishery, arranged by months. This graph should not be interpreted to signify that croakers are scarce or absent in the bay during the summer and autumn, for this does not appear to be true, as they are taken in fair numbers with hand lines at this time. A seasonal change in their habits is suggested

catch than had been taken during the preceding years and whether it should be regarded as an unusually large catch. Omitting the data for 1912 and calculating the decline for the remainder of the years by means of averages, it amounts to 27 per cent. From 1915 to 1922 an upward trend of 8.6 per cent took place, showing that during recent years no further decline has occurred in the catch at this fishery.

The starfish does not appear to have undergone a general decline. Comparatively large fluctuations have taken place, however. The largest yearly catch for the period 1912 to 1922, for which records are available, occurred in 1912. Here, as with the butterfish, it is impossible to know whether this is a "normal" catch as compared with immediately preceding years. The smallest catch for the entire period was made in 1916, and from the beginning of the period to that time the trend was decidedly downward, and thereafter it was definitely upward. A trend based on the average of the total catch for each half of the entire period shows a decline of 3.4 per cent. Determining a trend in the same way (omitting, however, the catch for 1912), an increase of 12 per cent is evident.

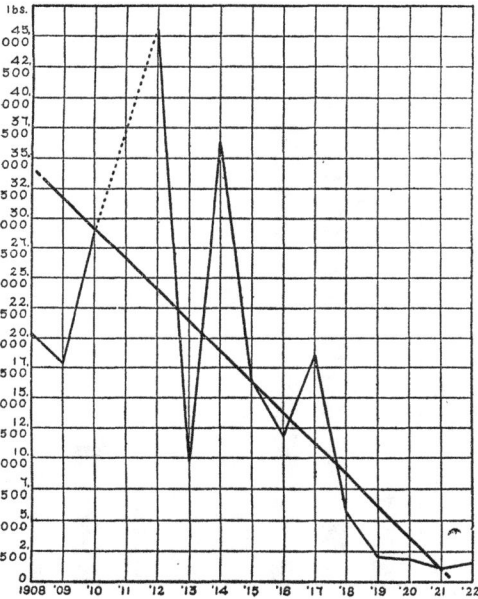

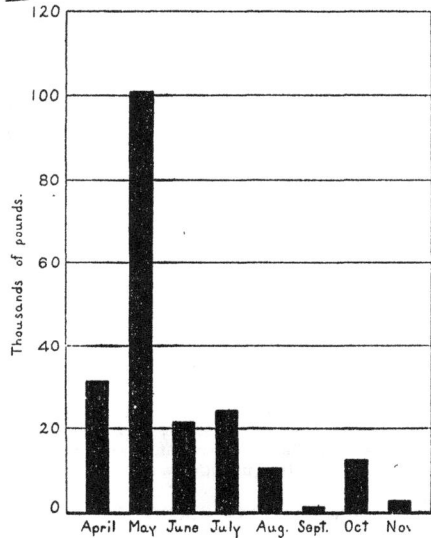

FIG. 21.—Graphic representation of the number of pounds of kingfish (*Menticirrhus americanus*, *M. saxatalis*, and *M. littoralis*) taken from 1908 to 1922 at the Buchanan Bros. fishery, arranged by months. This species usually is taken in commercial quantities first sometime during April.

0.—Graphic representation of the number of pounds of kingfish (*Menticirrhus americanus*, *M. saxatalis*, and *M. littoralis*) taken from 1908 to 1922 at the Buchanan Bros. fishery, arranged by years. The quantities of kingfish caught in seines, when they were operated, was rather insignificant, and in any event did not affect the catches made during the spring, when the largest quantities were taken. A pronounced decline in the abundance of the kingfishes, therefore, is certain and undeniable. The straight, heavy line shows the general trend in the quantities caught

Very large fluctuations have occurred in the catch of croakers. The catches for 1908, 1909, and 1913 were almost negligible. Later followed some very large catches, the largest being taken in 1916. An upward trend is evident from 1908 to 1916, and thereafter a decline took place. The increase for the entire period (1908 to 1922) for which statistics are at hand is 42.6 per cent, as shown by general averages of the catch arrived at as before.

The catch of summer flounders was quite stationary from 1912 to 1918. In 1919 a considerable decline took place. This small catch, however, was followed by large catches during the next three years. The increase of the catch of the second half over that of the first half of the period (1912 to 1922) for which records are at hand is 92.4 per cent. This large increase, as already indicated, resulted from the catches of the last three years of the period, and especially from the banner year 1921, when the catch was more than twice as large as for any other year for which records are available.

Other species taken in the pound nets in Lynnhaven Roads are bluefish, sheepshead, pompano, sturgeon, and sand perch. All of these were of minor importance in the fishery during the period covered by the records under consideration. Occasionally, also, small catches of mullets, pigfish, Spanish mackerel, and bonito are made. The last-named species are taken in such small quantities, however, that their value in the fishery does not justify any discussion. The decline in the bluefish

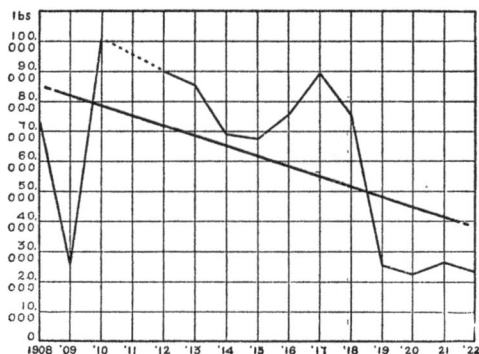

FIG. 22.—Graphic representation of the number of pounds of squeteague (*Cynoscion regalis*) taken from 1908 to 1922 at the Buchanan Bros. fishery, arranged by years. It is problematical whether the species will recover from the decline since 1918. The straight, heavy line shows the general trend in the quantities caught

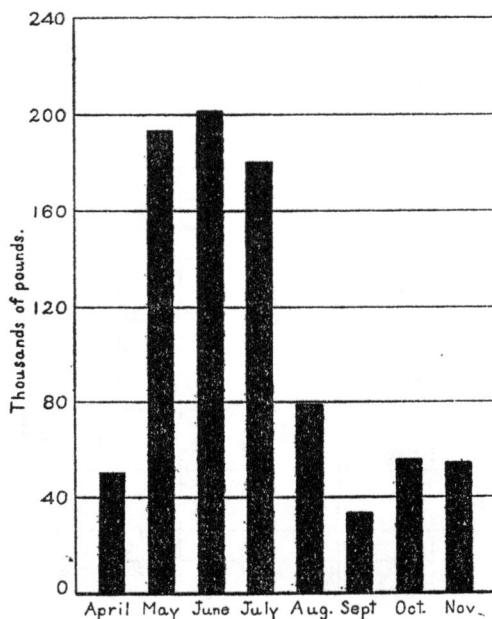

FIG. 23.—Graphic representation of the number of pounds of squeteague (*Cynoscion regalis*) taken from 1908 to 1922 at the Buchanan Bros. fishery, arranged by months. The species is not taken in commercial numbers during March

in Chesapeake Bay, according to all accounts, is quite general, and the catch unmistakably has declined at this fishery for the entire period (1908 to 1922) under consideration. A sharp drop occurred in 1916, and since that time a partial recovery is indicated. The sheepshead, too, is said formerly to have been much more numerous in Chesapeake Bay. The table presented herewith shows that at no time during the years covered by the records was this species of much importance in this fishery, and during recent years the catch has been negligible. The catch of pompano at this fishery warrants brief mention only because it is a highly prized food fish and because the small quantities taken bring a good price. Except for fairly large catches in 1913 and 1914, the species appears to have been rather stationary and uniformly scarce. The decline of the sturgeon is so well known that it does not require discussion. The catch at the Buchanan brothers' fishery was quite consistently low from 1916 to 1922, except in 1918, when it was more than twice as large as during any other year covered by the records. The sand perch is often taken in large numbers, and usually only the very largest individuals are retained for the market. The number retained, however, depends somewhat upon the abundance of more desirable species and market conditions.

CONCLUSION

It is evident from the foregoing discussion and the tables presented herewith that a number of important species in the fishery under discussion have declined during the period covered by the records at hand; one, at least, appears to have remained nearly stationary, and for two the catch has increased. It must be borne in mind, however, that a change in the gear used took place during the earlier years for which records of catches are at hand. The extent to which this change affected the trend, exclusive of the catch of shad and herrings (which was not influenced), is not known. Moreover, it has been shown that the change in the gear undoubtedly resulted in a somewhat larger catch, at least from 1912 to 1917. The calculated trend shown on the graphs, as well as the percentages of increase and decrease given in the preceding section, therefore, is subject to an error of unknown significance. Yet, it seems certain that for most of the species considered the decline was less rapid or the increase more pronounced than indicated, according to whether an increase or a decrease in the catch took place.

It is very interesting, and possibly significant, that the majority of the species discussed suffered a serious decline during about the middle of the period for which records are available, and that several species (shad, herrings, butterfish, starfish, spot, and flounder) during the last several years, when a set of two pound nets only was operated, showed a tendency to recover. The increase in the catches is regarded by the writers as a hopeful sign.

It is impossible to estimate the exact significance of these statistics in relation to the fisheries for the rest of the bay, as few records for the entire bay are available for comparison. Limited evidence has been produced to show that the records of this fishery of the catch of shad and herrings does reflect the status of these species for the entire bay, and the writers know of no reason why the same should not be true of the other important commercial species of this pound-net fishery. Inasmuch as no more reliable statistics are available, the present ones are offered for what they may be worth in this connection. Certainly, they are of interest as a local study and in showing when the species appear in the mouth of the bay in commercial numbers, the month or months during which they are the most abundant, and when they again become scarce.

Buchanan brothers' fishery

ACIPENSER OXYRHYNCHUS (STURGEON)

[Amounts given show the number of pounds of sturgeon taken at the Buchanan brothers' fishery from 1916 to 1922. It is evident that the sturgeon is of small importance in this pound-net fishery]

	1916	1917	1918	1919	1920	1921	1922	Average
April		285	150	150			175	109
May	40	185	275	100	100	100	240	148
June	100		230	25		125		68
July			260					37
August		40				50		13
September			50					7
October	50		290	40		100		68
November			50					7
Total	190	510	1,305	315	100	375	415	

Buchanan brothers' fishery—Continued

POMOLOBUS PSEUDOHARENGUS AND POMOLOBUS ÆSTIVALIS (HERRINGS)

[These species are not separated for the market and therefore are combined in the records under the name "herring." The entire catch (listed by pounds) for the period covered was taken in pound nets. Note that when a small catch was made in April it generally was followed by a larger catch than usual in May]

	1908	1909	1910	1911	1912	1913	1914	1915	1916	1917	1918	1919	1920	1921	1922	Average
March	9,380	12,950	10,250	1,575	2,725	3,500	3,750	850	1,065	1,000	3,165	1,190	1,815	4,750	1,950	3,994
April	26,850	36,850	10,400	6,950	4,110	8,100	16,885	5,345	1,885	6,135	13,390	6,525	1,565	2,810	4,600	10,160
May	1,550	1,100	815	3,525	4,190	5,225	1,025	660	850	2,485	3,365	200	1,435	450	1,600	1,898
June	------	------	------	------	------	275	325	100	------	75	100	------	------	------	300	78
Total	37,780	50,900	21,465	12,050	11,025	17,100	21,985	6,955	3,800	9,695	20,020	7,915	4,815	8,010	8,450	-------

ALOSA SAPIDISSIMA (SHAD)

[The entire catch of shad for the period covered was taken in pound nets and is listed by pounds. A few shad are caught early in March, as soon as the nets are set, and usually not many are caught after May 15. The largest single day's catch for the period covered was made on March 25, 1910, when 3,900 pounds were taken]

	1908	1909	1910	1911	1912	1913	1914	1915	1916	1917	1918	1919	1920	1921	1922	1923	Average
March	2,650	5,000	10,585	3,800	5,195	2,520	2,060	1,535	4,680	625	5,665	2,230	2,570	6,580	1,245	1,815	3,679
April	2,710	11,025	2,260	4,580	4,565	5,405	1,670	1,270	2,210	900	1,400	1,420	330	4,725	3,005	2,150	3,101
May	2,050	1,000	290	1,240	2,215	4,010	490	625	420	700	290	365	655	1,155	1,075	1,585	1,135
Total	7,410	17,025	13,135	9,620	11,975	11,935	4,220	3,430	7,310	2,225	7,355	4,015	3,555	12,460	5,325	5,550	-------

PARALICHTHYS DENTATUS (SUMMER FLOUNDER)

[Amounts are listed in pounds; those marked "b" were taken in part in a seine and in part in pound nets; all other amounts were taken in pound nets. The small catches during midsummer should not be interpreted to mean that this fish is scarce in the bay at that time, for it is taken in considerable numbers with hook and line. A seasonal change in habits is suggested]

	1912	1913	1914	1915	1916	1917	1918	1919	1920	1921	1922	Average
April	1,785	740	1,275	400	520	840	150	165	215	1,730	620	767
May	1,625	450	2,520	1,440	1,970	875	2,155	790	1,865	1,790	1,035	1,501
June	365	210	635	610	725	395	150	320	670	530	230	440
July	80b	75b	235b	100b	165b	60b	80	140	135	185	180	130
August	50b	50b	140b	75b	55b	60b	100	50	100	175	185	94
September	50b	135b	85b	75b	110b	45b	100	115	200	435	200	141
October	1,510b	3,000b	590b	1,275b	1,025b	1,390b	1,745	200	850	4,010	1,800	1,581
November	960	2,585	3,435	4,300	2,475	2,250	3,480	2,160	8,400	25,605	10,150	5,982
Total	6,425	7,245	8,915	8,275	7,045	5,915	7,960	3,940	12,435	34,460	14,400	--------

MUGIL CEPHALUS AND M. CUREMA (MULLETS)

[Mullets are not regularly caught in pound nets. The table, with amounts given in pounds, shows that only occasionally a school is trapped]

	1916	1917	1918	1919	1920	1921	Average
August	300	------	------	------	------	150	75
September	2,640	------	------	------	550	175	561
October	375	50	350	------	50	1,900	454
November	------	------	------	------	100	------	17
Total	3,315	50	350	------	750	2,225	--------

Buchanan Brothers' fishery—Continued

SCOMBEROMORUS MACULATUS (SPANISH MACKEREL)

[Amounts given show the number of pounds of Spanish mackerel taken by a set of two pound nets from 1918 to 1922. Blank spaces do not necessarily indicate that no fish of this species were taken, for daily catches of less than 10 pounds were not listed separately]

	1918	1919	1920	1921	1922	Average
May					100	20
June	375	680	105	1,125	205	498
July		1,400	175	1,005		516
August	125	1,150		50		265
September		550	300	100	300	250
October					40	8
Total	500	3,780	580	2,280	645	

SARDA SARDA (BONITO)

[Amounts given show the number of pounds of bonito taken from 1916 to 1922 at the Buchanan brothers' fishery. Blank spaces do not signify that no bonito were taken, as daily catches amounting to less than 10 pounds were not listed separately]

	1916	1917	1918	1919	1920	1921	1922	Average
May						15		2
June	30	25	25	20	140	20	25	40
July	25	20	30	55	55	30	35	36
August	20	85	20	105	25	30	30	45
September	15	10	15	15	55	10	25	21
Total	90	140	90	195	275	105	115	

PEPRILUS ALEPIDOTUS (STARFISH)

[Amounts are given in pounds; those marked "b" were taken in part in a seine, but mainly in pound nets; all other amounts were taken in pound nets. The first catches of the season generally are made from about May 10 to 25, the species apparently arriving about a month later than its relative, the butterfish]

	1912	1913	1914	1915	1916	1917	1918	1919	1920	1921	1922	Average
May	25,225	3,965	8,815	5,800	7,635	4,940	9,670	9,645	150	4,085	7,065	7,909
June	35,350	18,200	15,285	14,885	11,765	21,030	16,490	5,550	9,175	16,880	12,270	16,080
July	9,685b	11,185b	16,275b	20,000b	2,475b	6,855b	3,465	6,490	10,250	5,990	6,380	9,014
August	2,565b	1,380b	265b	13,120b	645b	2,920b	815	6,990	2,100	7,790	28,100	6,063
September	1,640b	185b	115b	745b	790b	1,080b	2,725	5,850	7,100	19,200	5,190	4,056
October	125b	145b	125b	115b	35b	425b	805	1,160	210	535	735	401
November									1,100			100
Total	74,590	35,060	40,880	54,665	23,345	37,250	33,970	35,685	30,085	54,480	59,740	

PORONOTUS TRIACANTHUS (BUTTERFISH)

[Amounts are given in pounds; those marked "b" were taken in part in a seine, but mainly in pound nets; all other amounts were taken in pound nets. The first catches of the season usually are made during the first half of April, or about a month before its relative, the starfish, is taken]

	1912	1913	1914	1915	1916	1917	1918	1919	1920	1921	1922	Average
April	15,435	210	160	4,000	570	70	290	5,275	315	605	2,400	2,666
May	100,910	1,805	36,260	24,215	9,080	3,920	14,600	15,770	600	15,890	4,410	20,678
June	132,600	26,750	44,190	8,425	23,120	18,095	17,810	21,990	9,850	13,520	12,630	29,907
July	35,840b	36,265b	64,050b	12,405b	7,800b	19,935b	7,000	16,860	17,000	5,830	5,790	20,798
August	10,075b	2,240b	1,070b	8,670b	1,115b	11,295b	2,645	23,480	4,200	5,200	10,010	7,273
September	415b	430b	475b	475b	215b	385b	990	13,300	6,500	4,840	3,950	2,907
October	510b	415b	515b	175b	125b	1,690b	2,020	1,935	240	1,120	585	848
November	125	665	635	2,020	1,705	1,455	2,460	250	850	1,445	910	1,138
Total	295,910	68,780	147,355	60,385	43,730	56,845	47,815	98,860	39,555	48,450	40,685	

Buchanan Brothers' fishery—Continued

TRACHINOTUS CAROLINUS (POMPANO)

[Amounts are given in pounds; those marked "a" were caught in a seine; those marked "b" were taken partly in a seine and partly in pound nets. All other amounts were taken in pound nets. The blank spaces signify that if any pompanoes were taken, the daily catches amounted to less than 10 pounds. This species is not taken in commercial quantities earlier than June]

	1908	1909	1910	1912	1913	1914	1915	1916	1917	1918	1919	1920	1921	1922	Average
June		50			300						20		100	80	43
July		210b		40b	65b	25b	165b	150b	275b		390		225	55	123
August	460a	100a			50b	30b	125b	475b	165b		25		150	115	108
September					60b	630b	35b				100				59
October	40a		50a		650b	1,805b	65b			50	25		35	25	196
Total	500	360	50	40	1,125	2,490	390	625	440	50	560	0	510	275	--------

POMATOMUS SALTATRIX (BLUEFISH)

Amounts are given in pounds; those marked "a" were caught in a seine; those marked "b" were taken partly in a seine and partly in pound nets; all other amounts were taken in pound nets. The blank spaces signify either that no fish at all or that less than 10 pounds were taken on any one day]

	1908	1909	1910	1912	1913	1914	1915	1916	1917	1918	1919	1920	1921	1922	Average
April		125		100			1,300			25				20	112
May	515	550	1,150	1,450	460	810	3,160	30	745	200	50	25	365	185	692
June	150	4,175	1,350	735	1,935	3,460	315	75	25	35	40	220	265	25	914
July	810b	1,975b	125b	910b	2,375b	2,340b	515b	50b	45b	25	150	35	100	200	690
August	2,490a	375a	75a	310b	680b	440b	1,635b	25b	40b	75	60	50	25	150	459
September	1,740a	325a	8,250a	2,430b	385b	375b	125b	20b	495b	130	150	625	50	350	1,103
October	740a	925a	7,850a	3,450b	1,435b	5,540b	635b	125b	400b	735	1,160	400	790	1,475	1,833
November		50a		135	560	3,925	175			100		50		375	383
Total	6,445	8,500	18,800	9,520	7,830	16,890	7,860	325	1,750	1,325	1,610	1,405	1,595	2,780	--------

ORTHOPRISTIS CHRYSOPTERUS (PIGFISH)

Amounts given show the number of pounds of pigfish taken from 1916 to 1922 at the Buchanan brothers' fishery. Blank spaces simply indicate that the daily catches amounted to less than 10 pounds during the periods covered]

	1916	1917	1918	1919	1920	1921	1922	Average
April	50	130	75					36
May	2,800	1,420	2,775					999
June	735	830	100					238
July								
August								
September		405						58
October	55	850	50					136
Total	3,640	3,635	3,000					

ARCHOSARGUS PROBATOCEPHALUS (SHEEPSHEAD)

[Amounts are listed in pounds; those marked "a" were taken in a seine; those marked "b" were taken in part in a seine and in part in pound nets; all other amounts were taken in pound nets. Blank spaces do not always signify that no sheepsheads were taken, as daily catches of less than 10 pounds were not listed separately. According to the fishermen, the sheepshead was an abundant fish "years ago" and was taken in large numbers. Its abundance must have diminished prior to 1908]

	1908	1909	1910	1912	1913	1914	1915	1916	1917	1918	1919–22	Average
April			100									7
May	65	350	350	2,660	325	100			25	140		287
June	25	50	50	400	25	60						43
July												
August												
September												
October		150a		220b								26
November							25					2
Total	90	550	500	3,280	350	160	25		25	140		--------

Buchanan Brothers' fishery—Continued

LEIOSTOMUS XANTHURUS (SPOT)

[Amounts are given in pounds; those marked "b" were taken in part with a seine and in part with pound nets; all other catches were made with pound nets]

	1912	1913	1914	1915	1916	1917	1918	1919	1920	1921	1922	Average
March	--------	2,360	1,100	310	500	90	--------	980	1,300	--------	330	118
April	--------	2,360	1,100	310	500	90	--------	980	540	300	330	591
May	1,320	7,250	12,050	3,085	5,605	1,950	5,360	2,040	1,615	980	7,560	4,438
June	6,700	5,500	8,820	4,585	14,080	6,015	5,815	2,350	4,580	1,530	3,585	5,778
July	18,400b	33,200b	7,960b	8,820b	18,025b	10,395b	3,980	4,310	3,745	9,615	2,985	11,039
August	45,675b	42,225b	16,175b	17,150b	4,400b	11,650b	8,370	3,485	6,200	17,850	3,130	16,028
September	62,950b	95,235b	28,915b	78,575b	57,860b	12,400b	8,420	7,515	4,750	25,655	5,200	34,898
October	41,400b	30,990b	27,410b	92,325b	26,700b	7,375b	20,815	25,970	23,400	37,900	36,250	33,678
November	1,425	285	1,735	8,165	785	1,540	2,095	15,175	5,200	6,185	1,045	3,967
Total	177,870	217,045	104,165	213,015	127,955	51,415	54,855	61,825	51,330	100,015	60,085	--------

BAIRDIELLA CHRYSURA (SAND PERCH)

[Amounts given show number of pounds of sand perch marketed. This species is taken in large numbers, particularly in the spring and summer, but the individuals generally are too small to market]

	1918	1919	1920	1921	1922	Average
April	175	570	85	175	205	242
May	565	110	210	750	260	379
June	215	110	180	165	110	156
July	110	70	85	65	120	90
August	130	50	150	475	50	171
September	275	150	200	745	270	328
October	2,435	3,655	650	3,540	3,755	2,807
November	1,055	300	1,500	1,085	935	971
Total	4,960	5,015	3,060	7,000	5,705	--------

MICROPOGON UNDULATUS (CROAKER)

[Amounts are given in pounds; those marked "a" were caught with a seine; those marked "b" probably were caught partly with a seine and partly with pound nets; amounts unmarked were caught in pound nets. The first catch of croakers usually is made sometime during the last half of March, when the fish arrive in large schools, the very first catches sometimes consisting of several thousand pounds]

	1908	1909	1910	1912	1913	1914	1915	1916	1917	1918	1919	1920	1921	1922	Average
March	1,975	--------	8,750	3,750	175	--------	25	--------	--------	24,180	4,100	12,725	28,870	80,075	11,759
April	2,895	140	27,175	151,320	675	116,010	100,200	288,285	56,100	171,180	24,825	23,870	12,950	22,035	71,261
May	400	--------	11,150	8,620	4,360	14,380	35,200	45,565	7,825	34,570	33,700	5,500	4,165	11,110	15,467
June	760	100	34,300	12,900	6,580	1,580	10,535	7,460	4,960	1,595	2,720	4,745	1,400	2,095	6,552
July	540b	100b	90,400b	12,745b	915b	6,180b	1,305b	4,645b	13,265b	170	210	2,860	31,400	450	11,800
August	510a	3,000a	510a	670b	260b	220b	1,080b	485b	5,665b	430	705	1,850	1,920	355	1,261
September	80a	570a	360a	18,480b	420b	470b	550b	210b	220b	140	400	200	3,830	3,855	2,127
October	20a	30a	80a	20b	210b	270b	1,440b	180b	220b	380	150	200	825	270	307
November	75a	55a	--------	--------	50b	1,150b	535b	200b	175b	435	310	350	130	535	285
Total	7,552	3,995	172,725	208,505	13,645	140,260	150,870	347,030	88,430	233,080	67,120	52,300	85,490	120,780	--------

Buchanan Brothers' fishery—Continued

MENTICIRRHUS AMERICANUS, M. SAXATALIS, AND M. LITTORALIS (KINGFISH)

[The three species of kingfish that occur in Chesapeake Bay are not separated in the market and therefore all were listed as kingfish in the records from which this table was compiled. However, *americanus* is the predominating species, and the quantities listed are chiefly of it. Amounts are given in pounds; those marked "a" were taken with a seine; those marked "b" were taken partly with a seine and partly with pound nets; all other amounts were taken in pound nets]

	1908	1909	1910	1912	1913	1914	1915	1916	1917	1918	1919	1920	1921	1922	Average
April	4,855	4,050	5,115	12,575	1,500	2,700	775	425	1,025	115	50	80	250	125	2,403
May	10,825	5,925	13,800	22,075	2,200	21,275	6,375	7,240	5,975	3,600	560	430	240	460	7,213
June	1,025	3,050	5,625	2,715	1,050	1,850	575	1,790	2,745	950	640	465	215	115	1,629
July	2,500b	2,300b	2,525b	5,075b	1,650b	5,575b	950b	780b	6,235b	210	140	60	120	150	2,019
August	400a	550a	1,900a	1,100b	1,325b	1,200b	1,275b	300b	1,380b	150	200	250	70	100	728
September	100a	175a	100a	320b	475b	400b	150b	50b	105b	70	40	50	50	50	152
October	450a	1,700a	200a	1,405b	1,250b	1,875b	4,350b	1,250b	1,000b	345	170	100	115	410	1,044
November	325a	100a		375	225	1,500	1,875	70	90	135	260	400	80	240	405
Total	20,480	17,850	29,265	45,640	9,675	36,375	16,325	11,905	18,555	5,575	2,060	1,835	1,140	1,650	--------

CYNOSCION REGALIS (SQUETEAGUE)

[Amounts are given in pounds; those marked "a" were taken in a seine; those marked "b" were taken partly in seines and partly in pound nets; all other amounts were taken in pound nets. The first catches in commercial quantities usually are made early in April]

	1908	1909	1910	1912	1913	1914	1915	1916	1917	1918	1919	1920	1921	1922	Average
April	7,775	2,800	11,785	3,440	8,450	1,175	5,250	675	360	245	175	120	1,650	1,125	3,216
May	37,450	5,300	23,060	18,140	7,800	10,755	17,350	14,695	19,515	12,420	7,475	3,660	6,630	8,585	13,774
June	5,200	7,650	45,700	12,200	28,310	13,825	5,100	27,135	30,055	20,780	6,400	4,160	2,955	2,100	15,112
July	7,450b	5,925b	17,550b	28,550b	27,215b	20,025b	17,950b	8,175b	25,215b	13,340	1,580	3,370	2,520	1,195	12,861
August	14,525a	2,365a	2,365a	10,960b	4,400b	5,350b	8,000b	12,305b	4,400b	8,910	1,510	1,000	1,555	870	5,609
September	1,110a	935a	185a	5,920b	2,975b	1,950b	3,725b	5,170b	1,355b	5,130	1,000	600	1,545	1,235	2,345
October	100a	235a	200a	7,485b	4,400b	5,125b	3,225b	5,560b	7,375b	10,590	630	3,600	3,895	2,380	3,914
November	565a	265a		3,500	1,925	10,725	6,975	1,795	925	4,075	6,575	6,000	5,770	5,585	3,905
Total	74,175	25,495	100,845	90,195	85,475	68,930	67,575	75,510	89,200	75,490	25,345	22,510	26,520	23,075	--------

SYSTEMATIC CATALOGUE OF THE FISHES OF CHESAPEAKE BAY

INTERPRETATION OF DESCRIPTIONS

Abbreviations used by many writers of ichthyological descriptions have been adopted. For example, the expression "head 3 to 3.5" signifies that the length of the head, measured from the tip of the upper jaw to the bony margin of the opercle (unless otherwise stated), is contained 3 to 3.5 times in the "standard length"— that is, in the distance from the end of the snout to the base of the caudal fin. Similarly, the expression "depth 2.5 to 3" signifies that the greatest depth of the body is contained 2.5 to 3 times in the standard length. Roman numerals are used for indicating spines and. Arabic numerals for soft rays. in giving fin-ray formulæ. For example, "D. VII–I, 15; A. III, 12" signifies that the dorsal fins are two in number, and that the first one consists of 7 spines and the second of 1 spine and 15 soft rays, and that the anal fin consists of 3 spines and 12 soft rays. If the dorsal fin had been single and had contained the same number of rays, the formula would have been written thus: D. VIII, 15. The number of scales given (unless otherwise stated) is the number of oblique rows that occur just above the lateral line from the upper angle of the gill opening to the base of the caudal. The terms used in the descriptions and keys in describing the external structure of a fish are largely indicated in the accompanying outline of the croaker.

USE OF KEYS

The keys have not been made with the view of showing natural relationships, but they are intended purely for the purpose of ready identification, and in preparing them only the characters applicable to the fishes of Chesapeake Bay have been taken into consideration. In using the keys, first determine to which of the major groups

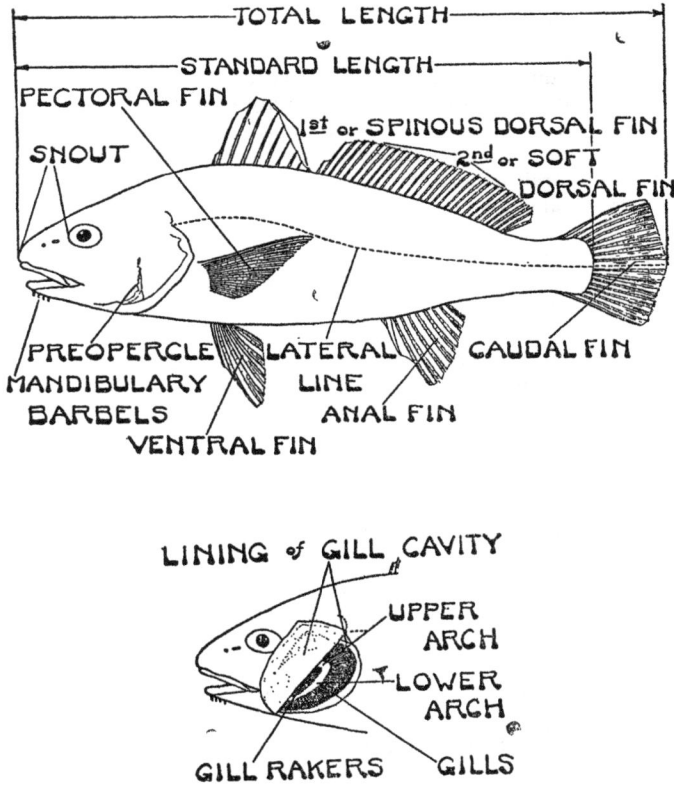

FIG. 24.—Diagram of a sciænid, explaining terms used in keys and descriptions

the specimen in hand belongs; then take up the regular order of letters under that group. If the characters of the specimen do not agree with those under the single letters, look under the double letters (occasionally triple letters are used), ignoring all intervening matter. By means of indentations, the order of subordination of the minor groups to the major groups is shown.

KEY TO THE FAMILIES

I. LEPTOCARDII: Amphioxi (the lancelets).—Skeleton a cartilaginous rod; brain and skull wanting; body elongate, compressed, translucent; mouth a longitudinal slit, surrounded by cirri; eyes and fins rudimentary_____*Branchiostomidæ* (lancelets), p. 42

II. MARSIPOBRANCHII: Hyperoartia (the lampreys).—Skeleton cartilaginous; brain and skull present; body eel-shaped; head not differentiated from the body; mouth circular, suctorial; seven small, round gill openings on each side_____*Petromyzonidæ* (lampreys), p. 43

III. ELASMOBRANCHII (Sharks, skates, and rays).—Skeleton cartilaginous; skull imperfectly developed; brain present; gill openings slitlike, five to seven on each side; skin with small, rough scales, spines, or tubercles, or naked; air bladder absent; jaws separable from the skull.

 1. Body elongate, usually more or less rounded, not greatly depressed and not forming a disk; gill openings all or partly lateral; pectoral fins not attached to the head.

  EUSELACHII (the typical sharks).

   a. Body typically fishlike; one or two dorsal fins present; anal fin present.

    b. Head normally shaped, not broad and expanded across the eyes.

     c. Nictitating membrane absent; each nostril with a cirrus or barbel; two or three gill slits over base of pectoral_____*Orectolobidæ* (nurse sharks), p. 44

     cc. Nictitating membrane absent; nostrils without a cirrus or barbel; gill slits all in advance of pectorals; mouth broad, mainly transverse
_____*Lamnidæ* (man-eater sharks), p. 45

     ccc. Nictitating membrane present, nostrils without a cirrus or barbel; last gill slit above base of pectoral; mouth narrow, crescent-shaped_____*Galeidæ* (gray sharks), p. 46

    bb. Head greatly expanded across the orbital region, more or less hammer-shaped
_____ *Sphyrinidæ* (hammerhead sharks), p. 49

  TECTOSPONDYLI (the dogfishes and angel sharks).

   aa. Body more or less depressed; two dorsal fins present; anal fin absent.

    d. Head and body not greatly depressed; each dorsal fin preceded by a spine; pectoral fins not greatly expanded_____*Squalidæ* (dogfishes), p. 51

    dd. Head and body notably depressed and expanded; dorsal fins without spines; pectoral fins large, greatly expanded_____*Squatinidæ* (angel sharks), p. 54

 2. Head and body much depressed; gill openings all inferior; pectoral fins greatly expanded, attached to the head; anal fin absent.

  BATOIDEI (skates and rays).

   a. Tail comparatively thick, bearing two dorsal fins and no caudal spine.

    b. Body elongate, depressed, but not forming a disk; snout produced into a long, thin, saw-like process, armed on each side with a series of large, strong teeth
_____*Pristidæ* (sawfishes), p. 55

    bb. Body broad, forming with the pectorals a rhomboidal or subcircular disk; snout more or less produced, not sawlike, and never armed with teeth.

     c. Disk rhomboidal; skin usually rough, bearing spines, prickles, or tubercles; no electric organs present_____*Rajidæ* (skates), p. 56

     cc. Disk subcircular; skin smooth, unarmed; an electric organ on each side of median line on head_____*Torpedinidæ* (electric rays), p. 61

   aa. Tail usually very slender; bearing one or no dorsal fins and usually one or more strong, serrated spines.

    d. Disk subcircular or rhomboidal; pectoral fins uninterrupted confluent around the snout
_____*Dasyatidæ* (sting rays), p. 63

    dd. Disk broad and angular; pectoral fins not confluent around the snout; head bearing one or a pair of rostral processes or cephalic fins.

     e. Head bearing one or a pair of rostral processes; teeth large, flat, largely hexagonal.

      f. Snout with a pair of rostral fins, joined together and forming a single rostral process_____*Myliobatidæ* (eagle rays), p. 68

      ff. Snout with two separate lobes, making the anterior margin of the snout concave
_____*Rhinopteridæ* (cow-nosed rays), p. 70

     ee. Head with a pair of cephalic fins, developed as two hornlike appendages; teeth small, numerous, in pavement_____*Mobulidæ* (sea devils), p. 71

IV. PISCES (The true fishes).—Skeleton usually bony, sometimes cartilaginous; skull with a well-developed system of bones; a single gill opening on each side; skin commonly with normally developed scales, sometimes with variously shaped bony plates and occasionally naked.

 1. GANOIDEI (ganoid fishes): Tail strongly heterocercal; arterial bulb muscular, with numerous valves.

GLANOSTOMI (the sturgeons).

    a. Skeleton cartilaginous; snout produced, with four flexible barbels; mouth underneath; teeth wanting; skin imperfectly covered with bony plates_____*Acipenseridæ* (sturgeons), p. 72

HOLOSTEI (the gar pikes).

    aa. Skeleton bony; both jaws greatly produced, armed with sharp teeth; no barbels; skin completely covered with rhombic plates_____*Lepisosteidæ* (gar pikes), p. 77

2. TELEOSTEI (nonganoid fishes): Tail homocercal or isocercal (not heterocercal); arterial bulb thin, with a pair of opposite valves.

  A. Ventral fins present, abdominal.

    a. Dorsal fin single; adipose fin present or wanting.

      b. Adipose fin wanting.

        c. Pectoral fins inserted low on side, below axis of body; lateral line, when present, normally placed; lower pharyngeal bones separate.

          d. Gill openings restricted, the membranes attached to the isthmus; jaws without teeth.

EVENTOGNATHI (suckers, carps, and carplike minnows).

            e. Maxillaries forming sides of margin of upper jaw; lower pharyngeal bones armed with a single row of comblike teeth_____*Catostomidæ* (suckers), p. 117

            ee. Premaxillaries alone forming margin of upper jaw; lower pharyngeal bones supporting one to three series of teeth, the teeth few in number

                _____*Cyprinidæ* (carps and minnows), p. 120

          dd. Gill openings not restricted, the membranes free from the isthmus; teeth in jaws present or absent.

            f. Head naked; dorsal fin more or less over the middle of the body; upper jaw not protractile; color silvery.

ISOSPONDYLI (the clupeoid and salmonoid fishes).

              g. An external bony plate present between the arms of the lower jaw; lateral line present.

                h. Scales comparatively small; pseudobranchiæ present, large; the last ray of dorsal not produced_____*Elopidæ* (10-pounders), p. 78

                hh. Scales very large, pseudobranchiæ absent; last ray of dorsal greatly produced, filamentous_____*Megalopidæ* (tarpons), p. 79

              gg. No bony plate between the arms of the lower jaw; lateral line absent.

                i. Body oblong or elongate; mouth small to moderate, terminal or slightly superior, oblique; stomach not gizzardlike_*Clupeidæ* (herrings), p. 81

                ii. Body rather short and deep; mouth small, inferior, terminal; stomach gizzardlike _____*Dorosomidæ* (gizzard shad), p. 106

                iii. Body elongate; mouth large; snout pointed, usually projecting far beyond mandible; stomach not gizzardlike

                _____*Engraulidæ* (anchovies), p. 108

            ff. Head scaly; dorsal fin commonly posterior in position; upper jaw protractile or not; color not silvery.

HAPLOMI (the pikelike fishes).

              j. Body very elongate; snout considerably produced, depressed; mouth large; maxillaries forming sides of upper jaw; size moderate to large_____*Esocidæ* (pikes and pickerels), p. 132

CYPRINODONTES (the killifishes and top minnows).

              jj. Body oblong or moderately elongate; snout not produced; mouth small; premaxillaries forming entire margin of upper jaw; size small.

                k. Anal fin similar to the dorsal and not modified in the male; species oviparous_____*Cyprinodontidæ* (killifishes), p. 134

                kk. Anal fin in the male modified, some of the rays produced, others short and more or less coalesced, the fin serving as an intromittent organ; species viviparous__*Pœciliidæ* (top minnows) p. 145

SYNENTOGNATHI (the gars, halfbeaks, and flying fishes).

cc. Pectoral fins inserted rather high on sides, on or near the axis of the body; lateral line usually placed abnormally low on the sides, frequently along the edge of the abdomen; body very elongate; vertebræ numerous (45 to 70).

  l. Snout not in the shape of a tube; body covered with scales.

    mm. Both jaws produced, forming a beak, each jaw with a band of sharply pointed teeth; pectoral fins normal.

      n. Dorsal and anal fins single, not followed by detached finlets_____*Belonidæ*, p. 147

      nn. Dorsal and anal fins followed by a series of four to six detached finlets_____*Scomberesocidæ*, p. 151

    mm. Upper jaw short, the lower much produced (in Cheasapeake specimens); pectoral fins normal__*Hemiramphidæ*, p. 152

    mmm. Jaws normal, neither produced (in adult); pectoral fins greatly enlarged, used as organs of flight _____*Exocœtidæ*, p. 154

  ll. Snout greatly produced, forming a long tube, terminating in a small mouth; scales wanting; bony plates on various parts of the body; caudal fin forked, the middle ray produced into a long filament_____*Fistulariidæ* (cornet fishes), p. 186

bb. Adipose fin present.

NEMATOGNATHII (the catfishes).

    o. Body without true scales (naked in Chesapeake specimens); anterior part of head with one or more pairs of whiskers; dorsal and pectoral fins each with a strong spine.

      p. Nostrils close together, neither with a barbel; ventral fins with 6 rays_____*Ariidæ* (sea catfishes), p. 127

      pp. Nostrils far apart, the posterior one with a barbel; ventral fins with eight or nine rays _____*Ameiuridæ* (horned pouts), p. 129

INIOMI (the lantern fishes).

    oo. Body with cycloid scales; head without whiskers; head and snout depressed; mouth very large; premaxillaries alone forming margin of upper jaw; fins without spines; caudal forked_*Synodontidæ* (lizard fishes), p. 130

aa. Two dorsal fins, the anterior with spines only, the posterior chiefly of soft rays; no adipose.

    q. Pectoral fins entire, no free rays.

      r. Head not pikelike; the jaws not produced; teeth small or wanting; lateral line obsolete.

        s. First dorsal with three to nine flexible spines; anal fin with a single weak spine _____ *Atherinidæ* (silversides), p. 187

        ss. First dorsal with four stiff spines; anal fin with three stiff spines (two in very young) _____*Mugilidæ* (mullets), p. 192

      rr. Head pikelike; the jaws produced; teeth strong; lateral line present _____*Sphyrænidæ* (barracudas), p. 197

    qq. The lowermost rays of pectorals free and feelerlike or barbellike ____*Polynemidæ* (threadfins), p. 199

AA. Ventral fins present, attached to the thorax or throat, under, anterior to, or slightly behind base of pectorals.
 a. Gill openings moderate or large, situated anterior to pectoral fins; carpal bones normally developed; the pectoral fins without a "wrist."
  b. Ventral fins always with I, 5 rays.
   c. Ventral fins separate and distinct, never united and never forming a part of a sucking disk.
    d. Suborbital without a bony stay; cheeks not mailed; pectoral fins entire, without detached rays.
     e. Anterior dorsal fin converted into a sucking apparatus, forming a disk at nape, consisting of several crosswise partitions and a single lengthwise septum
      --*Echeneididæ* (remoras), p. 328
     ee. Anterior dorsal fin normal, not converted into a sucking disk.
      f. Dorsal and anal fins followed by a series of detached finlets; anal fin not preceded by free spines; caudal fin broadly forked
       ------------------------------------*Scombridæ* (mackerels), p. 200
      ff. Dorsal and anal fins not followed by several detached finlets.
       g. Body elongate, spindle-shaped; head strongly depressed; snout broad; first dorsal with eight or nine free spines__*Rachycentridæ* (crab eaters), p. 234
       gg. Body not spindle-shaped; head never greatly depressed; snout not expanded.
        h. Anal fin preceded by two free spines (sometimes obsolete in very old, joined by membrane in very young); ventral fins present at all ages; œsophagus without teeth.
         i. Preopercle entire; caudal peduncle slender, frequently with lateral bony scutes; teeth, if present, small to moderate
          -------------------*Carangidæ* (crevallies, pompanos, etc.), p. 216
         ii. Preopercle serrate; caudal peduncle rather stout, never with bony scutes; teeth unequal, some of them enlarged
          --------------------------*Pomatomidæ* (bluefishes), p. 231
        hh. Anal fin not preceded by free spines.
         j. Oesophagus provided with lateral sacs containing teeth; anal fin long, similar to dorsal; ventral fins normal in young, sometimes reduced or wanting in adults_____*Stromateidæ* (butterfishes), p. 210
         jj. Oesophagus not provided with teeth.
          k. Lateral line extending to end of caudal fin; anal fin with one or two spines.
           l. Backbone typically with 10+14 vertebræ
            -------------------*Sciænidæ* (croakers and drums), p. 271
           ll. Backbone typically with 14+10 vertebræ
            --------------------------*Otolithidæ* (weakfishes), p. 296
          kk. Lateral line ending at base of caudal.
           m. Nape with a fleshy flap resembling an adipose fin; similar but smaller fleshy flaps on sides of lower jaw near angle of mouth; dorsal fin continuous____*Branchiostegidæ* (tilefishes), p. 305
           mm. No fleshy flap at nape or on lower jaw.
            n. Gills 4, a slit behind the fourth.
             o. Premaxillaries excessively protractile, their basal processes very long, entering a groove at top of cranium just underneath the skin; scales large; fin spines strong; color silvery_____*Gerridæ* (mojarras), p. 369
             oo. Premaxillaries only moderately protractile, or not protractile.
              p. Anal fin with one or two spines; dorsal fins separate, with about 8 to 16 spines; form elongate; fresh-water fishes.

q. Branchiostegals 7; preopercle serrate; air bladder present; fishes of moderate size
------------------------*Percidæ* (perches), p. 236

qq. Branchiostegals 6; preopercle entire; air bladder obsolete or nearly so; fishes of small size, the majority of the species not exceeding a length of 3 or 4 inches_____*Etheostomidæ* (darters), p. 237

pp. Anal fin with three to eight spines; dorsal fins separate or continuous; form various.

r. Teeth more or less bristlelike, or at least slender and close-set, movable; gill membranes attached to the isthmus; soft part of vertical fins completely covered with small scales; form short and deep.

s. Dorsal fins nearly or quite separate; teeth slender but scarcely bristlelike
------------*Ephippidæ* (spade fishes), p. 306

ss. Dorsal fin continuous; teeth numerous, very slender, bristlelike; color usually brilliant
------*Chætodontidæ* (butterfly fishes), p. 308

rr. Teeth not bristlelike, usually firmly attached to the jaws, not movable; gill membranes free from the isthmus; form usually elongate.

t. Pseudobranchiæ very small; anal fin with three to eight spines; dorsal fin continuous or notched, with 6 to 13 spines; form moderately short and deep to elongate, compressed; fresh-water fishes____*Centrarchidæ* (fresh-water basses and sunfishes), p. 238

tt. Pseudobranchiæ well developed; anal fin definitely with three spines; form elongate, generally more or less compressed; marine fishes.

u. Teeth on anterior part of jaws broad, incisorlike; form oblong or elongate, always notably compressed.

v. Teeth on sides of jaws molarlike; no teeth on vomer or palatines; vertical fins not densely covered with scales; intestinal canal of moderate length
------*Sparidæ* (porgies), p. 261

vv. Jaws without molar teeth; teeth present on vomer and palatines; vertical fins densely scaled; intestinal canal very long; species herbivorous
-----*Kyphosidæ* (rudderfishes), p. 269

uu. Teeth in jaws all pointed, not broad and incisorlike.

w. Vomer and palatines without teeth.

x. Body deep, strongly compressed; the back strongly elevated; preopercle with large serrations at angle; caudal fin round
-----*Lobotidæ* (triple-tails), p. 255

xx. Body elongate, only moderately compressed; the back not greatly elevated; preopercle entire or with fine serrations; caudal fin forked__*Pomadasidæ* (grunts),p.257

ww. Vomer and palatines with teeth.

y. Head and body much compressed; mouth very oblique to nearly vertical; eye very large; postorbital part of head short; scales small, very rough __*Priacanthidæ* (catalufas),p.253

yy. Head and body only moderately compressed; mouth moderately oblique to nearly horizontal; eye small to moderate; postorbital part of head not shortened; scales not excessively rough.

z. Maxillary for the most part slipping under preorbital; opercle without a spine; teeth in the jaws rather strong, unequal, some of them usually enlarged __*Lutianidæ* (snappers), p. 256

zz. Maxillary not, or only partly, concealed by the preorbital; opercle ending in a spine.

(a) Body elongate, compressed; maxillary without a supplemental bone; teeth pointed, fixed; two dorsal fins; scales of moderate size_____*Moronidæ* (white basses), p. 244

(aa) Body oblong, somewhat compressed; maxillary with a supplemental bone; dorsal fin continuous; scales quite small_____*Epinephelidæ* (groupers), p. 250

nn. Gills 3½, the slit behind the last small or wanting.

(aaa) Body rather robust; maxillary without a supplemental bone; teeth pointed, fixed; dorsal fin continuous; scales moderate or large_____*Serranidæ* (sea basses), p. 251

(b) Head and body more or less compressed; eyes lateral, moderately large; scales large; mouth horizontal to more or less oblique.

(c) Teeth in the jaws large, separate_____*Labridæ* (lipped fishes), p. 317

(cc) Teeth in the jaws coalesced, forming a continuous cutting edge _____*Scaridæ* (parrot fishes), p. 321

(bb) Head broader than deep, partly covered with bony plates; eyes very small, on top of head; mouth vertical, surrounded by fleshy fringes _____*Uranoscopidæ* (star-gazers), p. 329

dd. Suborbital with a bony stay; head inclosed in bony plates, bearing spines; pectoral fins long, winglike, with the three lowermost rays detached and free from each other, developed as feelers_____*Triglidæ* (sea robins), p. 312

cc. Ventral fins close together forming a sucking disk, or separate, with a sucking disk between them of which they form a part.

(d) Body short and thick, more or less triangular in cross section; skin with bony tubercles; suborbital stay present; opercles normally developed; gills 3½; ventral fins forming the bony center of a sucking disk _____*Cyclopteridæ* (lumpfishes), p. 311

(dd) Body oblong or elongate, roundish or more or less compressed; body with or without scales; no suborbital stay; opercle normally developed; gills 4; ventral fins close together, forming a sucking disk_____*Gobiidæ* (gobies), p. 322

(*ddd*) Body oblong, broad, and depressed anteriorly; skin naked; no suborbital stay; opercle reduced to a concealed spinelike projection; gills 2½ or 3; ventral fins far apart, with a sucking disk between them, of which they form a part --*Gobiesocidæ* (clingfishes), p. 339

aa. Gill openings reduced to small foramen, situated near the axils of pectorals; carpal bones greatly elongated, forming a "wrist."

(*e*) Mouth large, superior, very oblique to vertical; gill openings in or near lower axil of pectoral; oblique to vertical; two dorsal fins, the first dorsal with one to three detached tentacle-like spines on the head, the first spine expanded at tip, forming a lure or bait.

(*f*) Head and body very broad, depressed anteriorly; pseudobranchiæ present; mouth excessively large and broad; skin naked; head and sides with dermal flaps; size large--------------------------*Lophiidæ* (anglers), p. 351

(*ff*) Head and body compressed; pseudobranchiæ absent; mouth moderately large, not excessively broad; skin naked or with minute tubercles and dermal tentacles; size rather small----*Antennariidæ* (frogfishes), p. 353

(*ee*) Mouth small, inferior; gill opening above and somewhat behind axil of pectoral; a single short dorsal fin, consisting of soft rays only; a rostral process present; skin covered with bony tubercles and spines--*Ogcocephalidæ* (batfishes), p. 354

bb. Ventral fins not definitely with I, 5 rays.

(*g*) Form unsymmetrical, the eyes and color on one side, leaving the other side blind and colorless.

(*h*) Eyes large, usually separated; mouth moderate or large; teeth generally well developed; margin of preopercle not concealed by skin and scales ----------------------------*Pleuronectidæ* (flounders), p. 164

(*hh*) Eyes small, very close together; mouth small, twisted; teeth small or wanting; margin of preopercle concealed by skin and scales.

(*i*) Body oblong or ovate; eyes and color on the right side; caudal fin free from the dorsal and anal; right ventral on ridge of abdomen and continuous with the anal fin-----*Achiridæ* (broad-soles), p. 175

(*ii*) Body elongate; eyes and color on the left side; caudal fin joined to the dorsal and anal; ventral fins, if present, free from the anal ----------------------------*Cynoglossidæ* (tonguefishes), p. 177

(*gg*) Form symmetrical, the eyes and color not confined to one side.

(*j*) Tail isocercal, the vertebral column pointed behind, the last vertebræ very small; the fins all without spines.

(*k*) Ventral fins inserted almost on the chin, in advance of eyes, each developed as a long forked barbel; caudal fin confluent with the dorsal and anal; body more or less eel-shaped ----------------------------*Ophidiidæ* (cusk eels), p. 335

(*kk*) Ventral fins inserted posterior to the eyes, large or small; caudal fin separate and distinct from the dorsal and anal.

(*l*) Head elongate, shaped as in the pikes, its upper surface with an excavated area; no barbels; ventral fins normally shaped, well developed; dorsal fins 2, the first one short, the second one long----------------------*Merlucciidæ* (hakes), p. 162

(*ll*) Head not especially elongate and not shaped as in the pikes; chin with a barbel; ventral fins various, with two to seven rays; dorsal fins 1, 2, or 3, extending over most of the back ----------------------------*Gadidæ* (codfishes), p. 155

(*jj*) Tail not isocercal, truncate at base of caudal; at least some of the fins with spines.

(*m*) Head rough, bony, with spines, shields, and ridges.

    (*n*) Head rather high, compressed; interorbital space deeply concave; numerous fleshy cirri on head; pectoral fins moderately large, not especially produced and not divided into two sections__*Hemitripteridæ* (sea ravens), p. 309

    (*nn*) Head low, blunt, depressed, quadrangular, or nearly entirely covered with bony shields; interorbital not deeply concave; no fleshy cirri; pectoral fins divided into two sections, the inner one greatly produced, used as an organ of flight__*Cephalacanthidæ* (flying gurnards), p. 316

(*mm*) Head not especially bony, with or without a few spines, no bony shields.

    (*o*) Body robust, depressed anteriorly, compressed posteriorly; mouth large, broad; teeth short but very strong; scales wanting (in Chesapeake specimens); dorsal fins 2, the first with two or three low spines; ventral fins well developed, jugular, without a true spine _____*Batrachoididæ* (toadfishes), p. 337

    (*oo*) Body moderately or greatly elongate, more or less compressed; mouth usually small; teeth various; skin naked or with small scales; dorsal fin single, the anterior part and sometimes the whole fin with spines; ventral fins small, jugular, composed of I, 1 to 3 rays _____*Blenniidæ* (blennies), p. 332

    (*ooo*) Body elongate, somewhat compressed, tapering both anteriorly and posteriorly, the caudal peduncle being very long and slender; mouth moderate, oblique; skin naked or with vertically oblong plates on sides; middle or sides of abdomen shielded by the produced innominate bones; dorsal fin preceded by two or more free spines; ventral fins thoracic to subthoracic, with one strong spine and one or two rudimentary soft rays._____*Gasterosteidæ* (sticklebacks), p. 178

AAA. Ventral fins absent.

  a. Body very elongate, rounded, snakelike; premaxillaries rudimentary or wanting.

    b. Body covered with rudimentary, elongate, imbedded scales, placed at right angles to each other; lower jaw projecting; origin of dorsal far behind pectorals _____*Anguillidæ* (common eel), p. 111

    bb. Body scaleless; upper jaw projecting; origin of dorsal over or somewhat behind middle of pectorals_____*Congridæ* (conger eels), p. 116

  aa. Body not snakelike; premaxillary bones present.

    c. Gill membranes not joined to the isthmus.

      d. Body rather deep to very deep and strongly compressed; mouth small; caudal fin deeply forked; size rather small_____*Stromateidæ* (butterfishes), p. 210

      dd. Body very elongate, compressed, band-shaped, tapering posteriorly; head sharply pointed; mouth large, nearly terminal; teeth very large; scales wanting; dorsal fin beginning on head and extending over entire body; caudal fin wanting _____*Trichiuridæ* (cutlass fishes), p. 208

      ddd. Body moderately elongate, not compressed; upper jaw greatly produced, forming a sword; caudal fin large and forked; size very large _____*Xiphiidæ* (swordfishes), p. 209

    cc. Gill membranes broadly joined to the isthmus.

      e. Body inclosed in a bony armor composed of rings or polygonal plates.

 f. Snout tubular, bearing a small mouth at the tip; tail long, sometimes prehensile; body covered with bony rings

  _____*Syngnathidæ* (pipefishes and seahorses), p. 181

 ff. Snout not tubular; mouth small, terminal; tail of moderate length; body covered with boxlike shell, composed of polygonal plates

  _____*Ostraciidæ* (trunkfishes), p. 345

ee. Body not inclosed in a bony armor; the skin naked, with scales, or beset with prickles and spines of varying sizes.

 g. Teeth fused, forming a continuous cutting edge; body not compressed, somewhat globular in form and capable of considerable inflation; dorsal fin single.

  h. Teeth in each jaw anteriorly divided by a median suture; skin smooth or more or less prickly_____*Tetraodontidæ* (puffers), p. 346

  hh. Teeth in the jaws undivided, having no median suture; body covered with strong bony spines_____*Diodontidæ* (porcupine fishes), p. 349

 gg. Teeth separate, not fused and not forming a continuous cutting edge; body rather deep, compressed; two dorsal fins.

  i. First dorsal with three spines; scales rather large, bony, bearing spines or bony tubercles_____*Balistidæ* (trigger fishes), p. 340

  ii. First dorsal consisting of a single spine; scales small, bearing slender spines, making the surface of the body rough, velvety

  _____*Monacanthidæ* (filefishes), p. 342

Class LEPTOCARDII

Order AMPHIOXI

Family I.—BRANCHIOSTOMIDÆ. The lancelets

Body elongate, compressed, tapering gradually to both extremities; mouth a longitudinal slit surrounded by a fringe of cirri; eyes and fins rudimentary; color pale, translucent. A single genus is represented in United States waters.

1. Genus BRANCHIOSTOMA Costa. Lancelets

Reproductive organs present on both sides of the median line; anal fin present, with traces of rays; vertebral column not produced backward into a caudal process.

1. Branchiostoma virginiæ Hubbs. Amphioxus; Lancelet.

Amphioxus lanceolatus Rice, 1878a, p. 503; Andrews, 1893, p. 238.
Branchiostoma lanceolatum Jordan and Evermann, 1896–1900, p. 3, Pl. I, fig. 1.
Branchiostoma virginiæ Hubbs, Occ. Papers, Mus. Zool., Univ. Mich., No. 105, 1922, p. 8; Sewell's Point, Va.

"The lancelet of Chesapeake Bay appears to differ from the other American species of the genus in the increased number of myotomes. In this respect it resembles the European *B. lanceolatum*, from which, in turn, it is distinguished by the more posterior position of the anus in reference to the lower lobe of the caudal, the relatively shorter distance between this fin lobe and the atriopore, and the more numerous dorsal-ray chambers. It is more closely related to *floridæ* than to *lanceolatum*. All of the lancelets from the east coast of the United States, variously referred to *lanceolatum* or *caribæum*, are perhaps conspecific with the Chesapeake form. It seems not improbable that *virginiæ* and *floridæ* will be found to intergrade.

"Dorsal-ray chambers, 259 to 309 (average of five, 279); anal-ray chambers, 36 to 40 (average of six, 38). Dorsal-ray chambers about two or three times as high as long; dorsal fin about one-eighth as high as body. Anus near middle of lower caudal lobe; origin of this lobe about midway between tip of tail and atriopore. Postanal length, 8.5 to 11.5 in total. Preatrioporal length, 2.4 to 2.7 times postatrioporal length. Myotome formula: 36 to 40+14 to 16+9 to 12=60 to 64 (in type material); 36 to 38+13 or 14+11 to 15=61 to 64 (according to Andrews, 1893). Maximum length, 5.3 cm. (Andrews, 1893.)" (Hubbs, 1922.)

This curious little animal is not represented in our collection. It was first recorded from Chesapeake Bay by Rice (1880, p. 1), who followed European authors in considering the American and European species identical. Andrews (1893, pp. 238 to 240), after examining specimens from several localities, concluded that the specimens from Chesapeake Bay belonged to the European form, *B. lanceolatum*, rather than to the more southern American form, *B. caribæum*. Hubbs (1922, p. 8) found the Chesapeake Bay specimens to represent a new species—*B. virginiæ*—which differs from other American species in the more numerous myotomes.

These little animals were first made known to science in 1774 from specimens found upon the coast of Cornwall, England, and described by Pallas, who considered them a species of snail and gave them the name *Limax lanceolatus*.

The lancelets live principally in the sand. The young are often taken in plankton nets, but the adults that have been captured are reported either to have been dug out of sand along the shore or taken in dredges. Rice (1880, p. 8) states that live animals kept in glass containers swam much like tadpoles but different, in that the head, or anterior part of the body, moved from side to side as far and as vigorously as the tail. They swam about either on the side or on the abdomen and sometimes on the back but never backward.

The young did not "burrow," but the adults remained hidden in the sand (which was provided on the bottom of the containers) during the day, but at night they came near to the surface or emerged wholly or in part, indicating that the day is their rest period and that they feed at night.

Habitat.—Chesapeake Bay.

Chesapeake localities.—(a) Previous record: Fort Wool, Fortress Monroe, Willoughby Sandspit, and Sewell's Point. (b) Specimens in collection: None.

Class MARSIPOBRANCHII
Order HYPEROARTIA
Family II.—PETROMYZONIDÆ. The lampreys

Body eel-shaped, more or less cylindrical anteriorly, compressed posteriorly; head not differentiated from the body; mouth nearly or quite circular, suctorial, usually armed with teeth; eyes developed, at least in the adult; gill openings small, rounded, seven on each side, arranged in a row along the chest; dorsal fin notched or divided, its posterior part commonly continuous with the caudal and anal fins around the tail; intestine with a spiral valve.

2. Genus PETROMYZON Linnæus. Lampreys

Teeth present in mouth, arranged in concentric lines, pointed and rather close together, the teeth immediately anterior to mouth two or three in number; the lateral teeth bicuspid; dorsal fins 2, well separated. Of this genus, a single species is known, which lives in the sea but ascends rivers to spawn.

2. Petromyzon marinus Linnæus. Lamprey; Lamprey eel.

Petromyzon marinus Linnæus, Syst. Nat., ed. X, 1758, 230; European seas. Uhler and Lugger, 1876, ed. I, p. 194, ed. II, p. 164; Bean, 1883, p. 367; Jordan and Evermann, 1896–1900, p. 10, Pl. I, fig. 3; Smith and Bean, 1899, p. 180; Fowler, 1912, p. 51.

Body eel-shaped, somewhat depressed anteriorly, compressed posteriorly; head depressed, its length to first gill opening greater than the distance from the first to the last gill opening, 6.6 in total length; eye of moderate size, 6 in head; interorbital space broad, 3 in head; mouth, or buccal disk, large, its diameter about 2 in head; teeth on each side of mouth bicuspid, a series posterior to the mouth coalesced, the other teeth simple; the origin of the first dorsal distinctly behind the middle of the body, the distance from tip of snout to origin of dorsal 1.9 in total length; the second dorsal well separated from the first, continuous with the rounded caudal, with a depression posteriorly; anal fin represented by a mere fold.

Color in alcohol plain bluish-gray above, pale below. The color in life has been described as mottled brown or black above, occasionally plain bluish, with lower parts whitish or gray.

A single specimen, 158 mm. (6¼ inches) in length, is at hand and it forms the basis for the foregoing description. This lamprey is readily recognized by the bicuspid teeth on the sides of the mouth and by the divided and well separated dorsal fins.

The lampreys attach themselves to larger fish by means of the suctorial mouth, sucking their blood and making ulcerous sores, often producing death. Surface (1898, p. 212), in an account of the variety *P. marinus unicolor*, records that this lamprey destroyed large numbers of catfish, suckers, carp, etc., in Cayuga Lake, New York. Shad are sometimes taken with lampreys 6 to 14 inches in length hanging on their sides. Kendall (field notes, 1894) reports a 10-inch lamprey clinging to a menhaden only 6 inches in length. Bigelow and Welsh (1925, p. 20) report lampreys preying upon cod, haddock, and mackerel in Massachusetts Bay. At one time lampreys were said to be common in the Chesapeake during the early spring and to have destroyed many shad caught with gill nets. Within recent years, however, it has not been sufficiently abundant in Chesapeake Bay to be considered destructive of other fishes.

This lamprey is anadromus and ascends fresh-water streams in the spring to spawn, coming with the shad and branch herring. The number of eggs produced is large, as many as 236,000 having been found in one individual. The young differ considerably in appearance from the adults. They are blind and toothless and their mouths and fins are different in shape. They live in this state in fresh water for about three or four years and then undergo a transformation, after which they descend to the sea. When mature they return to fresh water to spawn but once and then die.

The young have been found to subsist on minute organisms. The stomachs of adults, while usually containing only blood, have been reported by Goode (1884, p. 677) to occasionally contain large numbers of fish eggs.

FIG. 25.—*Petromyzon marinus*

This species attains a length of 3 feet, although seldom exceeding 2½ feet. In the past, when it was more plentiful, it was used for food in parts of New England, while in Europe it has been considered a delicacy for many years. In Chesapeake Bay the lamprey is of no commercial value.

Habitat.—North Atlantic coasts of Europe and North America; on the American coast from Labrador south to Florida.

Chesapeake localities.—(*a*) Previous records: Potomac River and many points in the upper parts of the bay. (*b*) Specimens were taken during the present investigation (during April and May) at Havre de Grace, Md., and Lynnhaven Roads, Va.; also observed in the lower Patuxent River, Md., and Kendall reports (field notes, 1894) several from Hampton, Va.

Class ELASMOBRANCHII

Subclass SELACHII.   The sharks, skates, and rays

Order EUSELACHII

Family III.—ORECTOLOBIDÆ.   The nurse sharks

Body short and subcylindrical to moderately short and depressed; nostrils with a nasoral groove and with a cirrus or barbel; mouth transverse, with labial folds around angles; teeth compressed, with or without lateral cusps on each side of the median one; eyes very small, without nictitating membrane; spiracle minute and behind eye to large and more or less below it; gill slits small to medium, the posterior two or three above base of pectoral; caudal fin narrow, usually without exerted lower lobe; other fins short and broad, no fin spines; no caudal pits.

3. Genus GINGLYMOSTOMA Müller and Henle. Nurse sharks

Body moderately elongate, compressed posteriorly, depressed anteriorly; head broad; snout very blunt; nostrils near tip of snout, remote from each other, connected with the mouth by a groove, each anteriorly with a cylindrical barbel; mouth broad, little arched; teeth small, compressed, with a strong central cusp and one or more smaller lateral ones; several series functioning; spiracle minute and behind eye; gill slits moderate, the last two close together and above base of pectoral; dorsal fins rather close together, the first over the ventrals, the second somewhat in advance of anal.

3. Ginglymostoma cirratum (Bonnaterre). Nurse shark.

Squalus cirratus Bonnaterre, Tableau Encyclop., Method Nat. Ichthyol., 1788, p. 7; American seas.
Ginglymostoma cirratum Lugger, 1877, p. 90. Jordan and Evermann, 1896–1900, p. 26, Pl. IV, fig. 13; Garman, 1913, p. 54, pl. 7, figs. 4 to 6.

Body posteriorly compressed, head and anterior part of body broad, depressed; snout short, broadly rounded; mouth much in advance of eyes, broad; teeth small, with sharp median cusp and a shorter one at each side; nostrils nearly at margin of snout and connected with mouth by a groove, each with a barbel; eye very small, the greatest diameter a little shorter than the longest gill slit in young, proportionately much shorter in adult; spiracle situated just behind eye, very small; denticles on skin below base of dorsal irregular in size, triangular, slightly imbricate, one or three keeled; origin of first dorsal over ventrals; second dorsal a little smaller; caudal long, angles rounded, lower lobe not produced; anal smaller than second dorsal, its origin under middle of second dorsal; pectoral fins nearly as broad as long. Color grayish or yellowish brown above, somewhat paler below. The upper parts either with or without round black spots.

No specimens of this shark are at hand. The above description was compiled from published accounts.

Gudger (1921, p. 58), after examining specimens of this shark taken in southern Florida, with reference to stomach contents, says: "Its food, in keeping with its tooth structure, is mainly confined to invertebrates, squid, shrimp, the so-called crawfish (Palinurus), short-spined sea-urchins, small fish, and probably the more thick-bodied, succulent algæ. In short, the fish is more or less omnivorous."

The nurse shark, according to Gudger (1921, p. 59), is "ovoviviparous." The eggs are large, about 75 millimeters in diameter when they break through the walls of the ovary, and brownish, horny shells with blunted ends, bearing tendrils (as in some of the egg-laying sharks and rays) are later provided. These egg cases measure from 120 to 140 millimeters in length and 170 to 190 millimeters in circumference. The eggs then remain in the posterior part of the oviduct, where a "saddle-bag shaped" section is provided for them, until the young are hatched.

Habitat.—Tropical Atlantic and eastern Pacific; apparently not recorded from the Atlantic coast of America north of Chesapeake Bay.

Chesapeake localities.—(a) Previous records: "Southern part of Chesapeake Bay" (Lugger, 1877). (b) Specimens in collection: None; not seen during the present investigation.

Family IV.—LAMNIDÆ. The mackerel sharks; the man-eater sharks

Body robust; head conical; tail slender, the peduncle depressed, with lateral folds and caudal pits; nostrils oblique, near the mouth but not confluent with it; eyes without nictitating membrane; mouth broad; teeth large; spiracles small or wanting; gill slits wide, all in front of pectorals; first dorsal large; second dorsal and anal small; caudal lunate; pectorals large, falcate.

4. Genus CARCHARODON Müller and Henle. Man-eater sharks

Body very robust anteriorly; head conical; caudal peduncle strong, depressed; teeth large, compressed, serrate, triangular, the upper teeth broadest; first dorsal large, nearly midway between pectorals and ventrals; second dorsal and anal very small; pectorals large.

4. Carcharodon carcharias (Linnæus). Man-eater; Great white shark.

Squalus carcharias Linnæus, Syst. Nat., ed. X, 1758, p. 235; Europe.
Lamnidæ atwoodi Uhler and Lugger, 1876, ed. I, p. 191; ed. II, p. 161.
Carcharodon carcharias Jordan and Evermann, 1896–1900, p. 50; Garman, 1913, p. 32, pl. 5, figs. 5 to 9.

Body robust; head a little more than 4 in total length; depth about 5.5; snout conical, blunted at tip; eye above the front of the mouth; pupil vertical; nostrils small, far apart, nearer to the mouth than to tip of snout; spiracles minute, behind eye; mouth large, with labial folds; teeth large, triangular, serrated, in about 24 to 26 rows in each jaw; first dorsal moderate, its origin behind bases of pectorals, a little longer than high; second dorsal very small, its base entirely in advance of anal; caudal fin broad, the lower lobe produced, slightly shorter than upper; anal fin small, similar to second dorsal, its origin behind vertical from the base of that fin; ventral fins small, below middle of the interdorsal space; pectoral fins falciform, the front margin nearly twice the length of the inner margin; a well developed keel on each side of caudal peduncle; deep pit at base of caudal above and below.

FIG. 26.—*Carcharodon carcharias*

Color grayish, shading to white below; tips and edges of pectorals black.

This is one of the most ferocious of all sharks.

Uhler and Lugger (1876) writing in 1876, stated that this shark was common in Chesapeake Bay as far as the outer harbor of Baltimore. It is uncommon anywhere, however, even in the Tropics, and seldom strays on our Atlantic coast. None were seen during the present investigation, and we know of no record for the Chesapeake since 1876. It is believed, therefore, that the shark referred to by Uhler and Lugger was another species.

The man-eater grows to a length of 40 feet. The jaws of a specimen 36 feet long are in the British Museum.

Habitat.—Seas of the Temperate and Torrid zones; in the western Atlantic, rarely as far north as Nova Scotia.

Chesapeake localities.—(*a*) Previous records: Reported entering Chesapeake Bay by Uhler and Lugger (1876). (*b*) Specimens observed on present investigation: None.

Family V.—GALEIDÆ. The gray sharks

Body elongate; head and snout depressed; eyes lateral, with a more or less perfectly developed nictitating membrane; nostrils below the snout; spiracles present or absent; mouth crescent-shaped, inferior; teeth various; last gill slit above base of pectoral; dorsal fins 2, without spines, the first in advance of ventrals; anal fin present.

KEY TO THE GENERA

a. Teeth small, numerous, in pavement; spiracles present, small_____ Mustelus, p. 47
aa. Teeth not in pavement, compressed, more or less triangular, with a large cusp and usually with a broad base; spiracles wanting.

c. Labial folds wanting; teeth more or less serrate_____ Carcharpinus, p. 48
cc. Labial folds well developed, present on both jaws; teeth not serrate_____ Scoliodon, p. 49

5. Genus MUSTELUS Linck

Body and tail of about equal length, rather slender; head short, broad, depressed; snout long and flat; spiracles small, behind eyes; eyes with a nictitating membrane; mouth small, crescent-shaped; teeth small, many rowed, pavementlike; dorsal fins similar in shape, the first above the abdomen, the second above the anal; caudal fin not deep, the lower lobe feebly developed; pectoral fins large.

5. Mustelus mustelus (Linnæus). Smooth dogfish.

Squalus mustelus Linnæus, Syst. Nat., ed, X, 1758, p. 235.
Mustelus canis Jordan and Evermann, 1896-1900, p. 29.
Galeorhinus lævis Garman, 1913, p. 176.

Body long, slender; head narrow, depressed, flattened beneath, about 4 in length; snout moderate, tapering, its length greater than the width of mouth; nostrils large, placed about half as far from the mouth as from the tip of the snout; eye rather small, its length about equal to the pre-narial length of snout, the pupil elongate horizontally, a nictitating membrane present; mouth about twice as wide as long; teeth small, numerous, pavementlike, in about 10 rows, the upper ones with a short and blunt cusplike projection on the posterior margin, lower teeth similar, with less

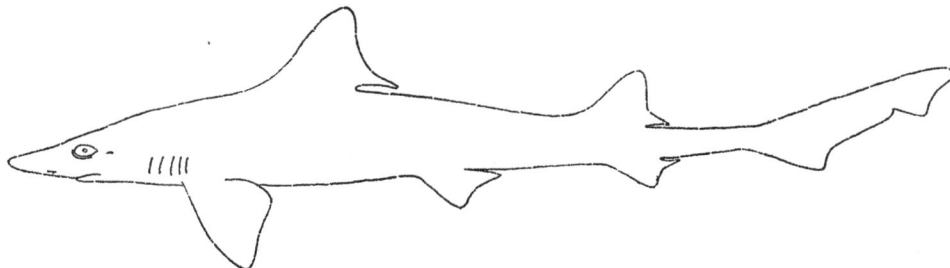

FIG. 27.—*Mustelus mustelus*

prominent cusps, no cusps on teeth near angles of mouth; the skin roughened by rather large, sharply pointed denticles, bearing two or four low keels; origin of first dorsal a little in advance of the posterior margins of the pectorals; second dorsal inserted in advance of the anal, about half as large as the first; caudal fin about 4.5 in total length, the lower lobe scarcely produced; anal fin notably smaller than the second dorsal and inserted under the middle of the base of the second dorsal; ventral fins rather small, inserted nearer the origin of the anal than the base of the anterior rays of the pectoral; pectoral fins of moderate size, about two-thirds as broad as long, the hinder margins only slightly concave.

Color usually uniform grayish, sometimes yellowish or olivaceous and with pale spots; pale underneath.

The smooth dogfish previously has not been recorded from Chesapeake Bay. The present record is offered on the authority of the following field note made by Lewis Radcliffe, at Gwynns Island, Va., May 6, 1915: "Among the fish brought in from pound nets in this locality and landed on the wharf was one smooth dogfish." The same investigator also reports having seen a specimen at Buckroe Beach, Va. The foregoing description is based upon published accounts of the species.

The food of the smooth dogfish consists mainly of the larger crustaceans. Field (1907, pp. 11–13) examined the stomachs of 388 fish caught around Woods Hole, Mass., and found the principal foods to be lobsters, rock crabs, lady crabs, spider crabs, hermit crabs, menhaden, squid, razor clams, and Nereis. Besides menhaden, various species of small fish are eaten indiscriminately.

"The eggs of this dogfish are fertilized internally, and the young are about 1 foot long when born. From 4 to 12 fish are produced at one time." (Smith, 1907, p. 33.) A female examined by Linton at Woods Hole, Mass., contained eight young, each 12½ inches long and ready to be born.

The smooth dogfish is particularly abundant along the coasts of New Jersey and Long Island, extending to Woods Hole, Mass.

The average length of this shark is 2 to 3 feet, but fish as long as 5 feet have been reported.

Habitat.—Cape Cod to Cuba, rarely straying to the Bay of Fundy; southern Europe.

Chesapeake localities.—(*a*) Previous records: None. (*b*) Specimens in present collection: None. This record is based upon a specimen observed at Gwynns Island, Va., May 6, 1915, and another at Buckroe Beach, Va., early in May, 1915, by Lewis Radcliffe.

6. Genus CARCHARHINUS Blainville

Body rather robust; head broad, depressed; snout produced; nostrils and mouth inferior; teeth compressed, more or less triangular, with large cusp and usually a broad base; eyes small, with a well developed nictitating membrane; spiracles wanting; first dorsal large, placed not far behind the pectorals; second dorsal small, wholly or partly above the anal; distinct pits at base of each caudal lobe. The embryos are attached to the uterus by a placenta.

6. Carcharhinus milberti (Müller and Henle).   Milbert's shark.[2]

Carcharias (*Prionodon*) *milberti* Müller and Henle, Plagiostomen, 1838, p. 38, Pl. XIX, fig. 3 (teeth); New York.
Carcharhinus milberti Jordan and Evermann, 1896–1900, p. 37; Smith and Bean, 1899, p. 180.

Body stout; head broad, strongly depressed; snout rather broadly rounded, its preoral part about 1.1 in its length to eye; mouth wide, its width equal to preoral length of snout; eye lateral, small, 4.1 to 5.1 in snout; nictitating membrane evident; interorbital space somewhat greater than length of snout; teeth in upper jaw triangular, the edges serrate, about 29 in outer series, teeth in lower jaw narrow, erect, with finely serrate edges, about 26 in outer series; longest gill slit 3.1 to 3.3 in snout; dermal denticles not overlapping, with three distinct keels; first dorsal with concave outer margin, inserted behind origin of pectorals, its base 2 to 2.15 in distance between dorsals; second dorsal small, its base 5.1 to 5.6 in distance between dorsals; upper lobe of caudal long, 4 to 4.15 in total length; anal opposite the second dorsal and only slightly larger, its outer margin deeply concave; ventral fins inserted at vertical from a point equidistant from the end of the base of the first dorsal and the origin of the second dorsal; pectoral fins longer than broad, 5.9 to 6.4 in total length.

Color in life, taken from two specimens—a male, 635 millimeters (25 inches), and a female, 620 millimeters (24⅜ inches)—bluish gray above, white below; highest part of both dorsals and upper extremity of caudal slightly dusky; tip of pectoral of one fish slightly dusky underneath.

This shark is represented in the collection by six specimens—five females and one male—ranging from 450 to 648 millimeters (17¾ to 25½ inches) in length. Although rather rare in Chesapeake Bay, it is perhaps more common than any other shark except the spiny dogfish. The only fish taken during the collecting of 1921 were caught off Janes Island, Crisfield, Md., where, on September 16, the catch was two, fishing one and one-half hours; on September 18 the catch was five, fishing six hours with hook and line at depths of 50 to 90 feet. During 1922 five sharks of this species were caught at Ocean View, Va., with seines, on October 6, 10, 17, and 18.

Like most sharks, this species feeds chiefly on fish. The stomachs of two specimens examined contained fragments of fish bones, and another had eaten one pinfish (*Lagodon rhomboides*).

The young on the coast of Long Island are born during June and July, from 8 to 14 at one time, and about equally, males and females (Nichols and Murphy, 1916, p. 16).

This is one of the medium-sized sharks, attaining a maximum length of about 8 feet. A fish 18 inches in length weighed 1¼ pounds; 24⅜ inches, 3⅜ pounds; 25 inches, 3½ pounds.

Habitat.—Middle Atlantic and middle eastern Pacific (Garman, 1913, p. 133); northward on the Atlantic coast of America to Woods Hole, Mass.

Chesapeake localities.—(*a*) Previous records: Fort Washington and Glymont, Md. (*b*) Specimens in collection or observed in the field: Crisfield, Md., September, 1921; Ocean View, Va., October, 1922.

---

[2] This shark is also known as the blue shark, but we discard this name in order to avoid confusion with *Galeus glaucus*, a shark of wide distribution and which for many years has been known to fishermen and whalers as the "blue shark."

7. Genus SCOLIODON Müller and Henle

This genus differs from Carcharhinus in the presence of labial folds, which extend some distance along the jaws from the angles of the mouth, and the teeth, which are never serrate.

7. Scoliodon terræ-novæ (Richardson). Sharp-nosed shark.

Squalus terræ-novæ Richardson, Fauna Bor. Amer. III, 1836, p. 289; "Newfoundland," where the species does not occur.
Scoliodon terræ-novæ Bean, 1891, p. 94; Jordan and Evermann, 1896–1900, p. 43; Garman, 1913, p. 115, pl. 2, figs. 1 to 4.

Body moderately robust; head rather broad; snout rather short, broadly rounded, preoral portion 1 to 1.05 in length to eye, its width at nostrils 1.05 in preoral length and 1.1 in length to eye; eye rather small, its diameter somewhat greater than width of nostril; interorbital area convex, 1.05 to 1.15 in snout; nostrils obliquely placed, the outer angles being notably in advance of the inner ones, the inner angles about two-thirds as far from the mouth as from tip of snout, narial valve with a sharply pointed lobe; distance from nostril to eye 3.1 to 3.2 in snout; internarial space two times diameter of eye; mouth rather strongly arched, its width at angles 1.2 to 1.25 in preoral part of snout; labial folds short, the upper one notably less than one-third the length of the jaw, about two-thirds the length of eye, 3.8 to 4.15 in preoral part of snout and 2.3 to 2.8 in internarial, the lower fold shorter, 6.35 to 6.75 in preoral part of snout; teeth not serrate, with broad bases and rather narrow cusps, the anterior ones erect, those of the sides directed inward and backward; gill slits rather narrow, the longest about 2.5 in internarial, 1.1 to 1.15 in distance from eye to outer angle of nostril; first dorsal rather large, its outer margin concave, the lower lobe pointed, its origin about two times diameter of eye behind vertical from axil of pectoral, its base 2.4 in distance between dorsal fins; second dorsal moderate, its origin over or a little behind middle of base of anal, its base 6.05 to 7.4 in distance between the dorsal fins; upper lobe of caudal very long, pointed, 3.85 in total length, the lower lobe broad, 6.4 to 6.75 in the upper lobe; anal fin with concave margin, its base 1.85 to 1.95 in distance from anal to base of caudal; ventral fins small, inserted equidistant from axil of pectoral and posterior margin of base of anal, the claspers about two-thirds the length of the fins in specimens 360 millimeters in length; pectoral fins moderate, the posterior margin little concave, reaching about opposite middle of base of dorsal.

Color bluish gray above; pale below.

This shark was not seen during the present investigation. It may be distinguished from the other sharks of this family known from Chesapeake Bay by the presence of folds in the lips, which extend forward from the angles of mouth, and by the smooth teeth.

The food of this shark is rather varied, consisting, however, largely of fish and crustaceans. The young, according to Smith (1907, p. 34), are born during the summer. The usual length attained is about 3 feet. This small shark, which is common on the South Atlantic coast, probably rarely enters Chesapeake Bay.

Range.—Cape Cod, Mass., to Brazil.

Chesapeake localities.—(a) Previous record: Cape Charles, Va. (b) Specimens in the collection: None.

Family VI.—SPHYRINIDÆ. The hammerhead sharks

This family resembles the species of the genus Carcharhinus, differing in the peculiar modification of the head, which is greatly depressed and broadly expanded, hammer-shaped. The eyes are far apart, being situated on the lateral margins of the expanded head; nictitating membrane present; no spiracles; nostrils remote from each other and distinct from the mouth; labial folds rudimentary; teeth compressed; first dorsal fin large, in advance of ventrals; second dorsal and the anal small, opposite; lower lobe of caudal prominent. A single genus is known.

8. Genus SPHYRNA Rafinesque

Body elongate, compressed; head much depressed, with a broad expansion on each side, more or less hammer-shaped; eyes far apart, placed on lateral edges of the broadly expanded head; nictitating membrane present; no spiracles; mouth inferior, strongly arched; labial folds rudimentary; teeth compressed, more or less triangular, with broadly expanded bases and a notch on posterior edge; first dorsal behind the origin of the pectorals and in advance of the ventrals; second dorsal over the anal; caudal pits present; lower lobe of caudal produced, upper lobe long.

KEY TO THE SPECIES

a. Head very broad, its greatest width about 3 in total length; anterior outline of head irregular, a deep concavity over each nostril _____zygæna, p. 50
aa. Head less broadly expanded, its width about 5 in the total length; anterior outline of head regularly convex, no concavity over nostrils _____tiburo, p. 51

8. Sphyrna zygæna (Linnæus). Hammerhead shark.

Squalus zygæna Linnæus, Syst. Nat., ed. X, 1758, p. 234; America.
Sphyrna zygæna Lugger, 1877, p. 88; Jordan and Evermann, 1896-1900, p. 45.
Cestracion zygæna Garman, 1913, p. 157, pl. 1, figs. 1 to 3.

Body elongate, compressed; head very broad, hammer-shaped, the front margin broadly and irregularly convex, with a deep concavity at each nostril; width of head at eyes from 3 to 3.25 in total length; nostril close to eye, with a long groove on margin of snout; mouth moderate, its width a little shorter than preoral length of snout; teeth similar in both jaws, oblique, cusps triangular, the lateral ones with a notch at base posteriorly; first dorsal high, its height greater than the length of its base, the outer margin concave, its origin a little behind axil of pectoral; second dorsal small, its posterior angle notably produced; upper lobe of caudal long, the lower lobe also produced, its length about 2.75 in the upper lobe; anal fin a little longer than the second dorsal, the outer margin deeply concave, its origin a little in advance of the second dorsal; ventral

Fig. 28.—*Sphyrna zygæna*

fins small, inserted slightly more than half as far from origin of anal as from base of pectoral; pectoral fins moderate, scarcely reaching to base of first dorsal, the lower angle not produced and the posterior margin of fin slightly concave.

Color of fresh specimen lead gray above, lower parts grayish white; tips of pectorals black; the tips of the other fins dark.

No specimens of this shark were preserved. The description herewith was compiled from published accounts.

Lugger (1877, p. 89) states that the hammerhead shark was so very common in the mouth of Miles River, Md., during the summer of 1876 that the fishermen were forced to abandon that ground. The species is not reported by other observers. During the present investigation only three individuals were seen. A hammerhead was taken on July 15 and another one on July 17, 1916, in pound nets in Lynnhaven Roads, and in the same locality a 2-foot specimen was caught with hook and line on June 26, 1921.

The food of this shark, according to stomach examinations made by investigators at Beaufort, N. C., consists of fish and crustaceans. Gudger (1907, pp. 1005-1006) took an almost perfect skeleton and many fragments of skeletons of the sting ray (*Dasybatus say*) from the stomach of a specimen of this shark, and he found imbedded in various parts of the shark numerous spines of the sting ray. In all, 50 spines were extracted, mainly from the mouth parts, and, according to this author, all that were present quite certainly were not recovered. This particular shark was harpooned while it was in pursuit of a sting ray and the evidence would suggest that this sting ray may form a considerable part of the food of this species of shark.

The hammerhead is viviparous, as many as 37 embryos having been taken from an 11-foot fish. Along the Atlantic coast fish 2 to 6 feet long are not uncommon, while larger examples are reported occasionally. The largest hammerhead of which we find record was 17 feet long, harpooned off Miami, Fla., on March 21, 1919.

Habitat.—Tropical and temperate seas; on both coasts of America from Cape Cod, Mass., and California southward.

Chesapeake localities.—(a) Previous record: Miles River, Md. (b) Specimens observed on present investigation: Lynnhaven Roads, Va., July, 1916, and June, 1921.

9. Sphyrna tiburo (Linnæus). Shovel-nose shark; Bonnet-nose shark.

Squalus tiburo Linnæus, Syst. Nat., ed. X, 1758, 234; America.
Sphyrna tiburo Jordan and Evermann, 1896–1900, p. 44, Pl. V, fig. 19.

Body moderately slender, compressed; head much depressed, expanded, the anterior margin semicircular, the posterior margins short, free, slightly concave, its greatest width quite equal to its length to first gill opening; eye small, lateral, 4 in preoral length of snout, nictitating membrane present; mouth moderate, its width 1.05 in preoral part of snout; teeth in jaws similar, with broad basal shoulders and a sharp, smooth cusp, the lateral teeth with a notch behind the cusp, upper jaw with about 30 teeth in a series, the lower with about 27; longest gill slit 2.1 in preoral part of snout; dermal denticles slightly imbricate, 3 and 5 keeled, the median keels projecting as sharp lobes; first dorsal rather short and high, elevated anteriorly, its origin slightly behind base of pectorals, the base 2.96 in distance between dorsals; second dorsal small, its posterior lobe elongate, pointed, the base 5.5 in distance between dorsals; upper lobe of caudal long, pointed, 35 in total length; anal fin notably longer than second dorsal and beginning farther forward, its base 1.05 in distance from anal to base of lower lobe of caudal; ventral fins moderate, inserted about equidistant from the origin of the first and second dorsals; pectoral fins rather small, 7.1 in total length.

Color grayish above, pale below.

This shark is represented in our collection by one small male specimen, 662 millimeters (26 inches) in length.

The food, as determined from specimens taken at Beaufort, N. C., consists of fish, crabs, shrimp, and other crustaceans.

This shark is viviparous, and as many as eight or nine young have been found at one time. (Smith, 1907, p. 35; Radcliffe, 1916, p. 266.)

This fish is comparatively rare in Chesapeake Bay, where only one specimen was observed. In the lower bay, between Ocean View and Cape Henry, however, fishermen said that it was occasionally taken in pound nets and they knew it well enough to give it the name "shovel-nose shark."

The maximum length attained is given as about 5 feet.

Habitat.—Tropical and temperate seas (Garman, 1913, p. 161); northward on the Atlantic coast of America to Long Island.

Chesapeake localities.—(a) Previous records: None. (b) Specimen in collection: Lynnhaven Roads, Va., pound net, June 9, 1921.

Order TECTOSPONDYLI

Family VII.—SQUALIDÆ. The dogfishes

Body elongate; head depressed; eyes lateral, no nictitating membrane; nostrils inferior, separate, remote from the mouth; mouth rather large, inferior, with labial folds and a deep groove at each angle; spiracles present; gill slits 5, all in front of pectoral; dorsal fins 2, each preceded by a spine; no anal fin.

9. Genus SQUALUS Linnæus

Body rather slender; head flattened below; snout produced, tapering; nostrils transverse, inferior, remote from mouth; spiracles behind eyes; mouth wide, little arched, with a deep groove and with labial folds at each angle; teeth compressed, alike in both jaws, with oblique cusps; dorsal spines not grooved on sides; first dorsal near the pectorals; second dorsal behind ventrals; caudal pits present; lower lobe of caudal produced.

10. Squalus acanthias Linnæus. Spiny dogfish; Spiked dogfish; Grayfish.

Squalus acanthias Linnæus, Syst. Nat., ed. X, 1758, 233; coast of Europe.
Squalus americanus Uhler and Lugger, 1876, ed. I, p. 194; ed. II, p. 163.
Squalus acanthias Jordan and Evermann, 1896–1900, p. 54, Pl. VII, figs. 24 and 24a.

Body moderately slender, somewhat depressed anteriorly; caudal peduncle laterally with a low dermal fold; head low; snout pointed, 2.5 to 2.8 in head to first gill slit; its preoral part 1.75 to 2.1 in head; mouth moderate, its width at angles 2.55 to 3 in head; eye lateral, elongate, 2.15 to 2.25 in snout; interorbital space 2.3 to 2.8 in snout; teeth similar in each jaw, the cutting edges transverse, each tooth with a sharply pointed cusp, outer series in upper jaw with about 27 teeth, the lower jaw with approximately 22; longest gill slit 3.2 to 3.6 in snout; spiracles behind eyes, prominent; dermal denticles not imbricate, situated in more or less definite rows, each with a quadrangular base and a high median keel ending in a triangular apex; first dorsal rather small, preceded by a spine, the outer margin very slightly concave, its base 3.4 to 4.35 in space between dorsal fins; second dorsal smaller, with a larger spine, its base 4.4 to 5.6 in space between dorsal fins; upper lobe of caudal produced, no notch in lower margin, 4.6 to 5.35 in total length; ventrals inserted about equidistant from axil of pectoral and base of lower lobe of caudal; pectoral fins moderate, the posterior margin notably concave, 6.2 to 7 in total length.

Color grayish above (occasionally brown), pale below, sides with small, roundish, pale spots in one or several rows, most prominent in young up to 14 inches, almost disappearing in largest fish.

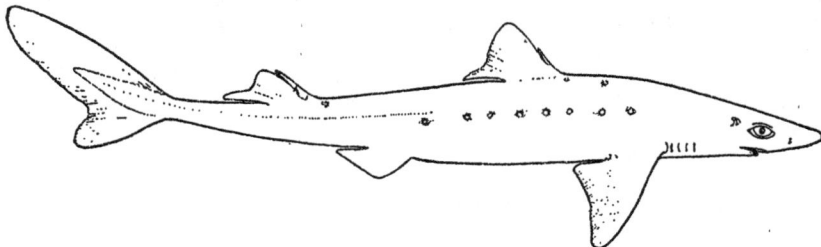

FIG. 29.—*Squalus acanthias.* (After Garman)

This shark is represented in our collection by six specimens—three males and three females—ranging in length from 560 to 800 millimeters (22 to 31½ inches).

The spiny dogfish is a very voracious feeder. The stomach of one contained a partly digested squeteague, probably *Cynoscion regalis*, approximately 7 inches in length, and that of another contained a mass of partly digested fish, from which eight menhaden, ranging from 3 to 5 inches in length, were recognizable. The other specimens, having been taken in pound nets, were not examined. One of several fish examined in the field by us had eaten crabs and a small croaker (*Micropogon*). Bigelow and Welsh (1925, p. 49) give the food of the spiny dogfish as all fish smaller than themselves, squid, worms, shrimps, prawns, crabs, and at times even ctenophores.

This shark is ovoviviparous. The large eggs, abundantly supplied with yolk, are at first in a horny capsule in the oviduct. Later the embryos break free, remaining in the oviduct or "uterus," with which they have no placental attachment. The period of gestation has been variously estimated, but it appears that 10 to 11 months, based upon the studies of Ford (1921, pp. 468–505), is correct.[3] Ordinarily, a female gives birth to three or four young at one time, but the number may be only one or as many as eight to eleven. Gudger (1912, p. 143) records a specimen at Beaufort from which three young were obtained. This fish was taken on May 23, but the size of the embryos was not given. Nichols and Murphy (1916, p. 32) state that spiny dogfish taken along the continental shelf off New York in late November contained well-developed young, the common number observed being three. They record a female taken near Gardiner's Island, N. Y., June 12, 1911, which gave birth to several young on the deck of the boat. We examined 12 specimens taken at Lynnhaven Roads, Va., April 4 to 8, 1922. Although selected at random, eleven were females and one was a male. The smallest fish (26 inches long) was the male. The length of the females and the number of embryos they contained are as follows:

---

[3] For an account of the embryology see Bigelow and Welsh, 1925, pp. 49–50.

Length, inches	Embryos or eggs	Length, inches	Embryos or eggs
28⅛	Immature eggs.	33¾	4 embryos.
31	2 embryos.	Do	1 large egg.
31½	Do.	34¼	4 embryos.
Do	1 embryo.	35¼	Do.
32	2 embryos.	35⅛	5 embryos.
32¼	4 embryos.		

This table agrees with the examinations of other investigators in that the number of young produced at one time usually is not more than four. It also suggests that larger fish produce more young than smaller fish, a fact noted also by Ford (1921, p. 473). The size of most of these embryos was 6 to 7 inches. At the time of birth dogfish are from 9 to 12 inches in length. Young appear to be born in the spring and autumn. If the period of gestation is 10 to 11 months, as probably is the case, a female can not give birth to young both in the spring and in the fall.

The spiny dogfish is generally common in the lower part of Chesapeake Bay, below the Potomac River, during the late fall and early spring. Nothing is known of its presence there in the winter, as fishing in the lower bay ceases entirely during this period. It is probable, however, that, due to the depleted food supply and the low water temperatures, it is scarce if not entirely absent during the winter. During the summer, at least from late May to October, it is entirely absent. Dogfish travel in schools, often appearing suddenly and irregularly. A set of two pound nets in Lynnhaven Roads, Va., caught spiny dogfish beginning with the first day's fishing—March 6, 1922. A few (perhaps less than 10) were taken nearly every day throughout the month. On April 4, when we began field operations in this locality, 25 were caught, followed by 8 on April 6 and 6 on April 8, all 2½ to 3 feet in length. It was of interest to note that all of these were taken by the pound net set in 32 feet of water, whereas the other net, placed in 12 feet of water and leading inshore directly from the deep-water net, caught none. On May 25, when these nets were again visited, no spiny dogfish were caught, nor had any been taken since early in May. At Cape Charles, Va., the fishermen reported this shark common in March and April. In the fall the spiny dog appears in apparently smaller numbers than in the spring. Our earliest record is November 15, 1922, when a 28½-inch fish was caught off Willoughby Spit, Va. Only stragglers are taken in pound nets late in November, or at the time when fishing ceases for the winter. Inquiries among the fishermen along the lower Potomac revealed that the spiny dogfish is not taken there, hence we can state with assurance that it is restricted to the lower parts of the bay, being most abundant near the capes.

Spiny dogfish are exceedingly abundant off the New England coast, at least from Nantucket Shoals to Cape Sable, Nova Scotia. They are present in this region from May until late October, or during the time when they are absent from Chesapeake Bay and points farther south. Along the New Jersey coast and western Long Island they appear suddenly in great numbers early in November, and are then regarded by fishermen as the forerunners of the cod. They soon disappear, however, and are not seen again until late April and early May, when they are present only a few weeks. Little appears to be known concerning the winter home of this dogfish. Their appearance south of New England directly after they leave and before they return would indicate a coastwise movement. Although they may occur as far south as the Carolinas, and to a more limited extent farther south, evidence produced by Bigelow and Welsh (1925, p. 47) indicates that the predominating migration is on and off shore rather than alongshore.

Spiny dogfish are usually from 2 to 3 feet long and attain a length of at least 3½ and possibly 4 feet. Females average somewhat heavier than males.

This shark is of no commercial importance in Chesapeake Bay and does not occur there in sufficient numbers to be regarded as a serious pest by the fishermen. Wherever abundant, it is destructive to other fish and fishing gear; because of this and its strong dorsal spines, with which it can inflict painful wounds, it is considered obnoxious by all fishermen.

Habitat.—On both coasts of the Atlantic; on the American Continents from Labrador to Uruguay, occasionally straying northward to Greenland.

Chesapeake localities.—(a) Previous record: Mouth of Chesapeake Bay. (b) Specimens in collection: Old Point Comfort, Va., beam trawl, depth 73 to 84 feet, December 2, 1915, April 2, 1921; Lynnhaven Roads, pound net, April 6, 1922, November 28, 1921; Willoughby Spit, Va., November 15, 1922; also seen at several other points in the southern sections of the bay.

Family VIII.—SQUATINIDÆ. The angel sharks

Body, head, and tail depressed and flat; snout obtuse; gill openings wide, partly inferior and partly hidden by the base of the pectorals; spiracles wide, crescent-shaped, behind the eyes; nostrils on the front margin of the snout, with skinny flaps; mouth terminal or nearly so; teeth rather small, far apart, erect; dorsal fins 2, small, subequal, situated on tail behind ventrals; anal fin wanting; pectoral fins very large, expanded in the plane of the body, but not attached to the side of the head, deeply notched at the base; ventral fins very large; caudal fin small.

This family of peculiar sharks is intermediate in both structure and general appearance between the sharks and rays.

10. Genus SQUATINA Duméril. Angel fishes

The characters of the genus are included in the family description. A single species is indigenous to the Atlantic coast of America.

11. Squatina dumeril Le Sueur. Nurse fish; Angelfish; Monkfish; Sand devil.

Squatina dumeril Le Sueur, Journ., Ac. Nat. Sci., Phil., I, 1818, p. 225, Pl. X; probably Florida.
Squatina squatina Jordan and Evermann (in part), 1896–1900, p. 58.

Body depressed throughout; head low, flat, its length to first gill slit 4.9 in total length; snout short and broad, the anterior outline slightly concave, 5.65 in head; eye small, 11.7 in head; spiracles crescent-shaped, at least as long as eye, situated behind eyes at a distance not quite equal to length of snout; interorbital very broad, concave, 2.4 in head; nostrils on anterior margin of snout, with skinny flaps, the interspace 3.5 in head; mouth only slightly behind anterior margin of snout, very broad, its width 1.5 in head; teeth, 18 in a series in each jaw, rather small, far apart, erect, with broad basal shoulders and a sharp median cusp; skin rough, with enlarged tubercles on head and snout and with sharp spines on outer margin of the pectoral fins; dermal dentacles in irregular rows rather far apart, of unequal size, each consisting of a low, strong, angled spine with a very broad base and a rather sharp point; gill slits 5, wide, all posterior to anterior angle of pectoral; pectorals broad, expanded, the anterior angle free from the body and not confluent with the head, the length of the outer anterior margin of fin 3.4 in total length, the outer posterior angle a right angle, the inner lobe of fin round; dorsal fins of about equal size, situated on the tail, far behind ventrals, the base of first dorsal 1.45 in distance between dorsals, the base of the second 1.55; caudal fin posteriorly truncate, both lobes pointed, the lower slightly the longer, 1.55 in head; ventral fins inserted opposite posterior margin of pectorals, very broad (the claspers in the male specimen at hand—that is, 42½ inches long—are 7½ inches in length).

Color grayish above, pale below. The abdomen, throat, and ventral fins with reddish spots in life.

A single male specimen, 1,080 millimeters (42½ inches) in length, occurs in the Chesapeake Bay collection. This peculiar fish, which has the combined characters of a shark and a ray, is a conspicuous form. Years ago it was said to be rather common on the Atlantic coast of Maryland. Within Chesapeake Bay, certainly, it is very rare, and none at all were seen or reported during the intensive collecting of 1921 and 1922. Lugger (1878, p. 122) says of this animal: "The not very inviting looks of this fish are not the only reasons why fishermen dislike it. It has, to some extent, the unpleasant habits of the snapping turtle, since it can open its mouth very suddenly, to an alarming extent, and not to play, either. In consequence of this biting propensity, it is called by the fishermen the 'sand devil,' and also the 'fair maid'; the first name not without any reason and the latter certainly not out of politeness."

Habitat.—Both sides of the Atlantic and on our Pacific shores, occurring sparingly northward on our Atlantic coast to Cape Cod, Mass.

Chesapeake localities.—(a) Previous records: None. (b) Specimen in collection from Lynnhaven Roads, Va., pound net, July 15, 1916.

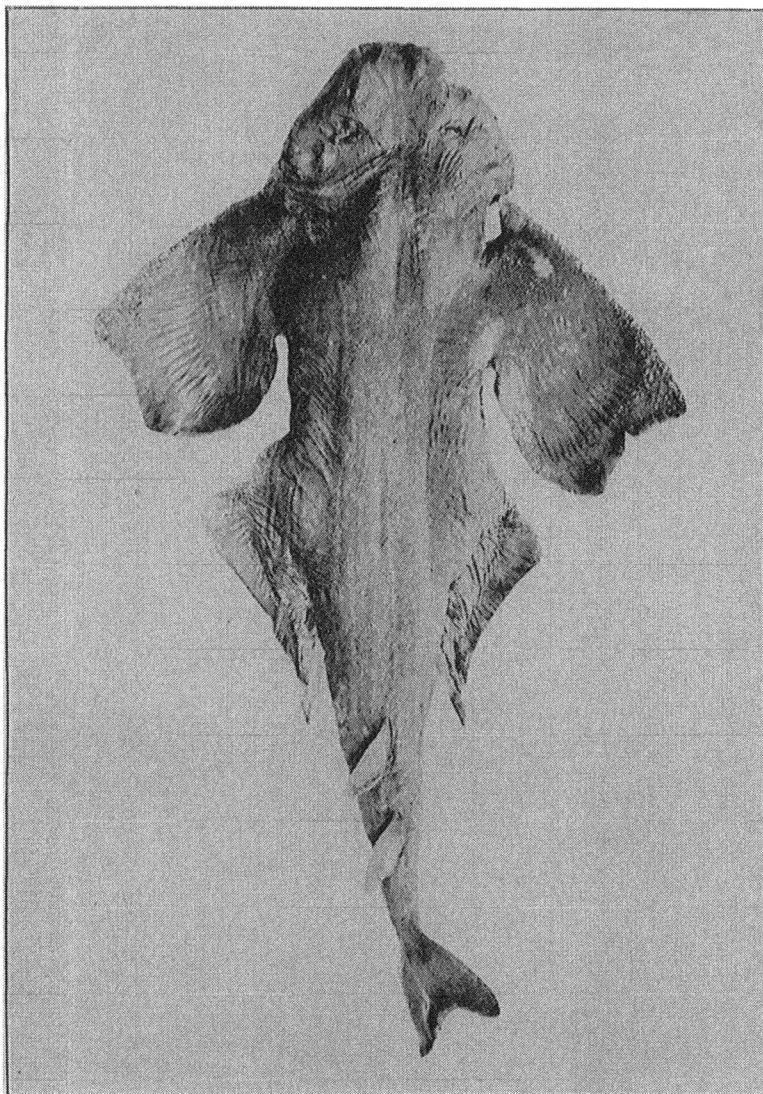

Fig. 30.—*Squatina dumeril*.   From a specimen 42.5 inches long

Order BATOIDEI. The skates and rays

Family IX.—PRISTIDÆ. The sawfishes

Body elongate, depressed; snout produced into a long, thin, flat process, armed laterally with a series of large, strong teeth; teeth in the jaws numerous, small, in pavement; gill slits moderate, inferior; spiracles wide, placed behind the eye; eyes without nictitating membrane; dorsal fins 2, large, the first nearly opposite ventrals; caudal fin well developed, bent upward; a fold along each side of tail; pectoral fins moderate, their front margins not extending to the head. A single genus is known. Viviparous.

11. Genus PRISTIS Linck. Sawfishes

The characters of the genus are included in the family description. A single species is known from the waters of the Atlantic coast of the United States. The sawfishes are bottom-dwelling animals. The large, sawlike rostrum probably is not used extensively as an offensive weapon, but it forms an effective defensive weapon, as the fish can strike from side to side with great force.

12. Pristis pectinatus Latham. Sawfish.

Pristis pectinatus Latham, Trans., Linn. Soc., London, II, 1794, p. 278, Pl. XXVI, fig. 2; "in the ocean." Jordan and Evermann, 1896–1900, p. 60, Pl. VIII, fig. 27; Garman, 1913, p. 262.
Pristis antiquorum Uhler and Lugger, 1876, ed. I, p. 190; ed. II, p. 160.

Body depressed, its depth between the dorsals about equal to its width at the same point; caudal peduncle depressed, provided with a lateral keel on each side; rostrum (or "saw") of moderate width, tapering, provided with 24 to 32 strong teeth on each edge, varying with age and among individuals; teeth on the jaws in pavement, in many rows; origin of first dorsal opposite or a little posterior to the origin of the ventrals; second dorsal scarcely smaller than the first; the lower lobe of caudal not produced; pectoral fins broad, the outer angles blunt, posterior margins nearly straight.

Color, dark gray or brownish above, pale yellow or white below.

The sawfish was not seen during the present investigation, but it was reliably reported by pound-net fishermen operating in the lower parts of Chesapeake Bay. The foregoing description was compiled from published accounts.

The prolongation of the snout, with its armature of teeth, at once identifies the sawfish from all other Atlantic fishes. Six species are known to exist. The only other species (*P. microdon*) found on this side of the Atlantic, chiefly in the Tropics, has 17 to 23 teeth along its snout, whereas the present species has 24 to 32 teeth. The number of teeth on each side of the snout may or may not be the same. A 14-foot fish taken by us at Key West had 28 teeth on the left and 27 teeth on the right side. Three 30-inch fish taken by us at Marco, Fla., had the following counts of rostral teeth: 24–24, 24–25, 25–25. The last-mentioned young fish were taken in the same locality on the same day. They were found swimming slowly along, parallel to and within 3 or 4 feet of the shore. Each was thrown ashore with a dip net. As they were exactly the same length (30 inches), it is quite certain that they were of the same age, and it is likely that they were recently born. This species gives birth to live young, as many as 20 being produced at one time. It is said to deliver its young in the summer, but as the three newly born fish mentioned above were found early in January, it is probable that young are born over an extended period, the period of reproduction varying in different sections with the climate.

The sawfish is only an occasional visitor in the lower Chesapeake.

Pound-net fishermen at Ocean View and Lynnhaven Roads report that it is rarely taken—sometimes one or two fish a year and sometimes none. The capture of a sawfish is long remembered by the fishermen, for it is very destructive of nets, from which it is removed with great difficulty.

This sawfish is said to attain a length of 20 feet. Examples 10 to 16 feet in length are not rare.

Habitat.—Caribbean Sea, Gulf of Mexico, and the east coast of the United States as far north as New Jersey.

Chesapeake localities.—(*a*) Previous record: "Occasionally enters Chesapeake Bay." (Uhler and Lugger, 1876.) (*b*) Specimens in collection: None. The species was not seen during the present investigation, but it was reliably reported by fishermen operating pound nets in the southern parts of the bay.

Family X.—RAJIDÆ. The skates

Body and head much depressed, united with the pectorals and forming a rhomboid disk; tail distinct, stout, rather long, with lateral folds; dorsal fins 2, small, both on the posterior half of the tail; eyes and spiracles superior; mouth inferior, small; teeth small, numerous, in pavement; skin usually more or less rough, with small spines and larger tubercles. The species are oviparous, the eggs being laid in large, leathery, four-angled cases, with two tubular "horns" at each end.

12. Genus RAJA Linnæus. Skates

Disk subquadrangular or subcircular; snout more or less produced, pointed, supported by a "rostral cartilage"; spiracles present, close to eyes; teeth small, varying from flat to sharp and pointed; pectoral fins not confluent around the snout; ventral fins deeply notched; dorsal fins 2; tail with a membranous fold on each side.

KEY TO THE SPECIES

a.   Snout very blunt, only the tip projecting beyond the general outline of the disk; median line of back and tail without a row of enlarged spines; tail with three lateral rows of spines on each side; teeth in about 74 series in each jaw_____ *diaphanes*, p. 56
aa.   Snout acute; median line of back and tail with a series of enlarged spines; tail with a single lateral row of enlarged spines on each side; teeth in fewer than 50 series in each jaw.
  b.   Dorsal surface mostly beset with bony prickles; snout only moderately acute; teeth in about 48 series in each jaw; dark markings on dorsal surface mostly elongate__ *eglanteria*, p. 58
  bb.   Dorsal surface largely smooth; snout very acute; teeth in 32 to 36 series in each jaw; dark markings on dorsal surface roundish_____ *stabuliforis*, p. 59
aaa. Snout moderate, more pointed than in *diaphanes* but less so than in *eglanteria* and *stabuliforis*; median line of back nearly or quite without tubercles; tail with two to four lateral rows of enlarged spines; teeth in about 50 series in each jaw_____ *erinacea*, p. 60

13. Raja diaphanes Mitchill.   Common skate; Spotted skate.

Raja diaphanes Mitchill, Trans., Lit. and Philo. Soc., N. Y., I, 1814, p. 478; New York. Garman, 1913, p. 339, pl. 22, fig. 1.
Raja ocellata Jordan and Evermann, 1896-1900, p. 68, Pl. X, fig. 30.

Disk broader than long, the anterolateral margin double concave, a slight concavity opposite snout and a very broad one opposite eyes and spiracles, the posterolateral margin broadly and evenly convex, length of disk 1.15 to 1.25 in its width, the width of disk 1.45 to 1.6 in total length; head to first gill slit 2.95 in width of disk; distance from snout to vent, 1.25; tip of snout projecting beyond the general outline of the disk, the length of snout 4.5 to 5.35 in width of disk; preoral length of snout 1.75 in head; interorbital (bone) 1.8 to 2.4 in snout; eye 5.55; spiracles immediately back of eyes, the longest diameter somewhat greater than the length of eye; nasoral groove extending to mouth; teeth in about 74 series in each jaw, each tooth with a roundish base, surmounted by a very low, blunt cusp, at least in the posterior, or newer, series; skin in the female on upper surface largely beset with prickles and spines; median part of head naked, also the snout, except the tip, which bears enlarged spines; the anterolateral margin of disk with a band of enlarged spines continued as intramarginal spines posteriorly; no definite spines or tubercles on median line of back or tail; three lateral rows of spines on each side, beginning on middle of back and extending backward on the tail, becoming larger posteriorly; the upper surface in the male somewhat less prickly, but with the spines on the margins of the disk larger; tail moderate, depressed, with dermal keel along lower ventral edges, 2.1 in total length; dorsal fins 2, close together; caudal fin represented by a dermal fold; ventral fins long, inserted somewhat in advance of posterior margin of disk, greatly thickened at the base, the fins rather deeply notched.

Color of upper parts brown, light brown, or grayish-brown, everywhere covered with irregular dark spots, variable in intensity; a white ocellated spot on pectoral somewhat in advance of its inner posterior angle. (This spot, according to Garman (1913, p. 339), may be present or absent.) White underneath.

This species is readily distinguished from the "clear-nose skate," *Raja eglanteria*, its nearest relative of the genus in Chesapeake Bay, by the shorter and less strongly pointed snout, the more numerous and larger prickles, and especially by the absence in the present species of an enlarged series of spines on the median line of the back and tail. The color, too, presents noticeable differences, the pair of white ocellated spots on the pectorals of *R. diaphanes*, when present, being very evident.

This skate feeds chiefly on rock crabs and squid. (Bigelow and Welsh, 1925, p. 61.) They take, also, small crustaceans, razor clams, and such fish as they can capture.

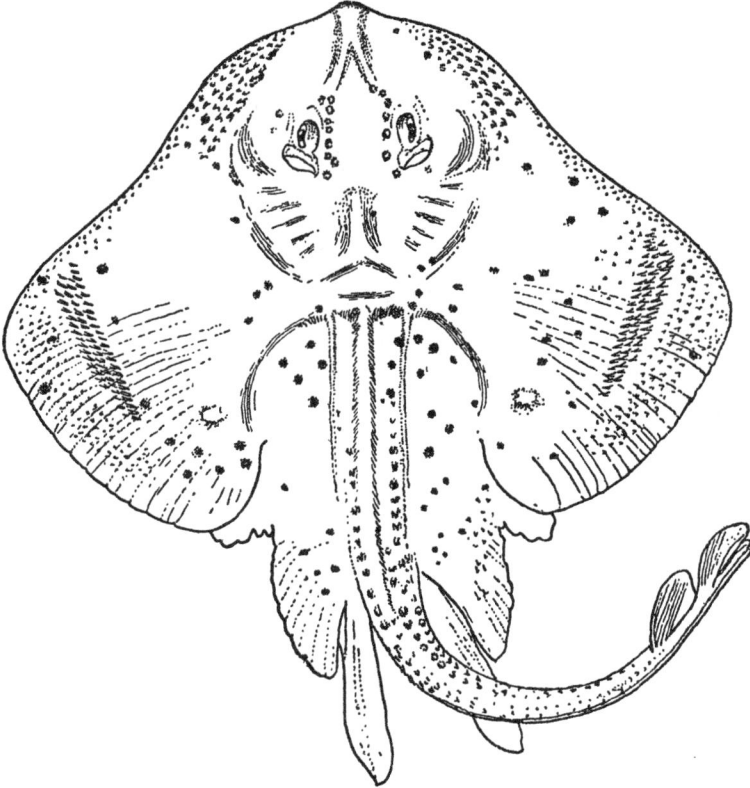

FIG. 31.—*Raja diaphanes*, male. (After Garman)

The breeding habits of this skate are unknown, except that, like all members of this family, the eggs are laid in leathery pouches.

A single specimen was preserved, but many were examined in the field, upon which notes and measurements were based, and these have been used in the foregoing description. The specimens examined ranged in length from 29 to 36 inches and the width of the disk varied from 19 to 25 inches. This skate was seen only in the southern parts of the bay, where it enters pound nets. Twelve to twenty-two individuals were taken each day in two pound nets located in Lynnhaven Roads, Va., from April 4 to 8, 1922, when the daily catches were observed; and one specimen was taken by the *Fish Hawk* near Cape Henry on January 16, 1914. This skate was not seen elsewhere in the bay, nor was it seen in the southern part of the bay on other dates than those previously mentioned, although observations of commercial catches and collections were made at nearly all seasons of the year. This species is recognized by the fishermen as distinct from the "clear-nose skate," but they do not appear to have a distinctive name for it, referring to it only as "skate." According to the local fishermen, this skate is taken only in the spring, when pound-net fishing is first resumed for the season, and at this time it is taken in considerable numbers.

The maximum length is about 6 feet.

Habitat.—Atlantic coast, from Virginia northward to Gulf of St. Lawrence.

Chesapeake localities.—(*a*) Previous records: None. (*b*) Specimen in present collection: Lynnhaven Roads, Va., pound net, April 4, 1922. Many others were observed during April at Lynnhaven Roads and also were taken near Cape Henry during January.

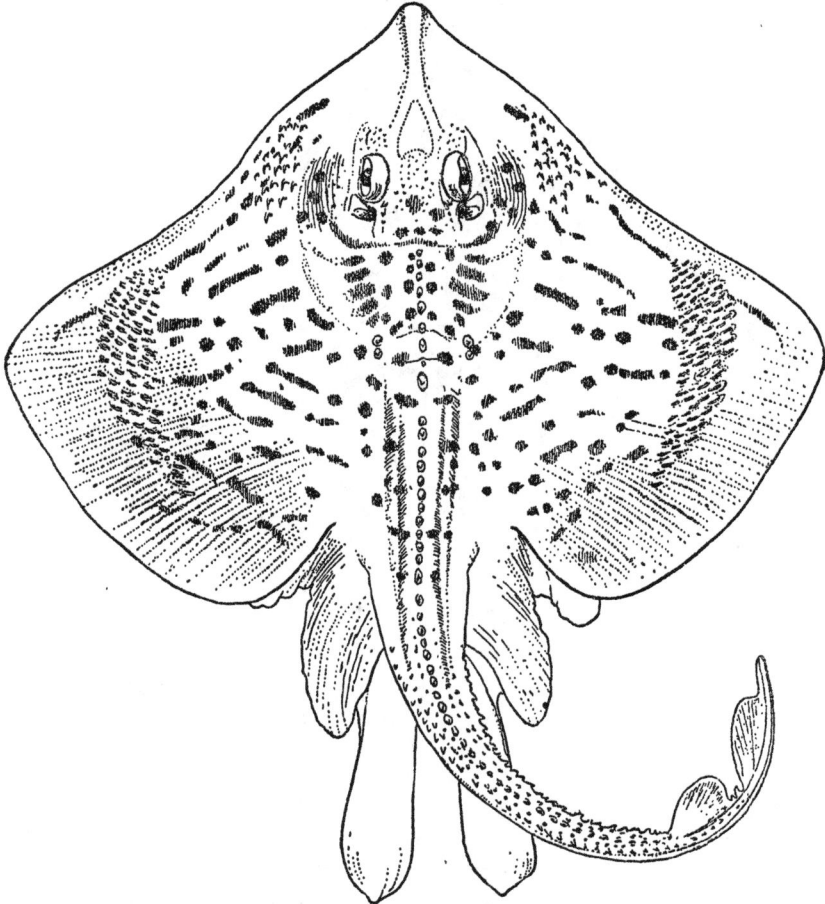

FIG. 32.—*Raja eglanteria*, male. (After Garman)

14. Raja eglanteria Lacépède.   Clear-nose ray; Brier ray.

Raia eglanteria Lacépède, Hist. Nat. Poiss., II, 1800, p. 109, Pl. IV, fig. 2; Charleston, S. C. Garman, 1913, p. 341, pl. 23.
Raja eglanteria Uhler and Lugger, 1876, ed. 1, p. 188, ed. 2, p. 159; Jordan and Evermann, 1896–1900, p. 71.
? *Raia lævis* Bean, 1891, p. 94.

Disk broader than long, anterolateral margins double concave, a slight concavity opposite snout and a much larger and broader one behind eyes, the margins meeting anteriorly in an angle a little greater than 90°, the outer angles rounded, the posterolateral margins broadly and evenly convex, length of disk 1.2 to 1.3 in the width; head to first gill slit 2.7 to 3.5 in width of disk; distance from snout to vent 1.15 to 1.37; snout projecting, apparently longer in females than in males, supported by a rather narrow cartilage with a large translucent area on each side, its length 1.45 to 1.57 in head; preoral length of snout 1.4 to 1.8; width of mouth 2.5 to 2.8; interorbital (bone) 4.8 to 5.25; eye 4.7 to 5.2 in snout; spiracles immediately behind eyes, slightly crescent-shaped, about as long as eyes; nasoral groove extending to mouth; teeth in about 48 rows in each jaw, each tooth with a large round or oval base, surmounted by a small pointed cusp on the posterior or newer teeth, the anterior or older teeth smooth, without pointed cusps; skin above largely beset with small bony prickles, these somewhat enlarged on tip of snout; a row of short, heavy spines on inner margins of eyes and spiracles; a few enlarged tubercles opposite median line of back on shoulders; a row of short, sharp spines on median line of back, extending from behind head to origin of first dorsal fin; tail with a row of enlarged spines on each side and with other prickles larger than those on the body; the male somewhat smoother than the female, with a patch of small recurved spines on disk opposite eyes and another intramarginal patch at widest part of disk; tail moderate, the base depressed, its length 1.9 to 2.25 in total length; dorsal fins 2, placed near the extremity of the tail, less than eye's diameter apart in four specimens examined, an eye's diameter in one specimen, and confluent in another; caudal fin represented by a dermal fold extending around the end of the tail; ventral fins long, beginning only a little in advance of the margin of the disk, greatly thickened at the base anteriorly, the fin deeply notched; the claspers in the male rather broad, not projecting far beyond posterior margins of the ventrals in adults.

Color varying from brownish to grayish above, with roundish and elongate dark markings on disk posterior to snout; lower surface white.

The foregoing description is based upon six preserved specimens—two males and four females—ranging in length from 457 to 672 millimeters (18 to 26½ inches); others were examined in the field. This ray is called "clear-nose" in allusion to the translucent snout and "brier ray" because of the numerous spines and prickles that beset the upper surface of the body and tail. This species, the sting ray (*Daybatus say*), and the sand skate (*Pteroplatea macrura*) are about equally common in the southern part of Chesapeake Bay. Early in April, in a set of two pound nets in Lynnhaven Roads, 7 to 15 brier rays were caught daily; and on May 25, when we again visited these nets, the catch was 25, all 1½ to 2½ feet in length.

This ray was found to feed chiefly on crustaceans and fish. Two stomachs examined in April contained crabs, shrimp, and fish; two examined in May contained shrimp and fish, and three in October the following: One had eaten a blue crab (Callinectes) 1 inch long; another a lizard fish (Synodus) 8 inches long; and the stomach of the third contained several blue crabs, 1 to 1½ inches long. The structure of the teeth suggests that mollusks and crustaceans probably form the principal foods.

Habitat.—Cape Cod to Florida; rarely to Cape Ann, Mass.

Chesapeake localities.—(a) Previous records: "Around the mouth of Chesapeake Bay" (Uhler and Lugger, 1876); Cape Charles City, Va. (b) Specimens in collection: Lynnhaven Roads, Va., pound nets, June 9, 1916, May 20 and September 27, 1921, and May 25, 1922. Numerous individuals also were seen and examined at Ocean View, Va., during the fall of 1922. It also was taken at 11 *Fish Hawk* stations, all made in southern sections of the bay during 1915 and 1916.

15. Raja stabuliforis Garman. Barn-door skate; Smooth skate.

Raja lævis Mitchill, Amer. Monthly Mag., II, 1818, p. 327; New York, not of Gronow. Jordan and Evermann, 1896-1900, p. 71; Uhler and Lugger, 1876, ed. I, p. 189; ed. II, p. 160.

Raia stabuliforis Garman, 1913, p. 341, pl. 22, fig. 2; pl. 44, figs. 4 to 6.

Disk broader than long, its width 1.5 in total length of fish, the length of disk 1.25 in its width; anterolateral margins slightly double concave; posterolateral margin broadly rounded; snout strongly projecting, acute; eyes small; spiracles as large as eyes; mouth large, the width more than half the length of the snout; teeth quite blunt in the female, sharper in males, in about 32 to 36 series in each jaw; upper surface comparatively smooth; tip of snout with small tubercles, and a narrow band of similar tubercles along anterolateral margin; small tubercles over the eyes and spiracle; a median row of compressed spines beginning on back and extending on tail; a similar row on each side of tail. The male, in addition to the armature already described, has a triangular patch of large sharp spines on disk opposite eyes, a large area of similar spines situated opposite the outer angle of the disk at about the beginning of the outer third of the disk.

Color of a fresh male specimen, 49 inches long, brownish above, with many scattered small dark spots of unequal size, the largest equal to the size of the eye; a pair of large, irregular, prominent, ocellated spots on the disk opposite the outer angles of pectorals; ventral surface nearly plain. The ocellated spots on the disk in the specimen described are sometimes wanting, and the lower surface is frequently marked, particularly in large individuals, with dusky or gray. The color of a female 46 inches long, taken with the above male, differed in having fewer spots and in the smaller size of the ocellated spots.

This species is readily recognized by the long, acute snout, by the smoothness of the skin on the upper surface, the prickles being much fewer than in related species, and, usually, by its large size.

The specimens of this species were too large to preserve conveniently. The above description is based upon several large specimens examined in the field and also upon published accounts.

The barn-door skate, like most other skates, feeds mainly on the bottom. Its food (Bigelow and Welsh, 1925, p. 67) consists of mollusks, crustaceans, fish and worms. It is regarded as more destructive of fish than any of the other skates.

The breeding habits of the barn-door skate are unknown.

The barn-door skate is frequently taken in early spring with pound nets in the southern part of Chesapeake Bay. The wings or "saddles," as they are called by the fishermen, are sometimes removed from the fish and shipped to New York, where there is a fair demand for them. In parts of Europe skate saddles are considered a delicacy, and it is the foreign population of New York and vicinity that furnishes most of the demand for them.

The barn-door skate reaches a length of 6 feet, and examples 4 to 5 feet long are not at all rare. The three largest seen by us in the Chesapeake were 46 to 49 inches in length.

Habitat.—Nova Scotia to Florida.

Chesapeake localities.—(*a*) Previous records: "Not uncommon in the ocean off Worcester County, but said to be scarce in Chesapeake Bay." (Uhler and Lugger, 1876, pp. 160 and 189.) (*b*) Specimens in present collection: None. A number of individuals, all large, were observed during the present investigation. They were taken in the spring with pound nets located at Ocean View and in Lynnhaven Roads, Va.

16. Raja erinacea Mitchill. Common skate; Little skate; Summer skate.

Raja erinaceus Mitchill, Amer. Journ. Sci. Arts, IX, 1825, p. 290; New York.
Raja erinacea Jordan and Evermann, 1896–1900 p. 68; Pl. IX, fig. 29; Garman, 1913, p. 337, pl. 20; pl. 55, fig. 5; pl. 68, fig. 1.

"Anterior margins of disk waved, convex opposite the eyes, concave opposite spiracles, outer and hinder angles and margins rounded. Snout short, longer than that of *R. diaphanes*, about one and one-half times the interspiracular width. Mouth strongly waved; teeth in about 50 rows. Back rough with strong, hooked spines over almost the entire surface on females, especially rough near and on head, on the snout, about the shoulders, on the hinder portions of the pectorals, and on the tail. A triangular patch of strong spines appears in front of the shoulder girdle; others are seen on each shoulder and in one to several rows at each side of the median line of the back. The ventral line is quite or nearly without tubercles; the tail has two to four rows on each side. Males have not so many tubercles as the females; their spines are more scattered, and smooth spaces exist on the middle of the back, over the gills, and above the abdomen; they have the band of erectile tenacula near the outer angle of the pectoral. * * *

"Back light grayish brown to very dark, clouded to uniform, usually spotted with small spots of darker, margins sometimes light. Color darker northward." (Garman, 1913.)

This species was not seen during the present investigation and it is not recorded from Chesapeake Bay. The species is included here on the authority of certain field notes by Dr. W. C. Kendall, made during an investigation in 1894, which he has kindly placed at our disposal. In these notes we find it stated that *R. erinacea* was taken in pound nets near Hampton, Va., on March 13 and 24 (one specimen on each date), and again near Cape Charles, Va., on March 21. Of the latter he says: "A few *R. erinacea* were brought in from the pounds."

"Little skates are omnivorous. Hermit and other crabs, shrimps, worms, amphipods, ascidians ('sea squirts'), bivalve mollusks, squid, small fishes, and even such tiny objects as copepods have been found in their stomachs." (Bigelow and Welsh, 1925, p. 59.)

Off the coast of southern New England, where the species is abundant, the eggs are taken from March to September, being most numerous during July and August. The egg, together with its case, is about 2 inches broad and 2½ inches long. The empty cases of these eggs frequently are seen washed upon the beach.

This is a common skate on our coast from Virginia northward. Its usual length ranges from 1 to 2 feet.

Range.—"Halifax to the Carolinas, abundant off New England and New York." (Garman, 1913.)

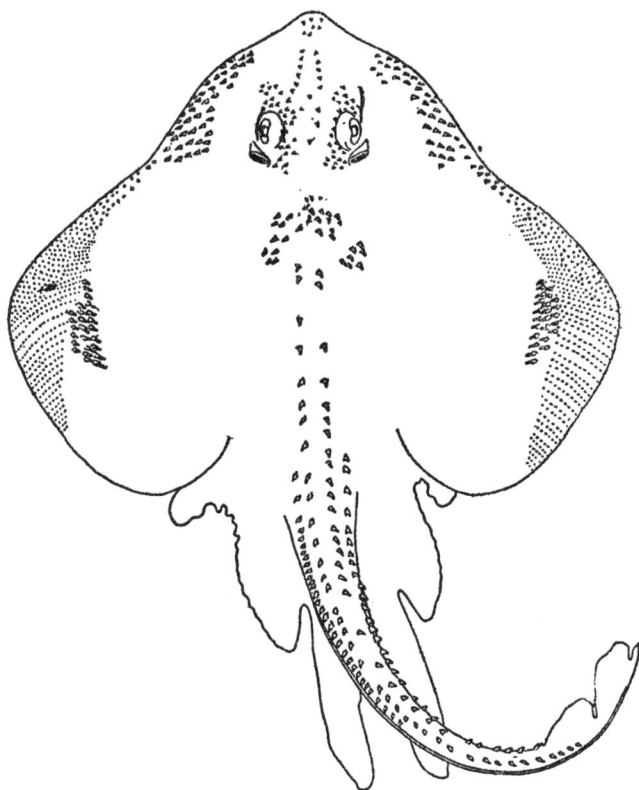

FIG. 33.—*Raja erinacea*, male. (After Garman)

Chesapeake localities.—(a) Previous records: None. (b) Specimens in the collection: None. The present record is based upon field notes by Dr. W. C. Kendall made during March, 1894, in which the capture of this species in pound nets near Hampton and Cape Charles, Va., is reported.

Family XI.—TORPEDINIDÆ. The electric rays

Head, trunk, electric organs and pectorals forming a depressed subcircular disk; tail short, rather stout, with or without a lateral membranous fold; spiracles present; gill slits small, between the electric organs and the head; electric organs composed of vertical cells, situated between the pectoral fins and the head; nasal valves confluent, forming a quadrangular lobe; skin smooth, unarmed; dorsal fins 2, 1, or none; caudal fin not lobed.

13. Genus TORPEDO Duméril.   Electric rays

Disk broader than long, subcircular; snout short, broad; tail short, distinct, with a large caudal fin and a low dermal keel on each side; spiracles moderate, placed at a short distance back of the eye, without fringes on the margins; mouth crescent-shaped, with a longitudinal fold on each side; dorsal fins 2; ventral fins separate and distinct. A single species is known from the Atlantic coast of the United States.

17. Torpedo nobiliana Bonaparte.   Torpedo; Electric ray; Crampfish

Torpedo nobiliana Bonaparte, Fauna Ital., 1832, fasc. 12; Italy.
Torpedo occidentalis Uhler and Lugger, 1876, ed. I, p. 188; ed. II, p. 159.
Tetronarce occidentalis Jordan and Evermann, 1896–1900, p. 77, Pl. XI, fig. 33.
Narcacion nobilianus Garman, 1913, p. 310, pl. 25, fig. 2; pl. 61, figs. 4 and 5.

Disk broader than long, the sides broadly rounded, the anterior margin slightly concave; tail short, thick, depressed, tapering abruptly, with a dermal fold on each side; spiracles close behind the eyes, their edges not fringed; mouth large, crescent-shaped, with a groove at each angle; teeth small, broad-based, with acute crowns on inner edges; skin smooth and unarmed; first dorsal about twice as large as the second, its origin in advance of the posterior edges of ventrals; caudal fin large, its posterior margin slightly rounded to slightly concave posteriorly.

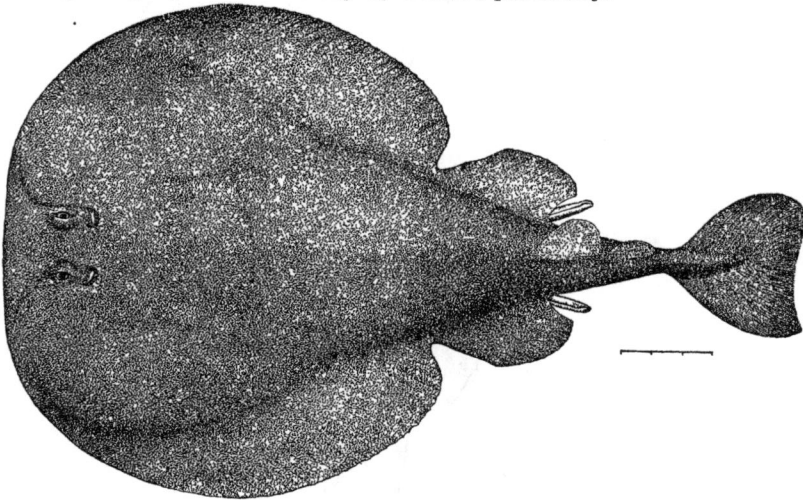

FIG. 34.—*Torpedo nobiliana*

Color above, uniform dark brown; mostly white underneath; the edges of the disk and the ventrals underneath purplish; caudal peduncle with irregular dark markings along ventral edges. Day (1880–1884, p. 331) records the color of European specimens as dull reddish gray or dull ash above, dashed with purple, and white below, sometimes with ill-defined blotches on the dorsal surface.

A single large (female) specimen, weighing about 100 pounds, was taken during the present investigation, upon which the following measurements are based:

	Inches
Total length	47. 6
Width of disk	35. 6
Length to base of first dorsal	31. 1
Distance between dorsal fins	2. 1
Interorbital (bone)	1. 6
Space between spiracles	2. 6
Length of base of first dorsal	3. 1
Length of base of second dorsal	1. 7
Height of first dorsal	3. 9
Height of second dorsal	2. 6
Greatest width of caudal fin	10. 5

The species is rare in Chesapeake Bay, where it is occasionally caught in the most southern parts. The fishermen do not seem to have a distinctive name for the animal, as those who saw the specimen merely called it a "ray." The species is readily recognized by its smooth, soft skin, dark brown color, broad disk, which is straight or slightly concave in front of eyes, and by the large caudal fin.

The torpedo is said to reach a weight of 200 pounds, and a specimen as heavy as 170 pounds has been recorded from Massachusetts Bay (Bigelow and Welsh, 1925, p. 69). A specimen has been recorded from Cape Lookout, N. C., which was 60½ inches in length and weighed 125 pounds.

It has long been known that the torpedo is capable of emitting strong electric shocks from large electric organs situated on each side just back of the head, a shock from a large fish being sufficient to knock a man down.

Little is known of the feeding habits of this ray along our coast, but Day (1880–1884, p. 331), working with European specimens, records from the stomach of one fish a 2-pound eel and a 1-pound flounder and from another a 4 or 5 pound salmon, all of which, he believes, may have been killed by the electric organs of the fish.

The species is viviparous, but little is known of its breeding habits.

Habitat.—Tropical and temperate parts of the Atlantic Ocean, from Maine to Cuba on the American coast and from the coasts of Great Britain to Madeira, including the Mediterranean Sea, on the European coast.

Chesapeake localities.—(a) Previous records: "Said to occur very rarely in the region near the entrance of Chesapeake Bay." (Uhler and Lugger, 1876). (b) Specimens seen or preserved during the present investigation: One large female, about 4 feet long, taken in a pound net in Lynnhaven Roads, Va., on May 25, 1922.

Family XII.—DASYATIDÆ. The sting rays

Body, head, and pectorals depressed, together forming a broad disk; the pectorals very broad and united around the snout; no supporting cartilage in snout; tail distinct from the disk, either long or short, and usually bearing one or more strong, serrated spines; spiracles large and near the eyes; skin smooth, or rough and with spines or tubercles, or both. The species are viviparous or ovoviviparous. The family contains numerous genera and species. Only two genera are represented in the Chesapeake Bay fauna.

KEY TO THE GENERA

a. Tail long and slender, whiplike, bearing one or more strong, serrated spines; disk more or less quadrangular to circular, not much broader than long_____Dasyatis, p. 63
aa. Tail short, the spine present or absent; disk much broader than long_____Pteroplatea, p. 67

14. Genus DASYATIS Rafinesque. Sting rays

Disk more or less quadrangular to circular, very strongly depressed; snout more or less prominent; tail long, whiplike, with one or more strong, serrated spines, with or without dermal fin folds or keels on the median line above, below, and behind the caudal spines; without lateral folds on the base; skin usually more or less spiny in adults; teeth small, paved.

KEY TO THE SPECIES

a. Tail without a dermal fin fold above, a rather broad fold below, its length more than twice the length of the disk; disk broader than long, its length about 1.25 in its width_____*centrura*, p. 64
aa. Tail with a dermal keel or fin fold above and below, its length less than twice the length of the disk.
  b. Tail with a low dermal keel above and a rather broad fold below; the color of the folds black; the disk quadrangular, its length about 1.2 in its width; middle of forehead with a small; round, light-colored spot_____*americana*, p. 64

49826—28——5

bb. Tail with a dermal fold above and below, both of about equal size, color of folds black; the disk rather narrower than in *americana*, little broader than long, its length about 1.1 in its width, the outline of the disk meeting at snout at an angle of about 120°; no light-colored spot on middle of forehead_____*say*, p. 66

bbb. Tail with a dermal fold above and below, the lower one the larger, the color of the folds brownish to yellowish, the lower one always of light color; disk still narrower, the length about equal to the width; snout more pointed than in related species, the outline of the disk meeting anteriorly at an angle of about 90°; no light-colored spot in the middle of the forehead_____*sabina*, p. 67

18. Dasyatis centrura (Mitchill). Sting ray; Stingaree.

Raja centrura Mitchill, Trans., Lit. Philo. Soc., N. Y., I, 1815, p. 479; New York.
Trygon centrura Uhler and Lugger, 1876, ed. I, p. 187; ed. II, p. 158.
Dasyatis centrura Jordan and Evermann, 1896–1900, p. 83.
Dasybatus marinus Garman, 1913, p. 382, pl. 33, figs. 1 and 2.

Disk quadrangular, notably broader than long, its length about 1.25 in its width, antero-lateral margins concave opposite the eyes, convex toward the slightly protruding snout, the outer angles rounded, the postero-lateral margins little convex; mouth arched forward, with five papillæ at base of lower jaw; teeth blunt, arranged in pavement; tail more than twice the length of the disk, bearing one or more strong, serrated spines, with a broad winglike expansion below but none above; young smooth, adults with conically pointed, broad-based tubercles on the middle and hinder parts of the back and on the top and sides of the tail, very old examples with still more numerous spines and tubercles on the back. Color dark brown above, pale underneath.

This ray was not taken in Chesapeake Bay during the present investigation and no specimens are at hand. The above description was compiled from published accounts. This species may be distinguished from all the others of the Chesapeake region by the entire absence of a fin fold on the dorsal surface of the tail, posterior to the large, serrated spine or spines, and by the prominent expansion of the fold below the tail. The tail appears to be longer than in related species, equaling more than twice the length of the disk.

The food of this ray no doubt is similar to that of related species, as it has the same type of teeth, which are suitable for crushing hard objects, such as the shellfishes. This ray appears to reach a larger size than the related rays, a length of about 12 feet from snout to tip of tail having been reported.

This ray probably is rare in Chesapeake Bay. It is included here on the strength of the following statement by Uhler and Lugger (1876, p. 187): "Common on the coast of Worcester County and *around the entrance to Chesapeake Bay.*"

Habitat.—Atlantic Ocean and Mediterranean Sea (Garman, 1913). On the American coast from Cape Cod to Cape Hatteras.

Chesapeake localities.—(*a*) Previous records: Around the mouth of Chesapeake Bay (Uhler and Lugger, 1876). (*b*) Specimens in the present collection: None.

19. Dasyatis americana sp. nov. Sting ray; Stingaree.

Dasibatis hastata Garman, in Jordan and Gilbert, Bull. U. S.Nat. Mus., XVI, 1882 (1883), p. 70. Not of De Kay, which herein is understood to be equivalent to *D. centrura* (Mitchill), and which, in turn, as understood by Garman (1913, p. 382), is identical with *D. marinus* Klein.
Dasyatis hastata Jordan and Evermann, 1896–1900, p. 83. Not of De Kay.
Dasybatus hastatus Garman, 1913; p. 391. Not of De Kay
Type No. 88378, U. S. National Museum; length of disk 15 inches; type locality, Crisfield, Md.

This ray is very similar to *say*, from which it may be distinguished, however, by the absence of a broad, winglike expansion on the upper side of the tail, which is replaced by a low, black keel; the cutaneous folds below the tail are identical in the two species; the disk appears to be slightly shorter in proportion to its width, and the ventral fins apparently project a little farther beyond the disk. This species bears a small, round, light-colored spot on the middle of the forehead, which is wanting in *say*. The following proportions were obtained from a male specimen having a disk 15 inches in length; length of disk in its width 1.2; distance from snout to vent 1.35 in width of disk; head to first gill slit 3.7; snout 1.6 in head; preoral length of snout 1.48; interorbital (bone) 3.43; width of mouth 3.1; eye 3 in snout; tail 1.63 in total length.

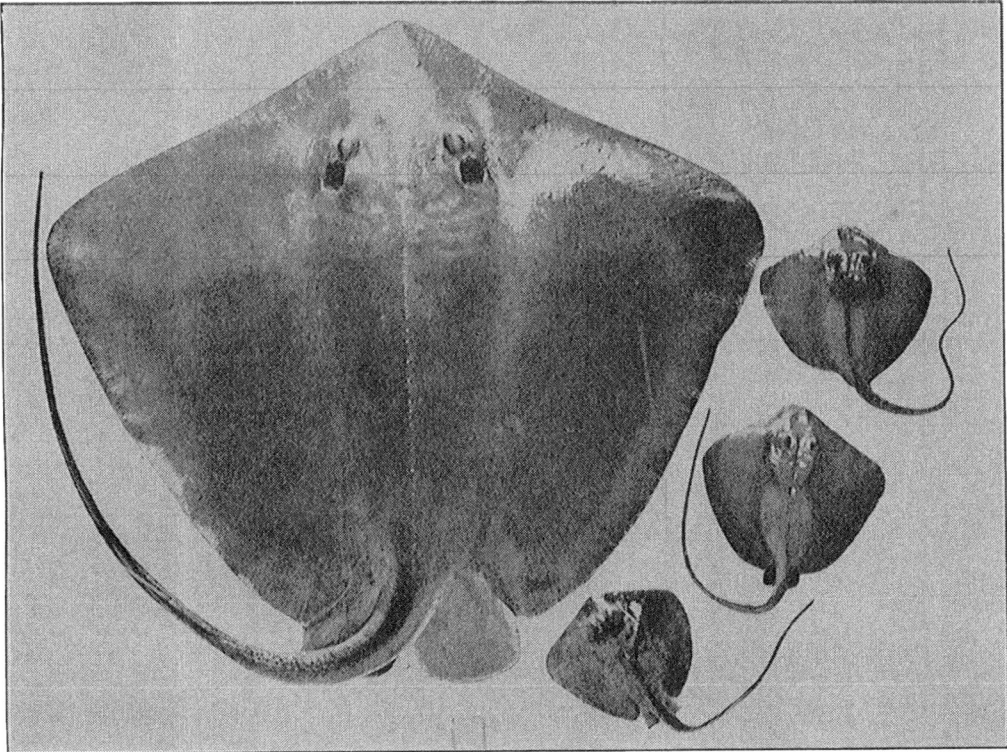

Fig. 35.—*Dasyatis americana* sp. nov.   From a female 62 inches long, and newly born young

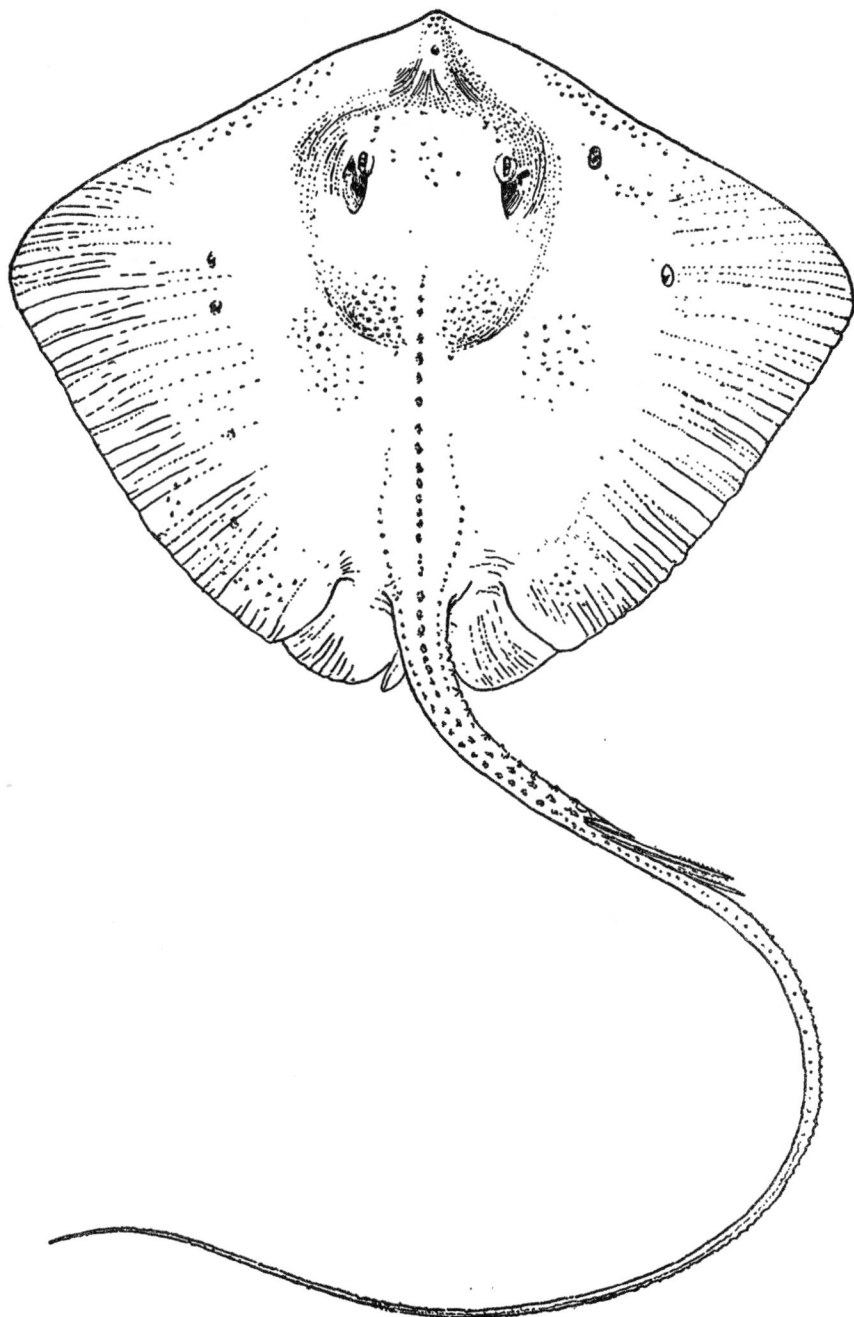

FIG. 36.—*Dasyatis centrura*. (After Garman)

A single (male) specimen with a disk 15 inches in length is present in the Chesapeake Bay collection. Nothing distinctive concerning food and reproduction can be said of this species.

Habitat.—Crisfield, Md., to Brazil.

Chesapeake localities.—(*a*) Previous record: None. (*b*) Specimen in collection: Crisfield, Md., hook and line, September 15, 1921.

20. Dasyatis say (Le Sueur). Sting ray; Stingaree.

Raia say Le Sueur, Journ., Ac. Nat. Sci. Phila., I, 1817, p. 42, with plate; New Jersey.
Dasyatis say, Jordan and Evermann, 1896–1900, p. 86.
Dasybatus say, Garman, 1913, p. 396.

Disk a little broader than long, the anterolateral margins nearly straight, meeting anteriorly in an obtuse angle of about 120°, the posterolateral margins broadly convex, the posterior angles rounded, length of disk 1.06 to 1.13 in its width; head to first gill slit 3.4 to 3.65 in width of disk; distance from snout to vent 1.27 to 1.32; snout 1.4 to 1.77 in head; preoral length of snout 1.6 to 1.75; width of mouth 3.2 to 3.75; interorbital (bone) 2.75 to 3.35; eye, 2.9 to 3.85 in snout; spiracles immediately behind eyes, elliptical, as long as eye; nasorial groove extending to mouth; teeth arranged in pavement, those of the male with an acuminate tip, those of the female smooth; 3 large papillæ at base of lower jaw, behind the teeth and a smaller one on each side back of the outer angle of the teeth; skin perfectly smooth in the young, large individuals with a row of short, blunt spines on median line of back, and sometimes one or two on each shoulder, the tail with spinules; tail long, slender, depressed anteriorly, round and whiplike posteriorly, bearing one or two long, sharply serrated spines, with a short cutaneous fold behind the spine above and a larger one below; tail 1.65 to 1.8 in total length; ventral fins broadly rounded posteriorly, not reaching far beyond end of disk.

Color grayish to brownish above, white below; the distal part of tail and the cutaneous folds on it black.

Eight male specimens with disks ranging in width from 230 to 290 millimeters (9 to 11.5 inches) are at hand and form the basis for the foregoing description.

The stomachs examined were void of recognizable foods. The teeth evidently are constructed for crushing hard objects, and Smith (1907, p. 45) says of this species, "It feeds largely on shellfish." Fish probably form a very small part of its diet. It is worthy of mention in this connection that although the individuals at hand were taken from a pound net, in which fish were easily available, they had not recently fed on them. No determination has been made as to whether or not a difference in the foods sought by the male and female (as suggested by the difference in the structure of the teeth) exists.

The sting ray appears to be common in the southern parts of Chesapeake Bay and at times even abundant. "Numerous and troublesome." (Moseley, 1877, p. 9.) In 1921 this ray was taken in large numbers in pound nets in Lynnhaven Roads, Va., during the latter part of September and early in October, as many as 40 individuals having been seen in one net at one time. The rays at this time probably were on their southward migration from the feeding grounds in the upper stretches of the bay. A southward migration of this species has been noted by observers elsewhere on the Atlantic coast.

The tail is used as a whip, and with the serrated spine the ray sometimes inflicts very painful wounds in the hands and feet of fishermen, who are generally of the opinion that a venom is injected with the spine. The difficulty experienced in healing such a wound, however, undoubtedly is due to septic infection.

Habitat.—New York to Brazil.

Chesapeake localities.—(*a*) Previous records: None. (*b*) Specimens in collections: Lynnhaven Roads, Va., pound nets, June 9, 1916, and September 26, 1921. Also seined at Cape Charles, Va., May 21, 1922, and observed numerous times in pound nets situated between Ocean View and Cape Henry.

21. Dasyatis sabina (Le Sueur). Sting ray.

Trygon sabina Le Sueur, Journ., Ac. Nat. Sci., Phila., IV, 1824, p. 109, Pl. IV.
Dasyatis sabina Jordan and Evermann, 1896-1900, p. 84, Pl. XIV, figs. 36 and 36a.
Dasybatus sabinus Garman, 1913, p. 397.

Disk little, if any, broader than long, the anterolateral margins distinctly concave in front of eye, meeting at an angle of about 90°, the posterolateral margins broadly and evenly convex, the outer angles of disk very broadly convex, the posterior angle much more sharply rounded, the length of disk 1 to 1.05 in its width; head to first gill slit 2.65 to 2.75 in width of disk; distance from tip of snout to vent 1.15 to 1.2; snout rather pointed, its length 1.4 to 1.5 in head; preoral length of snout 1.5 to 1.6; width of mouth 3.45 to 3.95; interorbital (bone) 3.8 to 4.3; eye 4 to 4.35 in snout; spiracles immediately behind the eyes elliptical, the longest diameter equal to length of eye; nasoral groove extending to mouth; teeth arranged in pavement, similar in both jaws and in the sexes; three large papillæ at base of lower jaw and two small ones at each side; skin almost perfectly smooth in our youngest examples with only a few spines on median line of back, the largest individual with a row of prominent, compressed spines on median line of back, extending from the occiput nearly to base of caudal spine, and two similar spines on each shoulder; the tail long and slender, depressed anteriorly, round and whiplike posteriorly, bearing one or two long, sharply-serrated spines; rather short cutaneous folds behind the spine, both above and below the tail, the lower fold a little broader than the upper; tail 1.6 to 1.65 in total length; ventral fins extending well beyond the disk, their posterior margins rounded, the outer angles sharper than the inner ones.

Color brownish on back, the winglike expansions paler; white underneath. The upper fin fold on the tail yellowish brown, the lower buff.

This ray is represented by four specimens—three males, respectively, 190, 215, and 275 millimeters (7½, 8½, and 10¾ inches) broad, and one female 185 millimeters (7¼ inches) broad. We also have the tail and the teeth of another specimen which measured 16 inches in width, evidently belonging to this species. This ray is rather closely related to *D. say*, from which it may be distinguished by the more pointed snout, the deeper concavity opposite the eyes, in the outline of the disk, and by the paler color of the fin folds on the tail.

The stomachs of the specimens at hand all contained fragments of crustaceans.

This sting ray apparently reaches a moderate size. The specimens previously mentioned, measuring 16 inches in width, appears to be among the largest taken to date. The species, although previously not recorded from Chesapeake Bay, probably is not rare there. It was not recognized in the field as distinct from *D. say*, and the fishermen do not distinguish it. The specimens were seclected at random from various catches, with the result that 5 of this species and 8 of *D. say* were preserved.

Habitat.—Previously recorded from North Carolina to Brazil. The range is now extended to Chesapeake Bay. The species enters fresh water.

Chesapeake localities.—(a) Previous records: None. (b) Specimens in present collection: Lower York River, Va., collecting seines, July 8 and October 6, 1921, and Ocean View, Va., commercial and collecting seines, October 2, 6, and 16, 1922.

15. Genus PTEROPLATEA Müller and Henle

Disk much broader than long, very strongly depressed, the anterior angle obtuse, the lateral angles acute; tail slender, shorter than body, with or without a serrated spine and without a fin. This skate reaches a rather large size. A single species inhabits the Atlantic coast.

22. Pteroplatea micrura (Schneider). Sand skate; Butterfly ray.

Raia micrura Schneider in Bloch, Syst. Ichth., 1801, p. 360; "Surinamo."
Pteroplatea maclura Bean, 1891, p. 94; Jordan and Evermann, 1896-1900, p. 86.
Pteroplatea micrura Garman, 1913, p. 414, pl. 23, figs. 3 and 4.

Disk much broader than long, anterolateral margins convex opposite the head, the median portion of the margins broadly concave, the outer angles rounded, the posterolateral margins broadly convex, length of disk 1.65 to 1.85 in the width; head to first gill slit 5.6 to 6.8 in width of disk; distance from snout to vent 1.95 to 2.1; snout at its tip projecting very slightly beyond the outline of the disk, its length 1.5 to 1.8 in head; preoral length of snout 1.37 to 1.62; width of mouth 1.67 to 2.05; interorbital (bone) 2.1 to 2.45; eye 3.6 to 5 in snout; spiracle in a quadrangular pit immediately back of eye, no tentacle on its posterior margin, the slit equal to diameter of eye; teeth in numerous rows, about 75 to 100 in each jaw, arranged in definite series like bricks in a pavement, the teeth slightly spear-shaped, each tooth with a broad base and an elongate, sharp cusp; skin entirely smooth in specimens at hand, large individuals are reported to bear caudal spines; tail very short, pointed, with a keel above and below, its length 4.07 to 4.95 in total length; ventral fins rather narrow, inserted notably in advance of the posterior margin of the disk, the posterior margins rounded; claspers of the male long and narrow, reaching half their length beyond posterior margin of ventrals in specimens 14½ inches broad.

Color variable, gray, brown, light green, or purple above, with vermiculations and punctulations of lighter and darker colors; the tail lighter than body, with three or four dark bars; the anterolateral margins of disk frequently with roundish spots; lower parts plain white, outer margin of wings sometimes grayish, dusky, or salmon.

The foregoing description is based on six specimens—four males and two females—ranging in length from 175 to 270 millimeters (7 to 10¾ inches) and in width from 265 to 375 millimeters (10½ to 14¾ inches), and the jaws with the teeth of a female, which, according to field measurements, was approximately 595 millimeters (23½ inches) long and 860 millimeters (34 inches) broad. This skate is characterized by the very broad body, the short tail, and by the absence of a tentacle behind the spiracle, or breathing hole, situated just behind the eye.

This skate is taken in Chesapeake Bay from May until November. In September and early in October of 1921 numerous specimens were taken in pound nets in Lynnhaven Roads, where the latest catch was made on November 28. As many as 40 individuals were frequently caught by one net during a 24-hour period. It is probable that this skate migrates southward at this season of the year, returning from the more northerly feeding grounds in the bay. Most of the individuals taken in pound nets were of small size, specimens ranging from 10 to 13 inches in width predominating. At Ocean View, Va., however, only three small specimens, 10½ to 10¾ inches in width, were taken in commercial and collecting seines from September 25 to October 27, 1922. This skate and the sting ray (*Dasyatis say*) appear to be about equally common.

Little is known of the food of this species, but it is known to feed on crabs (Sumner, Osborne, and Cole, 1913, p. 739) and no doubt also on other crustaceans. The teeth appear to be too weak to crush oysters and clams. The stomachs examined, taken from specimens caught in a pound net, were empty.

The species is viviparous and the normal number of young produced appears to be two, the greatest width of the young at birth being 6 inches (Smith, 1907, p. 45). The frequent presence of this species on sandy shores has caused it to be named "sand skate," and it is called "butterfly ray" in allusion to the very broad, winglike expansion of the pectoral fins. The "wings" of this species are utilized to a limited extent as crab bait in the crab industry on Chesapeake Bay.

This species is reported to reach a length of 4 feet (Smith, 1907, p. 45), also 15 to 18 feet (Uhler and Lugger, 1876). The largest individual mentioned in the field notes of the collectors of the present collection is the specimen previously mentioned measuring 23½ inches in length.

Habitat.—Cape Cod to Brazil.

Chesapeake localities.—(*a*) Previous records: Cape Charles; "vicinity of Norfolk, Va." (Moseley, 1877, p. 9.) (*b*) Specimens in collection or observed in the field were taken at Back River, Ocean View, Lynnhaven Roads, and Cape Charles, Va., during May, June, September, October, and November.

Family XIII.—MYLIOBATIDÆ. The eagle rays

Disk broad; pectoral fins not continued to the end of the snout, but ceasing at side of head; a pair of rostral fins joined in front of head, supported by fin rays; tail long and slender, bearing a dorsal fin and usually a strong serrated spine on its basal portion; eyes large, lateral; spiracles large, behind eyes; teeth broad, flat, tessellated, the median ones usually broader than the others.

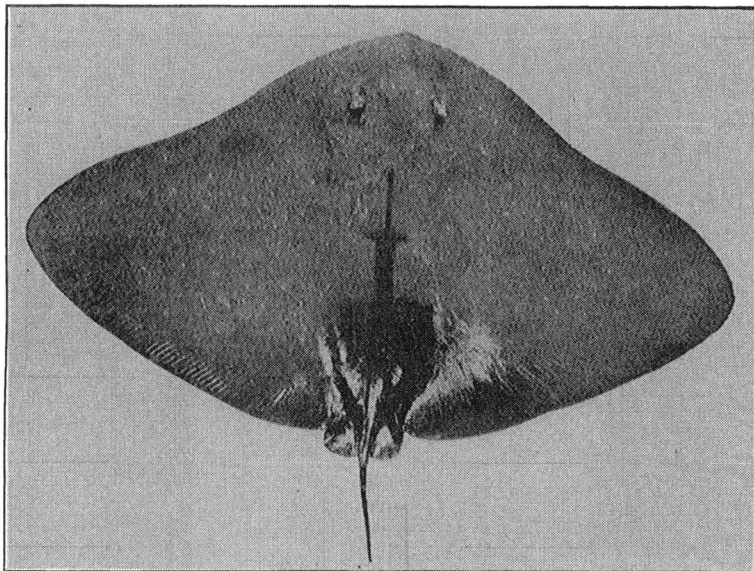

Fig. 37.—*Pteroplatea micrura.*   An adult female, 504 millimeters long

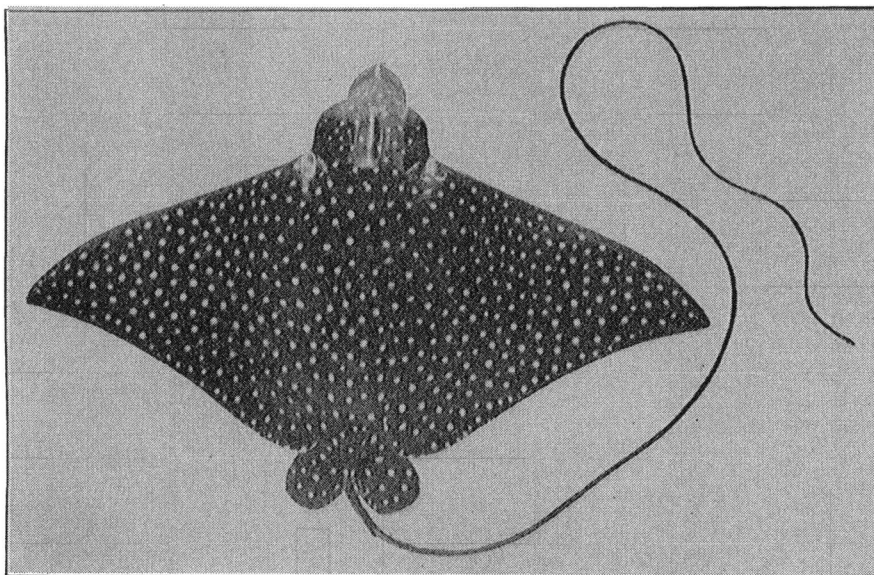

Fig. 38.—*Aëtobatus narinari.*   Showing coloration of young

KEY TO THE GENERA

a. Teeth in the jaws in several series; sides of head not entirely free from the pectorals; the rostral process and the pectoral fins narrowly confluent_____ __Myliobatis, p. 69

b. Teeth in one row in each jaw; pectoral fins not extending along the sides of the head; the rostral fins entirely distinct from the pectorals _____ _____Aëtobatus, p. 69

16. Genus MYLIOBATIS Cuvier. Eagle rays

Disk broad, the outer angles acute; rostral process narrowly confluent with the pectorals along the sides of the head; teeth in the jaws in 7 to 10 rows, tessellated, the median ones broader than the lateral ones; tail long, slender, bearing on its basal portion a dorsal fin and one or more serrated spines; skin smooth or nearly so.

23. Myliobatis freminvillii Le Sueur. Eagle ray; Bull-nosed ray; Sharp-nosed ray; Bull ray.

Myliobatis freminvillii Le Sueur, Jour., Ac. Nat. Sci., Phila., IV, 1824, p. 111; Rhode Island. Uhler and Lugger, 1876, ed I, p. 185; ed. II, p. 157.

Myliobatis freminvillei Jordan and Evermann, 1896–1900, p. 89.

Disk broader than long, the outer angles rather sharp, anterolateral margin slightly convex, the posterolateral margin broadly concave, length of disk 1.5 in its width; pectoral fins narrowly confluent below eyes with the rostral process, the latter broadly rounded with a slightly protruding median tip; head to first gill slit 4.85 to 5.1 in width of disk; distance from snout to vent 1.8 to 1.95; snout 2.1 to 2.35 in head; preoral length of snout 1.85; width of mouth 2.5 to 2.65; interorbital space 1.9; eye lateral, 1.95 in snout; spiracles quite as large as eyes and situated immediately behind them; nasorial groove extending to mouth; teeth in pavement about 9 transverse rows in upper jaw, 4 functioning, 10 transverse rows in lower jaw, 6 functioning; skin smooth; a prominent, serrated spine present behind dorsal; tail long, whiplike, 1.45 to 1.55 in total length; dorsal fin situated on tail, its origin at vertical from tips of ventrals; ventral fins rather broad, posteriorly convex, its base 1.75 in snout.

Color grayish above, white underneath, the outer tips of the disk becoming dusky. Sometimes reddish or reddish brown above, according to published accounts.

Two specimens—a male and female—355 and 368 millimeters (14 and 14½ inches) wide from tip to tip of disks, are at hand. The broad, pavementlike teeth obviously are constructed for crushing hard objects. The stomachs examined were empty, but, as suggested by the structure of the teeth, we learn from literature that the animal feeds on mollusks and various hard-shelled crustaceans (Sumner, Osborne, and Cole, 1913, p. 739).

This ray is not common. It is little known by the fishermen, who report that it is taken only occasionally in pound nets in the southern parts of Chesapeake Bay. In September and October, 1922, during five weeks collecting at Ocean View, Va., where numerous hauls were made with collecting seines and where 32 hauls of an 1,800-foot commercial seine were observed, only eight individuals were seen. The maximum width of the largest specimen observed was 34 inches and the total length was 5 feet. This appears to be about the maximum size attained by the species.

Habitat.—Cape Cod to Brazil.

Chesapeake localities.—(*a*) Previous record: "Chesapeake Bay" (Uhler and Lugger). (*b*) Specimens in collection: Lynnhaven Roads, Va., pound net, July 17, 1916, and June 25, 1921; also observed at Ocean View, Va., in the fall of 1922.

17. Genus AËTOBATUS Blainville. Spotted eagle rays

Disk broad, the outer angles acute; head prominent; snout narrower than head, produced; rostral fins separate from the pectorals and at a lower level on the sides of the head; teeth in a single row in each jaw, fused, the lower plate long; anterior nasal valves confluent; a median notch in the preoral flap; tail long, slender, bearing a dorsal fin and one or more serrated spines on its basal portion; skin smooth.

24. Aëtobatus narinari (Euphrasen). "Bishop ray"; Spotted eagle ray.

Raia narinari Euphrasen, Handl., K. Vetensk. Akad., XI, 1790, p. 217, Pl. X; Brazil.

Ætobatis narinari Uhler and Lugger, 1876, ed. I, p. 184; ed. II, p. 156.

Aëtobatus narinari Jordan and Evermann, 1896–1900, p. 88, Pls. XV and XVI, figs. 37 and 38; Garman, 1913, p. 441, pl. 49, figs. 1 to 3 (teeth); pl. 54, fig. 4 (pelvis); pl. 55, fig. 9 (vertebræ); pl. 57, fig. 4 (heart); pl. 73, fig. 4 (skeleton).

"Width of the disk nearly twice the length; anterior borders convex, posterior concave. Pectorals somewhat falciform, acute on the outer angle. Rostral fins distinct from the pectorals, joined in a single, produced, depressed, and pointed lobe. Cranium large, narrower toward the mouth, convex across the crown. Teeth in a single row on each jaw, broad and short, fused, upper wider; lower pavement flatter and more produced. Each tooth is curved or angled forward more or less in the middle, the amount varying in the individuals. Eyes prominent. Spiracles large, lateral, behind the eyes, partly visible from above. Ventrals narrow, elongate, nearly half extended behind the ends of the pectorals, rounded posteriorly. Dorsal small, rounded above, with a short, free margin and an angle behind the base, origin above the ends of the bases of the ventrals. Tail whiplike, very slender, more than four times the length of the body. In a specimen at hand the measurements are from snout to vent 13, from vent to end of tail 59, and across the pectorals 25 inches." (Garman, 1913.)

Color dark brown or black above, white underneath. The dorsal surface with white spots, which are somewhat variable in number, size, and shape, and usually are smaller anteriorly than posteriorly. Anteriorly the spots are round, but posteriorly they sometimes become elongate, ring-shaped, or they appear as incomplete rings, and occasionally two spots become more or less connected by a narrow isthmus. Tail plain black.

This ray was not taken in Chesapeake Bay during the present investigation. The color description offered herewith is largely from notes made by us from specimens examined at Beaufort, N. C. The species is readily recognized by the shovel-shaped snout, the broad disk, which is covered above with numerous white spots, and by the very broad, flat, platelike teeth.

The spotted eagle ray, according to several recent writers, subsists almost wholly upon clams, which it probably digs up with its shovellike snout. The large, flat, platelike teeth and strong muscular jaws are well adapted to crushing hard-shelled mollusks.

This ray has the habit of jumping high above the water and, according to Coles (1910, p. 340) and Gudger (1914, p. 301), it is during this leaping that the young are born. According to Coles, about four young are delivered at one time. The young, when born, are from 6 to 8 inches broad, and they are delivered rolled up lengthwise. The spotted eagle ray reaches a length of about 12 feet and a width of about 7½ feet.

This ray, if it occurs at all in Chesapeake Bay, is very rare, as it was not seen during the present investigation and no information concerning its occurrence could be obtained from the fishermen. It is here included because of the record by Uhler and Lugger (1876).

Habitat.—"Tropical parts of the Atlantic and Eastern Pacific" (Garman); ranging northward on our Atlantic coast to Virginia.

Chesapeake localities.—(a) Previous record: "Enters Chesapeake Bay from the ocean and is caught in seines near Norfolk, Va." (Uhler and Lugger, 1876.) (b) Specimens in collection: None.

Family XIV.—RHINOPTERIDÆ. The cow-nosed rays

Body, head, and pectorals united to form a broad disk; a pair of rostral fins present, not joined in front of the skull and not continuous at the sides with the pectoral fins; eyes prominent, lateral; spiracles large, behind the eyes, opening laterally; one dorsal fin present, situated on the base of the slender tail and just in front of one or more strongly serrated spines.

18. Genus RHINOPTERA Cuvier. Cow-nosed rays

Disk broader than long, but not as broad as in related genera; tail long, slender; head prominent; rostral fins detached from the pectorals, forming a free and detached lobe in front of each orbit but not produced in front of the middle of the head; dorsal fin present, followed immediately by one or more serrated spines. A single species is known from the Atlantic coast of the United States.

25. Rhinoptera quadriloba (Le Sueur).   Cow-nosed ray; Whipparee.

Raia quadriloba Le Sueur, Journ., Ac. Nat. Sci., Phila., I, 1817, p. 44, with plate; New Jersey.

Rhinoptera quadriloba Uhler and Lugger, 1876, ed. I, p. 184; ed. II, p. 156; Bean, 1891, p. 94; Garman, 1913, p. 444, pl. 37 figs. 1 to 5.

Rhinoptera bonasus Jordan and Evermann, 1896–1900, p. 90.

Disk about one-third broader than long; the tail very slender, less than twice as long as the disk; head short, as broad as long; snout deeply indented anteriorly between the rostral fins; teeth in pavement, mostly hexagonal, in seven to nine rows, the median row in each jaw the widest, the functioning teeth deeply pitted; skin smooth; one or two serrated spines immediately behind the dorsal fin; origin of dorsal a little behind the end of the ventral bases, the fin small, its lower angle sharp; caudal fin wanting; ventral fins more than half as wide as long, the posterior margins convex; pectoral fins longer than broad, the outer angles acute, the anterior margins nearly straight, the posterior margins broadly convex.

Color brownish above, pale underneath, with more or less brownish toward the outer angles of pectoral.

This ray was not seen during the present investigations, and although previously recorded from Chesapeake Bay it is evidently very rare.   The foregoing description is based upon published accounts of the species.   This ray is readily recognized by the broad, emarginate snout, the lateral eyes, and whiplike tail.

The following information concerning the species is submitted by Smith (1907, p. 47): "The species reaches a large size, some examples observed in Florida being 7 feet wide.   It feeds largely on mollusks, which it crushes with its powerful paired jaws; the razor clam and the oyster are favorite foods.   The young, numbering two or three, are born in spring and summer and are very active from birth.   The stout, barbed spine is usually covered with mucus, and the wounds which it inflicts are painful and often dangerous."

Habitat.—Nantucket, Mass., to Florida.

Chesapeake localities.—(a) Previous records: "Near the mouth of Chesapeake Bay" (Uhler and Lugger, 1876) and Cape Charles City, Va.   (b) Specimens in the present collection: None.

Family XV.—MOBULIDÆ.   The sea devils

Head, body, and pectorals forming a subrhomboid disk, broader than long; head broad and flat, bearing cephalic fins or processes, developed as two long hornlike appendages, separate from the pectorals; mouth large, transverse, terminal or inferior; teeth small, numerous, in pavement; tail long, whiplike, with a single dorsal fin at its base and with or without a serrated spine; eyes lateral; skin more or less rough; ventrals small, between the pectorals.

Some of the members of this family reach an enormous size.   It is said that individuals have been taken which were 20 feet wide and weighed more than 4 tons.

19. Genus MANTA Bancroft.   The devilfish

Disk broader than long, its exterior angles acute, the posterior margins concave; head broad, flat, truncate; cephalic processes long, turned forward and inward; mouth very wide, terminal; teeth on lower jaw only, very small, in numerous rows; skin rough, with small tubercles; tail long, whiplike; a small dorsal fin over the ventrals.

26. Manta birostris (Walbaum).   Devilfish.

Raia birostris, Walbaum, Artedi Piscium, 1792, p. 535.

Ceratoptera vampyrus Uhler and Lugger, 1876, ed. I, p. 185; ed. II, p. 157.

Manta birostris Jordan and Evermann, 1896–1900, p. 92, Pl. XVIII, fig. 39; Garman, 1913, p. 453.

"Disk nearly twice as wide as long; tail as long as the body, including the rostral fins.   Pectorals falciform, acute angles, anterior margin convex, posterior concave.   Teeth minute, rasplike, on the lower jaw only, occupying the entire width of the jaw, in about 100 rows separated by interspaces (in the young).   Base of the dorsal extending forward a little in front of the ends of the bases of the pectorals and backward to about the middle of the free inner margin of the same fins.   Ventrals small, hind margins rounded, not reaching to the ends of the pectorals.   Body and tail rough.   *   *   *   Back brown, darkening with age; white underneath."   (Garman, 1913.)

49826—28——6

The devilfish was not taken during the present investigation. This species and a related species (*Mobula hypostomus*), not as yet recorded from Chesapeake Bay, may be recognized at once by two hornlike appendages in front of the head, known as the cephalic or rostral fins. The present species differs from its relative in the entire absence of teeth on the upper jaw.

The devilfish, according to Gill (1910, p. 167), feeds chiefly on small crustaceans and young or small fish. This species, like the spotted eagle ray, also has the habit of leaping above the surface of the water, but it is not definitely known whether this habit is correlated with the delivery of the young, as reported by Coles (1910, p. 340) and Gudger (1914, p. 301) in the case of the spotted eagle ray. The devilfish has only one young at a time, according to Gill (1910, p. 172) and others.

No information concerning the occurrence of this species in Chesapeake Bay was obtained during the present investigation. It was unknown to the fishermen who were questioned. The species is included in this work because of the statement by Uhler and Lugger (1876)—namely, that it is occasionally seen near the entrance to Chesapeake Bay.

Habitat.—Warm waters of both coasts of America. On the eastern coast its range extends northward to Block Island, R. I.

Chesapeake localities.—(*a*) Previous record: "* * * near the entrance to Chesapeake Bay" (Uhler and Lugger, 1876). (*b*) Specimens in the present collections: None.

Class PISCES. True fishes

Superorder GANOIDEI

Order GLANIOSTOMI

Family XVI.—ACIPENSERIDÆ. The sturgeons

Body elongate, cylindrical; skeleton cartilaginous; body imperfectly covered with bony plates or shields; head with similar large plates; snout produced with four flexible barbles hanging from its lower surface; mouth underneath head, small, protractile, suckerlike; teeth wanting; eyes small; tail heterocercal; air bladder large. A single genus of these large fishes is known from Chesapeake Bay

20. Genus ACIPENSER Linnæus. The sturgeons

Bony plates not confluent, one series on back and a lateral and abdominal series on each side; ventral plates often deciduous; snout more or less conical, depressed; spiracles over eye; gill rakers small, pointed.

KEY TO THE SPECIES

a. Space between dorsal and lateral shields with stellate plates of rather large size, in 5 to 10 series; top of head with a smooth area in young and deeply concave; snout long, acute, rather narrow at base; D. 30 to 44; A. 23 to 30_____*oxyrhynchus*, p. 72
aa. Space between dorsal and lateral shields with minute spinules, in very many series; top of head without a smooth area in young and less deeply concave; snout short, proportionately broader at base; D. 33; A. 19 to 22_____*brevirostrum*, p. 76

27. Acipenser oxyrhynchus Mitchill. "Sturgeon"; Sharp-nosed sturgeon.

> *Acipenser oxyrhynchus* Mitchill, Trans., Lit. and Philo. Soc., N. Y., I, 1815, p. 462; New York. Uhler and Lugger, 1876, ed. I, p. 183, ed. II, p. 155; Bean, 1883, p. 367.
> *Acipenser sturio* Jordan and Evermann, 1896–1900, p. 105, Pl. XX, fig. 45; Smith and Bean, 1899, p. 181; Fowler, 1912, p. 51.

The sturgeon is a fish of variable characters. The following description has been compiled from published accounts, both of American and of European fish, and from an examination of specimens made by us.

Head 3.7 to 5; depth 7 to 10; D. 30 to 44; A. 23 to 30. Most authors give the dorsal rays between 30 and 40, but Ryder (1890, p. 235), who made an extensive study of the sturgeons of the Delaware River, counted 40 to 44 on the fish examined by him. The number of anal rays given by most authors is 23 to 27, but Ryder (loc. cit.) found 26 to 30. The body is elongate, somewhat hexagonal, tapering gradually to base of caudal; head flattened above; snout 2 to 3 in head, variable, pointed in young up to 3 or 4 feet but becoming blunt with age. Smitt (1892, p. 1058) states that the shortening of the snout in relation to length of fish during its growth is accomplished at the expense of its anterior part (the rostral cartilage), the distance from the anterior nostril to the tip of the snout being reduced with age from 47 to 28 per cent of the length of head. Ryder (1890, p. 235), too, is of the opinion that the snout of the common sturgeon undergoes actual shortening and loss of substance during growth. Eye small, elongate, about 5 to 7 in snout; interorbital about 2.7 to 3.2, somewhat concave; mouth underneath head small, protractile, suckerlike; premaxillaries passing around front of mouth; maxillaries small, lateral, articulated with premaxillaries and with palatines; two pairs of short, slender barbels placed in transverse line about midway between end of snout and anterior edge of mouth, never touching mouth when deflected; nostrils double, close together, in front of eye, the posterior pair larger than anterior; teeth wanting, except in young; gill rakers small, sparse; skin smooth, grandular, or covered with small osseous points; dorsal shields 10 to 16 (usually 10 or 11); lateral shields 25 to 36 (usually 26 to 29); ventral shields 8 to 14 (usually 9 to 11); preanal shields present; dorsal far back; caudal heterocercal, the upper lobe longest; anal beginning under posterior half of dorsal; ventrals inserted on a perpendicular beginning a little in front of dorsal; pectorals inserted low, near level of lower edge of gill cover.

Color olive green, bluish gray, or brownish above; pale below.

Two specimens, 7 and 9 feet long, were examined in the Museum of Comparative Zoology, Cambridge, Mass., which gave the following shield counts: Dorsal 10 and 11, lateral 25, ventral 9 and 11. American sturgeons are said to have fewer lateral plates (25 to 29) than the European fish, which usually have from 29 to 36.

FIG. 39.—*Acipenser oxyrhynchus*

The sturgeon feeds on the bottom, its food consisting of a large variety of animals and plants, perhaps chiefly mollusks, worms, and small fish. When ascending rivers to spawn the sturgeon feeds little or not at all.

Adult sturgeons, according to Smith (1907, p. 56), do not appear in the sounds and rivers of North Carolina until the latter part of April, when the main run of shad is over. Ryder (1890, p. 266) says: "As the season advances the spawning schools move upward from the salt waters of Delaware Bay and in the neighborhood of Fort Delaware and Delaware City, 45 miles south of Philadelphia, where they pass into brackish or nearly fresh water. From this point, southward 20 miles and northward as many more, it is probable that a large part of the spawning occurs." Records of catches of pound nets set in Lynnhaven Roads indicate that the sturgeon usually enters Chesapeake Bay during April. It later enters the rivers where the spawn is deposited. The eggs, when laid, are about 2.6 millimeters in diameter. They are demersal and adhesive, becoming attached to brush, weeds, stones, etc. The eggs hatch in about 1 week in water having a temperature of 64° F. The mature ovaries of the female, according to Smith (1907, p. 56), may constitute one-fourth of the total weight of the fish, and a total of 1,000,000 to 2,500,000 eggs may be produced by one female. The young fish, according to Ryder (1890, p. 267), are sometimes taken from under ice in the Delaware River in midwinter, indicating that they remain in fresh water the whole year.

The newly hatched fry is about 11 millimeters (⅖ inch) in length (Ryder, 1890, p. 268), and in a few days, when the yolk sac is absorbed, it reaches ¾ inch. The later growth has not been followed, but in Europe this sturgeon is said to reach a length of 4 to 5½ inches in two months. Sexual maturity is believed to occur when a length of about 4 feet has been attained.

Small, unmarketable sturgeon, less than 4 feet in length, are even yet taken in sufficient numbers in the Chesapeake to give promise that the present-day small catch of adults will at least hold its own, providing the fishermen in every instance return the immature fish to the water uninjured. From early March until April 8, 1922, in a set of three pound nets off Ocean View, Va., from 3 to

10 small sturgeons were taken each week; while in a set of two nets in Lynnhaven Roads during the same period the weekly catch was 3 to 6, the usual size being from 30 to 40 inches in length. Even in Lower New York Bay, where the adult sturgeon is almost extinct, we have reason to believe that young fish are present in small to fair numbers at the present time. We observed a sturgeon 575 millimeters in length (about 22½ inches) caught on December 21, 1923, off South Beach, New York, by being snagged in the side with a fish hook. A year later the same angler reported another small sturgeon caught in the same manner.

During 1920 the Chesapeake Bay catch of sturgeon amounted to 22,888 pounds, worth $5,353. In addition there was obtained 2,654 pounds of caviar, worth $7,618. The total value of the catch, therefore, was $12,971. In Maryland the sturgeon ranked nineteenth in quantity and sixteenth in value. The catch consisted of 714 pounds of fish, worth $172, and 20 pounds of caviar, worth $87. In Virginia it ranked eighteenth in quantity and tenth in value. The catch consisted of 22,183 pounds of fish, worth $5,181, and 2,625 pounds of caviar, worth $7,531. Of this amount, 90 per cent was caught in pound nets and 10 per cent in gill nets. According to the value of the fish and caviar, the leading counties were Norfolk, $3,518; Elizabeth City, $2,850; Mathews, $1,351; James City, $1,271; and Gloucester, $1,068.

At one time the sturgeon was caught in large numbers throughout Chesapeake Bay, but it has become scarce, and now it is seldom taken north of the mouth of the Potomac River. Fishing is done so intensively that very few are able to reach the headwaters of the bay.

A great decrease in the sturgeon catch occurred after the year 1897, followed by a further decline after 1904 (see table), since when it has never been taken in anything like its former abundance. In May, 1915, at Buck Roe Beach, Va., Radcliffe (field notes) stated: "Very few adults have been taken and few young observed. I saw fish caught on Buck Roe Beach 9 feet long, estimated weight 275 pounds, estimated weight of roe (prepared for shipment) 90 pounds. The owner had difficulty in marketing the fish. Roe worth 50 to 60 cents a pound." Inquiries around the bay during 1921 and 1922 elicited the fact that sturgeons were scarce everywhere and had been for many years. During April and May, 1921, there appeared to be a slight increase in the lower bay pound-net catch as compared with the previous few years. During April, in a set of five pound nets off Buck Roe Beach, six sturgeons of marketable size were caught. On May 16 a 225-pound fish was taken in Lynnhaven Roads. The roe of this fish, after being rubbed and salted, weighed 41 pounds and sold for $3.50 a pound. Other scattering fish were caught, of which we obtained no record. During 1922, in a set of three pound nets at Ocean View that fished from early March to April 8, one large female and two males (the latter weighing 90 and 100 pounds, respectively) were caught. The aggregate catch of these nets up to May 26 was 20 sturgeons over 4 feet in length, 13 of them males and 7 females. The largest amount of spawn from one of the females weighed 59 pounds. In a set of two pound nets operated in Lynnhaven Roads during the same period no adults were caught. The first marketable sturgeon taken in the last-mentioned nets in 1922 was a 40-pound male caught on May 25. At Buck Roe Beach only three sturgeons were reported in 1922 up to April 11. At Lewisetta, Va., on April 22, 1922, the fishermen reported that: "The sturgeon have been scarce this year but are occasionally taken." At Solomons, Md., on April 26, it was said: "A few sturgeon have been taken in this vicinity this spring; one large one was caught April 24." At Love Point, Md., no sturgeons were reported caught during the year 1921. At Havre de Grace, on May 9, the report was: "None caught this year nor for the past three years. At the end of May a few are sometimes taken."

Most of the sturgeons caught in the lower Chesapeake are taken during April and May. During this period large fish are taken, many of them containing eggs suitable for making caviar. Sturgeons are caught during the summer and fall, but usually these fish are rather small (less than 100 pounds) and contain immature roe. Records were obtained from a set of two pound nets located at Lynnhaven Roads, Va., giving the number of sturgeons caught from 1916 to 1922, both inclusive. The aggregate catch, by months, for this period is as follows: April, 9 fish; May, 15 fish; June, 9 fish; July, 4 fish; August, 2 fish; September, 1 fish; October, 9 fish; November, 1 fish.

In comparison with the present-day scarcity of sturgeons, the catches made in the following rivers during 1880 show that at one time this fish was abundant in the Chesapeake drainage: James River, 108,900 pounds; York River and tributaries, 51,661 pounds; Rappahannock River,

17,700 pounds; Potomac River, 288,000 pounds. The following table shows the tremendous decline in the catch and the corresponding increase in value of sturgeons caught in Chesapeake Bay.

The catch of sturgeons taken in Chesapeake Bay during certain years from 1887 to 1920

Year	Maryland				Virginia				Average price per pound received by fishermen	
	Sturgeon		Caviar		Sturgeon		Caviar		Sturgeon	Caviar
	Pounds	*Value*	*Pounds*	*Value*	*Pounds*	*Value*	*Pounds*	*Value*		
1887	7,800	$296	(?)	----------	(?)	----------	(?)	----------	$0.038	(?)
1888	7,350	312	(?)	----------	(?)	----------	(?)	----------	.042	(?)
1890	99,932	3,313	(?)	----------	814,400	$24,466	(?)	----------	.03	(?)
1891	72,445	2,343	(?)	----------	720,381	21,304	(?)	----------	.03	(?)
1897	141,069	4,788	1,594	$644	584,967	14,475	59,600	$17,717	.026	$0.30
1901	8,415	618	748	444	171,943	11,260	17,858	9,932	.065	.55
1904	8,705	552	913	621	153,865	13,429	19,904	13,977	.086	.70
1920	705	172	29	87	22,183	5,181	2,625	7,531	.23	2.87

It is a matter of common knowledge that at one time sturgeons were considered worthless and large numbers were destroyed annually by fishermen, who regarded them as a pest. Their value gradually became apparent, however, and a special fishery was inaugurated. Being a large, sluggish fish, it was easily captured in great numbers, with the result that each year the aggregate catch became smaller and smaller. The retail price of fresh sturgeon has advanced steadily from about 10 cents a pound during 1900 to 50 cents during 1921 and 1922. Smoked, it is considered a delicacy and is among the highest-priced fishes.

Even more phenomenal was the tremendous increase in the value of sturgeon eggs, from which caviar [4] is prepared. The wholesale price advanced from 30 cents per pound in 1897 to $2.87 in 1920 and $3.50 in 1922.

The sturgeon is mentioned in early American history. The first market for American sturgeon was established when (in 1628) the fish were cured near Brunswick, Me., and shipped to Europe, where they were much esteemed. Large quantities taken in the vicinity of Ipswich, Mass., about 1635, were likewise shipped to Europe. The Rhode Island Indians captured sturgeons with harpoons and prized them highly for food.

The vessels worked their way up the coast until Delaware Bay was reached about April 1, and operations were continued here until early in May. The fish caught in the south were sent to Savannah, where they were skinned, packed in ice, and forwarded to New York. The Delaware Bay and Chesapeake Bay fish were likewise shipped to New York, which seemed to be the only large market for sturgeon. At this time the fishermen received about 6 or 8 cents per pound. During 1880 about 3,000,000 pounds of sturgeon were smoked in New York City and were consumed mainly by the German population.

Preparing caviar from the eggs of the Atlantic sturgeon was attempted as far back as 1849 by a Boston firm operating at Woolwich, Me. Because of an alleged scarcity of fish, operations were discontinued in 1851 and were not revived until 1872. By 1880 sturgeon eggs were utilized at many places along the Atlantic coast, but at that time the fishermen received only about 7 cents per pound. At Cape Fear River, N. C., the eggs were discarded as being valueless.

The present-day method of preparing sturgeon for market is essentially the same as that used during 1880. The fish is bled by cutting off the tail, and later the head, viscera, and skin are removed. The carcass is then iced in a box or a barrel and is ready for shipment. The average weight of an adult sturgeon is about 150 pounds, and when a fish of this size is dressed the carcass weighs about 65 pounds.

At the present time most of the Chesapeake Bay sturgeons are caught incidentally in pound nets, but a few are taken in gill nets. After the fish are dressed they are shipped to Norfolk, Balti-

---

[4] The process of making caviar is explained in "Caviar: What it is and how to prepare it." U. S. Bureau of Fisheries Economic Circular No. 20, issued Apr. 19, 1916; revised edition, issued Oct. 28, 1925.

more, Philadelphia, or New York. The caviar, which is usually prepared by the fisherman himself, is shipped to New York exclusively.

The rapid decline in the abundance of the sturgeon has caused the enactment of laws for its protection. The Virginia law states that no sturgeon less than 4 feet long may be removed from the waters of the State. The Maryland law states that no sturgeon weighing less than 20 pounds may be caught or offered for sale, and that no sturgeons whatsoever might be taken during the 10-year period from 1914 to 1923.

When a survey of the fishery industries of the United States was made in 1880 it was found that the Atlantic coast sturgeon industry was of relatively large importance. The industry centered at Delaware Bay and Savannah, Ga. Schooners sailed from Delaware during January and commenced operations early in February on St. Mary's River, Ga.

This sturgeon attains a large size, a maximum length of 18 feet having been recorded from Europe and, many years ago, from New England. At the present day the maximum for American fish is more nearly 12 feet, with fish 7 to 9 feet long not at all uncommon. The males average considerably smaller than the females, rarely exceeding a length of 7 feet.

Habitat.—On the Atlantic coast of America from the St. Lawrence River to the Gulf of Mexico. Also, once recorded from Hudson Bay, on the northwestern coast of Europe, if the American and European sturgeons are considered identical.

Chesapeake localities.—(*a*) Previous records: Chesapeake Bay and virtually all tributary streams. (*b*) Specimens in collection: None. The species, however, was observed at Lewisetta, lower York River, Buckroe Beach, Ocean View, and Lynnhaven Roads, Va., during 1921 and 1922.

28. Acipenser brevirostrum LeSueur. Short-nosed sturgeon.

Acipenser brevirostrum LeSueur, Trans., Amer. Philo. Soc., I, new series, 1817, p. 390; Delaware River. Uhler and Lugger 1876, ed. II, p. 155; Jordan and Evermann, 1896–1900, p. 106, Pl. XXI, fig. 47.

Acipenser brevirostris Smith and Bean, 1899, p. 181.

Head 5 to 6; depth about 8; D. 33; A. 19 to 22. Body much like that of *A. oxyrhynchus;* snout, as compared with *A. oxyrhynchus* of about the same size, shorter, more blunt, and proportionately wider at base; eye small, somewhat elongate; interorbital 2.2 to 2.8 in head, somewhat concave; mouth one-sixth wider than in specimens of *A. oxyrhynchus* of same size; two pairs of barbels placed in transverse line about midway between end of snout and anterior edge of mouth; never touching mouth when deflected; nostrils double, close together, in front of eye, the posterior pair the larger; skin rather smooth, compared with *A. oxyrhynchus*, but with small osseous points on unarmed portion; dorsal shields 9 to 12; lateral shields 23 to 29; ventral shields usually 7 or 8, but occasionally fewer; with or without preanal shields; fins situated as in *A. oxyrhynchus.*

Color blackish, tinged with olive, or reddish brown above; sides reddish mixed with violet, sometimes with oblique black bands; white underneath.

The above description was compiled from published accounts and the examination by us of a specimen taken off Provincetown, Mass., and now in the Museum of Comparative Zoology, Cambridge, Mass. This latter specimen had the following fin and shield counts: D. 33; A. 19; dorsal shields 10; lateral shields 26; ventral shields 7; preanal shields 2; length of fish about 30 inches.

This comparatively rare species resembles rather closely *A. oxyrhynchus*, and it had frequently been thought that it was a variable form of the latter. However, the descriptions given by LeSueur (1817) and by Ryder (1890) of short-nose sturgeons taken in the Delaware River leave little doubt as to the validity of the species. Ryder (1890, p. 238) states that "The characteristic dark brown or brown color of the animal, its small size, width of mouth, comparatively smooth skin, and early maturity render it impossible to question the identification which is thus established. The color alone is diagnostic; none of the young of the common species are dark colored, while the characteristic dirty olive green or brownish, with a shade of green in it, is always markedly characteristic of the common species at all stages of its growth."

Five specimens came under Ryder's observation, the smallest 18 inches and the largest not exceeding LeSueur's 33-inch specimen. The sexual organs of four of these (roes and milts) were far more developed than specimens of *A. oxyrhynchus* of corresponding sizes; in fact, the sexes of the latter species of these lengths could not be determined with certainty.

This fish was not seen during the present investigation and if it really occurs in the Chesapeake region (concerning which there seems to be some doubt notwithstanding that it has been recorded from there by at least two authors) it is not recognized by the fishermen.

The maximum length attained by this sturgeon is about 3 feet.

Habitat.—The only definite locality records belonging to this species, rather than the young of *A. oxyrhynchus*, are Provincetown, Mass. (described herein); New York, Bean (1903, p. 68); Delaware Bay, LeSueur (1817, p. 390) and Ryder (1890, p. 236); and Chesapeake Bay, Smith and Bean, (1899, p. 181).

Chesapeake localities.—(*a*) Previous record: Potomac River. (*b*) Specimens in the present collection: None.

Order HOLOSTEI

Family XVII.—LEPISOSTEIDÆ. The gar pikes

Body very elongate, more or less cylindrical; jaws produced, more or less beaklike, both armed with sharp teeth of various sizes; external bones of skull very hard and rugose; eyes small; nostrils near the end of upper jaw; gills 4, a slit behind the fourth; an accessory gill on the inner side of opercle; branchiostegals 3; air bladder cellular, lunglike, somewhat functional; spiral valve of the intestine rudimentary; scales consisting of rhombic plates, more or less imbricated and placed in oblique series running downward and backward; tail heterocercal, produced as a filament extending beyond the caudal fin in young; dorsal and anal fins placed far back and nearly opposite each other; ventral fins abdominal.

21. Genus LEPISOSTEUS Lacépède. Gar pikes

The characters of the genus are included in the family description. A single species is reported from Chesapeake waters.

29. Lepisosteus osseus (Linnæus). Garfish; Gar pike.

Esox osseus Linnæus, Syst. Nat., ed. X, 1758, p. 313; "Virginia."
Lepisosteus osseus Uhler and Lugger, 1876, ed. I, p. 182; ed. II, p. 154; Jordan and Evermann, 1896–1900, p. 109; Smith and Bean, 1899, p. 181.

Head 3.22; depth about 8.5; D. 8; A. 9; snout produced, 1.55 in head, its least width about 12.5 in its length; eye 13.1 in head; interorbitals 6; mouth very large; teeth numerous, both jaws with an outer series of small teeth, followed by a series of large, sharp canines projecting into pits in the opposite jaw when the mouth is closed, smaller rasplike teeth following the large teeth and occupying the jaws, vomer, and palatines; tongue well developed, emarginate or with a shallow slit in the free tip; external bones of the head hard, rough; scales bony, rhombic, platelike, with sharp posterior cutting edges; dorsal fin placed on posterior part of body, its origin over middle of base of anal; caudal fin rounded, unsymmetrical, the upper median rays longest, the lowest ray shortest; anal somewhat larger than the dorsal; ventral fins placed on the median part of the abdomen, a little nearer the base of the pectorals than origin of the anal; pectoral fins rather narrow, elongate, 3.35 in head. Color dark grayish above, silvery underneath; the vertical fins with large black spots; the paired fins plain olivaceous.

A single specimen (a partial skin 30 inches in length) forms the basis for the above description. The gar is generally common in the fresh waters of the central and eastern States and at times it ventures into salt water. It is not common in Chesapeake Bay, however. Fishermen operating at the mouth of the York River did not know the fish. Those operating pound nets in Lynnhaven Roads stated that the "garfish" was seldom caught in that vicinity.

The garfish is very variable, the local variations having given rise to about 28 specific names. It is readily recognized, however, by its produced, beaklike snout, rough, bony head and by the quadrate bony plates that cover the body. The posterior edges of these plates are somewhat free and very sharp. Large individuals are capable of cutting the fishermen's hands severely while bending the body from side to side in their struggle to escape.

This gar pike reaches a length of about 6 feet, and with its long beaklike mouth, provided with sharply pointed teeth, it is popularly believed to be a terror among other fish; but stomach examinations made by various investigators have not fully borne out the reputation it has as a destroyer of other fish. It is essentially carnivorous, however, and no doubt feeds largely upon other fish. According to Smith (1907, p. 59), it spawns in the spring in shallow water. The species is nowhere valued as food, but in some localities, at least, the negroes smoke the meat to a limited extent for winter use.

Habitat.—From Vermont and the Great Lakes southward to the Rio Grande and west to Kansas and Nebraska.

Chesapeake localities.—(*a*) Previous records: Common in the brackish water of the Potomac and Patapsco (Uhler and Lugger, 1876); common in the Potomac River and tributaries (Smith and Bean, 1899); Havre de Grace, Md. (Bean, 1883); Elk River and Northeast River (Fowler, 1912); "vicinity of Norfolk, Va." (Moseley). (*b*) Specimens seen or captured during the present investigations: Bohemia River, Md., April, 1912, fyke net, length of specimen 33½ inches; lower York River, Va., July 8, 1921, pound net, salinity 1.0145, length of specimen 22 inches; Lynnhaven Roads, Va., May 19, 1921, pound net, salinity 1.015, length of specimen 30 inches.

Superorder TELEOSTEI.   The bony fishes

Order ISOSPONDYLI

Family XVIII.—ELOPIDÆ.   The ten-pounders

Body elongate, more or less compressed; mouth broad, the lower jaw projecting; maxillary extending beyond eye; premaxillaries protractile; an elongate bony plate between the branches of the lower jaw; villiform teeth on jaws, vomer, palatines, pterygoids, tongue, and base of skull; eye large, with an adipose eyelid; opercular bones with membranous border; gill membranes separate, free from the isthmus; branchiostegals numerous, 29 to 35; pseudobranchiæ present, large; lateral line present; scales small, wanting on head; dorsal fin inserted over or slightly behind ventrals, the last ray not produced, depressible in a scaly sheath; no adipose fin; caudal fin forked; axil of pectorals and ventrals each with a long accessory scale.

22. Genus ELOPS Linnæus.   Big-eyed herrings; Ten-pounders

Body elongate; opercular bones thin, with membranous borders; pseudobranchiæ present, large; lateral line straight, with simple tubes; scales thin, forming a very high sheath on dorsal and anal; axil of pectoral and ventral each with an excessively long accessory scale; dorsal fin anteriorly elevated, the last rays short; anal fin similar but somewhat smaller. The species of this genus are rather large fishes of wide distribution. The young are flat, ribbon-shaped, and they pass through a metamorphosis like the eels.

30. Elops saurus Linnæus.   Big-eyed herring; "Ladyfish"; "Jackmariddle", Ten-pounder.

Elops saurus Linnæus, Syst. Nat., ed. XII, 1766, p. 518; Carolina. Uhler and Lugger, 1876, ed. I, p. 154, ed. II, p. 131; Bean, 1891, p. 93; Jordan and Evermann, 1896–1900, p. 410, Pl. LXVII, fig. 178.

Head 4.15 to 4.35; depth 5.34 to 5.7; D. 22 to 24; A. 15 or 16; scales 114 to 116. Body quite elongate, compressed; the back not elevated; head low and long; snout moderate, a little depressed, its length 3.75 to 3.95 in head; eye in adult with well developed adipose lid, its diameter 5.2 to 5.75 in head; interorbital space 5; mouth large; terminal; maxillary reaching far beyond the eye in the adult, 1.3 to 1.8 in head; teeth all small, present on jaws, vomer, palatines, and tongue; gill rakers slender, 14 on the lower limb of first arch; scales rather small, with membranous borders, wanting on head, extending on base of caudal fin, and forming a broad sheath on base of dorsal and anal, an excessively large scale in the axils of the pectorals and ventrals; dorsal fin moderately elevated, its posterior margin deeply concave, its origin a little nearer base of caudal than tip of snout; caudal fin broadly and deeply forked; anal fin somewhat similar to the dorsal, but smaller, situated far behind end of dorsal, its origin a little nearer base of caudal than base of ventrals; ventrals rather small, inserted under origin of dorsal; pectorals rather small, similar to the ventrals, 1.8 to 2.2 in head. Color silvery, bluish on back, slightly yellowish below; dorsal and caudal dusky and yellowish; ventrals and pectorals yellowish with dusky punctulations.

A single specimen 565 millimeters (22¼ inches) in length was taken during the present investigations. This fish and a smaller one from the St. Johns River, Fla., form the basis for the above description. The big-eyed herring, like the common fresh-water eel (and other eels), passes through a metamorphosis. The young, or larvæ, are similar to the leptocephalus of the eels, being greatly compressed, more or less ribbon-shaped, and transparent.

The big-eyed herring is readily recognized by the elongate form, low head, large mouth, and the broadly forked tail. This fish, like the tarpon, has a gular plate. The Atlantic and Pacific coast forms of this genus were long regarded as identical, but comparatively recent investigations have shown that the Atlantic representatives constantly have fewer gill rakers on the lower limb of the first arch. The range for the Atlantic species in this respect is 11 to 14, while that for the Pacific form is 18 to 20.

The food and feeding habits of this fish have not been studied thoroughly. It undoubtedly is carnivorous. Smith (1907, p. 116) says: "A specimen examined at Beaufort in August, 1901, had in its stomach six large shrimp (Peneus)."

The spawning habits of this fish, too, are imperfectly known, but it is probable that spawning takes place out at sea and that the eggs are pelagic. The larvæ, like the leptocephalus of the eel, are pelagic. These more oceanic allies of the herrings do not migrate to fresh water to spawn, as already stated, but individuals are not infrequently taken in brackish water.

The maximum size attained by the big-eyed herring is 3 feet (Jordan and Evermann, 1896–1900, p. 410), but the average size probably does not exceed 20 inches. This fish is evidently rather rare in Chesapeake Bay, since only a single specimen was taken during the present investigation

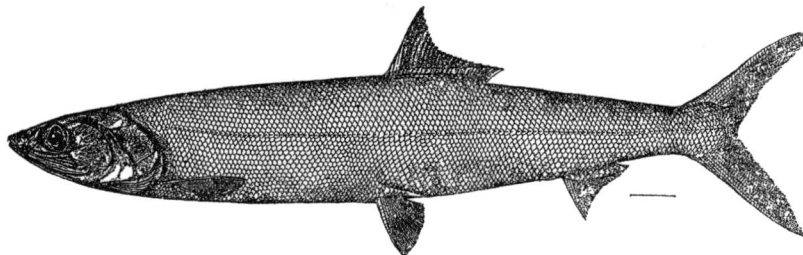

FIG. 40.—*Elops saurus*

and as fishermen report it as rare. "The species has no food value, the flesh being dry and bony." (Smith, 1907, p. 116.) Aside from the qualities of its flesh, it obviously is too rare in Chesapeake Bay to be of commercial importance in that vicinity.

Habitat.—The Atlantic coast, from Massachusetts to Brazil.

Chesapeake localities.—(a) Previous records: "Enters our large rivers from the salt waters of Chesapeake Bay, but seems to be quite common." (Uhler and Lugger, 1876); Cape Charles, Va. (Bean, 1891). (b) Specimen in collection: Lynnhaven Roads, September 17, 1921, taken in a pound net.

Family XIX.—MEGALOPIDÆ. The tarpons; The grande-ecailles

This family, as here understood, differs from the Elopidæ in the large scales, absence of pseudobranchiæ, and in the greatly produced (filamentous) last ray of the dorsal fin.

23. Genus TARPON Jordan and Evermann. The tarpon

Body oblong, rather strongly compressed; mouth large, very oblique, the lower jaw strongly projecting; maxillary broad, extending beyond eye; pseudobranchiæ wanting; lateral line decurved; scales very large, not forming a sheath on dorsal or anal; axil of pectoral and anal with a moderately large accessory scale; dorsal fin anteriorly elevated, the last rays of fin produced, filamentous; anal fin similar but larger, the last ray not notably produced; ventrals inserted well in advance of dorsal.

31. Tarpon atlanticus (Cuvier and Valenciennes).   Tarpon; Silverfish; Jewfish.

Megalops atlanticus Cuvier and Valenciennes, Hist. Nat. Poiss., XIX, 1846, 398; Guadeloupe, San Domingo, Martinique, Porto Rico.

Megalops thrissoides Lugger, 1878, p. 121.

Tarpon atlanticus Jordan and Evermann, 1896–1900, p. 409; 1900, Pl. LXVII, fig. 177.

Head 4.1 to 4.3; depth 3.4 to 3.85; D. 12 to 15; A. 20 to 23; scales 42 to 47.   Body elongate, rather strongly compressed; the ventral outline much more strongly curved than the dorsal; dorsal profile slightly concave over head; head moderate, notably compressed; snout short, broad, 4.8 to 5.1 in head; eye 3.9 to 4.65; mouth large, oblique, the jaws strongly curved; the lower jaw much in advance of the upper; maxillary reaching far beyond eye, 1.5 to 1.7 in head; teeth all small, in villiform bands; gill rakers slender, 32 to 36 on lower limb of first arch; lateral line decurved; scales very large, cycloid, wanting on head, present on base of anal but wanting on dorsal, the accessory scale in the axil of pectoral and ventrals less than half the length of fin; dorsal fin short, anteriorly notably elevated, the last ray filamentous, nearly equal to depth of body; caudal fin broadly forked, the lobes equal; anal fin deeply falcate, similar to the dorsal but longer, the posterior rays somewhat produced but not filamentous; ventral fins moderate, inserted well in advance of origin of dorsal; pectorals inserted low, under posterior margin of opercle, 1.1 to 1.16 in head.   Color uniform bluish-silvery above; sides and lower parts bright silvery; pectoral and ventral fins pale, the other fins more or less dusky.

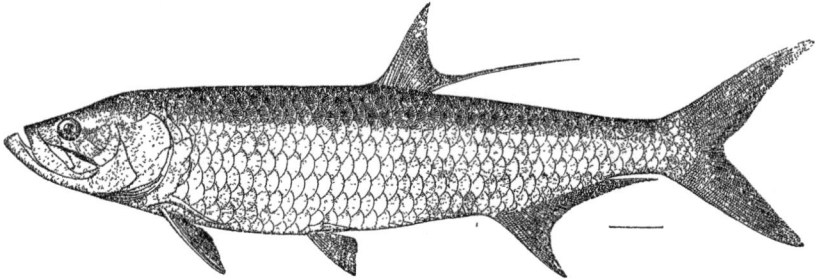

FIG. 41.—*Tarpon atlanticus*

The tarpon was not seen in Chesapeake Bay during the present investigation, but it was reliably reported by fishermen.   The species is readily recognized by the large silvery scales, decurved lateral line, and the small dorsal fin, which is smaller than the anal and which has the last ray produced into a long filament.

The tarpon feeds largely on small fish and at times it ascends fresh-water streams, presumably in pursuit of its prey.   It is a powerful and active swimmer and it has the habit of leaping entirely above water.   The purpose of these leaps remains unexplained, but it is generally supposed that this is a form of play.   The spawning habits of this species are little known.   The eggs and young of the American tarpon have never been found and the spawning grounds are unknown.   The young of the oriental tarpon pass through a stage of metamorphosis similar to Elops.   The smallest specimens of tarpon of which a record has come to the notice of the present writers are reported by Evermann and Marsh (1902, p. 80).   These specimens were collected at Fajardo, Porto Rico, and they ranged in length from 2¼ to 3¼ inches.   All the specimens reported from our coast were comparatively large individuals.   The tarpon reaches a maximum length of about 8 feet.   It is a game fish of considerable importance and is much sought by anglers.   Its flesh, however, is coarse and of little value.   The large, silvery scales are frequently sold as curiosities or as souvenirs, and at times they are used in ornamental work and in the manufacture of artificial flowers.

The species is too rare in Chesapeake Bay to be of economic importance, as it is taken only occasionally.

Habitat.—Massachusetts to Brazil; rarely as far north as Nova Scotia.

Chesapeake localities.—(a) Previous records: "Vicinity of Norfolk, Va." (Moseley, 1877, p. 9); Crisfield (Lugger, 1878, p. 121).   (b) Specimens in the present collection: None.   However, the species was reliably reported by fishermen operating in the southern parts of Chesapeake Bay.

Family XX.—CLUPEIDÆ. The herrings

Body oblong or elongate, more or less compressed; belly rounded or compressed, usually armed with bony serratures when compressed; mouth rather large, terminal, or more or less superior, with the lower jaw projecting; premaxillaries not protractile; teeth usually small, often feeble or wanting, variously arranged; adipose eyelid present or absent; gill rakers long and slender; gills 4, a slit behind the fourth; branchiostegals 6 to 15; pseudobranchiæ present; lateral line wanting; scales cycloid or pectinate; dorsal fin usually about median, rarely wanting; no adipose fin; ventral fins, if present, moderate or small; anal fin usually rather long; caudal forked, vertebræ 40 to 56.

KEY TO THE GENERA

a. Scales with their posterior edges round and smooth, or nearly so, never pectinate; cheeks and opercles not exceedingly broad; intestine of moderate length.
  b. Last ray of dorsal normal, not produced into a long filament; vertebræ 46 to 56.
    c. Vomer with a patch of permanent teeth; abdomen not strongly compressed; ventral scutes rather weak_____Clupea, p. 81
    cc. Vomer without teeth; abdomen rather strongly compressed; ventral scutes prominent.
      e. Cheeks as long as or longer than deep; jaws with minute teeth_____Pomolobus, p. 82
      ee. Cheeks deeper than long; jaws in the adult without teeth_____Alosa, p. 93
  bb. Last ray of dorsal produced into a long filament; vertebræ about 40 to 44__Opisthonema, p. 101
aa. Scales with the posterior edges nearly vertical and strongly pectinate; cheeks and opercles very deep; intestine very long_____Brevoortia, p. 102

24. Genus CLUPEA Linnæus. Herrings

Body long, compressed, with median line of abdomen armed with hard, bony scutes; maxillary with a broad supplemental bone; vomer with a permanent patch of teeth; vertebræ 46 to 56. A single species is known from the Atlantic coast, and it occurs in Chesapeake Bay only as a rare straggler.

32. Clupea harengus Linnæus. Sea herring.

Clupea harengus Linnæus, Syst. Nat., ed. X, 1758, p. 317; European seas. Lugger, 1877, p. 87; Jordan and Evermann, 1896–1900, p. 421, Pl. LXX, fig. 185.

"Head 4.5; depth 4.5; eye 4; D. 18; A. 17; lateral line 57; ventral scutes 28+13; vertebræ 56. Body elongate, compressed. Scales loose. Cheeks longer than high, the junction of the mandible and preopercle under middle of eye. Maxillary extending to middle of eye; upper jaw not emarginate, lower jaw much projecting. Vomer with an ovate patch of small permanent teeth; palatine teeth minute, if present; tongue with small teeth; jaws with or without minute teeth. Gill rakers very long, fine, and slender, about 40 on the lower part of the arch. Eye longer than snout. Dorsal inserted rather behind middle of body, in front of ventrals. Pectorals and ventrals short, anal low. Abdomen serrated in front of ventrals as well as behind, the serratures weak. Bluish; silvery below, with bright reflections. Peritoneum dusky." (Jordan and Evermann, 1896–1900.)

This species was not seen during the present investigation, and apparently it is not recorded from Chesapeake Bay in a published work. The species is included here on the authority of certain field notes by Dr. W. C. Kendall, which he has kindly placed at our disposal and which were made during an investigation in Chesapeake Bay in 1894. Doctor Kendall reports having taken one specimen, 12 inches long, on March 13 in a pound net near Hampton, Va., and he also states that according to the fishermen this herring is caught occasionally.

The sea herring may be recognized by the rather slender body, thin, deciduous scales, weak scutes on the ventral edge, and by the presence of a patch of teeth on the roof of the mouth.

The food of the sea herring consists of small organisms, chiefly copepods, and the larvæ of worms, mollusks, and other planktonic forms. It is stated in Bigelow and Welsh (1925, p. 103) that "the larvæ (European) feed on larval gastropods, diatoms, peridinians, and crustacean larvæ, but they soon begin taking copepods, and after they are 12 millimeters long depend on them exclusively for a time. * * * As they grow older they feed more and more on larger prey, turn-

ing to the larger copepods and amphipods, pelagic shrimps, and decapod crustacean iarvæ.''
Moore (1898, p. 402) examined a large number of adult herring taken near Eastport and found
them feeding solely on copepods and pelagic shrimps, while the young less than 4 inches long fed
only on the former.

Along the western North Atlantic coast the herring spawns during the spring, summer, and
fall, the spring and the fall being the chief seasons.   The fish spawns all along the coast from Nova
Scotia to Block Island within 25 miles of land and at depths usually not exceeding 75 fathoms.
The eggs are adhesive and demersal and they adhere to seaweeds and other objects on the bottom.
They are 1 to 1.4 millimeters in diameter, and an individual fish, according to size, deposits from
20,000 to 40,000.   The period of incubation ranges from 11 days at 50° F. to 40 days at 38.3° F.
According to Bigelow and Welsh (1925, p. 94), 10 to 15 days might be stated as an average for the
Gulf of Maine.   The larvæ are about 5 to 6 millimeters long at the time of hatching, and when a
length of 40 millimeters is attained adult characters are nearly developed

The sea herring is perhaps the most important food fish in the world.   Occurring in countless
numbers on both sides of the Atlantic, it is preyed upon by many species of fish, as well as whales.
One of the chief enemies of young herring is the squid.   This fact is known by many investigators,
but we had occasion to watch the wholesale destruction of 2 to 4 inch herring during June, 1925,
on the flats about Provincetown, Mass.   Schools of 10 to perhaps 50 squids circled around a school
of herring until they had bunched their prey into a compact mass.   Individual squids then darted
in and seized one, sometimes two, herring, ate only a small portion, and then darted back for more.
Along the beach there remained a silvery streak of dead herring.

During 1919 the catch of herring in the Gulf of Maine amounted to about 110,000,000 pounds
(Bigelow and Welsh, 1925, p. 105), of which about 80 per cent were sardines (young of about 3 to
5 inches) while the remainder were adults utilized as food and bait for cod and other banks fish.

The sea herring is principally a North Atlantic species and is very abundant on both coasts
of the Atlantic Ocean.   It is said to be the most important as well as the most abundant food fish
in the world.   This herring is not only an important article of food for man but it is of great importance
as food for the larger fish of the North Atlantic, such as the cod, haddock, halibut, bluefish, and
many others.   The herring is used in the fresh, salted, smoked, and canned state, and it is also
used as bait in the line fisheries for cod, haddock, etc.   The maximum length attained by the
species is given as 18 inches.

Habitat.—North Atlantic Ocean, on the coasts of Europe and America.   Recorded as far south
on the American coast as Cape Hatteras, and northward to northern Labrador.

Chesapeake localities.—(*a*) Previous records: None.   (*b*) Specimens in present collection: None.
The present record is based upon a field note by Dr. W. C. Kendall, made on March 14, 1894, in
which he reports having taken a specimen 12 inches long from a pound net near Hampton, Va.

25. Genus POMOLOBUS Rafinesque.   Alewives; River herrings

Body oblong, compressed; belly strongly compressed, serrate; mouth moderate, terminal, or
the lower jaw projecting; teeth weak, no patch on vomer; cheeks usually longer than deep; an
adipose eyelid present; dorsal fin short, nearly median; scales cycloid or with an emarginate membranous border, deciduous.   Three of the four American species are found in the Chesapeake
waters.

KEY TO THE SPECIES

a. Gill rakers rather few, 19 to 21 on the lower limb of the first arch; mandible strongly projecting
entering into the general dorsal outline of the head, no pronounced angle on the upper margin
near the median point of its length; tip of snout and lower jaw conspicuously dusky; peritoneum
pale _____*mediocris*, p. 83
aa. Gill rakers numerous, about 25 in very young, 40 to 50 in adults; mandible not strongly projecting, never entering into the general dorsal outline of the head, a very prominent angle
on the upper margin near the median point of its length, posteriorly very deep; the tip of
snout and lower jaw not conspicuously dusky.

b. Eye of moderate size, the diameter about equal to length of snout; color of back bluish green; peritoneum black_____*æstivalis*, p. 85
bb. Eye large, its diameter greater than length of snout at all ages; color of back grayish green; peritoneum pale_____*pseudoharengus*, p. 89

33. Pomolobus mediocris (Mitchill).   Hickory shad; Hick; Hickory jack; Bone jack; Freshwater tailor.

Clupea mediocris Mitchill, Trans., Lit. and Philo. Soc., N. Y., I, 1814, p. 450; New York.
Pomolobus mediocris Uhler and Lugger, 1876, ed. I, p. 159; ed. II, p. 136; Goode, in McDonald, 1879, p. 14.
Clupea mediocris Bean, 1883, p. 366.
Pomolobus mediocris Jordan and Evermann, 1896-1900, p. 425, Pl. LXXI, fig. 188; Smith and Bean, 1899, p. 183; Fowler, 1912, p. 51.

Head 3.5 to 3.9; depth 3.1 to 3.7; D. 15 to 18; A. 20 or 21; scales 45 to 50.   Body rather slender, compressed, dorsal profile straight over the head, gently convex from nape to dorsal fin; ventral outline more strongly rounded than the dorsal, without a prominent angle at the base of mandible; the margin of the abdomen compressed and provided with bony scutes; head long, not very deep; snout moderate, 3.75 to 4.2 in head; eye 4.2 to 5; interorbital 5.46 to 7; mouth rather large, superior; maxillary broad, reaching a little beyond middle of eye, 2.05 to 2.42 in head; cheek about as deep as long; mandible projecting very prominently, the tip not included in the upper jaw and entering

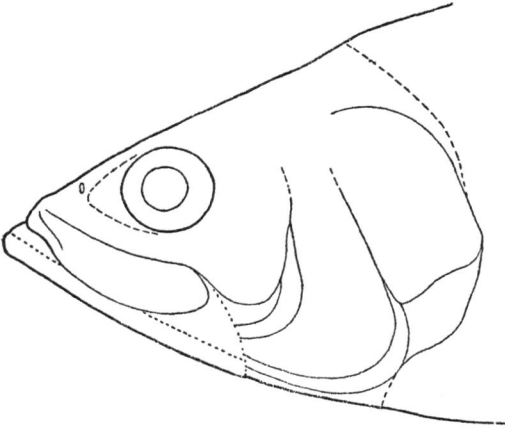

FIG. 42.—*Pomolobus mediocris.* Note strongly projecting mandible which enters into dorsal profile

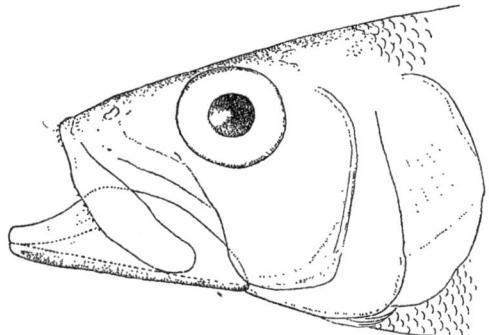

FIG. 43.—*Pomolobus mediocris.* Note that outline of upper margin of mandible bears no pronounced angle

into the dorsal profile, its upper margin without a prominent angle near the middle of its length (similar in outline to that of *Alosa sapidissima*); teeth very small, present on jaws, palatines, and tongue; gill rakers rather long and slender, 19 to 21 on the lower limb of the first arch; scales of moderate size, with emarginate membranous border, more or less deciduous; ventral scutes 20 or 21 in advance of ventrals and 14 to 16 behind ventrals; dorsal fin small, the outer margin straight or slightly concave, its origin equidistant from tip of snout and vertical from base of last anal ray; caudal fin forked, the lobes symmetrical and of equal length; anal fin longer than the dorsal but much lower, its origin a little nearer the base of ventrals than base of caudal; ventral fins rather small, inserted about equidistant from base of pectorals and origin of anal; pectoral fins of moderate size, similar to the ventrals but larger, 1.25 to 1.7 in head.

Color grayish green above, silvery on sides and below, more or less iridescent.   A dark shoulder spot, followed by several obscure dark spots; faint dark spots at base of scales on upper part of sides, forming longitudinal dark lines.   Nape green, opercle brassy, tip of snout dusky.   Dorsal and caudal dusky; anal and ventrals plain translucent; pectorals slightly dusky.   The dark lateral stripes are most conspicuous on specimens that have lost their scales, the black being on the skin underneath the scales and showing faintly through the somewhat transparent scales.   The dark lines along the rows of scales are not evident in the small specimens (155 to 190 millimeters) examined.   Peritoneum pale.

Fourteen specimens of this species, ranging from 155 to 328 millimeters (6⅛ to 13 inches) in length, form the basis for the above description.   This species is recognized by the strongly projecting lower jaw, which distinctly enters into the dorsal profile, and by the small number of gill rakers on the lower limb of the first arch.   It agrees with *Alosa sapidissima* in the general shape of the mandible, and it differs from the other species of Pomolobus in the absence of a prominent angle near the median point of its length.

Doctor Linton examined seven stomachs from specimens taken on five different dates and at three localities in Chesapeake Bay.   Four stomachs were virtually empty, but the small fragments found in washings indicated a fish diet.   The other three stomachs contained the remains of fish exclusively.   Two stomachs examined by us, taken from fish caught in April, were entirely empty.   Bean (1903, p. 198) states that specimens taken near New York City and examined by him had fed on sand launces, each stomach containing from 15 to 20 of these animals, ranging from 3½ to 5 inches in length.   Many stomachs of fish caught at Woods Hole, Mass., and examined by Vinal N. Edwards, contained, besides various species of small fish, squids, fish eggs, small crabs, and various pelagic crustaceans.

The habits of the hickory shad are even less perfectly understood than are those of the branch herring and the glut herring.   Jordan and Evermann (1896–1900, p. 425) and Fowler (1906, p. 95), state that this fish does not ascend fresh-water streams to spawn.   McDonald (1884, p. 609) says that no observations have been made on the breeding habits but that it is almost certain that the species spawns in the spring, and he thinks that it is "more than probable" that it spawns in fresh water under the same conditions as the shad but at a little earlier period.   Smith (1907, p. 121) says: "The species is common in the coast waters and rivers of North Carolina, coming in from the ocean in the late winter or early spring and ascending streams to spawn, going to the headwaters in company with the branch herring."   Not a single fish less than 155 millimeters (6⅛ inches) in length occurs in the present Chesapeake collection, and young previously reported from Chesapeake waters (so far as we have been able to secure the specimens for examination) were wrongly identified. Extensive collections of Clupeidæ were made in the Potomac River, in the fresh waters in the vicinity of Havre de Grace and at many points in the bay.   The collections in the Potomac were made chiefly during the summer and fall, those at Havre de Grace in the spring, late summer, and fall, and those in the bay were made at all seasons of the year in both shallow and deep water.   The fact that not a single young hickory shad (of less than 155 millimeters in length) was taken throughout these investigations shows rather conclusively that the hickory shad does not ascend the fresh waters of the Chesapeake region to spawn.   Five adults examined, taken during April and May (four females and one male), all had the roe somewhat developed but not ripe.   The information gathered during the investigation leads to the belief that the hickory shad leaves Chesapeake Bay to spawn.

A definite spring run and a somewhat less definite fall run of hickory shad takes place in Chesapeake Bay.   During the summer only stragglers are taken.   Hickory shad are taken with the opening of pound-net fishing in the lower bay early in March.   Like the shad and the alewives, the first fish appear sometime later in the upper reaches of the bay.   In the lower Potomac the bulk of the catch is taken late in March and early in April, agreeing in this respect with the Lynnhaven Roads region; but in the vicinity of Havre de Grace the run does not occur until late in April and early in May.   Most of these fish range in length from 14 to 18 inches.

This fish is taken during summer in all parts of the bay, at least as far north as Baltimore. The individuals are smaller fish than those of the spring and fall runs, measuring from 8 to 12 inches in length.   The number of fish taken in a set of pound nets during the summer (if, indeed, any at all are caught) usually ranges from one to six per day.

In the fall a definite but somewhat smaller run than that in the spring occurs.   The fall fish are taken mostly in the lower parts of the bay, from Solomons, Md., southward.   In a set of two pound nets in Lynnhaven Roads, fishing from November 1 to 16, 1921, from none at all to 100 pounds a day were caught; while in two nets fishing from November 16 to December 5 at Ocean View daily catches of 100 to 400 pounds were taken, the catch for the last day being 150 pounds. Virtually all fishing ceases by December 1, consequently we do not know at what date this fall run of fish ends, but in view of the catch made on December 5 it appears probable that a few fish, at least, remain after the nets are lifted.   The fall fish are of about the same size as those of the spring run. In the lower bay the hickory shad is often the principal species caught at the very end of the fishing season.

The maximum length attained by the hickory shad, according to published accounts, is about 2 feet. Uhler and Lugger (1876, p. 159) state that it attains a length almost equal to that of the shad. Observations made during the present investigations indicate that the maximum length now attained by this fish in Chesapeake Bay is about 18 inches, with a weight of 2 pounds. The average length of market fish, however, is only about 15 inches and the weight 1 pound.

The hickory shad has some commercial value, especially in the southern parts of the bay, where it is one of the first in the spring and one of the last fish in the fall to be caught in considerable quantities. During 1920 it ranked fourteenth among the fishes of Chesapeake Bay in quantity and fifteenth in value, the catch amounting to 218,620 pounds, worth $8,245. The bulk of the catch is taken in pound nets in March, after which a decline occurs and only stragglers are caught after April 15 in all sections of the bay except in the extreme northern stretches, where the spring run occurs later, as shown elsewhere. A smaller catch of fish is made in the late fall, and sometimes at the very end of the fishing season the hickory shad is the principal species caught. The following catches made by a set of two pound nets in Lynnhaven Roads, Va., in 1914 is somewhat typical of the hickory-shad catch made in the southern parts of the bay: March 10 to 31, 25 to 600 pounds per day; April 1 to 15, 10 to 100 pounds per day; April 16 to 30, less than 10 pounds per day; November 1 to 16 (end of season); none to 100 pounds per day.

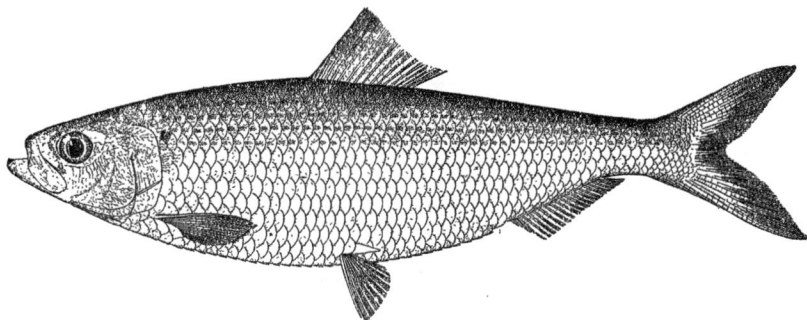

FIG. 44.—*Pomolobus æstivalis*. Male, 10.6 inches long

The fishermen separate the hickory shad from the alewives and shad, as the prices of each of these species differ widely. In April, 1922, run-boat buyers were paying 5 cents each for hickory shad, regardless of size. When the fish are packed in boxes and shipped direct to market they are sold by weight. The retail price in 1922 ranged from 10 to 15 cents per pound.

Habitat.—Maine to Florida, entering streams, except in New England.

Chesapeake localities.—(a) From virtually all streams tributary to Chesapeake Bay and from many localities within the bay. (b) The immature specimens in the collection, ranging from 155 to 255 millimeters (6⅛ to 10 inches) in length, are from Annapolis, Md., to Smith Point, Va., taken with the beam trawl at depths ranging from 16 to 27 fathoms from January 19 to March 18, 1914; Lynnhaven Roads, Va., June 9; Buckroe Beach, Va., June 22, 1921, taken in pound nets.

34. Pomolobus æstivalis (Mitchill). Herring; Glut herring; Blue herring; Greenback herring; Alewife.

Clupea æstivalis Mitchill, Trans., Lit. and Phil. Soc., N. Y., I, 1814, 456; New York. Bean, 1883, p. 366.
Pomolobus pseudoharengus Uhler and Lugger, 1876, ed. I, p. 158; ed. II, p. 135 (in part).
Pomolobus æstivalis Goode, in McDonald, 1879, p. 14; Jordan and Evermann, 1896–1900, p. 426, Pl. LXXI, fig. 190; Smith and Bean, 1899, p. 183; Evermann and Hildebrand, 1910, p. 158.

Head 3.33 to 4.5; depth 3.35 to 4.25 (average for 22 specimens, 3.6); D. 16 to 19; A. 18 to 21; scales 47 to 52. Body moderately elongate, compressed, slightly deeper in the adult than in the young; dorsal profile from snout to dorsal evenly and very gently convex; ventral outline more strongly convex than the dorsal, with a very slight angle at base of mandible; the margin of abdomen compressed, with sharp bony scutes; head moderate; snout rather long, 3.7 to 5 in head; eye small, about equal to length of snout, except in very young, 3 to 4.4 (average for 22 specimens 3.53) in head; interorbital 2.95 to 5.8; mouth moderate, oblique, slightly superior, but not entering into the dorsal profile; maxillary broad, reaching about opposite middle of eye, 2.2 to 2.6 in head; cheek broad, its width greater than its depth; mandible slightly projecting, the tip not included in the upper jaw, but not entering into the general dorsal outline, its outline as in *P. pseudoharengus;* teeth as in *P. pseudoharengus;* gill rakers long and slender, increasing in number with age, young of

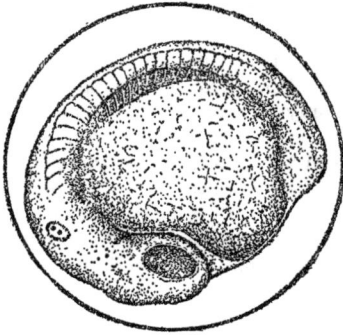

FIG. 45.—Egg with large embryo

40 to 50 millimeters in length, with 28 to 34 gill rakers on the lower limb of the first arch, adult specimens with 42 to 50 gill rakers; scales moderate, more or less deciduous: ventral scutes 19 to 22 in advance of ventrals and 13 to 16 behind ventrals, total number of scutes 33 to 36; dorsal fin rather small, its outer margin concave, the origin at least an eye's diameter nearer tip of snout than base of caudal; caudal fin forked, the lobes about equal; anal fin a little longer than the dorsal, but lower, its origin about equally distant from base of ventrals and base of caudal; ventral fins small, inserted equidistant from the base of pectorals and the origin of the anal; pectoral fins similar to the ventrals, but larger, 1.3 to 1.85 in head.

Color bluish above, sides silvery; upper rows of scales with more or less distinct dark lines in the adult; a dark spot at shoulder (this is rarely present in specimens less than 100 millimeters (4 inches) in length). Fins all plain, sometimes slightly yellowish or greenish in life. Peritoneum black.

Numerous specimens, ranging from 20 to 295 millimeters (⅘ to 11⅝ inches) in length, have been examined. This species is similar to the branch herring (*P. pseudoharengus*), the most outstanding difference being the color of the peritoneum, which is black in the present species and pale or silvery in the branch herring. Externally, the glut herring differs from the branch herring in being a more slender and elongate fish. It has a somewhat smaller eye, and the color of the back

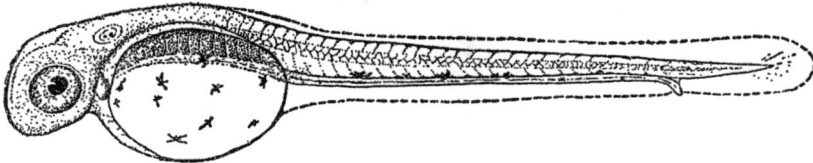

FIG. 46.—Newly hatched larva, 3.5 millimeters long

is bluish rather than grayish green, as in the branch herring. This difference in color is recognized by the fishermen and gives rise to the local names "blue herring" and "gray herring." All of the external differences mentioned, however, appear to vary, and occasionally intermediate specimens are found, which are difficult to separate without examining the peritoneum. The difference between the young and the adults of this species are not especially pronounced and are not unusual. The sexes are so similar that they are not readily distinguished externally.

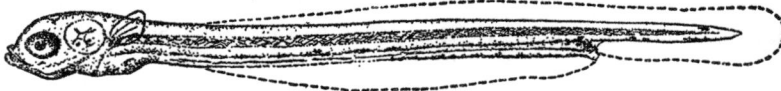

FIG. 47.—Larva 4 days old, 5.2 millimeters long

The habits of this fish are similar to those of the branch herring, and the remarks regarding the latter in general also apply to this fish. The glut herring, however, enters fresh water several weeks later than the branch herring. In the lower bay a few are caught with pound nets in March, the catch increasing toward the end of the month. In the first week of April, 1922, the pound nets at Lynnhaven Roads and Ocean View were catching the two species in the following ratio: Branch herring 60, glut herring 40. The peak of the catch of glut herring is usually taken between

April 1 and 20 in the lower bay. The numbers decrease throughout May, until after June 1 only stragglers are caught. At Havre de Grace notes made by the late William W. Welsh in 1912 record the first catch of glut herring on April 11 and the height of the run on about April 27. Of course, it is well known that the time of arrival and the height of the run vary somewhat from year to year, but in general the glut herring is expected in the lower Chesapeake region the first half of April and in the upper reaches of the bay during the last half of April. This species does not

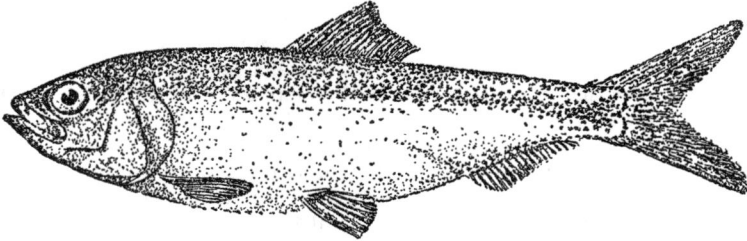

Fig. 48.—Young, 30 millimeters long

ascend fresh-water streams as far as the branch herring, and spawning takes place at a shorter distance from the sea. The greater part of the young, as in the branch herring, appear to pass through Chesapeake Bay and out to sea upon the approach of cold weather, but a few stop in the deeper waters of the bay during their first winter and very few apparently remain there for the second winter.

The rate of growth [5] in the young of this species appears to be somewhat more rapid than in the branch herring. The size attained at a given age is quite uniform, as no difficulty was expe-

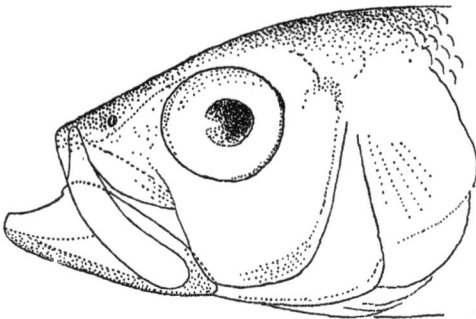

Fig. 49.—*Pomolobus æstivalis*. Note deep mandible and sharp angle on its upper margin

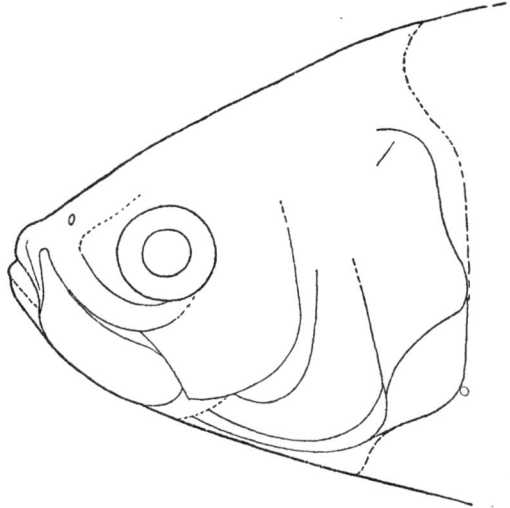

Fig. 50.—*Pomolobus æstivalis*. Note that mandible scarcely projects and that it does not enter dorsal profile

rienced in separating young fish into year groups. Specimens taken in March, for example, clearly fall into two separate lots, one group consisting of individuals that are in their first year and the other group comprising those in their second year. The uniformity in size of the young of a certain age suggests a short spawning period. The young of the branch herring, on the other hand, vary greatly in size, and much difficulty was experienced in separating into year groups catches taken

[5] The eggs, embryology, and larval development of the glut herring are described by Kuntz and Radcliffe (1918, pp. 123 to 126, figs. 87 to 100).

in the bay on various dates. This large variation in size of young taken on the same date suggests a rather protracted spawning period.

A large number of young glut herring taken in fresh water, principally in 1912, had attained an average length of 28 millimeters (1⅛ inches) on July 1, 46 millimeters (1¹³⁄₁₆ inches) on September 1, and 64 millimeters (2½ inches) on December 1. A decided upward jump takes place in the growth curve between the last lots taken in fresh water and those taken in the salt water of the bay. The reasons for this sudden upward curve are not definitely known and the subject needs further investigation. The two possible reasons that have occurred to us are (1) that growth is greatly exhilarated when the fish enters salt water; this explanation is rendered somewhat unlikely because of the cold winter weather; (2) it seems probable that the smaller individuals of the season's brood remain in fresh water later than the larger ones, and therefore the lots taken late in the fall in fresh water consist of fish that are either "runts" or hatched late in the season, whereas the catches in January and February from the bay consist of fish of more average growth. It seems altogether unlikely that the fish from the bay belong to a different year class. A limited number of specimens taken in the deeper waters of the bay had reached an average length of 82 millimeters (3¼ inches) by February 1 and 90 millimeters (3½ inches) by April 1. A few individuals taken in Chesapeake Bay, which probably were in their second year, had attained an average length of 174 millimeters (6⅞ inches) on February 1.

The maximum length attained by the glut herring is about 380 millimeters (15 inches) and the weight 13 ounces. The average length of market fish, however, is only about 280 millimeters (11 inches) and the weight 7 ounces.

Length frequencies of 2,035 glut herring, Pomolobus æstivalis

[Measurements in millimeters, grouped in 5-millimeter intervals]

Total length, millimeters	June		July		Aug.		Sept.		Oct.		Nov.		Dec.		Jan.		Feb.		Mar.		Apr.		May	
	1-15	16-30	1-15	16-31	1-15	16-31	1-15	16-30	1-15	16-31	1-15	16-30	1-15	16-31	1-15	16-31	1-15	16-28	1-15	16-31	1-15	16-30	1-15	16-31
20–24		22																						
25–29		4																						
30–34				2																				
35–39		1		3	2																			
40–44				13	34	10	2		1	1														
45–49				10	38	33	25	19	4															
50–54				2	12	5	19	36	47	7	3													
55–59				1	5	2	4	11	71	21	9													
60–64					2			2	33	44	29													
65–69								1		29	31				1	2		1	2					
70–74								1	2	5	3				3	4	1	2	4					
75–79															25	16	9	17	39					
80–84															76	21	56	67	128					
85–89															81	5	68	123	139					
90–94															52	7	40	87	56					
95–99															18	1	30	25	45				2	
100–104															16	1	15	11	14				2	
105–109															2		4	5	7				1	
110–114															1			1	1				1	
115–119															1			1					2	
120–124																								
125–129															2									
130–134																								
135–139																								
140–144																			1					
145–149																								
150–154																		1						
155–159																		1	1					
160–164															2			2	2					
165–169															1			3	1					
170–174															3			1						
175–179																		4						
180–184																			1					
185–189															1									
190–194															1			1						
195–199																		1	1					
205–209																			1					
Total		27		31	93	51	50	70	162	106	75				286	57	238	346	435				8	

The glut herring and the branch herring are not separated for the market, and the data and remarks concerning the commercial importance of the branch herring, therefore, also include the present species.

Habitat.—Nova Scotia to St. Johns River, Fla.

Chesapeake localities.—(*a*) Previous records: Many parts of the bay and virtually all streams tributary to the bay. (*b*) The numerous young in the present collection, ranging in length from 20 to 119 millimeters (⅘ to 4⅔ inches), are from the following localities: Beam-trawl catches in many parts of the bay from Annapolis, Md., to Old Point Comfort, Va., including the Potomac River below Mathias Point, at depths ranging from 5 to 28 fathoms, January 15 to March 24, 1914, January 16 to March 12, 1916, January 22 to 26, 1921, February 14 to 19, 1922. Taken with seines in the Potomac River from Bryans Point, Md., to Lewisetta, Va., October 14 to November 11, 1911, June 17 to December 3, 1912, October 29, 1914, August 8, 1921; in the bay at Havre de Grace, Md., May 10, 1922, August 26, 27, 1921; Baltimore, May 4, 1922; Annapolis, Md., May 3, 1922; Love Point, Md., September 5, 1921; Buckroe Beach, Va., April 10, 1922.

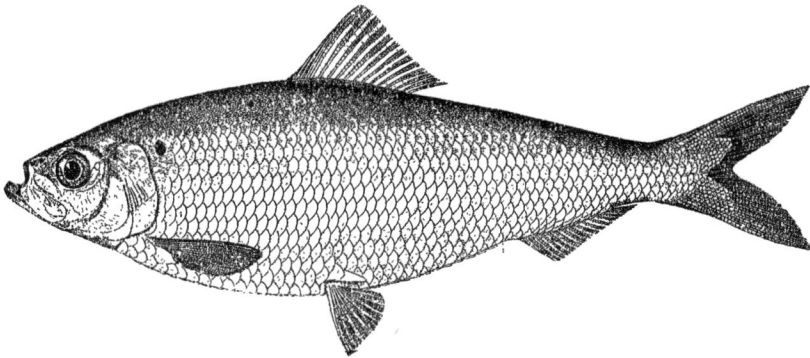

FIG. 51.—*Pomolobus pseudoharengus.* From a specimen 11.5 inches long

35. Pomolobus pseudoharengus (Wilson). Alewife; "Branch herring"; Big-eyed herring; "Herring"; "Gray herring"; "White herring".

Clupea pseudoharengus Wilson, Rees's Cyclopedia, IX, no pagination and no date, about 1811; Philadelphia.
Pomolobus pseudoharengus Uhler and Lugger, 1876, ed. I, p. 158; ed. II, p. 135 (in part).
Pomolobus vernalis Goode, in McDonald, 1879, p. 14.
Clupea vernalis Bean 1883, p. 366.
Pomolobus pseudoharengus Jordan and Evermann, 1896–1900, p. 426, Pl. LXXI, fig. 189; Smith and Bean, 1899, p. 183; Evermann and Hildebrand, 1910, p. 158; Fowler, 1912, p. 51.
Pomolobus mediocris Evermann and Hildebrand, 1910, p. 158 (not of Mitchill).

Head 2.9 to 4.3; depth 2.8 to 4.15 (average for 22 specimens 3.23); D. 15 to 19 (usually 16 or 17); A. 17 to 21; scales 46 to 49. Body rather deep, compressed, slightly deeper in the adult than in young; dorsal profile from snout to dorsal fin gently and nearly evenly rounded; ventral outline more strongly convex than the dorsal, with a slight angle at base of mandible; the margin of the abdomen compressed and provided with strong, bony scutes; head rather short and deep; snout rather blunt, 3.5 to 5 in head; eye large, longer than snout, 2.6 to 4.15 (average for 22 specimens 3.12) in head; interorbital 4 to 6.45; mouth moderate, slightly superior; maxillary broad, reaching about opposite middle of eye, 2 to 2.65 in head; cheek broad, its width greater than its depth; mandible slightly projecting, the tip not included in the upper jaw but not entering into the general dorsal outline, its upper margin strongly elevated, with a prominent angle near the middle of its length; teeth very weak, present on premaxillaries and tip of lower jaw in the young, sometimes persisting in the adult; gill rakers rather slender, of moderate length, increasing in number with age, young 30 to 58 millimeters in length with 22 to 29 gill rakers on the lower limb of the first arch, specimens ranging from 158 to 284 millimeters with 33 to 40 gill rakers; scales of moderate size, cycloid, more or less deciduous; ventral scutes 19 to 22 in advance of ventrals and 11 to 15 behind ventrals, total number of scutes 30 to 35; dorsal fin rather small, its outer margin very slightly concave, the origin usually slightly nearer tip of snout than vertical from end of base of anal; caudal fin forked, the lobes nearly symmetrical; anal fin a little longer than the dorsal, but lower, the anterior rays only slightly longer than the posterior ones, its origin nearly equally distant from base of ventrals and base of caudal; ventral fins rather small, pointed, inserted about midway between the base of the pectorals and the origin of the anal; pectoral fins moderate, similar to the ventrals but larger, 1.2 to 2 in head.

Color grayish-green with metallic luster above, sides silvery; a dark spot at shoulder (rarely developed in the young of less than 100 millimeters (4 inches) in length). Rows of scales with indistinct dark lines, which are present only in the adult, appearing somewhat later in life than the dark shoulder spot. Fins all plain, slightly greenish or yellowish in life; the dorsal and caudal with dusky punctulations; peritoneum pale.

Numerous specimens, ranging from 30 to 284 millimeters in length, have been examined. This species is recognized by the large eye (which is longer than the snout at all ages), by the deep body and by the pale peritoneum. The young of this species do not differ greatly from the adults, except that the body is scarcely as deep, the eye is proportionately larger, and the gill rakers are fewer in number. The sexes are very similar, but the dorsal fin in the male appears to be a little higher. The difference in size of the sexes, as shown by a limited number of weights and measurements, is not pronounced. The female, however, appears to reach a slightly greater length and weight.

In 12 stomachs examined, taken from fish ranging in length from 83 to 178 millimeters (3¼ to 7 inches) examined by Linton, the principal food of the smallest specimens consisted of copepods; in the medium-sized and in the largest ones it consisted of Mysis. In 7 large fish, 2 stomachs were empty, 2 had fed wholly on Mysis, and 3 wholly on fish. The authors examined 6 stomachs of specimens taken in fresh water during the summer, ranging in length 50 to 70 millimeters, and found 2 stomachs empty, 2 fish had fed wholly on copepods, one contained a worm, and the smallest had fed on ostracods. Stomachs of five larger specimens, taken in salt water in the bay during March, ranging from 90 to 178 millimeters in length, contained Mysis only, and these had been eaten in great numbers.

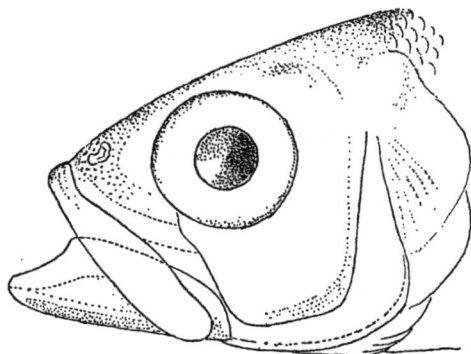

Fig. 52.—*Pomolobus pseudoharengus*. Note broad cheek bone and deep mandible with pronounced angle on its upper margin

In this species, as in the common shad, the cœca, connected with the intestine near the stomach, are very small or wanting in the young and become greatly developed with age.

This fish enters streams in the spring to spawn, and when this purpose is accomplished the adult again returns to the ocean, where most of its life is spent. The young remain in fresh water throughout the first summer of their lives, but with the approach of cool weather they gradually migrate to salt water, but they do not all leave the fresh or brackish water (of the Potomac River) until late in the fall, as specimens have been taken as late as November 11 at Bryans Point, Md., and at Riverside, Md., as late as December 3. The majority of the young evidently pass through Chesapeake Bay without stopping and migrate directly to the ocean, but we have specimens taken in the deeper waters of the bay throughout the winter months, indicating that at least some of them do not enter the ocean until they are a year or more old. Measurements of specimens indidate that a few fish may stay in the bay until they are 2 years old. The movements and the life history of the branch herring during the period or periods spent in the ocean are very imperfectly understood. "During the summer months enormous schools of full grown but sexually immature alewives migrate along the coast." (Bean, 1903, p. 201.) Further information concerning the fish after it enters the sea is wanting.

The branch herring generally reach fresh-water streams in the spring three or four weeks earlier than the glut herring and they also precede the first run of shad. In the Chesapeake drainage they usually arrive some time in March. This fish runs far upstream to spawn, frequently entering small brooks only a few feet wide and a few inches deep. "The alewives are very prolific. In the Potomac River, 644 female branch herring yielded 66,206,000 eggs, an average of 102,800 per fish; and probably 100,000 eggs may be taken as a fair average for the species. The eggs are 0.05 inches in diameter, and are very glutinous when first laid, adhering to brush, ropes, stones, piling, and other objects. The hatching period is six days in a mean water temperature of 60° F." (Smith, 1907, p. 123.)

The young grow rapidly, reaching an average length of about 55 millimeters (2⅙ inches) by July 1, 65 millimeters (2½ inches) by September 1, and 70 millimeters (2¾ inches) by December 1. The individuals that stop in Chesapeake Bay during the first winter of their lives appear to grow very rapidly (possibly the explanations given on p. 88 for *P. æstivalis* apply to this species also) after entering the salt water, as specimens taken in the bay had attained an average length of about 105 millimeters (4⅛ inches) by February 1 and 120 millimeters (4¾ inches) by May 1. Fish taken in Chesapeake Bay, which apparently were in their second year, had reached a length of about 140 millimeters (5½ inches) by October 1 and 165 millimeters (6½ inches) by March 1. It would appear that if this rate of growth is maintained the branch herring may reach its average maximum length of 11 inches in about four years.

Length frequencies of 1,967 branch herring, "Pomolobus pseudoharengus"

[Measurements in millimeters, grouped in 5-millimeter intervals]

Total length, millimeters	June 1-15	June 16-30	July 1-15	July 16-31	Aug. 1-15	Aug. 16-31	Sept. 1-15	Sept. 16-30	Oct. 1-15	Oct. 16-31	Nov. 1-15	Nov. 16-30	Dec. 1-15	Dec. 16-31	Jan. 1-15	Jan. 16-31	Feb. 1-15	Feb. 16-28	Mar. 1-15	Mar. 16-31	Apr. 1-15	Apr. 16-30	May 1-15	May 16-31
30-34	1	2																						
35-39	2	6																						
40-44	1	5																						
45-49	1	2																						
50-54	1	3		25	6	4		3																
55-59	1	1	4	25	14	8	8	9	15	6	2													
60-64				11	6	14	14	33	14	27	5	6	1											
65-69			1	10	2	13	18	28	11	21	21	37	3						1					
70-74				2	2	5	7	10	5	10	1	96	9								1			
75-79						1		4	4	4	1	37	6			1								
80-84						1			3	1		4				6		2	2	2	1			
85-89										1		3				15		2	1	6				
90-94								2								28		11	9	19				
95-99																36		14	25	31			1	
100-104												2				63		38	32	67	1			
105-109												1				44		35	29	48	1	2		
110-114																29		38	20	52	1			1
115-119										1						25		33	20	42			3	
120-124										1					1	25		16	12	25		3	1	
125-129										3					13			12	14	13	1	1	1	
130-134											1	1			11	8		10	14					1
135-139											1	2			9	7		10	14		1			
140-144										2		1				5		8	5	6	1	1		
145-149										1	1	1				3		6	5	12	1	1		
150-154										1		1				2		7	2	15			1	
155-159										1		1				3		8	2	8			1	
160-164										2						4		12	3	9	1			
165-169		1								2		1				4		9		3	1			
170-174										1		1				3		12		8				
175-179										1						2		5		4				1
180-184										1		1						2		1	1			
185-189										1		1						1		3				
190-194		1							1			1								1				
195-199										1						1				1				
220-224																				1				
Total	7	21	5	74	31	49	47	87	54	85	41	188	19		34	301		285	202	404	9	13	8	3

The maximum length attained by the branch herring is about 380 millimeters (15 inches) and the weight about 14 ounces. The average length of market examples, however, is only about 11 inches and the average weight is about 8 ounces.

The branch and glut herrings are both very abundant species in the Chesapeake region, probably occurring in about equal numbers. Since the branch herring arrive earlier than the glut herring, the earliest catches consist wholly of the former species, which is gradually replaced by the latter as the season advances. The species are not separated for the market and are sold either as river herring or alewives; therefore, the available statistics include both species, and their relative abundance is judged only from the observations made of various catches. Observations made at Lynnhaven Roads, Va., from April 4 to 8, 1922, showed that the catch taken in pound nets consisted of 27 per cent branch herring and 43 per cent glut herring. At Lewisetta, Va., from April 24 to 28, 1922, the catch taken with pound nets consisted of 29 per cent branch herring and 61 per cent glut herring. The largest catches of herrings are made in the southern sections of the bay between March 20 and April 20, whereas in the vicinity of Havre de Grace, Md., at the head of the bay, the principal fishing season usually extends from April 10 to May 10.

Throughout Chesapeake Bay, during 1920, the alewives ranked first in quantity and second in value, the catch being 22,986,158 pounds, worth $416,968.

In Maryland the alewives ranked first in quantity and third in value, the catch being 6,604,891 pounds, worth $163,544. Of this amount, 87 per cent was caught in pound nets, 9 per cent in haul seines, 3 per cent in gill nets, and 1 per cent with other apparatus. The five leading counties, with respect to the pounds of alewives caught, were Talbot, 1,506,865; Cecil, 1,170,780; Dorchester, 595,482; St. Marys, 534,888; and Harford, 453,840.

In Virginia they ranked first in quantity and fourth in value, the catch being 16,381,267 pounds, worth $253,424. Of this amount, 90 per cent was caught in pound nets and 10 per cent with seines, gill nets, fyke nets, and slat traps. The five leading counties, with respect to the pounds of alewives caught, were Northumberland, 5,726,586; Mathews, 3,057,900; Lancaster, 2,060,353; Elizabeth City, 1,120,000; and Gloucester, 1,068,800.

Somewhat over half of the herring catch is salted. In 1920, 1,456,300 pounds of salt herring were marketed by fishermen, and an additional 7,696,420 pounds were put up by wholesale and canning firms, making a total production of 9,152,720 pounds, valued at $291,948. Salting fish and canning roe is engaged in only during the height of the run. As a rule, vegetable canneries are utilized, as very little added equipment is necessary to handle the fish. The greater part of the salting and canning is done at Havre de Grace and Oxford, Md., and Lewisetta and Gwinns Island, Va. One cannery at Havre de Grace during most of April, 1922, utilized about 125,000 herrings per day.

The prices that the fishermen receive fluctuate considerably from year to year and during the same season. During the 1920 season the average price was slightly less than 2 cents per pound, or about $8 for 1,000 fish. During 1922 the prices were lower and the salting houses paid $5 per thousand fish during the earlier part of the season, but this price had dropped to $1.50 by the middle of April. All the salting establishments employed run boats for collecting the fish from the fishermen, and many fishermen preferred to dispose of their catches in this way, as it obviated packing and shipping to market. The market prices always were higher than the price paid by run-boat operators. The difference in the prices, however, was somewhat offset by the cost of packing and shipping.

Comparison of lengths and weights

ADULT FISH

Locality	Date	Sex	Length, inches	Average weight, ounces
Havre de Grace, Md.	Apr. 30, 1912	Females	11⅜ to 12¼ (5 fish)	9.6
		Males	10½ to 11¾ (7 fish)	8.0
	May 9, 1922	Undetermined	10¼ to 13 (14 fish)	7.5
Lewisetta, Va.	Apr. 24, 1922	Female	11¼ (1 fish)	9.6

YOUNG FISH

Locality	Date	Sex	Length, inches	Average weight, ounces
Solomons, Md.	Oct. 28, 1921	Undetermined	4½ to 4¾ (2 fish)	0.55
			5 to 5⅜ (2 fish)	.8

Habitat.—Nova Scotia to North Carolina. Landlocked in Lakes Cayuga and Seneca, N. Y., and also present in Lake Ontario. "In Lake Ontario, since the introduction there of the shad, the alewife has become so plentiful as to cause great difficulty to fishermen, and its periodical mortality is a serious menace to the health of people living in the vicinity. The belief is that the fish were unintentionally introduced with the shad." (Bean, 1903, p. 200.) It is supposed to have reached Cayuga and Seneca Lakes in a natural way.

Chesapeake localities.—(*a*) Previous records: From virtually all streams tributary to Chesapeake Bay. (*b*) The numerous young in the present collection, ranging in length from 30 to 165 millimeters, are from the following localities: Beam-trawl catches in many parts of the bay, from Annapolis, Md., to Old Point, Va., including the Potomac River below Cedar Point, at depths ranging from 5 to 28 fathoms, January 15 to April 28, 1914, January 16 to April 25, 1916, January 22 to January 27, 1921, and February 17 and April 20, 1922. Taken with seines in Potomac River from Washington, D. C., to Lewisetta, Va., September 21, 1911, June 7 to December 3, 1912, and October 24 and 25, 1921; and in the bay from Havre de Grace, Md., to Lynnhaven Roads, Va., June 22 to November 21, 1921, and April 8 to October 27, 1922.

Comparative statistics of the alewife product of Maryland and Virginia for various years from 1880 to 1921

Years	Pounds	Value	Years	Pounds	Value
1880	16,129,372	$215,967	1901	27,660,601	$206,732
1887	15,463,905	118,858	1904	29,088,836	228,715
1888	17,964,779	150,660	1908	66,690,000	328,000
1890	30,408,692	235,467	1909	51,425,300	283,874
1891	28,432,335	225,150	1915	28,621,710	297,720
1896	29,864,922	189,074	1920	23,736,788	436,448
1897	30,828,969	194,294	1921	25,339,009	390,529

NOTE.—The catch of alewives in these States, outside of the Chesapeake Bay, is included for some years but is practically negligible.

26. Genus ALOSA Linck. The shad

The genus Alosa is described as differing from Pomolobus in having the cheeks deeper than long, the upper jaw deeply indented anteriorly, and the toothless jaws of the adult. These differences separate the genus very satisfactorily from *P. æstivalis* and *P. pseudoharengus*, but in *P. mediocris*, with the exception of the toothless jaw of the adult Alosa, these differences become very slight or disappear; for in *P. mediocris* the cheeks are at least as deep as long and an indentation in the upper jaw is distinctly present. It is the opinion of the present authors, therefore, that the genus Alosa is scarcely tenable.

36. Alosa sapidissima (Wilson). Shad.

Clupea sapidissima Wilson, in Rees's New Cyclopedia, IX, no pagination and no date (about 1811); Philadelphia. Bean, 1883, p. 366.
Alosa sapidissima Uhler and Lugger, 1876, ed. I, p. 157; ed. II, p. 133; Jordan and Evermann, 1896–1900, p. 427, Pl. LXXII, fig. 191; Smith and Bean, 1899, p. 184; Fowler, 1912, p. 51.

Head 3.2 to 4.3; depth 2.7 to 3.9; D. 17 to 19; A. 19 to 23 (usual number 21 or 22); scales about 52 to 64. Body elongate, compressed, deeper in adult than in young, average depth in length to base of caudal of young of 35 to 100 millimeters, 3.5, adult females about 2.75; dorsal profile nearly straight on head, gently convex from nape to dorsal, ventral outline gently and evenly rounded, the abdomen compressed, with sharp ventral edge, provided with scutes; head rather small, low, and comparatively long; snout slightly tapering, 3.2 to 4.7 in head; eye 3 to 5.95; interorbital 3.95 to 5.85; mouth rather large, terminal; maxillary broad, reaching middle of eye in young (50 millimeters long), to or a little beyond posterior margin of eye in adults, 1.85 to 2.7 in head; cheek deeper than long, narrower below than above; mandible not projecting, included in upper jaw and not entering into the dorsal profile, its upper margin rather gently elevated, without a prominent angle near the middle of its length; teeth in the adult wholly wanting, the young with small, weak teeth on the anterior part of the jaws; gill rakers rather numerous, long and slender, increasing greatly in number with age, specimens 35 to 70 millimeters in length with 26 to 31 gill rakers on the lower limb of the first arch, specimens 110 to 180 millimeters long with 34 to 41, adults 413 to 580 millimeters in length with 62 to 76 gill rakers; scales of moderate size, deciduous in young and to a lesser extent in the adult; ventral scutes, 20 to 24 in advance of ventrals (usual number 21 or 22) and 12 to 16 behind ventrals (usual number 14 or 15), total number of ventral scutes 32 to 39 (usual number 35 to 37); dorsal fin rather small, its outer margin slightly concave, the origin considerably nearer tip of snout than base of caudal; caudal fin deeply forked, both lobes pointed; anal fin somewhat longer than the dorsal, the anterior rays only slightly longer than the posterior ones, its origin at least twice the diameter of the eye behind vertical from the end of the dorsal; ventral fins rather small, pointed, inserted a little in advance of the vertical from middle of base of dorsal; pectoral fins much larger than the ventrals but similar in shape, 1.4 to 1.7 in head.

Color greenish, with metallic luster above, sides silvery; a dark spot at shoulder, occasionally followed by smaller ones, rarely with a second parallel row somewhat above the median line of side. Fins all pale to slightly greenish, the dorsal and caudal somewhat dusky in the larger specimens, darkest at tips. Peritoneum pale.

Leim (1924, p. 224), who made an exhaustive study of the shad in the Bay of Fundy, gives the following counts for rays, scutes, and vertebræ (the predominating numbers are placed in parentheses): Dorsal rays 15 to 19 (17 or 18), 676 fish; anal rays 18 to 24 (20 to 22), 317 fish; pectoral rays 14 to 18 (15 to 17), 287 fish; pelvic (ventral) rays 8 to 10 (9), 277 fish; anterior ventral scutes 19 to 23 (20 to 22), 315 fish; posterior ventral scutes 12 to 19 (16 or 17), 653 fish; vertebræ 51 to 59 (56 or 57), 170 fish. These counts vary somewhat from those taken of Chesapeake Bay

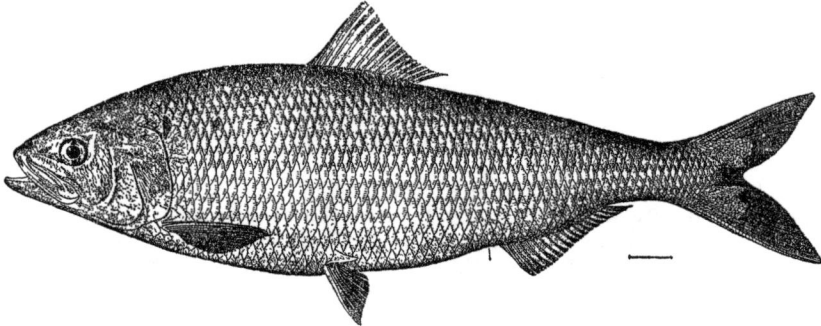

FIG. 53.—*Alosa sapidissima*

shad. This, however, is only to be expected, as Leim, who examined a large number of fish, found a slight variation even among the several localities of the Bay of Fundy where his specimens were obtained.

Numerous small specimens, 21 to 115 millimeters (⅚ to 4⅛ inches) in length, taken in fresh water, are at hand. We also have 26 specimens contained in various small lots, of different localities and dates, taken in salt or brackish water during the fall and winter, ranging from 97 to 243 millimeters (3⅞ to 9½ inches) in length, and three adult females. The young shad is not readily distinguished from the young of the genus Pomolobus. It is particularly close to the hickory shad, (*P. mediocris*), from which the young are difficult to separate. In the common shad the lower jaw, however, is included in the upper; it does not protrude and it does not enter into the dorsal profile. In the hickory shad the lower jaw projects strongly and the tip of it enters into the dorsal profile. Another and more pronounced difference is evident in the number of gill rakers supported by the lower limb of the first arch. The number of gill rakers increases greatly with age in at least some of the Clupeidæ, but in our series there is no overlapping, the common shad always having more gill rakers than the hickory shad. The range in the number of gill rakers on the lower limb of the first arch for the common shad in specimens ranging from 1½ to 23 inches in length is 26 to 71. In the hickory shad in specimens 6¼ to 12 inches long it is from 18 to 22. The difference is more evident when specimens of the same size are compared. A common shad 160 millimeters long, for example, has about 40 gill rakers on the lower limb of the first arch, whereas a hickory shad of the same length has only about 18. From the other species of Pomolobus the young shad may be separated by the much narrower and proportionately deeper cheek and by the much lower and broader angle near the middle of the sides of the upper margin of the mandible. The young shad does not differ greatly from the adult, except that the body is more slender, the gill rakers much fewer, and the dark spot at the shoulder is undeveloped. The sexes are very similar, except that the female reaches a larger size than the male.

The young shad, according to published accounts, feed on small crustaceans, insects, and insect larvæ, as well as on small fish. This statement was verified through the examination of 14 stomachs. Stomachs of specimens ranging from 1¾ to 2⅜ inches in length, taken in fresh water, had fed mainly on adult insects but also on ostracods. Specimens somewhat larger, ranging from 4 to 6 inches in length, taken in salt or brackish water, had fed almost wholly on small crustaceans (Mysis), but one stomach contained a small amount of plant tissue and another contained fragments of a small fish.

Little or no food has been found in the shad while they were migrating up rivers. Various investigators, however, have examined the stomachs of adult fish caught in the sea or at the mouths of estuaries and have found food. An adult female taken in the southern part of Chesapeake Bay early in December, 1921, and examined by us, had the stomach gorged with parts of plants, consisting not only of the softer parts but also of hard stems. Fragments of a molluscan shell also were present. Perley (1851, p. 139) found that the shad in the Bay of Fundy fed on shrimp and "shad worm." Mordecai (1860, p. 278) examined shad stomachs from the vicinity of Savannah, Ga., and as a result states that "shad feed and fatten on marine fuci and microscopic organisms that are parasitically attached." Leidy (1862, p. 2) obtained a shad in a market and upon opening it found in its stomach nine small fish, which were identified as follows: Three *Hydrargyra swampina;* five *Pœcilia latipinnis,* and one *Cyprinodon ovinus.* As these species inhabit fresh or slightly brackish water, and as the shad was probably received in Philadelphia, it is likely that it was caught in Delaware Bay. Leidy (1868, p. 228) examined a shad caught in the fall, probably off the coast of New Jersey or in Delaware Bay, and found in its stomach 30 sand launces (*Ammodytes americanus*), 2 to 4 inches long. Baird (1874, p. LVIII) says that in the sea the food of the shad consists "of worms,

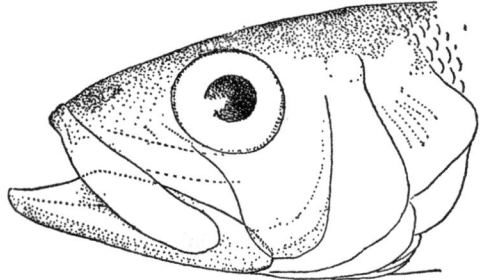

FIG. 54.—*Alosa sapidissima.* Note narrow, deep cheek bone and rather long, slender mandible without a pronounced angle on its upper margin

small fishes, and most largely of minute crustaceans, especially of the genus Mysis." Huyler (1876, p. 233) examined the stomachs of 15 shad caught in the North River near Fort Lee, N. J., on May 5, 1874, and found them containing many young shrimp about half an inch long. One of these stomachs contained several hundred shrimp. Prime (1876, p. 138) reports the capture of several shad (at least one with full roe) with artificial flies in the Connecticut River, thus indicating that at times the shad will feed just prior to spawning. It is of interest to note that these fish were caught on July 1, at the very end of the spawning season for that region. Smith (1896, p. 405) quotes Seth Green as saying that sand fleas [Gammarus?] are the principal food of the old shad in the Atlantic. Bean (1903, p. 207) says that the shad coming in to spawn will sometimes take the artificial fly and live minnows. Willey (1923, p. 313) examined many shad from the Nova Scotian coast of the Bay of Fundy and found the stomachs to contain chiefly copepods (Acartia, Temora, and other genera), mysid shrimp, and the larval stages of barnacles. Leim (1924) examined the stomachs of about 350 shad caught in Scotsman Bay, Bay of Fundy, during 1920 and 1922, consisting of mature and immature fish. Copepods formed the chief food for the smaller shad, but were of lesser importance in the diet of fish longer than 40 centimeters (16 inches). Mysids, however, while eaten sparingly by the younger fish, formed the chief constituent of the food of the adult shad. Copepods and mysids together formed about 90 per cent of the food of the shad of all sizes. Among the foods of lesser importance were ostracods, amphipods, isopods, decapod larvæ, insects, mollusks, algæ, fish eggs, and fish. Bigelow and Welsh (1925, p. 117) found adult shad taken in the Gulf of Maine in summer packed full of copepods (chiefly Calanus).

The fish examined by Bigelow and Welsh were taken far enough out in the Gulf of Maine to be removed from the influence of bay or river, and it is probable that their stomach contents show more truly than any of the other records given in the foregoing list the sort of food eaten by the shad at sea. These stomach examinations, therefore, justify Bigelow and Welsh's statement that the shad is primarily a plankton feeder.

The fact that adult shad in the sea are known to feed partly on mysid shrimps (bottom dwellers), on bottom-dwelling amphipods, etc., and on algæ is of considerable importance, as it suggests that part of a shad's life, perhaps a considerable part, is spent near the bottom of the sea. This may explain why so few shad are caught in the open sea south of Cape Cod. In the Gulf of Maine, however, large catches of shad are often made near the surface with purse seines, and it may be that in this region an abundant supply of food in the upper stratum of water lures the fish within reach of the nets. A change in the diet of the shad with age is suggested by the presence of numerous long cœca connected with the intestine near the stomach, which are very small or wanting in the young.

The life history of the shad is not well understood. It enters bays and rivers of the Atlantic coast of the United States in the spring, reaching the southern streams much earlier than the northern ones, and it ascends to fresh water for the purpose of spawning. The shad may spawn anywhere, but it appears to prefer shallow flats in rivers near the mouths of creeks. The fish are paired, swimming side by side, while spawning. The eggs are cast loose in the water, quickly sinking to the bottom, where many doubtlessly find unfavorable ground and fail to hatch.

The shad enters Chesapeake Bay in March, the date of arrival varying from year to year, and, with the exception of a few stragglers, they are gone again by the 1st of June. In 1921 a few shad were taken in pound nets in Chesapeake Bay throughout June and July and again during the latter part of November and early in December. These small and rather unusual fall runs, which occur only occasionally, naturally excite considerable interest. Whether such shad have remained in the inshore waters since spring, or whether they represent a new run from the sea is not definitely known; but inasmuch as extensive collecting in the inshore waters during the summer months has never brought an adult shad to light, it appears more reasonable to believe that they come from the sea. A female 17¼ inches long, taken on December 5, 1921, was preserved and critically examined in the laboratory and found to agree in all respects with the individuals of the spring run, except that it was smaller than the average size of the spring run of roe shad. This particular shad was full of eggs, which evidently were nearly mature. Whether or not the shad that constitute the fall runs usually are gravid is not known to the authors, but if this were the case the reason for their migration toward fresh water evidently would be the same as for the spring run—viz, for the purpose of reproduction. When the shad first arrive the males, or "buck shad," are greatly in the majority, but later in the season the females, or "roe shad," are the more numerous.

Spawning takes place soon after the fish reach fresh water. The eggs are relatively large after impregnation, measuring about 3 millimeters (⅛ inch) in diameter. The average number of eggs produced by a female, according to Smith (1907, p. 127), is from 25,000 to 30,000, but there are records of over 100,000 and in one case of 156,000 eggs having been taken from one fish. The period of incubation varies from six days and four hours in an average water temperature of 57.2° F. to a little less than three days in an average temperature of 74° F. (Ryder, 1884, p. 796). It is not considered an advantage, however, either to retard or to hasten hatching unduly, as a very long period of incubation may result in a proportionately smaller hatch and a very short incubation period yields weak fry. The average hatching period at temperatures that prevail during normal shad seasons varies from about six to ten days.

Although shad usually deposit their spawn in fresh water, recent studies by Leim (1924, p. 264) in the Bay of Fundy have shown that "the optimum conditions for the development of the eggs and larvæ up to the end of the period of yolk-sac absorption were a temperature of about 17° C., a salinity of about 7.5 per mille, and darkness." This is a most important finding, as virtually all shad hatching heretofore has been done in fresh water and largely in comparatively bright light.

Leach (1925, p. 485) reared shad successfully in a fresh-water pond at Washington, D. C., and in October, at the age of 5 months, transferred them to brackish-water aquaria (increasing the density of the water gradually from 1.005 to 1.018), in which they continued to thrive. Others of the same lot, transferred to fresh-water aquaria, died within three days.

The young when hatched are scarcely 10 millimeters (⅖ inch) in length, but they grow rapidly, reaching an average length of about 47 millimeters (1⅞ inches) during the first half of July (computation based on 74 specimens taken from the Potomac River at Bryans Point, Md., on July 2 and 9, 1912); 66.5 millimeters (2⅝ inches) by the last half of August (computation based on 5 specimens taken at Havre de Grace, Md., August 19, 1921, and also 30 specimens taken at Bryans Point, Md., August 23 and 30, 1912); and 70 millimeters (2¾ inches) by the last half of October (computation based on 138 specimens taken in the Potomac River at Bryans Point, Md., on the following dates: October 20, 1909, October 21, 1910, October 21, 22, and 28, 1911, and October 19 and 23, 1912).

The migration of the young shad to salt water begins with the approach of cool weather, but it is not until near the end of November or the beginning of December that all the young shad have left the fresh waters. The foregoing statement is based on investigations made on the Potomac River by Lewis Radcliffe and the late William W. Welsh. It is evident from the collections at hand that some of the young shad may remain in the salt water of the bay until they are a year or more of age, as specimens were taken on the following dates: January 15, 17, and 19, 1916, February 18 and 21, 1916, and March 6 and 21, 1916. Eight shad, taken in January, 1916, had an average length of 108 millimeters (4¼ inches); 3 taken in February, 1914, average 114 millimeters (4½ inches); 10 specimens taken in March, 1916, average 142 millimeters (5⅝ inches); and 2 specimens taken in May, 1922, average 152 millimeters (6 inches). Two specimens taken in November (one in 1912 and the other in 1921), having lengths of 149 millimeters (6⅛ inches) and 243 millimeters (9½ inches), may be in their second year, as they are very much larger than other specimens taken during the same month, although we are unable to observe anything in the structure of the scales that suggests a "winter ring."[6] We are aware that there are published accounts of shad having attained, under especially favorable circumstances, a length of 6 or 7 inches, or in one instance of 9 inches (Smith, 1907, p. 127), at the age of 7 months. At Washington shad placed in ponds with an abundant supply of young carp for food attained a length of 6 inches by early November, but shad kept in an aquarium at the Bureau of Fisheries in Washington attained a length of less than 4 inches at the age of 1 year. Bean (1903, p. 208) states that "Nets set offshore in Gravesend Bay in the fall frequently inclose large quantities of young shad, sometimes a ton and a half at one time, during the migration seaward * * *. The fish are usually about 6 to 8 inches long." Bean, no doubt, assumed that these fish were the young of the last spawning season. However, in view of the fact that Chesapeake fish are only about 3 inches long in late fall, it is our opinion that these 6 to 8 inch shad probably were in their second year. As there are no "connecting links" between our two individuals (6⅛ and 9½ inches) already mentioned and 140 other individuals at hand, taken during November, the largest of which is only 117 millimeters (4½ inches) long, the question naturally rises as to whether or not these specimens belong to an older class. The growth of the shad apparently is not noticeably retarded by the approach of cold weather, and it appears to continue throughout the winter. The limited number of specimens taken during the winter probably are too few to permit of definite conclusions, but if the rate of growth should be even slightly more rapid during the winter than during the summer, as indicated by the specimens at hand, an explanation for the apparent absence of definite "winter rings" on the scales of adult shad becomes evident. (The suggestions on p. 88, relative to the sudden upward bend in the growth curve of *P. æstivalis*, may apply to the shad also.) No specimens of young shad, except fry, were taken during the present investigations from April to October, both inclusive. The extensive collecting that was done shows quite conclusively that a few of the young shad spend the first winter of their lives in the salt water of the bay, and furthermore, that if any at all remain until they are more than a year old it is only a rare straggler.

---

[6] For recent works on the age determination of the shad see "Age of shad (*Alosa sapidissima* Wilson) as determined by the scales," by N. Borodin. Transactions, American Fisheries Society, fifty-fourth annual meeting, Quebec, Canada, Sept., 1924 (1924), pp. 178–184, 6 figs. Hartford. Also, "A confirmation of Borodin's scale method of age determination of Connecticut River shad," by R. L. Barney. *Ibid.*, pp. 168 to 177, 4 figs.

Length frequencies of 807 shad, " Alosa sapidissima "

[Measurements in millimeters, grouped in 5-millimeter intervals]

Total length, millimeters	June		July		Aug.		Sept.		Oct.		Nov.		Dec.		Jan.		Feb.		Mar.		Apr.		May	
	1–15	16–30	1–15	16–31	1–15	16–31	1–15	16–30	1–15	16–31	1–15	16–30	1–15	16–31	1–15	16–31	1–15	16–28	1–15	16–31	1–15	16–30	1–15	16–31
20–24		5																						
25–29	2	12																						
30–34	11	14	3																					
35–39	5	15	11	8																				
40–44	4	26	17	5																				
45–49		16	12	11	3		2																	
50–54		3	17	20	8		4			8														
55–59		8	15	9	10	4	8	4	1	9														
60–64		6	1	2	5	10	22	5	11	17	1	1												
65–69		1	1	4	3	8	11	8	32	21	7	1	1											
70–74				2	2	12	6	8	21	40	21	2												
75–79					1	1	1	2	17	21	36	9												
80–84							1	3	5	9	23	6												
85–89										1	7	12				5		2					1	
90–94								1	1	1	1	3												
95–99										1	1	1												
100–104										2	1					2				1				
105–109										3	1	1				1				1				
110–114										1	1	1				1				3				
115–119											1	1				1				1				
120–124											1					1				1				
125–129											2													
130–134																								
135–139																		1		1				
140–144																								
145–149											1									1			1	
150–154											1												1	
155–159																				1				
240–244													1											
Total	22	106	77	61	32	35	53	31	88	137	103	37	2			7		3		10			3	

It is still a mystery, at least south of the New England coast, in what part of the sea the shad spends its life after it leaves the rivers and bays and until it again returns to spawn. To our knowledge no shad ever have been captured in the open sea off Chesapeake Bay, although menhaden purse seines are used in this region throughout the summer and fall. In the Gulf of Maine, however, adult shad not only are present (at times in abundance) from October until into December, but, according to Bigelow and Welsh (1925, p. 116), schools of immature shad from 1 foot long and one-half pound in weight up to 2 or 2½ pounds in weight are reported every year at Provincetown for a short period in June. · These authors report the capture of numbers of shad about 14 inches in length in the traps at Magnolia and Beverly from June 20 to July 6, 1921.

Atkins (1887, p. 684) reports large numbers of immature shad feeding about the bays and mouths of rivers along the coast of Maine during the summer after the main body of spawning fish had ascended the rivers. These immature so-called sea shad belonged to the group ranging from about one-half to 2½ pounds in weight. Up to the present day these immature shad are caught every year along the New England coast, although the quantities taken now, in keeping with the reduced numbers of adult fish, are much smaller than they once were. Our only knowledge of the shad from the time (late in the fall of the year) the young leave the rivers and bays in which they were hatched until they return as mature spawning fish is obtained from the immature fish, probably 2 to 3 years old, that are found during the summer, as already stated, in fairly large numbers along the shores of the Gulf of Maine and in smaller numbers south of Cape Cod.

Shad make their first appearance on the Atlantic coast in the St. Johns River, Fla., where they are first seen late in November and remain until March. It is extremely improbable that adult shad present in the Gulf of Maine in November and December could migrate to Florida and arrive there within the short period that would be necessary if they were to form part of the early winter catch of the St. Johns River. Stevenson (1899, p. 106), who made a study of the shad fisheries of the Atlantic coast, believed that the shad have a bathic migration rather than one toward and away from the Equator. The theory that shad migrate north and south along the Atlantic coast appears not to be as tenable as that they probably move off into deeper water of suitable temperature as winter approaches and remain somewhere near the general vicinity of the rivers in which they were hatched and which they will ascend to spawn.

The shad received attention early by fish culturists, but it was not until 1867 that a hatching apparatus that proved successful was perfected. The United States Fish Commission first began hatching shad on the Connecticut River at South Hadley Falls, Mass., in 1872. In the following year this commission hatched 95,000 young shad on the Potomac River at Washington, D. C. This was the beginning of shad-hatching operations in the Chesapeake drainage. In 1875 the fish commission of Maryland also undertook the hatching of shad. Similar action was taken by Virginia a few years later, and the hatching of shad has been continued from year to year by the Federal Government and more or less intermittently by Maryland and Virginia.

Many millions of young shad have been hatched and liberated in the streams flowing into Chesapeake Bay, but this fish has not been reestablished in its former abundance. In fact, it is evident from statistics that a more or less fluctuating decline is taking place. (See table of comparative statistics.) In interpreting the statistics it is necessary to remember that larger and more effective fishing apparatus has been used from year to year and that, therefore, the reduction in the abundance of shad very probably is greater than the figures given would indicate. This decline in the abundance of shad while millions of young were being liberated no doubt is attributable mainly to overfishing and to pollution in the streams. Many of the gravid shad are taken in the bay before they reach fresh water, and those that are successful in entering streams must follow a maze to escape the numerous nets set in the rivers and in order to reach their spawning grounds. A boat trip on the lower part of Chesapeake Bay during the shad season will convince the most skeptical that it is astonishing that any shad at all reach their spawning grounds. In some of the rivers, at least, there is great danger that pollution is so great that the eggs produced will fail to hatch, or, if they do hatch, that the fry may not be able to survive. Further restrictions concerning the use of nets, the placement of obstructions in rivers, and the discharge of refuse and wastes into streams are undoubtedly necessary if the shad is to be maintained as an important commercial species.

Many experiments in transplanting the shad to waters in which it was not native were made by the United States Fish Commission when hatching operations were first undertaken.

Fry were liberated in various streams in the Mississippi Valley, also in several lakes, including the Great Lakes, in Bear and Jordan Rivers (both tributary to Great Salt Lake, Utah), and in the Sacramento River, Calif., from whence they descended to the Pacific Ocean. Only the last-mentioned introduction has proved successful, and large numbers of shad annually ascend the streams of the Pacific slope of the United States. However, on the Pacific coast this fish is not as highly regarded for food as it is on the Atlantic, and it is being shipped to eastern markets (including Baltimore), where it finds a ready sale.

The shad is the most valuable food fish caught in Chesapeake Bay, its value in 1920 being $1,482,294, or more than the combined value of the four next most important species—namely, alewives, croakers, squeateagues, and striped bass. It ranked third in number of pounds caught (9,074,333), being exceeded only by the alewives and the croaker.

In Maryland it ranked second in quantity and first in value, the catch being 593,573 shad, weighing 1,816,346 pounds, worth $344,110. Of this amount 53 per cent was taken in pound nets, 43 per cent in gill nets, 3½ per cent with seines, and one-half of 1 per cent with other apparatus. The three counties having the largest catches were Dorchester, with 348,883; Talbot, with 328,543; and Kent, with 307,300 pounds.

In Virginia it ranked third in quantity and first in value, the catch being 2,199,390 shad, weighing 7,257,987 pounds, worth $1,138,184. Of this amount, 76 per cent was taken in pound nets, 23 per cent in gill nets, and 1 per cent in seines and fyke nets. The three counties credited with the largest catches were Mathews, with 2,295,730; Northumberland, with 1,291,488; and Lancaster, with 526,129 pounds.

In the Southern parts of the bay pound-net fishing is begun about March 1, and by March 15 virtually all the nets are set for the expected run of shad. Small quantities are taken early in March. The heaviest catches, however, are made between March 20 and April 20. The catch then declines, and after May 10 only small numbers are taken. However, it is not unusual to catch a few stragglers until late in June. In the vicinity of Havre de Grace, Md., the heaviest run occurs during April. At this place pound nets are set only for shad and herring, the runs of these fish being over by the end of May, when fishing usually ceases for the year. Two of the best catches made by a set of nets in 1922 occurred on April 10, one at Ragged Point, Va., lower Potomac River, where 3,650 shad were taken in 2 nets, and the other at Cheapeake Beach, Md., where 4,600 shad were taken in 3 nets.

Shad are always packed and shipped in the fresh state, and facilities are available for bringing the fish to the market and the consumer in a remarkably short period of time. Fishing is done early in the day, usually at slack tide, and the fish at once are brought ashore to be packed in boxes and shipped on the first outgoing boat or train. Some fishermen dispose of their catch to "run boats," which anchor in convenient localities and which are fully equipped for this kind of trade. The fish are paid for in spot cash after being counted or weighed, and are packed loosely in the hold according to sex, the roe shad and buck shad being separated. Unless the catch is very small a run boat seldom waits for a second day's fishing but makes a rapid run to its home port, where the fish are properly packed and forwarded to the various markets. Many fishermen prefer to trade with the run-boat buyers as this relieves them of the trouble of packing and shipping their catches and brings prompt payments.

The maximum weight attained by shad on the Atlantic coast is about 12 pounds, but on the Pacific the shad is said to average 1 pound heavier, and a maximum weight of 14 pounds has been reported. A series of 21 female (roe) and 35 male (buck) shad from the Chesapeake region was measured and weighed. The female averaged 576 millimeters (22¾ inches) in length and 6 pounds 5 ounces in weight, while the males averaged 500 millimeters (19¾ inches) and 3 pounds 11 ounces, showing that the females of this lot averaged 2 pounds 10 ounces heavier than the males. The general run of female shad in the Chesapeake region weigh from 4 to 5½ pounds and the males from 2½ to 4 pounds.

Weight of various-sized fish, according to sex

Number of specimens measured	Length [1]	Weight [2]		Number of specimens measured	Length [1]	Weight [2]	
MALE	Inches	Pounds	Ounces	FEMALE	Inches	Pounds	Ounces
2	14	2	4	3	18	3	11
2	15	2	4	4	20	4	9
13	16	2	3	8	21	4	12
12	17	2	5	4	22	5	8
10	18	2	14	9	23	6	4
8	19	3	5	5	24	6	9
5	20	3	9	3	25	7	14
5	23	5					

[1] The fish measured have been grouped by inches for convenience; that is, if the specimen was nearer 14 inches in length than 15 inches, it was considered a 14-inch specimen.
[2] The weights given are the average of all fish of any one length group.

Habitat.—Gulf of St. Lawrence to Florida. (The shad occurring in the Gulf drainage is here considered as a separate and distinct species.)

Chesapeake localities.—(a) Previous records: From virtually all streams tributary to Chesapeake Bay. (b) The numerous young in the present collection, ranging in length from 21 to 243 millimeters (⁵⁄₈ to 9½ inches), are from the following localities: Beam-trawl catches, Sandy Point, Md., November 18, 1912; Barren Island, Md., to Smith Point, Va., January 19 to March 22, 1914; Oxford, Md., November 16, 1921; Hampton Roads, Va., January 15 and March 6, 1916, at depths ranging from 9 to 27 fathoms. Taken with seines: Havre de Grace, Md., August 19, 1921; Sassafras River, May 10, 1922; Potomac River, Bryans Point, Md., to Riverside, October 19 and 20, 1909; September 13 to November 11, 1911; June 7 to December 3, 1912. Two fish were caught with hook and line, using dough for bait, at Bryans Point, Md., October 21, 1910.

Comparative statistics of the shad product of Maryland and Virginia for various years from 1880 to 1921

Years	Pounds	Value	Years	Pounds	Value
1880	6,946,379	$275,422	1901	10,083,393	$486,805
1887	7,855,946	319,223	1904	10,332,148	599,397
1888	11,924,908	498,289	1908	11,251,000	733,000
1890	14,393,693	471,806	1909	9,282,888	761,205
1891	12,723,115	418,969	1915	6,168,669	849,527
1896	16,712,018	473,606	1920	9,161,001	1,500,323
1897	17,329,037	463,813	1921	8,716,250	1,546,990

NOTE.—The catch of shad in these States, outside of the Chesapeake Bay, is included for some years but is practically negligible.

27. Genus OPISTHONEMA Gill. Thread herring

Body elongate, compressed; the abdomen strongly compressed, armed with about 33 prominent scutes; lower jaw projecting; upper jaw somewhat emarginate; dorsal inserted in advance of ventrals the last ray greatly produced, filamentous; vertebræ about 42. A single species of this genus is known from the Atlantic coast of America.

37. Opisthonema oglinum (LeSueur). "Hairy-back"; Thread herring; "Shad herring."

Megalops oglina LeSueur, Journ., Ac. Nat. Sci., Phila., I, 1817, p. 359; Newport, R. I.
Opisthonema thrissa Uhler and Lugger, 1876, ed. I, p. 158; ed. II, p. 134.
Opisthonema oglinum Bean, 1891, p. 93; Jordan and Evermann, 1896–1900, p. 432.

Head 3 to 4.3; depth 2.6 to 2.9; D. 18 or 19; A. 22 to 24; scales 70 to 77. Body moderately deep, compressed; abdomen compressed, with sharp scutes on ventral edge; head rather small; snout moderate, 3.7 to 4.2 in head; eye with adipose eyelid, 3.6 to 4.2; interorbital 3.8 to 4.2; mouth nearly terminal; the lower jaw projecting a little; maxillary reaching anterior margin of pupil, 2.4 to 2.6 in head; teeth wanting in the jaws, small ones present on median line of tongue; gill rakers long and slender, numerous, 70 to 77 on the lower limb of the first arch; scales rather large, cycloid, loosely adherent; ventral scutes 17 or 18+15 or 16; lateral line wanting; dorsal fin rather small, somewhat elevated anteriorly, the last ray greatly produced in the adult, reaching nearly or quite to base of caudal, origin of dorsal in advance of ventrals and much nearer tip of snout than base of caudal; caudal fin forked, the lower lobe slightly the longer; anal fin long and very low, its origin nearer base of caudal than base of ventrals; ventral fin small, inserted under middle of base of dorsal; pectoral fins moderate, inserted a little in advance of margin of opercle, 1.2 to 1.3 in head.

Color in alcohol, bluish gray with a metallic luster above; lower part of sides silvery; tip of snout black; a more or less distinct dark shoulder spot; indefinite dark lines along the rows of scales on the back; fins chiefly plain translucent, the dorsal and caudal with black tips.

No small individuals were taken. Eight specimens of adult fish, ranging from 198 to 230 millimeters (7¾ to 9 inches) in length, were preserved. This species is readily recognized (except the very young) by the greatly produced posterior ray of the dorsal fin, which reaches nearly or quite to the base of the caudal fin. It is from this long, threadlike ray that the fish has received the name "hairy-back" and "thread herring."

The food of this fish appears to consist largely, if not wholly, of small organisms, which it strains from the water by means of its long gill rakers. Doctor Linton examined the contents of three stomachs and found copepods exclusively.

The hairy-back is essentially a tropical fish and as a rule it is not abundant in Chesapeake Bay. Its spawning habits are almost wholly unknown. This herring reaches a size of about 12 inches, but its flesh is bony and of little value as food. Its commercial importance among the fishes of Chesapeake Bay is slight, as it is rarely used for food. However, it is utilized along with the menhaden in the manufacture of fertilizer and oil when taken in sufficient quantities. The fish usually makes its appearance about the middle of May, and it leaves the bay during October. It is taken in comparatively small quantities in pound nets throughout the summer in the southern parts of the bay, the catch rarely exceeding 100 pounds a day for one set of nets. The hairy-back appears to visit mainly that section of the bay that lies southward from the mouth of the Rappahannock River. The fish taken in the spring, among the specimens at hand, are very thin and poor, but those collected during the fall are fat and have broad, round backs.

Habitat.—Middle Atlantic States, southward to Brazil, and occasionally straying northward to Massachusetts.

Chesapeake localities.—(a) Previous records: Tributaries of Chesapeake Bay in the salt water (Uhler and Lugger, 1876); Cape Charles City (Bean). (b) Specimens seen or taken during the present investigation: York River, Va., July 8, 1921; Buckroe Beach, June 22, 1921; Lynnhaven Roads, May 25, 1922, and September 26, 1921.

28. Genus BREVOORTIA Gill. Menhadens

Body elongate, compressed, tapering posteriorly; head large; cheeks notably deeper than long; abdomen compressed and provided with bony scutes; mouth large; lower jaw included; teeth wanting; gill rakers long, thin, and numerous; scales deeper than long, closely imbricated, strongly pectinate; alimentary canal long; peritoneum black; vertebræ 46 to 49; fins small. A single species is known from Chesapeake Bay.

38. Brevoortia tyrannus (Latrobe). Menhaden; Skipjack; Bunker; Moss bunker; Alewife; Fatback; Bugfish.

Clupea tyrannus Latrobe, Trans., Amer. Phil. Soc., Phila., V, 1802, p. 77, Pl. I, Chesapeake Bay.
Brevoortia menhaden Uhler and Lugger, 1876, ed. I, p. 156; ed. II, p. 133.
Brevoortia tyrannus Bean, 1883, p. 366; Bean, 1891, p. 93; Smith, 1892, p. 64; Jordan and Evermann, 1896–1900, p. 433. Pl. LXXIII, fig. 195; Smith and Bean, 1899, p. 184; Evermann and Hildebrand, 1910, p. 158; Fowler, 1912, p. 52.

Head 2.9 to 3.4; depth 2.4 to 3.8; D. 18 to 20; A. 20 to 22; scales in oblique series along median line of side 48 to 56. Body elongate, compressed, the ventral outline much more strongly curved than the dorsal; abdomen compressed, with sharp scutes on the ventral edge; head large,

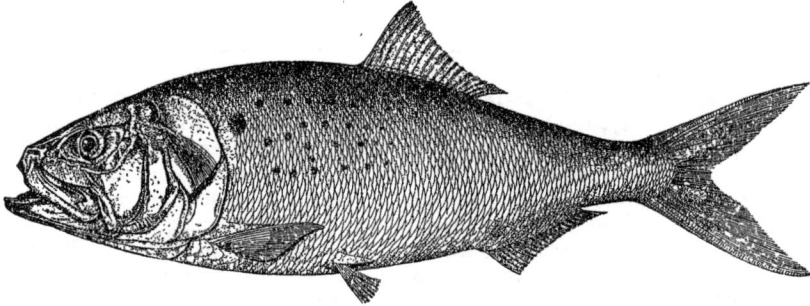

FIG. 55.—*Brevoortia tyrannus.* From a specimen 12 inches long

compressed; snout rather blunt, 3.8 to 5.5 in head; eye 4 to 5.9; interorbital 3.9 to 5.8; cheeks and opercles very deep, the upper part of opercle in adult with strong radiations, less prominent ones on the lower part of the preopercle; mouth moderate, terminal, the lower jaw largely included in the upper, the latter with a deep notch anteriorly; maxillary broad, rounded posteriorly, reaching past eye, in the adult, to middle of eye in young about 50 millimeters long; teeth in the jaws in the very young weak, disappearing entirely early in life; gill rakers extremely long, slender, close-set and exceedingly numerous; scales very closely imbricate, strongly pectinate, irregularly placed on upper part of sides, but in more definite series below median line of sides, the posterior margins nearly vertical instead of rounded, as in the herrings; lateral line wanting; dorsal fin rather small, somewhat elevated anteriorly, with a sheath of scales at base, except in the very young, origin of fin about equidistant from tip of snout and base of caudal; caudal fin rather deeply forked, the lower lobe somewhat the longer; anal fin rather long and low, slightly elevated anteriorly, its origin under tips of last rays of dorsal; ventral fins small, inserted slightly behind vertical from origin of dorsal; pectoral fin moderate, inserted slightly in advance of posterior margin of opercle, 1.7 to 2.1 in head.

Color of back dark green to bluish; sides brassy; a round, black, humeral spot present (except in the young of less than about 70 millimeters in length) and with or without a variable number of smaller dark spots on sides behind it; fins mostly pale yellow, some of them often more or less punctulate with dusky.

Many specimens of this species, ranging from larvæ 20 millimeters to adults 370 millimeters in length, were examined, and a large series was measured for the purpose of determining the range of variation within the species. The menhaden is so well known to those who live on the seashore within the range of the species that it is recognized at sight by old and young. The chief recognition marks of the species are the rather deep body, the compressed abdomen, deep cheeks, broad opercles, deeply emarginate upper jaw, strongly pectinate scales with posterior margins nearly vertical, and the greenish and brassy coloration. A pronounced variation in the depth of the body takes place within the species, which appears to be correlated, to a large degree, with the state of nourishment of the individual fish, the well-nourished fish being deeper than the poorer specimens. Similarly, a great variation in the width of the back also exists. When the fish is in a well-nourished state the back is very broad and layers of fat lie underneath the skin. The common name "fatback" is very appropriately applied to fish in this condition. A large crustacean parasite (*Cymothoa prægustator*) is commonly found inside the mouth of menhaden, giving rise to the name "bugfish."

The sexes are not distinguishable externally, so far as known to the writers, and the size attained appears to be nearly equal.

The menhaden feeds on small organisms, which it strains from the water by means of its long, slender, and very numerous gill rakers. The feeding and movements of schools of fish, as observed in the Patuxent River from aboard the *Fish Hawk* by the junior author, are described as follows in his field notes:

The fish swam swiftly in circles, like the dust driven by a whirlwind; then suddenly formed in a straight line, continually rising and falling at various depths. Each time they rose their mouths were wide open, but it was not possible to see whether or not their mouths were open when they swam downward. The fish near the shore seldom "broke water," but those observed in the open swam in compact schools, causing ripples at the surface; at times hundreds of them swiftly darted a few inches out of the water, causing a noise that could be heard easily at a distance of 300 feet. One large school was seen to divide into two parts. Some schools swam against the tide and then suddenly turned back with the tide. No general direction seemed to be maintained.

Doctor Linton examined the contents of the alimentary canal of 44 specimens taken in Chesapeake Bay and found that in most cases they consisted of sandy mud, vegetable débris (mostly algæ), and some diatoms, and in a few cases they consisted principally of copepods. He gives (from his notes as follows) the contents of the alimentary canal of a specimen taken in the lower part of the Patapsco River, November 7, 1921, as typical of the lot examined:

Gizzard full of yellowish mud, which, under high magnification, is resolved, as in previous cases, into vegetable silt with a little very fine sand. The vegetable material is reduced to a pulp, but vegetable cells can be distinguished, evidently of algal origin, material which makes up the vast majority of the food. Diatoms were present in considerable numbers, but do not constitute a large percentage of the food; very small, in fact, much less than 1 per cent. * * * Intestine filled with the same material.

Peck (1894, p. 113) gives the food of the menhaden as unicellular organisms, both vegetal and animal, together with the smaller Crustacea and other free-swimming forms.

Concerning the spawning habits of the menhaden, Kuntz and Radcliffe (1918, p. 119) state:

Observations on the movements of the schools and examination of the reproductive organs lead to the belief that in New England spawning takes place in late spring or early summer and that from Chesapeake Bay southward the season is late fall or early winter. Some reasons have been advanced for believing that in the Chesapeake region, at least, there are two spawning seasons.

The present writers have secured no information that suggests two spawning periods in Chesapeake Bay during one year. The evidence at hand, however, indicates that spawning takes place during the fall, as fish with well-developed (although not ripe) roe were taken only during that season of the year. The size and development of the young taken during the winter and spring furthermore suggest that they were hatched during the fall. Fourteen larvæ caught during January had attained an average length of 27.7 millimeters; 6 taken during February averaged 33.5 millimeters; 5 taken during March averaged 27.3 millimeters; 4 taken during April averaged 33 millimeters; and 137 taken during May averaged 46 millimeters. The number of larvæ caught from January to April, of course, is too small to show the rate of growth during the winter months, but at any rate the indications are that it is very slow. These fish all bear large chromatophores, the majority of them still possess indications of fin folds, and none of them have developed scales, all of which shows that the fish are very young. No larval menhaden were taken during any other season of the year. Further evidence that the breeding season may vary in different latitudes is produced by Bigelow and Welsh (1925, p. 122), for these authors state that in the vicinity of Woods Hole, Mass., spawning takes place chiefly in June, and that it continues well into October, and they add that the menhaden is equally a summer spawner in the Gulf of Maine, where spent fish and others approaching maturity have been reported during July and August.

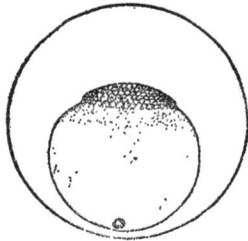

FIG. 56.—Egg in advanced stage of cell division

The eggs of the menhaden (Kuntz and Radcliffe, 1918, p. 119) are highly transparent, spherical in form, and they have a diameter of 1.4 to 1.6 millimeters. The period of incubation is given as "not over 48 hours," and the newly hatched larvæ have a length of approximately 4.5 millimeters. When the young fish reaches a length of about 33 millimeters all of the fins are well formed and scales are beginning to appear, but the body remains very slender. Large, black chromatophores are present on the head and nape, along the base of the anal, on the base of the caudal, and on the caudal peduncle posterior to the dorsal fin. Young fish 45 millimeters in length are fully scaled, and they have assumed the adult characters to such a degree that anyone familiar with the adult fish would recognize the young of this size. At one year of age the menhaden has reached a length of about 130 millimeters (5⅛ inches), and at two years of age it is 215 millimeters (8½ inches) long. Sexual maturity probably is reached during the third or fourth year.

The menhaden, as already indicated, is present in Chesapeake Bay throughout the year, although much less common during the winter than during the summer. The specimens caught during the winter were taken with a beam trawl in the deeper waters of the bay. During March, however, the fish again is common in the shallow waters and is taken in pound nets and haul seines. Very large schools of the migrating fish do not appear, as a rule, to enter Chesapeake Bay, and the abundance of menhaden does not seem to be affected by spring, summer, and fall "runs," as is the case along the outer shores of the middle Atlantic States.

The menhaden industry [7] in the Chesapeake is of considerable importance. The amount of fish utilized in 1920 was 366,-

FIG. 57.—Newly hatched larva, 4.5 millimeters long

379,425 pounds, valued at $2,158,518. It is not known how much of this amount was taken within the bay, but a large percentage was caught outside the capes by purse-seine boats and brought in to the various factories in Virginia. Pound nets are credited with 6,233,920 pounds, worth $22,114, almost the entire catch being confined to Virginia.

The menhaden is utilized almost entirely for fertilizer (fish scrap) and fish oil. In 1920, 18 factories were in operation, employing about 900 persons. These factories were supplied by 42

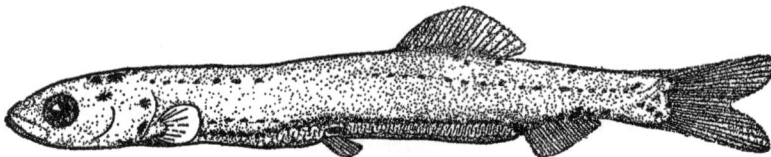

FIG. 58.—Larva 23 millimeters long

steam vessels, on which over 1,500 persons found employment. The industry is confined to Virginia, the chief centers being Northumberland and Lancaster Counties.

In many parts of the bay this fish is not utilized by the pound-net fishermen, but is separated from the catch of more valuable species and thrown away. In some localities it is sold to farmers at a small price and is used to enrich the soil. Within the vicinity of the factories, the pound-net fishermen sometimes dispose of a catch (when sufficiently large) by bringing the menhaden directly

[7] For a detailed account of the menhaden industry see "The Menhaden Industry of the Atlantic Coast" by Rob Leon Greer. Report, U. S. Commissioner of Fisheries, 1914 (1915), Appendix III, 27 pp., Pls. I-VII. Bureau of Fisheries, Document No. 811. Washington, 1915.

to the factory.   During much of the fishing season the daily catch for a pound net is about 1 bushel of menhaden, an amount too small to market.   Fishermen generally do not record the catch of menhaden, and for this reason the amount taken by pound nets probably is larger than that given in the statistics.

This species is taken during the major part of the fishing season—from March until late November.   It ranges from the capes to the head of the bay and is very common as far north as Baltimore.

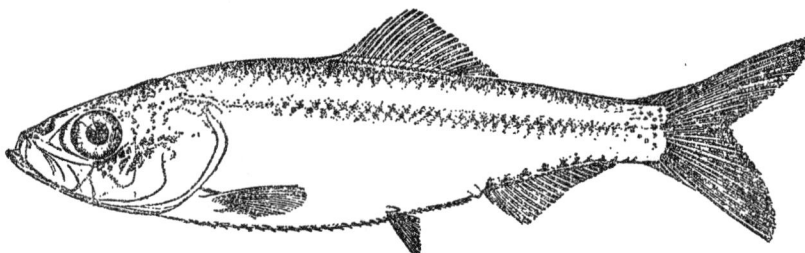

FIG. 59.—Young fish 33 millimeters long

Habitat.—"Nova Scotia to Brazil." (Jordan and Evermann, 1896–1900.)

Chesapeake localities.—(a) Previous records: Many parts of the bay, in salt and brackish waters.   (b) Specimens in collection: From nearly all sections of the bay.   Taken in shallow water during the summer and in deep water during the winter.

Comparison of weights and measurements of menhaden

Number of fish weighed and measured	Length	Average weight	Number of fish weighed and measured	Length	Average weight
	Inches	*Ounces*		*Inches*	*Ounces*
1	4	0.35	11	8.5	4.10
35	4.5	.58	9	9	4.65
17	5	.74	4	9.5	6.05
23	5.5	1.05	8	10	7.15
30	6	1.34	3	10.5	8.00
20	6.5	1.60	1	11.5	8.50
16	7	2.13	1	13	16.60
4	7.5	2.52	2	14	17.10
12	8	3.59			

For convenience the fish were divided into groups varying one-half inch in length.   For example, the 4.5-inch group is composed of fish ranging in length from 4.25 to 4.74 inches, and the 5-inch group is composed of individuals ranging from 4.75 to 5.24 inches in length, etc.   The weights given constitute the average weight for all fish weighed and measured falling within a group.

The weight of menhaden, with respect to size, varies according to season, the fat fish being heavier in the fall than corresponding sizes of spring-caught fish.   Apparently weights of fish taken in the same season may vary from year to year, depending upon the amount and kind of food available.   The following table illustrates the difference in weight of menhaden from lower Chesapeake Bay caught during October, 1921 and 1922.   These fish were weighed by the same metric scale.

Number of fish weighed and measured	Length	Weight	Number of fish weighed and measured	Length	Weight
1921			1922		
	Inches	*Ounces*		*Inches*	*Ounces*
7	5.1	0.78	2	5.1	1.07
5	5.5	1.01	4	5.5	1.15
2	5.9	1.19	8	5.9	1.36
4	6.3	1.44	7	6.3	1.55
1	6.7	1.57	7	6.7	1.93
2	7.1	1.95	8	7.1	2.21
6	7.9	3.59	3	7.9	3.27
3	8.3	3.52	7	8.3	4.26
2	8.7	3.77	9	8.7	4.64
1	9.1	4.22	2	9.1	5.17

Family XXI.—DOROSOMIDÆ. The gizzard shads

Body rather short and deep, strongly compressed; head small, short; mouth small, inferior; gill rakers numerous, slender; no lateral line; scales thin, cycloid, deciduous; anal fin long and low; the stomach rounded and very muscular, developed into a "gizzard." Mud-eating fishes.

29. Genus DOROSOMA Rafinesque

This genus is readily recognized by the prolongation of the last ray of the dorsal fin. A single species is recognized from the United States.

39. Dorosoma cepedianum (LeSueur). Gizzard shad; "Toothed herring"; "Oldwife"; "Mud shad."

Megalops cepedina LeSueur, Journ., Ac. Nat. Sci., Phila., vol. 1, 1818, p. 361; Delaware and Chesapeake Bays.
Dorosoma cepedianum Uhler and Lugger, 1876, ed. I, p. 160; ed. II, p. 136; Bean, 1883, p. 367; Jordan and Evermann, 1896–1900, p. 416, Pl. LXIX, fig. 183.

Head 3.3 to 4.6; depth 2.25 to 2.8; dorsal 14 or 15; anal 30 to 34; scales 56 to 64; ventral scutes 29 to 31. Body rather deep (with depth quite variable), compressed, the abdomen compressed,

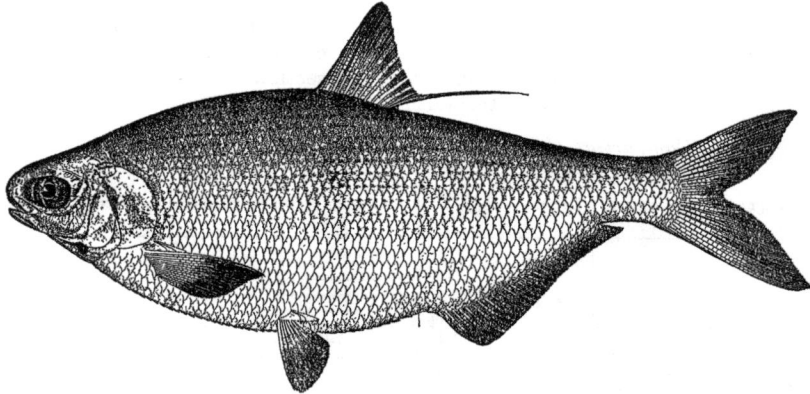

FIG. 60.—*Dorosoma cepedianum.* From a specimen 12⅝ inches long

with sharp scutes on ventral edge; head rather small (variable); snout blunt, projecting beyond mouth, 4.5 to 5.1 in head; eye with adipose eyelids, 3.45 to 5.25; interorbital 3.3 to 4.3; mouth inferior, rather small; maxillary reaching about opposite anterior margin of pupil, 3.1 to 3.75 in head; teeth wanting in the adult; gill rakers long and slender, numerous, about 135 on lower limb of first arch; scales rather large, reduced scales extending on base of caudal fin; lateral line wanting; dorsal fin rather small, somewhat elevated anteriorly, the last ray produced, sometimes nearly as long as head, origin of the fin somewhat nearer tip of snout than base of caudal; caudal fin rather deeply forked; anal fin very long, longer than head, 2.35 to 3.6 in length of body, its origin well behind the end of base of dorsal; ventral fins small, inserted about equidistant from base of pectorals and origin of anal, 1.65 to 2.3 in head; pectoral fins larger than ventrals, 1.15 to 1.3 in head.

Color of adult plain metalic blue above, silvery on sides; fins all more or less dusky. The color of immature fish, 107 to 127 millimeters in length, sea-green above, silvery below, frequently with a small black spot at shoulder; fins plain, dorsal and anal sometimes slightly dusky.

The Chesapeake collection contains 31 specimens ranging from 3¾ to 13 inches in length. No very young individuals were seen in brackish water. "The young are extremely different from the adult, slender and minnowlike in shape, and with a row of fine teeth on upper jaw, although the mouth of the adult is entirely toothless and smooth. The internal structure of the young also differs remarkably from that of the full-grown fish, especially in the much greater simplicity of the digestive apparatus, the intestine in specimens not more than an inch long passing almost directly back from the stomach to the vent." (Forbes and Richardson, 1908, p. 47.) The young also differ from the adult in having a large dark spot on the shoulder.

This species is readily recognized by the inferior mouth, the produced posterior ray of the dorsal, and by the very long anal fin. The adults also are characterized by the numerous, slender, close-set gill rakers, by the greatly thickened walls of the stomach, from which it derives the name "gizzard shad," and by the long convoluted intestine and numerous cœca.

The size of the head and the depth of the body vary greatly among specimens, as shown in the description. The dark shoulder spot, always present in the young, appears to persist much longer in some specimens than in others, and occasionally it probably never is lost. These variations form the basis for the descriptions of several nominal species. Only one species, however, is now recognized, and variations occur even among individuals taken in the same school.

The food consists almost exclusively of small organisms derived from mud, upon which it feeds. For the purpose of extracting these organisms from the mud, the fish is provided with a very effective straining apparatus in its gillrakers, which have already been described. Linton examined 10 stomachs taken from Chesapeake Bay specimens and found about 20 per cent of the "gizzard" content to consist of sand and mud and about 80 per cent of vegetable débris. One copepod was recognized and two Foraminifera. The intestine in this lot contained the same material, but with a rather larger proportion of sand. One Foraminifera, one Diffulgia, and one diatom were recognized.

Spawning occurs during the early summer. The species is very prolific. The gizzard shad is more fresh-water in habit than are the true shad and herrings, as it is found in fresh water at all seasons; in some instances it has become landlocked, under which conditions it is thriving. It has, in fact, become so fresh-water in its habits that it frequents only fresh and brackish water and is rarely seen in strictly salt water. Nevertheless, it appears to make certain migrations, at least in Chesapeake Bay, as there is a fall "run" in September and October; but we have no evidence that there is a corresponding spring "run", as one would expect and as reported for North Carolina by Smith (1907, p. 119).

We know comparatively little of the rate of growth of the gizzard shad in Chesapeake Bay, but the following total catches, and therefore unselected fish, show that a length of about 4 to 5 inches is attained by October. These fish all were collected at Ocean View, Va., except one specimen, which was taken from the Patuxent River on November 8.

Date	Number of fish taken	Range in size		Average length
		Millimeters	*Inches*	*Inches*
Sept. 25, 1922	5	101–125	4–5	4.3
Oct. 20, 1922	1	109	4.3	4.3
Oct. 18, 1922	14	107–126	4.2–5	4.5
Oct. 25, 1922	20	109–160	4.3–6.3	5.3
Nov. 8, 1921	1	101	4	4

The gizzard shad is a bony fish of rather poor quality and it commands a low price in the market. In the Chesapeake region it sells fairly well to a class of trade that demands a cheap fish. The retail price in the Baltimore market in 1921 was about 5 cents a pound. This fish is not taken in large quantities in Chesapeake Bay and it does not command a separate fishery, but at times when "fishing is bad" catches are made that are very helpful to the fishermen. During 1920, among the various Chesapeake Bay fishes, it ranked fifteenth in quantity and twentieth in value, the catch being 72,852 pounds, worth $2,013.

The importance of this fish among the commercial species, however, must not be judged from the quantity that is marketed and the price received. The food that the gizzard shad furnished for other fish, without itself eating foods utilized by most species, is no doubt of great economic importance. This point is well stated by Forbes and Richardson (1908, p. 46) in speaking of its importance among the fishes of Illinois:

This immensely abundant species, although little esteemed as a food fish, is one of the most useful in our waters because of the almost exhaustless food supply which it offers to all the game fishes of our larger streams and lowland lakes. Living itself mainly upon food derived from the muddy bottoms of our very muddy rivers and lakes, it serves as a means of converting this mere waste of nature into the flesh of our most highly valued fishes.

The maximum length attained by the gizzard shad is given as 15 inches by Jordan and Evermann (1896–1900, p. 416), and the *average weight* is given as 1½ to 2 pounds by Smith and Bean (1899, p. 183) and others. This weight is quite certainly the maximum instead of the average weight. The average length of the adult of Chesapeake Bay, at least, appears to be between 11 and 12 inches, and fish of this length, as shown by the accompanying table, weigh less than 1 pound.

The gizzard shad, as already indicated, appears to be common in Chesapeake Bay only during the fall months, when it is taken principally in brackish water near the mouths of fresh-water streams. In the rivers of the Chesapeake region it is common or even abundant throughout the year.

Habitat.—Fresh and brackish waters of the Atlantic coast, from Massachusetts to Mexico, and the Mississippi Valley and Great Lakes. Also landlocked in many ponds and lakes.

Chesapeake localities.—(a) Previous records: Chesapeake Bay (Le Sueur, 1817); Baltimore docks, Potomac, Patapsco, and other rivers (Uhler and Lugger, 1876); head of Chesapeake Bay (Bean, 1883); Potomac River (Smith and Bean, 1899). (b) Specimens seen or taken in brackish or salt water during the present investigation: Hawkins Point, Baltimore; mouth of Severn River; Chesapeake Beach; Blackistone Island, Md.; Lewisetta; Ocean View and Lynnhaven Roads, Va. Greatest salinity, 22.63 per mille.

The fish upon which the following weights are based with few exceptions were caught off Ocean View, Va., from September 25 to October 25, 1922.

Number of fish weighed	Length	Weight	Number of fish weighed	Length	Weight
	Inches	*Ounces*		*Inches*	*Ounces*
2	4.00	0.36	1	6.00	1.18
1	4.12	.40	1	6.25	1.42
7	4.25	.38	1	7.00	2.00
1	4.37	.43	1	[1] 8.00	4.50
3	4.50	.46	1	9.33	5.60
1	4.62	.54	1	[2] 9.63	4.57
4	4.75	.56	1	10.00	7.68
3	5.00	.59	1	[2] 11.75	7.42
1	5.25	.78	2	[1] 12.75	15.00
1	5.37	.82	1	[2] 13.00	12.00
1	5.75	.96			

[1] Caught off Chesapeake Beach, Md., in October, 1921.
[2] Caught October, 1921, on Blackistone Island, Potomac River, in a pond nearly landlocked. Note that the three fish from this locality are all below normal weight, due, perhaps, to the fact that they lived in a pond where the food supply was not abundant. All the remaining fish were taken in the open bay.

Family XXII.—ENGRAULIDÆ. The anchovies

Body elongate, more or less compressed; abdomen frequently compressed, forming a slight keel; snout pointed, usually projecting far beyond mandible; mouth large; maxillary usually reaching far past eye; premaxillaries not protractile; teeth usually small but sometimes uneven and caninelike; gill membranes separate or joined, free from the isthmus; gill rakers long and slender; pseudobranchiæ present; lateral line wanting; scales thin and cycloid, usually deciduous; dorsal usually about median in position; no adipose fin; caudal fin forked. A single genus of the family occurs in Chesapeake Bay.

30. Genus ANCHOVIELLA Fowler. Anchovies

Body elongate, compressed; abdomen usually compressed; snout conical, projecting prominently beyond the mandible; mouth large; the maxillary usually reaching far beyond eye; teeth very small, pointed; gill membranes separate and free from the narrow isthmus; gill rakers long and slender; scales rather large, thin, and usually deciduous.

KEY TO THE SPECIES

a. Anal fin with 24 to 27 rays, the origin of the fin under middle of dorsal base; silvery lateral band more or less diffuse; length about 3 inches_____*mitchilli*, p. 109
aa. Anal fin with 20 or 21 rays, the origin of the fin under the last rays of the dorsal; silvery lateral band very bright and well defined; length about 4½ inches_____*epsetus*, p. 110

40. Anchoviella mitchilli (Cuvier and Valenciennes). Anchovy.

Engraulis mitchilli Cuvier and Valenciennes, Hist. Nat. Poiss., XXI, 1848, p. 50, New York; Carolina and Lake Pon-chartrain, La.
Engraulis vittatus Uhler and Lugger, 1876, ed. I, p. 161; ed. II, p. 137.
Stolephorus mitchilli Bean, 1891, p. 93; Jordan and Evermann, 1896–1900, p. 446; Smith and Bean, 1899, p. 184; Evermann and Hildebrand, 1910, p. 159.
Anchovia mitchilli Fowler, 1912, p. 52.
Anchoviella mitchilli Jordan and Seale, 1926, p. 405.

Head 3.9 to 4.45; depth 4.1 to 5.1; D. 13 to 15; A. 24 to 27; scales 37 to 40. Body strongly compressed; ventral outline much more strongly convex than the dorsal; the margin of abdomen compressed, forming a rather sharp edge; head moderate; snout conical, projecting notably in advance of lower jaw, 4.6 to 7 in head; eye 2.6 to 4.2; interorbital 4.1 to 5.9; mouth large, slightly oblique; maxillary long and sharply pointed posteriorly, reaching nearly or quite to margin of opercle, 1.1 to 1.38 in head; teeth pointed, present on both jaws; gill membranes largely separate

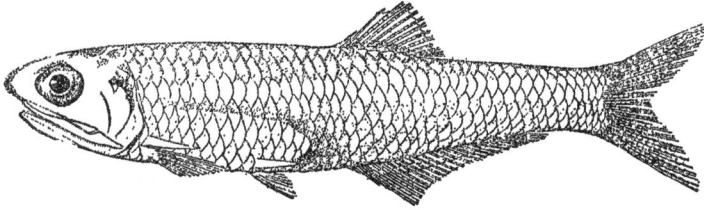

FIG. 61.—*Anchoviella mitchilli*, adult

and free from the isthmus; gill rakers rather long and slender, about 25 on the lower limb of first arch; scales thin, cycloid, deciduous, extending on the base of the fins; dorsal fin small, its origin notably nearer base of caudal than tip of snout, distance from tip of snout to dorsal 1.6 to 1.77 in body; caudal fin well forked; anal fin long and low, its origin near vertical from middle of base of dorsal; ventral fins small, inserted nearer origin of anal than base of pectorals; pectoral fins inserted low, 1.7 to 3.2 in head.

Color largely translucent, silvery; sides with a silvery lateral band, narrower than eye; back along base of anal and lower margin of caudal peduncle with dusky punctulations; cheeks and opercles silvery; fins pale or yellowish and usually with dark dots.

Many specimens of various sizes were preserved. The anchovies are readily recognized by their generally soft, delicate, more or less translucent appearance, large mouth, the prominently projecting, conical snout and the usually brilliant, silvery, lateral band. The present species differs from *A. epsetus* (the only other anchovy known from Chesapeake Bay) in the smaller size, narrower and less brilliant silvery lateral band, slightly longer anal, and in the relative position of the dorsal and anal fins. In *A. mitchilli* the origin of the anal is about under the middle of the base of the dorsal, whereas in *A. epsetus* the origin of the anal is only a little in advance of the base of the last ray of the dorsal.

A considerable variation in the depth of the body occurs among individuals of the same size, and a similar variation is especially great among individuals of various ages. In general the body becomes deeper with age. The larvæ are extremely slender, as the depth of specimens of about 16 millimeters in length is contained about 12 times in the body, 9 times in specimens 20 millimeters long, and 5.5 times in specimens 25 millimeters long. The range of variation in the depth of adult fish is shown in the foregoing description. The young, furthermore, differ from the adults in having a terminal mouth, a short rounded maxillary (which does not reach the margin of the opercle), and in the absence of a definite silvery lateral band. The fish does not acquire all the characters of the adult until a length of about 60 millimeters is reached.

The food of this anchovy (according to an examination made of 44 stomachs taken from specimens collected during the months of January, February, April, May, July, August, October, and November) consisted almost wholly of Mysis and copepods. The former appeared to be the principal food of the adult and the latter the sole food of the young. Other foods consisted of two small anchovies (indicating cannibalism), three small gastropods, and one isopod. No changes in the foods taken at different seasons of the year are apparent.

The spawning season, as shown by field observations, egg collections, laboratory dissections, and by the widely separated dates upon which very young specimens were taken is a prolonged one, extending through the months of May, June, July, and August. The eggs, according to Kuntz (1914, p. 14), are slightly elongate, the major axis being 0.65 to 0.75 millimeter and the minor axis is from 0.1 to 0.3 millimeter shorter. The eggs are pelagic and almost perfectly transparent. The period of incubation at summer temperatures is about 24 hours. The larvæ, when hatched, are only 1.8 to 2 millimeters in length. The rate of growth of the young fish is extremely difficult to follow as it is impossible to separate collections into age groups by lengths. This almost perfect gradation of size among the young no doubt largely results from the protracted spawning season.

The maximum size attained by this anchovy, as shown by measurements made of Chesapeake collections, is a little less than 4 inches, for the largest fish obtained were 97 millimeters long (weight, one-third ounce). The average length of this fish for Chesapeake Bay is about 3 inches. This anchovy occurs in schools, and it is the most abundant species of fish, with the probable exception of the silverside (*Menidia menidia*), that inhabits the bay. It is present at all seasons of the year. During cold weather it appears to frequent chiefly deep water, but during the summer it is generally common along the shores and even in muddy coves, and it also ascends fresh-water streams. It is sometimes taken in the Potomac River in fresh water near Bryans Point, about 12 miles below Washington.

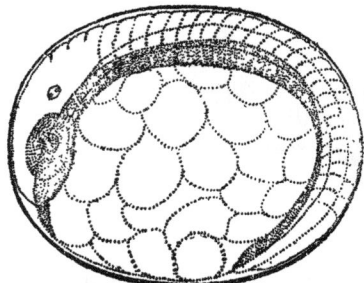

FIG. 62.—Egg with large embryo

In several Old World countries the anchovies are preserved like sardines and in various other ways. In America, however, they are much more important as food for other fish than as food for man. This species is not at all utilized by man in the Chesapeake region, yet it undoubtedly is of very great indirect commercial importance, as it appears to enter into the food of the larger predatory species more frequently than any other one species.

Habitat.—Atlantic and Gulf coasts, from Massachusetts to Texas; rarely northward to Maine.

Chesapeake localities.—(a) Previous records: Tolchester Beach, Riverside, Somerset Beach, lower Potomac, St. Jeromes, Md., and Cape Charles, Va. (b) Specimens seen or taken during the present investigation: From virtually all parts of the bay, from Havre de Grace, Md., to Lynnhaven Roads, Va.

41. Anchoviella epsetus (Bonnaterre). Anchovy.

Esox epsetus Bonnaterre, Ichthy., 1788, p. 175.
Stolephorus brownii Bean, 1891, p. 93; Jordan and Evermann, 1896–1900, p. 443.
Anchoviella epsetus Jordan and Seale, 1926, p. 396.

Head 3.6 to 4; depth 4.5 to 5.2; D. 14 to 16; A. 20 or 21; scales 38 to 40. Body moderately compressed; the ventral and dorsal outlines about evenly curved; the margin of abdomen little compressed; head moderate; snout conical, strongly projecting, 4.3 to 5.6 in head; eye 3.6 to 4.4; interorbital 3.8 to 5; mouth large, slightly oblique; maxillary long and sharply pointed, reaching nearly or quite to margin of opercle, 1.2 to 1.58 in head; teeth in the jaws small, sharply pointed; gill rakers rather long, about 20 on the lower limb of the first arch; scales thin, deciduous, extending on the base of the fins; dorsal fin small, its origin a little nearer base of caudal than tip of snout; caudal fin forked; anal fin of moderate length, its origin under the last rays of dorsal; ventral fins small, inserted equidistant from base of pectorals and the origin of the anal; pectoral fins moderate, 1.65 to 1.95 in head.

Color of fresh specimens pale gray and somewhat iridescent; the scales on back with dusky punctulations; sides with a broad, bright silvery band, a little narrower than eye; upper surface of head green and yellow; cheeks and opercles iridescent, silvery; fins mostly plain, the dorsal and caudal with more or less dusky.

Many specimens ranging in length from 46 to 150 millimeters were preserved and have been made use of in writing the foregoing description. The marks of distinction between this anchovy and *A. mitchilli*, the only other anchovy known from Chesapeake Bay, are indicated in the discussion following the description of the last-mentioned species.

The very young of this species, as in *A. mitchilli*, are much more slender than the adults. The great variation in the depth of the body among adults, noticed in *A. mitchilli*, however, is not apparent in the present species. The larvæ of both species are much alike, but those of *A. brownii* have the vent located correspondingly farther posteriorly, and as soon as the dorsal and anal fins have become differentiated the young of the present species may be recognized by the more posterior origin of the anal fin, which is under the base of the last rays of the dorsal, whereas in *A. mitchilli* it is under the middle of the dorsal base.

Nothing distinctive concerning the food of this anchovy can be said, as the examination of 16 stomachs shows that it is identical with that of *A. mitchilli*, consisting almost wholly of small crustaceans.

The spawning season of the present species appears to be identical with that of *A. mitchilli*. The eggs and embryology for *A. brownii* have not been described, and therefore such differences as may exist between the two species can not be given. The statements concerning the rate of

Fig. 63.—Larva 10 millimeters long

growth of the young fish, given in the discussion of *A. mitchilli*, appear to apply equally as well to *A. brownii*.

The maximum size attained by this anchovy, as shown by measurements made of Chesapeake collections, is 6 inches (weight, 1 ounce), and the average length is about 4½ inches (weight, one-half ounce). This anchovy, like *A. mitchilli*, occurs in schools. However, it is much less abundant in the bay as a whole than is *A. mitchilli*. *A. brownii* is common and at times very abundant in the southern parts of the bay. North of the mouth of the Rappahannock River it is comparatively rare. No specimens were taken during winter months, indicating that the species withdraws from the bay during cold weather.

The species has no direct commercial value in the region of the Chesapeake, but indirectly it must be of considerable importance because of the large numbers of these fish that are found in the food of the larger predaceous fishes.

Habitat.—Massachusetts to Uruguay.

Chesapeake localities.—(*a*) Previous record: Cape Charles, Va. (*b*) Specimens seen or taken during the present investigation: Annapolis, August 17, 1921, and Crisfield, Md., September 14, 1921; Lewisetta, August 4–8, 1921; lower Rappahannock River, July 25, 1921; Cape Charles, September 20–22, 1921; Buckroe Beach, October 5–10, 1921; Lynnhaven Roads, July 16, 1916, and September 27–30, 1921, and Ocean View, Va., September and October, 1922. All caught with collecting seines.

Order APODES. The eels

Family XXIII.—ANGUILLIDÆ. The common eels

Body very elongate, compressed posteriorly; head conical; opercles and branchial apparatus well developed; tongue distinct; teeth small, in cardiform bands on jaws and vomer; gill openings vertical; lateral line present; scales rudimentary, imbedded and placed at right angles to each other; dorsal and anal fins confluent around the tail; pectoral fins well developed.

31. Genus ANGUILLA Shaw.   Common eels

Mouth large, the lower jaw projecting; nostrils well separated, the anterior one tubular; dorsal and anal fins long, the origin of the dorsal not near the head.   A single species is known from American waters.

42. Anguilla rostrata (LeSueur).   Common eel; Fresh-water eel.

Muræna rostrata LeSueur, Jour., Ac. Nat. Sci., Phila., V, 1817, p. 81; Lakes Cayuga and Geneva, N. Y.
Anguilla bostoniensis Uhler and Lugger, 1876, ed. I, p. 181; ed. II, p. 153.
Anguilla rostrata Bean, 1883, p. 367; Bean, 1891, p. 94.
Anguilla anguilla Smith, 1892, p. 69.
Anguilla chrysypa Jordan and Evermann, 1896–1900, p. 348, Pl. LV, fig. 143; Smith and Bean, 1899, p. 183; Fowler, 1912, p. 52.

Head 7.4 to 8.5 in total length; depth variable 1.65 to 2.65 in head.   Body elongate, round anteriorly, compressed posteriorly; head of moderate length; snout rather pointed, 4 to 5.5 in head; eye 9.5 to 12; interorbital 6.5 to 8; mouth large, slightly oblique, reaching opposite middle of eyes; lower jaw projecting; anterior nostril situated on upper lip, provided with a tube; lateral line well developed, complete; scales small, imbedded, linear, arranged in groups, mostly at right angles to each other; origin of dorsal from 1.5 to nearly 2 times the length of head behind the gill slit; predorsal length of body 2.8 to 3.2 in total length; the dorsal and anal fins low, continuous with the caudal fin, which is round; pectoral fins moderate, proportionately longer in the adult than in the young; posterior margin round, the median rays longest, 2.65 (in adults) to 6 (in young) in head.

Color uniform greenish brown to yellowish brown above; white below.

FIG. 64.—*Anguilla rostrata*

Numerous specimens, all of the adult form, ranging from the glass stage, 48 millimeters long to adults of 740 millimeters (1⅞ to 29⅛ inches), are represented in the Chesapeake collection. The young or larval form, known as the leptocephalus, has not yet been taken in Chesapeake Bay nor within the immediate vicinity of the shore anywhere along the American coast.

The eel is an omnivorous feeder.   It is reported to be very destructive of other fish and even of one another and of the spawn of shad, herring, etc.   Stomachs of 31 Chesapeake Bay specimens, ranging from 14¼ to 29¼ inches in length, examined by Linton, had fed on crustaceans, annelids, fish, echinoderms, mollusks, and eel grass, named in the order of the abundance in which they were found in the stomachs examined, beginning with the most common one.   Thirteen stomachs of small examples, 2 to 8 inches in length, from various sections of the bay, examined by us, had fed mainly on amphipods and isopods.   Three stomachs also contained fragments of a segmented worm bearing bristles; one contained the siphon of a mollusk, another a portion of a tunicate, and three specimens contained plant leaves or stems or both.

The life history of the eel is very complicated but extremely interesting.   Although the female fresh-water eel spends most of its adult life in fresh water, it runs far out to sea to spawn.

Exactly where its spawning grounds are probably is not yet definitely known, although, with reference to the European and American eels, Dr. Johannes Schmidt is quoted [8] as saying:

---

[8] Fisheries Service Bulletin, Aug. 2, 1920, No. 63, p. 3 (United States Bureau of Fisheries).   For an extensive account of the life history of fresh-water eels see Johs. Schmidt, IV.—The Breeding Places of the Eel.   Philosophical Transactions, Royal Society of London, series B, vol. 211, 1922 (1923), pp. 179 to 208, pls. 17–18.

I think I am now able, after so many years' work, to chart out the spawning places of the European eel. The great center seems to be about 27° N. and 60° W. [southwest of Bermuda], a most surprising result, in my opinion. The American eel seems to have its spawning places in a zone west and south of the European, but overlapping. The larvæ of both species appear to pass their first youth together, but when they have reached a length of about 3 centimeters the one species turns to the right, the other to the left.

Neither is it definitely known, as far as the writers are aware, whether the eggs are pelagic or at what depth they are laid. The larvæ of such sizes as have been taken live at the surface, and it is now supposed that the eggs are pelagic.

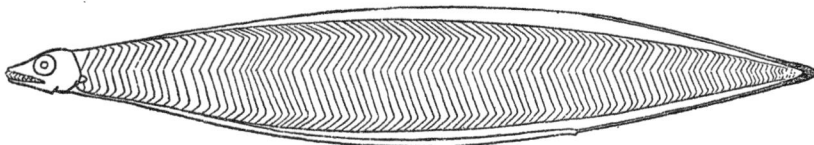

FIG. 65.—Leptocephalus stage, 49 millimeters long

The larva, or leptocephalus, is nearly as different in form from the adult as the caterpillar is from the butterfly. It was not until about 1895, or approximately 40 years after the leptocephalus was first described, that it was definitely determined that the leptocephalus was a young eel. The

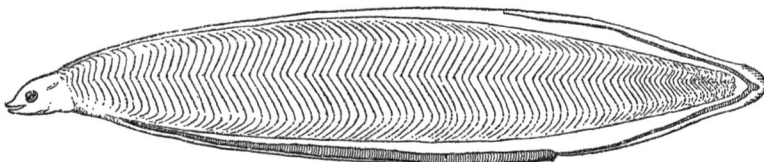

FIG. 66.—Leptocephalus stage, 55 millimeters long

larvæ are flat, ribbon shaped, tapering toward both ends. They are transparent, being entirely devoid of pigment, except in the eyes, and are readily overlooked in the collecting net. They have a well-developed mouth with very large teeth. In the process of metamorphosis the creature loses in length and depth but gains in width until the adult stage is attained.

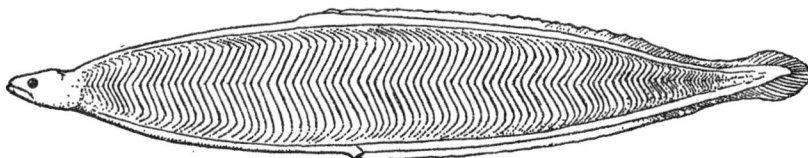

FIG. 67.—Leptocephalus stage, 58 millimeters long

The growth of the larvæ and metamorphosis take place while the young migrate from mid-ocean toward the shores. The smallest larvæ are taken nearest the spawning ground in mid-ocean and larger ones nearer the shores. By the time the eel reaches fresh water a complete metamorphosis has taken place. The length of fully developed larvæ, according to Schmidt (1912, p. 8),

FIG. 68.—Transition stage, 61 millimeters long

ranges from 60 to 85 millimeters, the length being reduced from 82 to 53 millimeters in the process of metamorphosis. The large larval teeth are lost, but they are replaced immediately by permanent ones. The dorsal and anal fins are produced farther forward; the pigment, however, is still largely wanting, as is indicated by the name "glass eel," and it forms very gradually. The following "glass eels" were collected at the surface in Chesapeake Bay by the *Fish Hawk* and the *Albatross*.

Date	Number	Size	Date	Number	Size
		Millimeters			*Millimeters*
Jan. 24, 1921	1	57	Jan. 17, 1916	14	50–60
Jan. 26, 1921	1	52	Jan. 18, 1916	2	48–50
Mar. 27, 1921	3	52–59	Feb. 19, 1922	2	54–57
Mar. 28, 1921	1	57			

Little is known of the rate of growth of the American eel or the approximate age at which it reaches sexual maturity and returns to salt water. Some attempts have been made by European investigators to determine the rate of growth and age of the European eel by an examination of the otoliths, the centra of the vertebræ, and the scales. Information derived from these studies indicate that the life of the eggs and larvæ may be two years (the American eel, however, is said to require only one year to pass through the metamorphosis), and that they probably have an average length of 7½ inches when 6 years old, and that at the age of 9 years the average length of the male is 14 inches and that of the female 15½ inches. Some evidence has also been obtained which indicates that maturity is reached at from 7½ to 9½ years of age. What becomes of the eel after it has spawned in mid-ocean also is not known, but it is generally supposed that it dies.

It is impossible to segregate eels into age groups based upon size, for all sizes are well represented in the catch of collecting seines. Commercial seines and nets have a mesh of such size that the smaller eels escape, so that data from this source are worthless as a means for determining rate of growth. However, our finest-meshed seines have caught enough very small eels so that some idea of the early growth may be had. It has already been pointed out that the "glass eel" that reaches our coast during the winter (January to March off the Chesapeake; as late as April in the Gulf of Maine) has a length of from 48 to 60 millimeters and is about 1 year old. The following catches of young eels that have passed the glass stage and possess the pigment of the adult have been made in Chesapeake Bay:

Date	Number measured	Inches	Date	Number measured	Inches
Apr. 26	2	2.4–3.1	Sept. 30	1	3
Aug. 8	3	2.3	Dec. 20	1	2.9
Sept. 1	1	2.5	Apr. 18	4	4.6–5.5
Sept. 19	3	2.7–3.9	Apr. 24	6	4.4–6.5
			July 23	3	6.9

Assuming that the fish have been grouped correctly in the foregoing table, the increase from one April to the next is from about 2½ to 5 inches, the greater length being attained when the eel is a little more than 2 years old.

The fresh-water eel is very common in the Chesapeake region, and in many places it is abundant in brackish water at the mouths of rivers and creeks.

The eel was considered so destructive of other fish that the legislature of Maryland, in 1888, passed an act and appropriated funds providing for the destruction of this fish. In 1892 and 1893 one-fourth of the funds appropriated for the use of the State fish commission was set aside for the destruction of the eel (Sudler and Browning, 1893, p. 27). The oak-split eel pot, baited with "fresh offal of any kind," was utilized in the capture of the eels. According to the report of the commissioners of fisheries of Maryland for 1892 and 1893 (p. 27), $3,413.25 were expended during these two years for destroying eels. A total of $80.77 was realized from the sale of the eels thus taken. No information concerning the number of eels destroyed or marketed is given. The work was discontinued in the following year. The effects upon the abundance of the eel and other fish, if any, which were brought about through the attempted destruction of eels, is not stated.

The following incident, which appears to be worthy of note on account of the difficulty with which an eel, because of its "slipperiness," is captured and retained after capture, was made by the junior author, whose field notes we quote:

WEEMS, RAPPAHANNOCK RIVER, VA., *July 25, 1921.*

An interesting incident was observed when a fish hawk caught an eel fully 20 inches long in several feet of water. The bird dropped the fish twice, recovering it each time, and several times it almost lost it. The hawk circled about several times with the eel before flying away. The fish could be seen plainly, squirming frantically to get away.

The eel reaches a large size in the Chesapeake region. Individuals 3 and 4 feet in length are seen occasionally in the markets. These large individuals are females, for the male probably does not exceed a length of 2 feet (Smith, 1907, p. 109). The flesh of the eel is firm and well flavored, but, owing to its resemblance to a snake, many people will not eat eels. In Europe this prejudice, if ever it existed, has been much more generally overcome, and the eel fisheries are of much greater importance than in America.

The eel is one of the important food fishes taken in the Chesapeake and its tributaries. During 1920 it ranked eleventh in quantity and tenth in value, the catch being 318,008 pounds, worth $33,704. Among the Maryland fishes it ranked seventh both in quantity and value, the catch being 197,293 pounds, worth $21,395. Of this amount, 77 per cent was taken in eel pots, 10 per cent in fyke nets, 7 per cent in pound nets, 4 per cent with spears, and 2 per cent with seines. In Virginia it ranked fourteenth in quantity and twelfth in value, the catch being 120,715 pounds, worth $12,309. Of this amount, 63 per cent was caught in eel pots, 21 per cent in fyke nets, and 16 per cent in pound nets.

This fish is taken principally in the vicinity of Rock Hall, Love Point, Oxford, and Crisfield, Md., and all western-shore rivers. A special fishery is conducted by means of eel pots in many of the tributaries of the bay. The majority of the pots are cylindrical in form with a conical entrance and are constructed of fine-meshed chicken wire. Sometimes many eel pots are attached to one cable, which may be from 500 to 2,000 feet long, similar to the gear used by lobster fishermen. The traps also are attached to the stakes of pound nets, for in such localities eels are attracted by the presence of dead fish.

Habitat.—Atlantic slope of North America from the Gulf of St. Lawrence to Panama, and in the West Indies, ascending fresh-water streams east of the Rocky Mountains.

Chesapeake localities.—(a) Previous records: Virtually all tributary streams. (b) Many specimens were taken during the present investigation from all parts of the bay and its tributaries.

Comparison of lengths and weights of Anguilla rostrata

[Actual lengths and weights of individual fish are given]

Inches	Ounces	Inches	Ounces	Inches	Ounces
2.7	0.03	20.8	11.1	23	12.4
3.1	.04	20.9	10.4	23	12.0
3.9	.07	21	11.3	23	14.1
7.9	.4	21.1	11.6	23	15.5
12.4	1.7	21.3	10.1	23.4	13.6
13	2.3	21.8	8.5	23.7	16.9
15.1	3.4	21.9	12.0	24.2	17.1
16.3	4.0	22	10.2	24.3	14.4
16.4	5.5	22.2	10.7	24.3	19.9
17.1	5.1	22.5	12.4	24.4	20.0
19	7.5	22.6	14.2	24.8	19.9
20.2	8.1	22.6	14.2	30	44.5
20.7	8.0	22.8	14.8	30.1	33.9

It will be noted that eels of the same length vary considerably in weight, due to the fatness of the individual. Thus, two fish, each 24.3 inches in length, differ in weight 5½ ounces, and two 30-inch fish differ by over 10 ounces. The 44½-ounce fish was abnormally fat. The sex of these eels was not determined, but it is not believed that a marked difference in weight due to sexual difference would occur between fish of the same length. As the male eel is said to reach a length of only 2 feet, the two largest fish of about the same length in the above table apparently were females and at the same time varied considerably in weight.

Family XXIV.—CONGRIDÆ. The conger eels

Body elongate; tongue largely free anteriorly; posterior nostril remote from the upper lip and placed near the eye; lateral line present; scales wanting; dorsal and anal fins confluent around the tail; pectoral fins well developed. A single genus of this family is represented in the fauna of Chesapeake Bay.

32. Genus CONGER Houttuyn. Conger eels

Mouth large, the upper jaw projecting; nostrils remote from each other, the anterior near tip of snout and tubular, the posterior near the eye; origin of the dorsal over or behind middle of the pectorals. A single species occurs in Chesapeake Bay.

43. Conger conger (Linnæus). Conger eel; Sea eel; Silver eel.

Muræna conger Linnæus, Syst. Nat., ed. X, 1758, 245; Mediterranean.
Conger oceanica Uhler and Lugger, 1876, ed. I, p. 180; ed. II, p. 153.
Leptocephalus conger Jordan and Evermann, 1896–1900, p. 354, Pl. LVII, fig. 148.

Head 6.35 to 7.3 in total length; depth 2.3 to 2.85 in head. Body elongate, anguilliform, round anteriorly, compressed posteriorly; head moderate; snout long, somewhat compressed, projecting beyond the mouth, 3.9 in head; eye 7.2 to 9; interorbital space 5.35 to 6.65; mouth inferior, slightly oblique; the gape reaching opposite posterior margin of pupil; anterior nostril situated on upper lip, provided with a short tube; lateral line complete, well developed; scales wanting; origin of dorsal over tips of pectorals, 0.4 to 0.7 length of head behind gill slit; predorsal length of body 4.6 to 5.15 in total length; dorsal and anal fins rather low, continuous with the caudal fin, which is narrowly rounded; pectoral fins moderate, round, 2.9 to 3.07 in head.

Color bluish gray, white beneath, dorsal fin with outer edge black, center light blue, dusky at base; anal pale with outer edge black; pectorals blue gray.

This eel is represented in the present collection by two specimens, 320 and 685 millimeters (12½ and 27 inches) in length. The conger eel is very similar in shape to the common fresh-water eel, from which it may be separated, however, by the projecting snout, the absence of scales, and by the very long dorsal fin, which has its origin about 0.4 to 0.7 the length of head behind the gill slit or over the tips of the pectoral fins, whereas in the fresh-water eel the origin of the dorsal is 1½ to 2 times the length of the head behind the gill slit and far behind the tips of the pectorals.

The conger eel feeds chiefly on fish, but it also takes other animal foods. (Smith, 1907, p. 112.) We have observed congers caught on the New Jersey coast on hooks baited with crab (Cancer) and clams (Macra, Mya). Cut fish is given as another bait.

The conger eel deposits its eggs at sea but evidently not as far from the shores as the fresh-water eel, for what were presumably conger eggs were collected by the *Grampus* 30 miles south of Nantucket Lightship, off the southern Massachusetts coast. These eggs were pelagic and about one-tenth inch (2.4 to 2.75 millimeters) in diameter when fertilized. (Eigenmann, 1902, p. 40.) "The number produced by a single eel is enormous, exceeding 7,000,000 in certain large European specimens. A conger in the Berlin aquarium, weighing 22.5 pounds, had ovaries weighing 8 pounds, which contained over 3,000,000 eggs (estimated)." (Smith, 1907, p. 111.)

The young, like the common eel, pass through a ribbonlike or leptocephalus stage. At this period the larvæ are recognized by the number of vertebræ and muscle segments, having 153 to 159 or more, whereas the American fresh-water eel has about 107 and the European fresh-water eel 114. The conger leptocephalus reaches a length of about 6 inches, while the American fresh-water eel reaches a length of only about 2½ inches and the European fresh-water eel only about 3 inches.

The conger eel seldom is caught in nets and nearly the entire catch is taken with hook and line. This eel is caught along our entire coast at least as far north as Woods Hole, where fish weighing up to 12 pounds are sometimes fairly common. The conger eel is a regular visitor along the Long Island and New Jersey coasts where from early summer to fall fish from 3 to 7 feet long and weighing up to 18 pounds are not uncommon. The usual length in the last-mentioned locality is 3½ to 6 feet, with a weight of 5 to 12 pounds. A Chesapeake specimen 27 inches in length weighed 1 pound 7½ ounces.

Uhler and Lugger (1876, p. 181) say of this fish for the Chesapeake region: "Common in the lower Potomac and in the parts of rivers within the reach of tide. Brought to our markets in large numbers and find a ready sale." At the present time, however, the conger is only a straggler in the bay, for many of the fishermen did not know the fish and we observed and collected only two specimens during 1921 and 1922. Because of its rarity, the conger obviously is of no commercial importance in the Chesapeake. Smith (1907, p. 112) remarks: "Although an excellent food fish, extensively sought and eaten in Europe and Asia, the conger supports no fishery in the United States and is sparingly utilized."

The conger eel attains a very large size in Europe, from whence a specimen of 128 pounds is recorded, and fish up to 60 pounds are not unusual. On our coast about 20 pounds appears to be the maximum. Only the female grows large and the male is thought to reach only 2½ feet in length and only several pounds in weight.

Habitat.—All warm seas except the eastern Pacific, inhabiting the Atlantic coast of America from Massachusetts to Uruguay.

Chesapeake localities.—(a) Previous records: Lower Potomac and within parts of rivers within the reach of tide (Uhler and Lugger, 1876). (b) Specimens in collection: Cape Henry, Va., February 19, 1922, beam trawl, depth 48 feet; Ocean View, October 11, 1922, 1,800-foot haul seine.

Order EVENTOGNATHI

Family XXV.—CATOSTOMIDÆ. The suckers

Body oblong or elongate, usually more or less compressed; head somewhat conical; nostril double; no barbels; mouth variable in size, usually protractile and with fleshy lips, jaws toothless; lower pharyngeal bones falciform, armed with a single row of numerous comblike teeth; branchiostegals 3; gill membranes somewhat connected with the isthmus, restricting the gill openings to the sides; gills 4, a slit behind the fourth; scales cycloid, wanting on the head; the fins without true spines; adipose fin wanting; ventral fins abdominal. The suckers comprise a large family of freshwater fishes. Only a few of the species venture into brackish water and none of them enter salt water.

KEY TO THE GENERA

a. Lateral line interrupted or wanting; scales large, 35 to 50 in a lateral series.
  b. Lateral line entirely wanting; species small_____Erimyzon, p. 117
  bb. Lateral line present, at least in adults, more or less interrupted; species larger
  _____Minytrema, p. 118
aa. Lateral line complete and continuous; scales small, 55 or more in a lateral series.
  _____Catostomus, p. 119

33. Genus ERIMYZON Jordan. Chub suckers

Body oblong, compressed; mouth subinferior; upper lip protractile; lower lip large, plicate, V-shaped; gill rakers long; pharyngeal bones weak, with small slender teeth; lateral line wanting; scales rather large, somewhat crowded anteriorly; dorsal fin short, with about 11 rays; the anal fin still shorter; caudal fin scarcely forked, but usually more or less concave. A single species of wide distribution in fresh and slightly brackish water is known.

44. Erimyzon sucetta (Lacépède). Chub sucker; "Mullet"; "Horned sucker."

Cyprinus sucetta Lacépède, Hist. Nat. Poiss., V, 1803, p. 606; South Carolina.
Moxostoma oblongum Uhler and Lugger, 1876, ed. I, p. 165; ed. II, p. 141.
Erimyzon sucetta Jordan and Everman, 1896–1900, p. 185, Pl. XXXVI, fig. 89; Smith and Bean, 1899, p. 181.
Erimyzon sucetta oblongus Fowler, 1912, p. 53.

Head 3.5 to 4.1; depth 3.1 to 3.9; D. 9 to 12; A. 7 or 8; scales 36 to 45. Body oblong, compressed, back elevated; head rather short; snout short, 2.5 to 3.2 in head; eye 3.8 to 5.8; interorbital space 2.2 to 2.6; scales large, closely overlapping, at least anteriorly, 13 to 15 in a transverse series; dorsal fin a little higher than long, situated over the ventrals; caudal fin with a more or less concave posterior margin; anal fin very small, higher than long, its origin slightly nearer the base of caudal than the base of ventrals; ventral and pectoral fins moderate and of about equal size.

Color variable with age and environment; adults nearly uniform brownish olive above, intermixed with pinkish anteriorly, and everywhere with a coppery luster; pale underneath; fins all more or less dusky, sometimes reddish. The young with a black lateral band, later becoming broken into blotches, forming transverse bands and disappearing entirely with age.

This fish was not taken in brackish water during the present investigation, but it is reported from brackish water from the vicinity of Baltimore by Uhler and Lugger and for that reason the species is included in the present work. The chub sucker is readily recognized by the small dorsal and anal fins, the absence of the lateral line, and the thick lower lip, which contains many folds and the halves of which meet anteriorly in a V-shaped angle. The young, in general appearance, are very similar to some of the cyprinoid minnows. The males of this species, like many of the cyprinoid minnows, develop tubercles on the snout during the breeding season.

The chub sucker is a bottom feeder and largely herbivorous, yet it bites readily at a small hook baited with a piece of meat or earthworm. Spawning takes place in the spring. The species reaches a length of only about 10 inches; its flesh is bony and not of good flavor. It is common, although not abundant, in the fresh waters of the Chesapeake region. During cold weather, according to Smith and Bean (1899, p. 181), it ascends streams to the head waters, where it is taken and considered a good winter fish for the table.

Habitat.—Great Lakes, the Mississippi Valley, and seaboard streams from Maine to Texas.

Chesapeake localities.—(a) Previous records: Many fresh-water streams and in brackish water of the Patapsco River. (b) Specimens in the present collection: None. We have records of specimens taken near Havre de Grace, Md., in April, May, October, and December. The headwaters of Chesapeake Bay are slightly brackish from late fall until late winter.

34. Genus MINYTREMA Jordan.   Spotted suckers

Body elongate, compressed; mouth inferior; upper lip freely protractile; lower lip plicate, the halves forming an acute angle anteriorly; air bladder in two parts; lateral line interrupted in adults, wanting in young; scales rather large, about 43 to 47 in a longitudinal series; dorsal fin high and short, with about 12 rays; caudal fin moderately forked.

45. Minytrema melanops (Rafinesque).   Spotted sucker; Striped sucker.

Catostomus melanops Rafinesque, Ichthyologia Ohiensis, 1820, p. 57; Ohio River.
Minytrema melanops Jordan and Evermann, 1896–1900, p. 187, pl. XXXVI, fig. 90.

Head 4.4 to 4.9; depth 3.7; D. 15 or 16; A. 9 or 10; scales 43 to 46. Body elongate, compressed; upper anterior profile evenly and gently convex; head rather small; snout conical, 1.2 to 2.6 in head; eye 5.8; interorbital space 2.1 to 2.5; mouth inferior; the lips with strong folds, the lower lip much broader than the upper; scales large, cycloid, 12 longitudinal rows between the origin of dorsal and base of ventrals; lateral line present, complete; dorsal fin a little higher than long, its origin about equidistant from tip of snout and end of base of anal, its outer margin gently concave; caudal fin forked, the lobes pointed; anal fin much higher than long, its origin slightly nearer base of caudal than base of ventrals, the fourth or fifth ray the longest, the posterior rays decreasing rapidly in length; ventral fins moderate, inserted under the end of anterior third of base of dorsal; pectoral fins inserted less than an eye's diameter behind margin of opercle, 1.1 to 1.2 in head. Color of preserved specimens bluish-gray above, pale below; scales on sides with dark areas at base, which are deeper than long in large individuals, roundish in medium-sized individuals, and indistinct in young; dorsal and caudal slightly grayish, with darker margins; other fins plain, colorless.

A specimen 420 millimeters (16½ inches) long, weighing, when fresh, 1¾ pounds, taken in brackish water, and four small specimens, all of equal size, 85 millimeters (3⅜ inches) long, taken in fresh water, occur in the Chesapeake collection. We have compared these fish with specimens from Indiana and Texas. It was noticed that the body becomes much more compressed and deeper with age and size, the folds on the lips become more pronounced, and the dark spots on the scales on the sides of large specimens are much less distinct than they are in specimens 6 to 10 inches in length. On the large example at hand the dark spots are much deeper than long; in the smaller specimens they are roundish. In specimens 5 inches and less in length these spots are indistinct or absent. The lateral line is not always complete and at times it is wanting. This character, however, does not appear to be correlated with age, as has been supposed. The spotted sucker usually is readily distinguished from all other suckers by the presence of dark spots on the scales, forming dark longitudinal lines. It also differs from related suckers in having the outer margin of the dorsal fin concave, and in the reduced number of longitudinal rows of scales on sides between the dorsal and ventral fins. This fish is known to reach a length of 18 inches.

The striped sucker evidently is rare in the Chesapeake Bay vicinity, as we are unable to find a record of its previous capture and the specimens in hand are the only ones seen in the field by the collectors. The species appears to be mainly a creek and small-river fish. However, the large specimen at hand was caught in brackish water in the narrows off Spesutie Island. "From the little that is known of its food we may surmise that it lives largely on mollusks and insect larvæ." (Forbes and Richardson, 1908, p. 83.) We are unable to find anything in the literature on the spawning and breeding habits of this sucker, and it is probable that nothing distinctive is known about it.

Habitat.—"Great Lakes region to North Carolina (Cape Fear River) and west to Texas; rather common westward." (Jordan and Evermann, 1896–1900.)

Chesapeake localities.—(*a*) Previous records: None. (*b*) Specimens in collection: From Spesutie Island near Havre de Grace, Md., 300-foot seine, Nov. 12, 1921, salinity, 1.53 per mille; Susquehanna River, Havre de Grace, Md., 30-foot seine, Aug. 27, 1921, water fresh.

35. Genus CATOSTOMUS LeSueur. Fine-scaled suckers

Head somewhat elongate; mouth inferior, the upper lip thick, protractile, papillose, lower lip greatly developed, incised behind, forming two lobes; scales small, 50 to 115 in a lateral series; lateral line well developed, air bladder with two chambers; dorsal fin with 14 to 19 rays.

46. Catostomus commersonii (Lacépède). Common sucker; White sucker; Mud sucker; Black mullet.

Cyprinus commersonii Lacépède, Hist. Nat. Poiss., V, 1803, p. 502; locality unknown.
Catostomus communis Uhler and Lugger, 1876, ed. I, p. 162; ed. II, p. 138.
Catostomus teres Bean, 1883, p. 367.
Catostomus commersonii Jordan and Evermann, 1896–1900, p. 178, Pl. XXXIV, fig. 83.

Head 4.08 to 4.35; depth 4.45 to 4.82; D. 14 or 15; A. 8; scales 63 to 67. Body elongate, little compressed; head quadrate, a little deeper than broad; snout conical, 1.9 to 2.15 in head; eye 4.4 to 6; interorbital 2.4 to 2.55; mouth inferior; lips papillose, the lower one broader than the upper; scales rather small, reduced in size anteriorly, about 20 longitudinal rows on sides between the dorsal and ventral fins; lateral line complete; dorsal fin about as long as high, the outer margin nearly straight, its origin a little nearer tip of snout than base of caudal; caudal fin moderately forked; anal fin much shorter, but higher than the dorsal, its origin about equidistant from base of ventrals and base of caudal; ventral fins short, inserted under middle of base of dorsal; pectoral low, 1.4 in head.

Color bluish-green above, pale below; dorsal and caudal fins more or less dusky, the other fins more or less orange. Spring males with a more or less distinct rosy lateral band. Young darker gray, mottled with black; the blotches sometimes more or less confluent and occasionally forming a lateral band.

The Chesapeake collection contains three specimens, respectively, 215, 222, and 235 millimeters (8½, 8¾, and 9¼ inches) in length, which were taken in slightly brackish water. These three and some smaller specimens from fresh water form the basis for the above description. This sucker is distinguished from all other suckers of the vicinity by the small scales, of which there are from 63 to 67 in a lateral series and about 20 longitudinal rows on the side between the dorsal and ventral fins. The scales are reduced in size anteriorly and appear crowded.

The alimentary canal is long and somewhat convoluted, without a sharp differentiation between the stomach and intestine. A specimen 8½ inches in length had an alimentary canal 17½ inches long. The food of this sucker, according to Smith (1907, p. 73), consists of insects, mollusks, worms, and "other animals." Forbes and Richardson (1908, p. 85) point out that the thick pharyngeal jaws with a relatively small number of pharyngeal teeth, the lower ones of which are much thickened and expanded at the crowns, constitute a crushing and grinding apparatus strongly suggesting that a molluscan diet prevails. The specimens at hand had fed abundantly on plankton, consisting mainly of Cladocera, copepods, and Ostracoda. No insects or insect larvæ were noticed. The earthworm is the commonly used bait for hook and line fishing for this sucker.

Referring to the spawning habits of this sucker, Uhler and Lugger (1876, ed. I, p. 162, and ed. II, p. 138) say: "In early summer these fish build their nests of piles of sand and stones, and shortly afterwards their dead bodies may sometimes be found in dozens along the shores of streams such as Gwynns Falls, Md." The death of adult fish after spawning is not reported by other observers. Smith (1907, p. 73) states that in North Carolina spawning occurs in spring in the headwaters of small streams. According to Forbes and Richardson this sucker prefers riffles or swiftly flowing water for depositing the spawn. The writers have seen this sucker ascend small creeks in the spring in schools, when the splashing of water on the shallow riffles could be heard from a distance. It is then frequently possible to approach quietly with a torch, and when the light once is over the fish they become quiet and may be gigged easily.

This fish, although quite bony, is generally considered a fairly good food fish. Uhler and Lugger (1876, ed. I p. 162, and ed. II, p. 138), however, say: "The rank taste of the flesh renders it distasteful to many persons, but in the interior sections of the western shore (Maryland) it is generally eaten by the people."

The sucker is found in the fresh waters of the Chesapeake Bay region throughout the year, and according to Smith and Bean (1899, p. 181) it is taken in the Potomac and its tributaries, chiefly in winter, with seines and fyke nets. This species reaches a length of about 2 feet and a weight of about 5 pounds.

Habitat.—"Streams and ponds from Quebec and the Great Lakes to Montana, Colorado, and southward to Missouri and Georgia. * * * Excessively abundant from Massachusetts west to Kansas." (Jordan and Evermann, 1896–1900.)

Chesapeake localities.—(a) Previous records: Apparently all from strictly fresh water. (b) Specimens in collection: From Susquehanna River, Havre de Grace, Md., 300-foot seine, November 9, 1921, salinity 1.53 per mille.

Family XXVI.—CYPRINIDÆ. The minnows and carps

Body more or less elongate, compressed or rounded; margin of upper jaw formed only by the premaxillaries; lower pharyngeal bones supporting one to three series of teeth, the teeth few in number and sometimes differing in number on the two sides; snout sometimes with two to four small barbels; gill membranes joined to the isthmus; pseudobranchiæ present; branchiostegals 3; body scaly; head naked; dorsal fin short; ventral fins abdominal. During the breeding season the males often develop tubercles on the snout, and in some of the species they become brightly colored.

KEY TO THE GENERA

a. Mouth with four barbels; dorsal and anal each with three spines, the third in each fin enlarged and serrated behind_____Cyprinus, p. 121
aa. Mouth without barbels; dorsal and anal fins without spines.
  b. Body in adult much compressed; belly behind ventrals compressed to a keel; lateral line strongly decurved; anal fin long, with about 14 to 16 rays; origin of dorsal behind ventrals_____Notemigonus, p. 123
  bb. Body not greatly compressed; belly rounded; lateral line only slightly decurved; anal fin short, with about 8 to 10 rays.
    c. Peritoneum black; alimentary canal long, more than three times the length of body_____Hybognathus, p. 124
    cc. Peritoneum pale; alimentary canal short, less than twice the length of body__Notropis, p. 125

36. Genus CYPRINUS Linnæus. Carps

Body robust, compressed; mouth moderate, inferior, with four barbels; snout blunt; scales large (wanting in the leather carp); lateral line complete; dorsal fin long, with three spines; anal fin short, also with three spines; the third spine of dorsal and of anal serrated behind.

47. Cyprinus carpio Linnæus. The carp.

Cyprinus carpio Linnæus, Syst. Nat., ed. X, 1758, p. 320; Jordan and Evermann, 1896, p. 201; Fowler, 1912, p. 53.

This fish has been domesticated and as a consequence it is subject to much variation; numerous varieties have resulted, which vary greatly in the depth of the body, the relative length of the head, the length of the fins, and especially in the number and arrangement of the scales. One variety has only a few scales on the back or is wholly naked and possesses a thick, soft skin. This variety is known as the "leather carp." Another variety has enlarged scales on the sides, often in only a few rows. Such fish are known as "mirror carp." A third variety is fully and normally scaled. This variety is the "scale carp" and probably is most like the original "wild" species.

Owing to these great and numerous variations, no attempt is made to offer a technical description. In general, the body is elongate and compressed, the back being elevated. The head is rather low and small. The dorsal fin is long, consisting of three spines and usually from 20 to 23 soft rays. The anal fin, too, has three spines, and it has only about six soft rays. The third spine in both fins is enlarged and has a rough posterior edge. The upper jaw has two barbels on each side, which readily distinguish the carp from all American forms.

The carp is a native of the temperate parts of Asia, especially of China, from whence it was introduced into Europe, Java, and also into America. Exactly when the first carp were brought to America has been a subject for discussion. It is claimed that they were introduced into the Hudson River many years before they were brought in by the United States Fish Commission in 1877, but this report apparently never was definitely verified. A few specimens of scale carp were brought from Germany by a Mr. Poppe, of Sonoma, Calif., some years before they were introduced by the United States Fish Commission. In the Chesapeake vicinity, however, the carp was first introduced in 1877, when 227 leather and mirror carp and 118 scale carp were brought directly from Germany by a representative of the United States Fish Commission and placed in ponds especially prepared for their reception in Druid Hill Park, Baltimore, Md. About a year later several carp ponds were constructed in Washington, and a part of the brood stock originally placed in Druid Hill Park was transferred to Washington. Other small lots were imported in 1879 and 1882 and placed in the aforementioned ponds. Young fish were shipped from these sources to various applicants, resulting in the general distribution of the carp to all suitable waters of the United States.

The expectations from the introduction of the carp were great. Prof. S. F. Baird, Commissioner of Fisheries, stated at the time of introduction (1879, pp. 41 and 42):

I have for a long time attached much importance to the introduction of carp into the United States of America as supplying an often-expressed want of a fish for the South, representing the more northern trout and capable of being kept in ponds. In the carp this desideratum is amply met, with the additional advantage that the same water will furnish a much larger amount of fish food in the aquatic plants, roots, seeds, etc., to be found, while feeding may be accomplished by means of leaves, seeds, pieces of cabbage and lettuce, by crumbs of bread, or by boiled corn and potatoes or other cheap substances. * * * There is no ditch, or pond, or milldam, or any muddy, boggy spot capable of being converted into a pond of more or less size that will not answer for this fish. Except for unforeseen casualties, I fully believe that within 10 years to come this fish will become, through the agency of the United States Fish Commission, widely known throughout the country and esteemed in proportion.

Prof. Baird's expectations concerning the multiplication and distribution of this fish have been fully met, but the fish is not esteemed in the same proportion. In the markets of the Chesapeake Bay region, as elsewhere, it is considered an inferior food fish. During recent years, however, it has gained in favor, and the demands for it are increasing. Throughout the Mississippi Valley it is commercially one of the most important food fishes.

It has attained a small commercial importance in the lower Potomac, where the pound-net fishermen catch them in April and May. Fishermen of Lewisetta, Va., brought in six on one April 25, the largest weighing 25 pounds. At Love Point, Md., haul seiners consider it one of their most profitable fish during May. It is also taken, although sparingly, in the lower Patuxent and Choptank Rivers.

The carp is omnivorous, but its principal food probably consists of plants.   Hessel (1878, p. 865) says: "The carp lives upon vegetable food as well as upon worms and larvæ of aquatic insects, which it turns up from the mud with the head; it is very easily satisfied and will not refuse the offal of the kitchen, slaughterhouse, and breweries, or even the excrement of cattle and pigs."   Three stomachs examined by Linton from specimens taken at Havre de Grace, Md., contained only vegetable matter, mostly the fruit of eelgrass.

In this country the carp has not infrequently been accused of destroying our native fishes. In some localities this has become a popular belief, but investigators have been unable to find much incriminating evidence.   It is a well-known fact that since the introduction of the carp our native fishes have become fewer in many streams, and that the carp is becoming abundant.   A very natural and logical conclusion, with such evidence alone at hand, is that the carp is responsible for the decrease of the native species.   It must be remembered, however, that a similar decrease has taken place in many of our marine fishes.   For such decrease other causes must be sought, as no introduction of foreign species of strictly marine fish has taken place.   It is not argued that the carp does not at times, through the uprooting of vegetation, destroy nests of other fish, nor that it at times eats the spawn of other fish, or that it destroys some of the young fish of other species; on the other hand it must be remembered that our native species, too, prey upon each other and upon the carp, very probably to a much greater extent than the carp preys upon them.   The carp, being largely herbivorous, gains much of its sustenance from plants; whereas many of our native fishes are strictly carnivorous, requiring animal foods, and where the young carp is present it not infrequently furnishes a considerable portion of the food of the carnivorous fishes.   From this standpoint the presence of carp appears to be a distinct advantage.   It seems necessary, in the light of our present knowledge, to seek the reason for the decline in our fresh-water food fishes elsewhere.   For a complete and admirable account of the carp and the various accusations that have been made against it in America see the report of the Bureau of Fisheries for 1904, pages 523 to 641, under the title "The German carp in the United States," by Leon J. Cole.   Overfishing, fishing during the spawning season, the construction of obstructions in streams (prohibiting the free passage of fish to and from their natural spawning and feeding grounds), and, most important of all, the pollution of streams are undoubtedly the important factors in bringing about the diminution of our native food fishes.

The carp prefers rather quiet waters that support an abundance of vegetation, but it is not limited thereto, as it is not infrequently taken in rather swiftly flowing streams.   Although the carp is essentially a fresh-water fish, it does enter brackish water, and in the Old World, according to Hessel (1878, p. 869), it even frequents salt water.   In the Chesapeake the carp is found abundantly in fresh water, sparingly in brackish water, but not at all in the salter parts of the bay.

Spawning takes place in the spring and may extend over a considerable period of time.   The eggs are deposited among vegetation; they are adhesive and usually adhere to vegetation in lumps. Field notes made by Lewis Radcliffe state that the ovaries of 4 to 5 pound carp contain from 400,000 to 500,000 eggs and that a 16½-pound fish contained ovaries weighing 5 pounds with over 2,000,000 eggs.   During warm weather the eggs hatch in from 12 to 16 days.   Under favorable conditions the young grow rapidly.   Hessel (1878, p. 873), in speaking of carp culture in Europe, says: "The normal weight which a carp may attain to in three years, whether it be scale carp, mirror carp, or leather carp, is an average of from 3 to 3¼ pounds; that is, a fish which has lived two summers, consequently is 18 months old, will weigh 2¾ to 3¼ pounds the year following." The carp is said to attain a great age—100 to 150 years—and a weight of 80 to 90 pounds, but such statements generally are based upon insufficient evidence.   Hessel (1878, p. 874) says: "It is a well-known fact that two large carps, weighing from 42 to 55 pounds, were taken several years ago on one of the Grand Duke of Oldenburg's domains in northern Germany."   Smith (1907, p. 106) makes the following statement: "The carp attains a relatively large size, examples weighing upward of 60 pounds being known in Europe and fully 40 pounds in the United States, although full sexual maturity is attained by the second or third year, when the fish weigh only 3 or 4 pounds."

The following weights were secured: Length, 17½ inches, 2 pounds 12 ounces; 20 inches, 4 pounds 8 ounces; 22½ inches, 6 pounds 5 ounces; 26 inches, 9 pounds 3 ounces.

Habitat.—Temperate Asia; introduced into Europe, Java, England, United States, Canada, Mexico, Costa Rica, Ecuador, the Hawaiian Islands, etc.

Chesapeake localities.—(*a*) Previous record: Apparently none from salt or brackish water. (*b*) Specimens in present collection: From brackish water, Spesutie Island, Havre de Grace; Love Point; and Blackistone Island, Md. Highest recorded salinity, 15.66 per mille.

37. Genus NOTEMIGONUS Rafinesque. Roaches

Body strongly compressed; back and belly curved; belly behind ventrals forming a keel; head small, conic; mouth small, oblique; barbels wanting; pharyngeal teeth 5—5; alimentary canal short, not much longer than the body; scales moderate; lateral line complete, decurved; dorsal origin behind ventrals; anal fin rather long, with 13 or more rays.

48. Notemigonus crysoleucas (Mitchill).   Golden shiner; Shiner; "Dace"; Chub; Bream.

Cyprinus crysoleucas Mitchill, Rept., Fish., N. Y., 1914, p. 23; New York.
Stilbe americana Uhler and Lugger, 1876, ed. I, p. 171; ed. II, p. 145.
Notemigonus crysoleucas Bean, 1883, p. 367; Smith and Bean, 1899, p. 182.
Abramis crysoleucas Jordan and Evermann, 1896–1900, p. 250, Pl. XLV, fig. 111; Fowler, 1912, p. 52.

Head 4 to 4.75; depth 2.85 to 4.25; D. 9 or 10; A. 14 to 16; scales 46 to 52.   Body in adult deep, rather strongly compressed, the back elevated and the ventral outline strongly decurved, more elongate and not as strongly depressed in young; head small, somewhat depressed above; snout short, blunt, its length 3.55 to 4.6 in head; eye 2.55 to 4.1; interorbital 2.25 to 2.9; mouth very oblique, the lower jaw slightly in advance of the upper; maxillary failing to reach anterior margin of eye; pharyngeal teeth in one row, usually with five teeth, occasionally with only four, each tooth with a prominent, nearly right-angled hook at the tip; scales moderate, rather deep in adult, 23 to 25 rows in advance of dorsal; lateral line complete, decurved; dorsal fin rather small, the anterior rays longest, reaching past the posterior rays when deflexed, the origin of fin a little nearer upper anterior angle of gill opening than base of caudal; caudal fin forked, both lobes pointed; anal fin rather long, the outer margin concave, its base 1.2 to 1.55 in head; ventral fin inserted nearly an eye's diameter in advance of dorsal, reaching origin of the anal in the young, proportionately shorter in the adult; pectoral fins pointed, the upper rays longest, 1.05 to 1.3 in head.

Color in adult bluish-green above, with metallic luster, gradually merging into bright silvery on lower part of sides; upper surface of head brownish; fins plain or sometimes yellowish and occasionally dusky.   A gravid male, 6 inches long, had a pale yellow dorsal and caudal and bright yellow anal, ventral, and pectoral fins.   Smith (1907, p. 89) describes a fish 7¾ inches long as having, in addition to the yellow color, crimson ventrals and the anal dull orange with a black margin. The young have less of the metallic luster and they have a distinct black lateral band, extending from the eye to the base of the caudal.

Many specimens of this species were preserved, ranging from 33 to 215 millimeters (1⅝ to 8½ inches) in length.   This minnow is locally very abundant, occurring in the tide waters principally in the upper parts of Chesapeake Bay, whether fresh or brackish, and on various kinds of bottom, but more usually where vegetation is present.   The adult of this minnow is readily recognized by the very oblique mouth, deep, compressed body, the long anal fin, the strongly decurved lateral line, and by the bright golden and silvery colors.   The young, however, are not so readily distinguished, for they are not much deeper than other minnows of related genera, and they have a black lateral band like many of the species of this family.   The strongly oblique mouth and the long anal fin serve as the most reliable characters in separating the young from related minnows. The scales in advance of the dorsal fin are somewhat reduced, from 22 to 25 rows crossing the back in front of the origin of the dorsal.   In most of the related minnows the scales are larger, and fewer rows cross the back in advance of the dorsal fin.   The peritoneum in this species is silvery with dusky punctulations.   The air bladder is large and has a constriction a little in advance of the middle of its length, from which arises a very small tube, which extends forward to the throat. The alimentary canal is about as long as the total length of the fish.

The food in six specimens examined consisted of algæ, fragments of higher plants, and débris. Many grains of sand, probably taken by accident, also were present in some of the stomachs examined.   Linton examined five stomachs and found amphipods, mollusks, and débris.

Spawning takes place during the spring.   Gravid fish were taken at Havre de Grace, Md., on May 8 to 10, 1922.

This species reaches a larger size than the other minnows of this family occurring in the Chesapeake vicinity. The maximum length given in various publications is 1 foot, but the largest individual taken in the Chesapeake was 8½ inches long. This minnow is considered excellent bait in the South for black bass and pike or pickerel. The large individuals are used for home consumption and are said to make good pan fish. When confined in cisterns or shallow wells the golden shiner feeds on mosquito larvæ and successfully prevents mosquito production. The weights of Chesapeake Bay fish were as follows:

Length, in inches	Weight, in ounces	Length, in inches	Weight, in ounces
5	0.7	6¾	2.0
5¼	.8	7	2.3
5½	.9	7¼	2.7
5¾	1.3	7½	3.1
6	1.4	8	3.7
6¼	1.6	8½	4.8

Habitat.—Nova Scotia, west to the Dakotas and south to Florida and Texas on both sides of the Alleghanies, frequenting weedy ponds and sluggish streams.

Chesapeake localities.—(a) Previous records: "Maryland" (Uhler and Lugger, 1876), Havre de Grace, Md. (Bean, 1883), Little Bohemia Creek, Bohemia Mills, Bohemia Bridge, Elk Neck, North East, Stony Run, Conewingo, Susquehanna River, and Broad Creek (Fowler, 1912). (b) Specimens in collection: From Havre de Grace, Baltimore, Annapolis, Love Point, Solomons Island, Md., and Lewisetta, Va., taken with 30 and 300 foot collecting seines and in one instance with a pound net from April to November. Highest salinity 14.4 per mille.

38. Genus HYBOGNATHUS Agassiz. Shiners; Gudgeons

Body elongate, somewhat compressed; mouth horizontal; the jaws normal, the lower one with a slight protuberance in front, the upper one protractile; no barbels; pharyngeal teeth 4—4; alimentary canal elongate, three to ten times the length of body; peritoneum black; scales large; lateral line complete; dorsal fin inserted in advance of ventrals; anal short.

49. Hybognathus nuchalis Agassiz. "Gudgeon"; Silvery minnow.

Hybognathus nuchalis Agassiz, American Jour. Sci. and Art., 1855, p. 224; Qunicy, Ill.
Hybognathus regius Uhler and Lugger, 1876, ed. I, p. 177; ed. II, p. 150.
Hybognathus nuchalis Jordan and Evermann, 1896–1900, p. 213; Smith and Bean, 1899, p. 182; Fowler, 1912, p. 52.

Head 4.1 to 5; depth 3.56 to 4.95; D. 9 or 10; A. 9 or 10; scales 37 to 40. Body rather slender, compressed; caudal peduncle moderate, its depth 1.5 to 2.6 in head; head rather long and low; snout conical, 3 to 3.5 in head; eye 3.05 to 3.35; interorbital space 2.35 to 3.35; mouth small, a little oblique, slightly inferior; maxillary not quite reaching eye; pharyngeal teeth in one row, consisting of four teeth; scales moderate, 13 or 14 rows crossing the back in advance of dorsal fin; lateral line complete, slightly decurved; origin of dorsal a little nearer tip of snout than base of caudal, the anterior rays of fin longest, reaching past the posterior ones when deflexed; caudal fin moderately forked, the lobes of about equal length; anal fin similar to the dorsal, its origin about 1.5 times diameter of eye behind the end of base of dorsal; ventral fins moderate, inserted a little behind vertical from origin of dorsal; pectoral fins pointed, the upper rays longest, 1.05 to 1.35 in head.

Color greenish above, sides silvery, lower parts pale. Some specimens have a slight indication of a plumbeous lateral band, at least posteriorly. The fins pale, the dorsal and caudal slightly dusky.

Nine specimens of this species, ranging from 70 to 157 millimeters (2¾ to 6¼ inches) in length, were taken in brackish water in the upper part of Chesapeake Bay. The adults of this species are very similar to *Notropis hudsonius amarus*, from which, however, they may be distinguished by the black peritoneum and the long convoluted intestine. *N. hudsonius amarus*, furthermore, usually has an indication of a black spot at base of caudal, which is never present in *H. nuchalis*.

The food of this species consists of plants. Only fragments were present in stomachs examined. This fish spawns early in the spring. Large specimens taken in November already have the ovaries somewhat distended with eggs easily visible to the unaided. eye.

This minnow reaches a somewhat larger size than *N. hudsonius amarus*, the largest specimen at hand being 6¼ inches in length, which is probably the maximum size attained. This minnow is used to a limited extent for food and also for bass bait. The food it provides for the larger predatory fishes, however, constitutes its chief economic importance. It is said to be abundant in the fresh-water streams of the Chesapeake region and is taken in company with *Notropis hudsonius amarus*.

Habitat.—New Jersey and southward to Texas and in the Mississippi Valley northward to the Dakotas.

Chesapeake records.—(*a*) Previous records: None definitely from brackish water. (*b*) Specimens in collection from the vicinity of Havre de Grace, Md. (Northeast River, Susquehanna River, and Spesutie Island), 30-foot seine, August 27 to 31 and November 10 to 12, 1921; highest salinity 2.23 per mille.

39. Genus NOTROPIS Rafinesque. Shiners

Body elongate, subcylindrical or compressed; abdomen rounded; mouth terminal or slightly inferior; no barbels; pharyngeal teeth in one or two rows, the main row with four teeth on each side; lateral line present and usually complete; scales rather large; vertical fins short; the dorsal situated over or posterior to the ventrals. The shiners comprise a large genus of fresh-water fishes, only a few of which venture into brackish water and none of which enter salt water.

KEY TO THE SPECIES

a. Lateral line complete; scales 37 to 41; no dark lateral band, except in very young; base of caudal usually with a dark spot_____ *hudsonius amarus*, p. 125
aa. Lateral line incomplete, usually extending only to end of base of dorsal fin; scales 33 to 36; a prominent dark lateral band extending around tip of snout to base of caudal

_____ *bifrenatus*, p. 126

50. Notropis hudsonius amarus (Girard). Spawn-eater; Silver-fin; Shiner; "Gudgeon."

Hudsonius amarus Girard, Proc., Ac. Nat. Sci., Phila., 1856, p. 210; Chesapeake Bay.
Hybopsis hudsonius Uhler and Lugger, 1876, ed. I, p. 175; ed. II, p. 149.
Notropis hudsonius amarus Jordan and Evermann, 1896–1900, p. 270; Smith and Bean, 1899, p. 182; Fowler, 1912, p. 52.

Head 3.8 to 4.8; depth 3.6 to 5.8; D. 9 or 10; A. 9 or 10; scales 37 to 41. Body rather slender, compressed; caudal peduncle quite long and slender, its depth 2 to 2.8 in head; head rather long; snout conical, 3.05 to 4.2 in head; eye 2.5 to 3.4; interorbital space 2.4 to 3.5; mouth somewhat oblique, terminal or nearly so in young, slightly inferior in adults, lower jaw included; maxillary not quite reaching anterior margin of eye; pharyngeal teeth usually in two rows, the second row sometimes wanting, with one or two teeth when present, the main row usually with four, rarely with only three teeth, the teeth in the main row rather large and prominently curved near the tips; scales moderate, 14 to 16 rows crossing the median line of back in advance of the dorsal fin; lateral line complete, somewhat decurved; origin of dorsal slightly nearer tip of snout than base of caudal, the third and fourth rays longest, reaching past the succeeding rays when the fin is deflexed, about equal to length of head; caudal fin forked, the lobes of about equal length; anal fin similar to the dorsal; but the rays not quite as long, its origin more than an eye's diameter behind the end of base of dorsal in large examples, less than an eye's diameter behind end of dorsal base in young, somewhat nearer base of caudal than tips of pectorals in adults, equidistant from base of caudal and base of pectorals in very young; ventral fins inserted a little behind vertical from origin of dorsal, reaching to or a little past origin of anal in very young, not nearly reaching anal in large individuals; pectoral fin rather pointed, the upper rays longest, 1.05 to 1.4 in head.

Color greenish above, sides silvery, lower parts pale. Young with a dark, plumbeous lateral band extending forward through eye and across snout and ending in a dark caudal spot. The lateral band and finally the caudal spot, also, almost wholly disappear with age. The fins are all plain translucent.

Numerous specimens ranging from 26 to 125 millimeters (1 to 5 inches) in length were preserved. Most of the specimens were taken in fresh water, but some of them were found where the water was slightly brackish. The young are very similar in color to the young of *Notropis bifrenatus*, both species having the black lateral band, but in the present species, in specimens of 2 inches and upward in length, the lateral band is not nearly as black, having become more plumbeous. This fish is said to differ from the typical *hudsonius* in having a longer and more obtuse head and in the faint or absent caudal spot. (The latter is a character that applies only to adult fish.) This minnow may be distinguished from other species of this genus of the Chesapeake by the rather small scales in advance of the dorsal, 14 to 16 rows crossing the median line of back, the slightly inferior mouth, and by the anterior rays of the dorsal and anal, which are long, projecting beyond the posterior rays when the fin is deflexed.

The peritoneum is silvery, but sometimes with few and at other times with numerous dark punctulations. The air bladder is large and it has a constriction a little in advance of the middle of its length, from which arises a small tube that extends forward to the throat. The alimentary canal is a little longer than the total length of the body.

The food in 13 stomachs examined consisted mainly of insects, a few small mollusks, and various forms of plants.

Spawning evidently occurs very early in the spring, as specimens taken in November have the ovaries somewhat distended with eggs, which are plainly visible to the unaided eye. This shiner, although locally abundant, is of little commercial value because of its small size. It is considered good bait for black bass, however, and the food it furnishes for larger fish is probably of considerable importance. The largest individual in the Chesapeake collection measures 5 inches in length, which is probably the maximum length attained, as it is an inch longer than the greatest length given in current works. All specimens at hand except one were taken in rather quiet, shallow water, and usually on grassy bottom. One, however, was taken off the mouth of the Sassafras River, one-half mile from shore, with a beam trawl hauled at depths of 36 to 54 feet. It is not known whether this fish was caught on the bottom or at the surface as the trawl was being hauled up, but the capture is unusual because this species is typically an inhabitant of brooks and rivers and rarely strays from the immediate vicinity of the shore.

Habitat.—"Delaware and Potomac Rivers." (Jordan and Evermann, 1896–1900.)

Chesapeake records.—(*a*) Previous records: "Chesapeake Bay" (Girard); Patapsco River; Potomac River; East River and Octoraro Creek, Md. (*b*) Specimens in collection: From the vicinity of Havre de Grace, Md. (Susquehanna River to Spesutie Island), 30-foot seine, August 26 to September 1 and November 9 to 12, 1921; Howell Point, mouth of Sassafras River, beam trawl, depth 36 to 54 feet, May 11, 1922; Baltimore, Hawkins Point and Bear Creek, 30-foot seine, May 4, 1922; highest salinity 6.54 per mille.

51. Notropis bifrenatus (Cope). Bridled minnow; "Minnie."

Hybopsis bifrenatus Cope, Trans., Amer. Philo. Soc., XIII, 1869, p. 384; Schuylkill River, Conshohocken, Pa.
Notropis bifrenatus Jordan and Evermann, 1896–1900, p. 258; Fowler, 1912, p. 52.

Head 3.7 to 4.25; depth 4 to 4.9; D. 9 or 10; A. 8; scales 33 to 36. Body rather slender, compressed; caudal peduncle long and slender in young, less so in larger individuals, its depth varying from 2.2 to 3.7 in head; head of moderate size; snout blunt, its length 3.35 to 4.6 in head; eye 2.8 to 3.25; interorbital space 2.5 to 3.3; mouth very oblique, upper anterior margin of gape slightly below median line of eye; maxillary reaching about to anterior margin of eye; pharyngeal teeth 0, 2—2, 0 to 1, 4—4, 1, the larger teeth prominently curved at the tips; scales rather large, 12 or 13 rows crossing the median line of back in advance of dorsal; lateral line incomplete, curved somewhat downward, extending nearly to or somewhat beyond vertical from origin of dorsal; origin of dorsal a little nearer tip of snout than base of caudal; caudal fin forked, the lower lobe slightly the longer; anal fin with slightly concave outer margin, its origin slightly behind vertical from end of base of dorsal, about equidistant from insertion of pectorals and base of caudal; ventral fins inserted under or a little in advance of the dorsal, reaching beyond origin of anal; pectoral fins inserted on ventral edge, 1.35 to 1.6 in head.

Color greenish brown above, pale silvery below; the scales in upper part of body with brownish punctulations, densest on the margins, making brownish edges; sides with a prominent black band, extending around the tip of snout but not involving the lower lip, through the median part of the eye, to base of caudal, where it ends in a dark spot; a somewhat darkened vertebral stripe, at least in advance of the dorsal; fins pale, the dorsal and caudal a little darker, the first rays of the dorsal and pectorals and the outer rays of the caudal more or less dusky.

Numerous specimens of this species ranging from 25 to 60 millimeters (1 to 2⅜ inches) in length are at hand, all taken in slightly brackish water. This species is very close to *Notropis procne* (Cope), from which it differs, however, according to Fowler (1906, p. 140), in having a shorter caudal peduncle and tail, larger dark edges on the dorsal scales, and a more plumbeous lateral band. According to the same author, there is much variation in *N. bifrenatus* in the development of the lateral line, which, he says, barely extends to the origin of the dorsal in some specimens and is nearly complete in others. In 48 specimens examined by us in regard to the development of the lateral line, we find comparatively little variation, as only one specimen has a few scattered pores posterior to the end of the base of the dorsal fin, and in only a few specimens the lateral line fails to reach opposite the origin of the dorsal.

This fish is known only from coastwise streams, but we find no previous mention made of its occurrence in brackish water. The alimentary canal is short, not as long as the body. The food in six stomachs examined consisted wholly of vegetable matter, ranging from the lowest forms of algæ to the higher plants. Except that it furnishes food for larger fish, the species is of no commercial importance.

Nothing appears to be known concerning the spawning habits of this fish. According to Fowler (1906, p. 140), this minnow prefers the smaller creeks with deep water having a gentle current. The specimens at hand were taken in tidal currents, on a grassy bottom, and in shallow water.

Habitat.—In coastwise streams from Massachusetts to Maryland.

Chesapeake localities.—(a) Previous record: "Tributaries of the Big Bohemia Creek," Md. (Fowler). (b) Specimens in collection: Havre de Grace, Md., May 10, 1922; Baltimore, Hawkins Point, and Bear Creek, May 4, 1922, and August 24, 1921; Annapolis, lower Severn River, May 2 and 3, 1922, and August 19, 1921; Love Point, May 12, 1922; highest salinity 11.80 per mille.

Order NEMATOGNATHII

Family XXVII.—ARIIDÆ.   The sea catfishes

Body naked; gill membranes united, forming a fold across the isthmus; mouth terminal; nostrils usually close together, without barbel; maxillary and one or two pairs of mandibular barbels present; dorsal fin anterior, with a spine; adipose fin present; anal short or of moderate length; ventral fins with six rays.

40. Genus FELICHTHYS Swainson.   Gaff-topsail catfishes

Body elongate, little if at all compressed; head depressed; snout very broad, projecting; mouth large; teeth all villiform, in more or less distinct bands on jaws, vomer, and palatines; a large fontanel; barbels 4, maxillary barbel long, broad, bandlike; pectoral spines and usually the dorsal spine with a long bandlike filament; caudal fin deeply forked; anal fin more or less emarginate.

A single species is known from the Atlantic coast of the United States.

52. Felichthys felis (Linnæus).   Gaff-topsail catfish; Sea catfish.

Silurus felis Linnæus, Syst. Nat., ed. XII, 1766, p. 503; Charleston, S. C.
Ælurichthys marinus Uhler and Lugger, 1876, ed. I, p. 177; ed. II, p. 150.
Felichthys marinus Jordan and Evermann, 1896–1900, p. 118, Pl. XXIII, fig. 52.

Head 3.66 to 4.2; depth 4.35 to 5.4; D. I. 7; A. 22 to 24. Body robust, depressed anteriorly, compressed posteriorly; head low and broad; snout very broad, 2.4 to 3.5 in head; eye 5.35 to 7.2; interorbital space 1.42 to 1.64; mouth very broad, the cleft extending nearly or quite to eye; maxillary 2.32 to 2.38 in head; teeth small, in villiform bands on the jaws, vomer, and palatines; two pairs of barbels present, the maxillary barbel flattened, ribbon-shaped, reaching from vertical below middle of base of dorsal nearly to base of ventrals; mandibular barbels small, reaching nearly or quite to the gill covers; dorsal spine bearing a filament, varying in length, frequently reaching to or past adipose fin; adipose fin rather small, inserted over or a little behind the middle of the base of the anal, its base 6.1 to 7.75 in head; caudal fin deeply forked, the upper lobe slightly the longer; anal fin moderate, its outer margin rather deeply concave, its base 1.43 to 1.6 in head; ventral fin inserted about equidistant from the tip of the snout and the base of caudal; pectoral spine bearing a compressed filament, reaching nearly or quite opposite the origin of the anal, the spine 1.15 to 1.48 in head.

Color, top of head and back uniform steel blue, blending into bronze; sides silvery; underneath white; dorsal fin white or bluish, adipose blue, caudal dusky or gray, anal white or pale blue, ventrals plain white or slightly dusky, pectorals more or less dusky.

The Chesapeake Bay collection contains five female specimens, ranging in length from 325 to 565 millimeters (12¾ to 22¼ inches). The gaff-topsail catfish is characterized by the reduced number of barbels or whiskers, only two pairs—the maxiliary and mandibular barbels—being present. Other characters that readily distinguish this fish from all others of the Atlantic coast of America are the long, flat, ribbon-shaped filaments borne by the dorsal and pectoral spines. The filament on the dorsal often projects far above the surface of the water as the fish swims, and it is from this character and habit that the fish has received the name "gaff-topsail catfish."

The stomachs of the specimens at hand were not examined for food content, as the fish were taken from a pound net where the usual foods may not have been available and where other foods probably were taken. According to Gudger (1918, p. 39), the principal food of the gaff-topsail

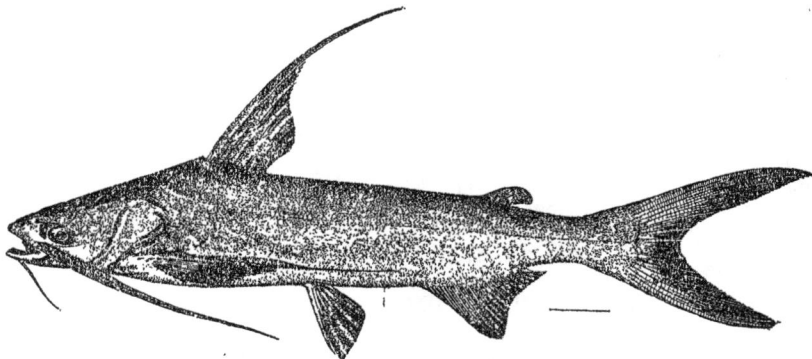

FIG. 69.—*Felichthys felis*

catfish at Beaufort, N. C., consists of crabs, supplemented by an occasional shrimp or fish or both. According to the same author, while carrying eggs and young in the mouth the male fish does not feed at all.

The sexual organs of our specimens taken on May 17, 1921, at Lynnhaven Roads, Va., are completely collapsed, as if the fish had spawned shortly before being captured. The ovaries of two fish caught in the same locality on May 25, 1922, contained eggs in various stages of development, as follows: A fish 557 millimeters in total length contained 5 eggs about 20 millimeters in diameter, 51 eggs 10 to 12 millimeters, 16 eggs 8 to 9.5 millimeters, 30 eggs 5.5 to 7.5 millimeters, and 50 eggs 3 to 5 millimeters, all opaque. In addition many undeveloped, translucent eggs, from less than 3 to 6 millimeters in diameter also were present. A fish 537 millimeters long contained 25 eggs 10 to 12 millimeters in diameter, 9 eggs 7.5 to 9 millimeters, 17 eggs 5 to 7 millimeters, and 21 eggs 3 to 4 millimeters, all of which were opaque. In addition about 22 translucent eggs, 2 to 4 millimeters in diameter, were present. A fish 476 millimeters in total length, taken on June 25, 1924, in the lower Potomac, contained numerous immature translucent eggs 2 to 4 millimeters in diameter.

The breeding season at Beaufort, N. C., according to Gudger (1918, p. 30–32), occurs during the last half of May and to a lesser extent in June. The eggs of the catfish are very large; Gudger (1918, p. 35) gives the size as varying from 15 to 25 millimeters (three-fifths to 1 inch) in diameter. After they are laid and fertilized the eggs are transferred in some mysterious manner to the mouth of the male, where they are held until hatched and where the young are retained for some time after hatching. The largest number of eggs found in the mouth of one fish by Gudger (1918, p. 36) was

55, which were taken from a male 22 inches long. The size of the young soon after hatching is probably about 1¾ inches, and according to Gudger (1918, p. 37) they are 4 inches in length before they are released from the shelter of the paternal mouth.

This catfish usually is not used as food in the United States, but on September 23, 1921, about 10 individuals were observed in a lot of "mixed" fish in the Baltimore wholesale fish market, which were said to have been taken in Chesapeake Bay. In Panama marine catfishes are seen in the markets daily and form an important food. In southern Florida this fish, together with a related species, *Galeichthys milberti*, is very abundant, causing considerable damage to the nets of mullet fishermen.

This catfish is known to reach a length of at least 22½ inches. It is not abundant in Chesapeake Bay, but a few individuals are taken from time to time during the spring and summer from the lower Potomac River to the mouth of the bay. "It is said to have become less common than formerly." (Uhler and Lugger, 1876, p. 177.)

Habitat.—Cape Cod to the Isthmus of Panama.

Chesapeake localities.—(*a*) Previous records: "Chesapeake Bay" (Uhler and Lugger, 1876); "vicinity of Norfolk, Va." (Moseley). (*b*) Specimens in collection: Lynnhaven Roads, Va., pound net, May 17, 1921, and May 25, 1922; Rock Point, Md. (Potomac River), June 25, 1924.

Family XXVIII.—AMEIURIDÆ. The horned pouts

Body naked; gill membranes separate or notched, free or at least forming a free fold across the isthmus; nostrils far apart, the posterior with a barbel; dorsal fin anterior, with a spine; adipose fin present; anal fin short or of moderate length; ventral fins with 8 or 9 rays.

41. Genus AMEIURUS Rafinesque. Horned pouts

Body moderately elongate, robust anteriorly; caudal peduncle compressed; head large, wide, supra-occipital extending backward, terminating in a more or less acute point, entirely separate from the second interspinal buckler, making the bony bridge from snout to dorsal incomplete; mouth large; teeth in broad bands on the jaws; those of the upper jaw without backward extensions at angle of mouth; adipose fin short, inserted over the posterior half of the anal; anal fin varying in length, with 15 to 35 rays.

53. Ameiurus catus (Linnæus). White cat; "Channel cat."

Silurus catus Linnæus, Syst. Nat., X, 1758, p. 305; "northern part of America."

Pimelodus lynx Girard, 1859 (1860), p. 160.

Amiurus catus Uhler and Lugger, 1876, ed. I, p. 179; ed. II, p. 152; Bean, 1883, p. 367; Jordan and Evermann, 1896–1900, p. 138; Smith and Bean, 1899, p. 181; Fowler, 1912, p. 53.

Amiurus lynx Uhler and Lugger, 1876, ed. I, p. 180; ed. II, p. 152.

Ameiurus albidus Bean, 1883, p. 367.

Head 3.38 to 3.9; depth 3.75 to 4.5; D. I, 6; A. 22 or 23. Body rather robust, somewhat compressed; head depressed and broad; snout very broad, 2.06 to 2.57 in head; eye 6.9 to 7.84 (young 4.25 to 4.8); interorbital space 1.46 to 1.67 (young 1.89 to 1.92); mouth very broad, the cleft short, not extending to eye; maxillary 2.19 to 2.66 in head (young 2.87 to 3); teeth small, in villiform bands on jaws; four pairs of barbels present, two on chin, the longest about equal to or shorter than snout, one at angle of mouth, slightly greater than interorbital space, and one at posterior nostril equal to about twice diameter of eye; the barbels of young fish are generally longer; margin of dorsal rounded, longest ray 1.67 to 1.82 in head (young 1.2 to 1.36); adipose moderate, inserted about over middle of base of anal; caudal fin moderately forked; the lobes about equal, rounded; anal fin moderate, its outer margin gently rounded, its base 1.35 to 1.5 in head (young 1.05 to 1.1); ventrals inserted a little nearer base of caudal than tip of snout in adult, in the young this proportion is reversed; pectoral spine stout, not as long as longest soft rays, 2.22 to 2.41 in head (young 1.53 to 1.8); humeral process very rough.

Color of fresh specimen, grayish on back and sides, head olive gray; underneath white; dorsal, adipose, and caudal grayish; anal whitish, edged with gray; ventrals and pectorals plain, with trace of gray.

The above description is based on eight specimens taken in Chesapeake Bay, ranging in length from 35 to 330 millimeters (1⅜ to 13 inches).

This catfish is seldom taken in the Chesapeake proper, except at the head of the bay, where the water is usually fresh. It is a common species in the deeper rivers, particularly the Potomac, where it occasionally strays into brackish water. The species is included in the present report upon receipt of an adult specimen taken from a pound net at Rock Point, Md., on June 25, 1924. The water in this part of the Potomac is decidedly brackish, and salt-water species such as weakfish, croakers, etc., are commonly caught there throughout the summer. The channel catfish, as its name implies, is frequently found in river channels and deep holes. The channel cat of the Mississippi is a different species (*Ictalurus punctatus*).

This catfish spawns in the summer. The parent fish are said to build a nest of gravel and to guard the eggs and the young until some time after hatching.

The channel catfish is considered a good food fish, being superior to several other species of cats. At Washington and Baltimore it is an important market species. It is known to reach a length of about 2 feet and a weight of 5 pounds.

This catfish is caught chiefly with hook and line and in pound nets and fyke nets. Large numbers are caught in the Potomac, in Back River near Baltimore, and in the Havre de Grace region of the bay.

Habitat.—Coastwise streams from New York to Texas.

Chesapeake localities.—(a) Previous records: Havre de Grace, Baltimore, Patapsco, and Potomac Rivers. (b) Specimens in collection: Rock Point, Md., pound net, June 25, 1924; Spesutie Island, Md., August 26, 1921; Elk River, Md., May 8, 1922; Northeast River, Md., August 29, 1921, collecting seines. Definite density readings are not available; however, it is known certainly to enter brackish water.

Order INIOMI

Family XXIX.—SYNODONTIDÆ. The lizard fishes

Body elongate, more or less cylindrical; mouth large; premaxillaries very long, forming entire margin of the upper jaw; maxillaries long and slender, closely adhering to the premaxillaries; teeth sharp, present on jaws, palatines, and tongue; gill membranes separate and free from the. isthmus; branchiostegals usually numerous; pseudobranchiæ present; gill rakers small or obsolete; lateral line present; scales cycloid, rarely absent; adipose fin present; dorsal fin single, consisting of soft rays only; caudal fin forked; anal fin moderate or long; pectoral and ventral fins present; air bladder small or wanting; intestinal canal short. A single genus of this family is represented in the fauna of Chesapeake Bay.

42. Genus SYNODUS (Gronow) Scopoli. Lizard fishes

Body elongate; cylindrical; head depressed; snout pointed; mouth very large; premaxillaries long, not protractile; teeth rather large, present on jaws, palatines, and tongue; teeth in the jaws compressed, very sharp; branchiostegals 12 to 16; gill rakers small, spinous; scales cycloid, present on body, cheeks, and opercles; upper surface of head naked; dorsal fin short, placed well forward; adipose fin small, situated over the anal; caudal fin forked; ventral fins moderately large, the inner rays longest; pectorals rather small. A single species is known from Chesapeake Bay.

54. Synodus fœtens (Linnæus). Lizard fish; "Providence whiting"; "Scarpen fish."

Salmo fœtens Linnæus, Syst. Nat., ed. XII, 1766, p. 513; South Carolina.
Synodus fœtens Uhler and Lugger, 1876, ed. I, p. 152, ed. II, p. 129; Lugger, 1877, p. 85; Bean, 1891, p. 93; Jordan and Evermann, 1896–1900, p. 538, Pl. LXXXVIII, fig. 236; Evermann and Hildebrand, 1910, p. 159.

Head 3.8 to 4.4; depth 5.8 to 7.3; D. 10 to 12; A. 12; scales 6–60 to 65–7. Body elongate, more or less cylindrical, about as broad as deep; head depressed, broader than deep; snout pointed, projecting beyond tip of mandible, 3.4 to 3.7 in head; eye 8.5 to 9.6; interorbital broad, concave, 5.3 to 5.9; mouth very large, the gape extending far beyond eyes; maxillary long and narrow, 1.6 to 1.7 in head; teeth present on jaws, palatines, and tongue, those in the upper jaw sharp and compressed, in two series, the inner and larger series depressible, the teeth in the lower somewhat smaller and in a narrower band, the teeth on the tongue and palatines rather prominent in bands;

scales moderate, rather thin, cycloid, seven rows on cheeks; dorsal fin rather high, the anterior rays not reaching tips of the posterior ones when deflexed, origin of fin a little behind base of ventrals and about equidistant from the adipose fin and middle of eye; adipose fin small, its base about as long as pupil of eye, situated over anterior half of anal base; caudal fin forked, the lobes of about equal length; anal fin rather long and low, the median rays shortest, its origin about equidistant from end of base of dorsal and base of caudal; ventral fins long, the inner rays about 1.5 times as long as the longest rays of the pectoral, inserted about equidistant from tip of snout and vent; pectoral fins rather small, 2 to 2.1 in head.

Color of two specimens 9¼ and 13 inches in length, brownish or olivaceous above, lower sides and below silvery white; operculum yellowish above; chin white; dorsal plain or pale yellow; adipose pale with dusky spot posteriorly; caudal dusky or yellowish, lower lobe darkest; anal white; ventrals white, pale yellow at base; pectorals plain, yellowish or light green. The young usually have more or less distinct dark crossbars on the back.

Six large specimens of this species, ranging from 234 to 330 millimeters (9¼ to 13 inches) in length, form the basis for the foregoing description. The young of this fish was not seen during the investigation. This species is the only one of the genus known from Chesapeake Bay and therefore is easily distinguished by its elongate form, depressed head, low, pointed snout, very large mouth, and the presence of an adipose fin.

In regard to the food of this fish, Smith (1907, p. 139) says:

"The lizard fish has a formidable mouth and it is a voracious feeder; small fish constitute its principal food, but crabs, shrimp, worms, and other animals are also eaten."

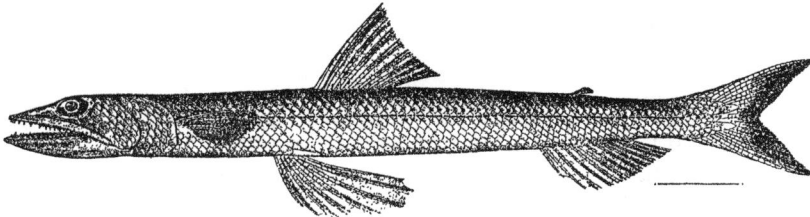

FIG. 70.—*Synodus fœtens* (lizard fish)

The contents of five stomachs of specimens taken in Chesapeake Bay consisted exclusively of small fish, as many as three being contained in one stomach. Three of these small fish, taken from three different stomachs, could be recognized as young weakfish (*Cynoscion regalis*). The others were too fragmentary to be identified.

The lizard fish is found on sandy shores and it is a bottom species. In some localities, because of this habit of living on the sand, together with the general shape of the body, this fish is known as the "sand pike."

Its spawning habits appear to be unknown.

This species is not common in Chesapeake Bay. During an entire season (April to November, 1921) of shore collecting not a single specimen was taken. From September 23 to October 27, 1922, at Ocean View, Va., 18 fish were caught in 11 of a total of 32 hauls of 1,800-foot seines on 10 different dates, the highest catch in one haul being 5. Nearly all these fish had gilled themselves far out on the wings of the seine and very few were taken in the bunt. These specimens were 11 to 13¼ inches in length. The only other lizard fish taken during the present investigation (a fish 9¼ inches long) was caught with hook and line on September 1, 1922, far up the bay at Chesapeake Beach, Md. Fish of the following lengths and weights were secured: Eleven inches, 5½ ounces; 12 inches, 7½ ounces; 12½ inches, 8½ ounces; 13 inches, 10 ounces. The lizard fish is reported to reach a length of 2 feet. As a food fish it has no value.

Habitat.—Massachusetts to Brazil.

Chesapeake localities.—(*a*) Previous records: Tide waters of the Potomac and along the shore of the southern end of the eastern peninsula, St. Charles Island, St. Jeromes, and Cape Charles City, Va. (*b*) Specimens in collection: Chesapeake Beach, Md., September 1, 1922, hook and line; Ocean View, Va., September and October, 1922, 1,800-foot seine.

Order HAPLOMI

Family XXX.—ESOCIDÆ.   The pikes, pickerel, and muskallunges

Body very elongate, not elevated, and not much compressed; head long; snout long and broad, depressed; mouth very large; mandible projecting; margins of upper jaws formed by maxillaries, provided with a supplemental bone; teeth present on jaws, vomer, palatines, and tongue; gill slits wide; gill membranes not united and free from the isthmus; gill rakers small, tuberclelike or toothed; branchiostegals numerous; lateral line weak, obsolete in young; scales small, wanting on upper surface of head and snout, cheeks and opercles partly or completely scaled; dorsal placed far back, similar to and opposite the anal; caudal fin forked; no adipose fin; air bladder simple. This family consists of a single genus.

43. Genus ESOX Linnæus.   Pikes; Pickerels; Muskallunges

The characters of the genus are included in the family description.

KEY TO THE SPECIES

a. Dorsal rays 13 or 14 (counting branched rays only); anal 12 or 13; branchiostegals 14 to 16; scales 122 to 126; adults with numerous lines, forming reticulations on the sides_____ _____ _____*reticulatus*, p. 132

aa. Dorsal rays 11 or 12; anal 11 or 12; branchiostegals 11 to 13; scales about 105; sides usually with black vertical bars; no reticulations_____*americanus*, p. 134

55. Esox reticulatus LeSueur.   Eastern pickerel; "Chain pickerel"; "Pike."

Esox reticulatus LeSueur, Journ., Ac. Nat. Sci., Phila., I, 1818, p. 414; Connecticut River, Adams, Mass.; Philadelphia. Uhler and Lugger, 1876, ed. I, p. 145; ed. II, p. 124; Bean, 1883, p. 366; Evermann and Hildebrand, 1910, p. 159; Fowler, 1912, p. 54. *Lucius reticulatus* Jordan and Evermann, 1896–1900, p. 627; Smith and Bean, 1899, p. 184.

Head 2.7 to 3.3; depth 5.3 to 6.6; D. 13 or 14 (counting divided rays only; 18 or 19 including rudiments); A. 12 or 13 (counting divided rays only; 16 or 17 including rudiments); scales 122 to 126. Body rather slender, somewhat compressed, deepest near the middle; head large, depressed above, the profile a little concave over snout; snout long and broad, equal to or a little longer than postorbital part of head, 2.1 to 2.8 in head; eye 5 to 10; interorbital 5.5 to 12; mouth large, nearly horizontal; the lower jaw projecting; maxillary scarcely reaching eye in very young, to or slightly past anterior margin of pupil in large specimens, 2 to 2.8 in head; teeth present on jaws, vomer, palatines, and tongue; the lateral teeth on lower jaw and those on vomer enlarged; branchiostegals 14 to 16; scales small, covering entire cheek and opercle; dorsal fin mostly opposite the anal, its origin about an eye's diameter in advance of origin of anal; caudal fin forked, the lower lobe the larger; anal fin similar to the dorsal; ventral fins rather small, inserted a little nearer the origin of the anal than base of pectorals; pectoral fins similar to the ventrals, 2.5 to 6.6 in head.

Color greenish above; pale underneath; scales above with golden luster; sides in adult reticulated with dark lines and streaks.   These reticulations are most evident in the largest specimens and entirely wanting in the very young.   A dark vertical bar under the eye and the young also with a dark longitudinal bar extending from the tip of the snout, through the eye, to margin of opercle. Fins plain, the dorsal and caudal darker than the others.

Many specimens of this pickerel, ranging from 28 to 490 millimeters (1⅛ to 19¼ inches) in length, were preserved, and these fish form the basis for the foregoing description.   The chief diagnostic differences between this species and the banded pickerel are shown in the key to the species.   The principal change that takes place with age is in color.   The reticulations that are characteristic of the species, according to specimens at hand, are not well defined until the fish reaches a length of about 12 inches.   The very young (1 to 2 inches long) are grayish in color in spirits, and the upper parts everywhere bear dusky punctulations.   A black bar extends from the tip of the snout, through the eye, to the margin of the opercle.   When the fish exceeds a length of 2½ inches dark areas appear on the sides, which for some time become more prominent with age.   As the fish increases in length, narrow, pale, vertical bars appear along the upper parts of the sides.   These pale bars form the first suggestions of the reticulations that are to appear later. In specimens upwards of 10 inches in length the pale bars still persist, and along the lower part of the sides are numerous, elongate, pale blotches. It is these pale blotches that are destined soon to be inclosed by darker lines, forming the reticulations. In the adults, 12 inches or more in length, the reticulations usually are well formed; the pale vertical bars along the upper parts of the sides are very indefinite or entirely wanting. The dark bar, which in the young passes from the snout through the eye and over the opercle, has entirely disappeared, but instead a dark bar extending downward from the eye has developed. It will be seen from the description of the pronounced changes in color that take place with age that a general description of the color markings can not be relied upon in classifying specimens of this fish of varying sizes. The most reliable diagnostic characters are the number of rays in the dorsal and anal fins and the number of branchiostegals (the riblike rays under the lower edge of the gill cover), as shown in the description and in the key to the species.

The eastern pickerel, according to Kendall (1917, p. 27), feeds principally upon other fish, although it includes many other animals in its diet, such as frogs and other batrachians, and in fact any living thing moving in the water within reach and which it can capture and manage. Of 6 stomachs from specimens taken in the tide waters of Chesapeake Bay, which we examined, 1 was empty, 4 contained fish only, and 1 contained fish and shrimp. The fishes (which could be recognized among the food) consisted of silversides, sticklebacks, and killifishes.

The usual haunts of the pickerel are weedy streams and bays or coves of lakes. It is characteristically found among weeds, with the head slightly projecting. It often remains very quiet in this position for a long time, and upon the approach of small fish or other small animals it "shoots" forth from its hiding place with great rapidity in an effort to capture its prey.

Spawning takes place early in the spring. Specimens taken in Severn River early in November already had the sexual organs somewhat developed. Welsh (field notes) took a ripe female at Havre de Grace, Md., on April 11, 1914. Kendall (1917, p. 28) writes that the eggs are laid in glutinous strings of a yellowish-white color, which often form large masses and have been seen clinging to submerged bushes in great mats or long strings. Strings of eggs averaging from 2 to 9 feet in length have been reported. The eggs of the pickerel are said to hatch in about a week to 10 days. The larvæ are reported to be very small when hatched, but under favorable conditions and with a sufficient food supply growth proceeds fairly rapidly. Nearly all the young collected during the investigation were taken in a brackish, marshy pond near Solomons, Md., on April 28, 1922. These specimens, 90 in number, ranged from 28 to 90 millimeters (1⅛ to 3½ inches) in length.

As a food fish it is variously esteemed, being regarded by some as an excellent fish and by others as decidedly inferior. (Kendall, 1917, p. 29.) In the Chesapeake drainage it is regarded with much favor.

This species is comparatively common in the tide waters of the Chesapeake, particularly at the head of the bay and in the lower Chester and Severn Rivers, where it is common in brackish water.

The pickerel is one of the important food fishes of the Chesapeake drainage, where, during 1920, 76,818 pounds, worth $16,591, were caught. The greater part of the catch was taken in fresh or slightly brackish water in the numerous tributaries of the Chesapeake. In Maryland, 62,208 pounds were caught, and in Virginia 14,610 pounds. The largest part of the catch was taken in seines and fyke nets, followed by pound nets, gill nets, and trammel nets.

The greater part of the catch is taken from October to April. During November it is one of the principal species found in the large fish markets of Baltimore and Norfolk. It commands a good price and sells well.

It is reported (Kendall, 1917, p. 29) that in Massachusetts this species has reached a weight of 4 or 5 pounds in three years when kept in a large, warm pond, covered with lily pads and well stocked with young alewives. The largest pickerel of which we have a record weighed 9 pounds and was caught during 1909 in a New York lake. Fish weighing more than 5 pounds are rare. In Chesapeake Bay and its tributaries the average size of the pickerel, as caught by anglers and fishermen, is from 1 to 2 pounds. The following weights were secured: 10½ inches, 3.7 ounces; 11 inches, 4.3 ounces; 11½ inches, 5.1 ounces; 12 inches, 5.8 ounces; 16½ inches, 12.6 ounces; 19¼ inches, 1 pound 8.4 ounces.

Habitat.—"It is believed originally to have been restricted to the fresh waters of the Atlantic seaboard, being commonly found anywhere east of the Allegheny Mountains, from southwestern Maine to Florida. Aided by man, its range has been extended throughout the southern half of Maine and even farther north, into the lower waters of the St. John River, into New Brunswick, and elsewhere." (Kendall, 1917, p. 24.)

Chesapeake localities.—(*a*) Previous records: Many places from the tide waters and streams tributary to Chesapeake Bay. (*b*) Specimens in collection or observed in the field: Havre de Grace, Md., April, 1912; May 8–19, 1922; Aug. 26–31, 1921; November 9–12, 1921; December 20, 1911. Baltimore, Md., fish market, November 4–8, 1921. Annapolis, Md., May 1–3, 1922; August 17–19, 1921; November 1–3, 1921. Solomons, Md., April 26–28, 1922. Love Point, Md., May 11, 1922. Norfolk, Va., fish market, November, 1921, and November, 1922. The greatest salinity in which the species was taken was 12.61 per mille.

56. Esox americanus Gmelin. Banded pickerel; "Pike."

Esox lucius americanus Gmelin, Syst. Nat., 1788, p. 1390; Long Island, N. Y.
Esox umbrosus Uhler and Lugger, 1876, ed. I, p. 144; ed. II, p. 123.
Esox niger Uhler and Lugger, 1876, ed. I, p. 146; ed. II, p. 124.
Lucius americanus Jordan and Evermann, 1896–1900, p. 626; Smith and Bean, 1899, p. 184.
Esox americanus Fowler, 1912, p. 54.

This species is much less common than *E. reticulatus* in the tide waters of Chesapeake Bay. A single small specimen occurs in the collection. This little species may be distinguished from the chain pickerel (the only other species of the genus known from Chesapeake waters) by the somewhat shorter dorsal and anal fins, fewer branchiostegals, and by the color. These differences are shown in the key to the species.

The food of this fish, according to Bean (1903, p. 294) and Smith (1907, p. 143), consists principally of minnows. Its breeding habits have not been specifically described, but Kendall (1917, p. 37) wrote that they probably were very similar to the eastern pickerel.

This pickerel is of small size; according to Smith (1907, p. 143) and other authors it rarely exceeds a foot in length and is of less importance than the eastern pickerel as a food fish.

This species is reported from brackish and salt water from New York and New Jersey (see Kendall, 1917, p. 36), but it appears to be rare in the brackish waters of Chesapeake Bay, and it is of no commercial value.

Habitat.—East of the Allegheny Mountains, from Vermont to Alabama.

Chesapeake localities.—(*a*) Previous records: Streams in the vicinity of Havre de Grace, Md.; Rappahannock River; Potomac River. (*b*) Specimen in collection: Havre de Grace, Md., September 1, 1921. This fish is known to enter brackish water, but it was not taken under these conditions in Chesapeake Bay during the present investigation.

Order CYPRINODONTES

Family XXXI.—CYPRINODONTIDÆ. The killifishes

Body elongate, compressed (at least posteriorly); mouth small, usually terminal; premaxillaries protractile; teeth pointed in Fundulinæ, incisorlike in Cyprinodontinæ; gill membranes united, free from the isthmus; gill rakers short and thick; scales large, cycloid; no lateral line; dorsal fin single, composed of soft rays only; caudal fin posteriorly square or rounded, not forked; anal fin somewhat similar to the dorsal, not modified in the male; ventral fins abdominal; species oviparous.

KEY TO THE GENERA

a. Body short and deep, compressed; head not notably depressed; teeth in a single series, incisorlike, tricuspid_____Cyprinodon, p. 135
aa. Body elongate and less strongly compressed; head more or less depressed; teeth in a single series or in bands, pointed.
  b. Teeth in a single series, all pointed; head scarcely depressed_____Lucania, p. 136
  bb. Teeth in bands, all pointed; head depressed, flattened above_____Fundulus, p. 137

44. Genus CYPRINODON Lacépède. Short minnows

Body short and deep; back elevated; mouth small; teeth in a single series, incisorlike, tricuspid; opercle superiorly fused with the shoulder girdle; scales large; males larger than females; oviparous. A single species of the genus occurs in the brackish waters of Chesapeake Bay.

57. Cyprinodon variegatus Lecépède. Variegated minnow; Sheepshead minnow.

Cyprinodon variegatus Lacépède, Hist. Nat., Poiss., V, 1803, p. 486; South Carolina. Lugger, 1877, p. 84; Smith, 1892, p. 64 Pl. XVIII; Jordan and Evermann, 1896–1900, p. 671, Pls. CXI and CXII, figs. 296 and 296a; Evermann and Hildebrand, 1910, p. 159; Fowler, 1912, p. 54.

Head 2.65 to 3.3; depth 2 to 2.65; D. 11 or 12; A. 10 or 11; scales 24 to 27. Body compressed, short and deep, becoming deeper with age, especially in the male; back elevated; upper profile gently and evenly elevated in the young and in females, with a concavity at occiput in adult males; head short; snout blunt, its length 3.3 to 4.7 in head; eye 2.5 to 4; interorbital 2.8 to 3.8; mouth rather small, terminal; premaxillaries strongly protractile; teeth in the jaws in a single

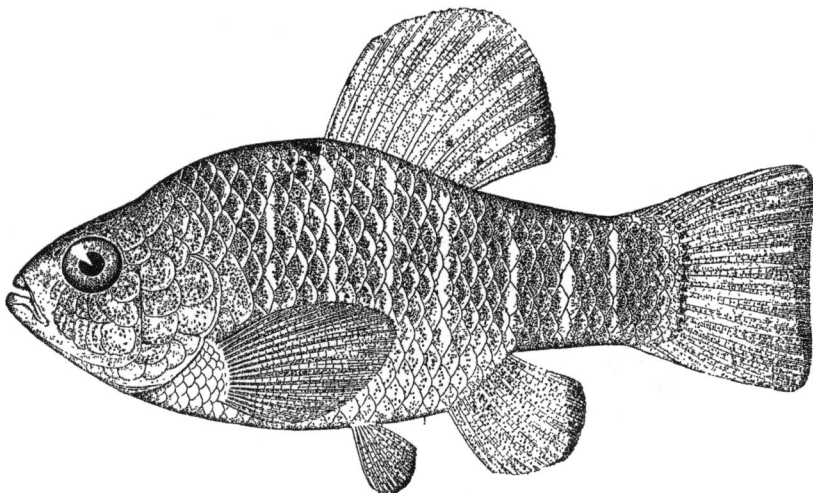

FIG. 71.—*Cyprinodon variegatus*, adult male

series, rather large, tricuspid, the median cusp the longest and broadest; scales large, the one placed just above the base of pectoral excessively enlarged, 4 or 5 oblique rows between the upper angle of gill opening and origin of dorsal; fins moderate, the dorsal particularly proportionately much larger in adult males than in females, inserted much in advance of the anal in both sexes; caudal fin with almost straight margin; anal fin smaller than the dorsal, particularly in adult males; ventral fins rather small, inserted equidistant from tip of snout and base of caudal or slightly nearer the latter; pectoral fins moderate, rounded, proportionately longer in adult males than in females, 1 to 1.5 in head.

Color of female brassy on the back and sides, with dusky blotches, usually forming bars on the lower part of the sides; yellowish or white below; dorsal olive or dusky, with a black blotch on the middle of the posterior rays; caudal greenish to dusky, with a dark bar at base; anal and ventrals pale yellowish with white margins; pectorals dusky to orange. Males darker, with bluish reflections on upper parts in advance of dorsal, sometimes brilliant blue along entire back; abdomen, at least during the breeding season, deep orange; dorsal bluish to dusky anteriorly, edged with pink or orange; caudal olive, with a very narrow dusky bar at base and a black margin; anal, ventrals, and pectorals orange, with bluish black margins.

Many specimens of this species were preserved. The above description is based on specimens ranging in length from 20 to 65 millimeters (⅘ to 2½ inches). The species is readily recognized by the short, deep body and the variegated color.

49826—28——10

The food probably consists largely of vegetable matter. In 20 stomachs examined, only sand, mud, débris, and filamentous algæ were found. The intestine is convoluted and more than twice as long as the body, which is a further indication that the species is chiefly herbaceous. In the aquarium this fish readily feeds on fish and other meats. It is very ferocious in confinement, waging constant fights with other fishes. Whether they are of its own kind or of another species does not appear to make a particle of difference. It frequently kills fishes larger than itself by making repeated attacks and inflicting wounds here and there with its sharp tricuspid teeth until the victim succumbs from exhaustion or from the attack of disease. It then proceeds to devour its prey by tearing off piece after piece at any convenient place.

Spawning takes place throughout the spring and summer (Hildebrand, 1917, p. 13), one female laying eggs several times during one season. The eggs are spherical and about 1.2 to 1.4 millimeters in diameter (Kuntz, 1916, pp. 410–414). They are slightly heavier than sea water and adherent, being held together by minute adhesive threads. The period of incubation at ordinary summer temperatures in the laboratory extended over five to six days. The newly hatched larva is quite plump and about 4 millimeters in length. At a length of 9 millimeters the young fish already has many of the characters of the adult, and at a length of 12 millimeters virtually all of the diagnostic characters of the full-grown fish are developed.

The males of this species, contrary to the more usual rule among fishes, especially in related genera, are notably larger than the females, the average difference in length being about 12 millimeters. The sexual differentiation in color takes place when the fish are about 30 millimeters long. The young of 8 millimeters and less in length are almost entirely unmarked, but when the fish becomes a little larger, spots and bars appear and all the young (males and females) assume the color of the female.

This minnow is very common in all brackish waters of Chesapeake Bay, from Cape Henry to Love Point and Annapolis, but none at all were found from Baltimore to Havre de Grace, and none were taken in strictly fresh water. It is especially abundant in coves, bays, ponds, and creeks, and less common along the open beaches. In one instance, while collecting near Buckroe Beach, one haul with a 30-foot bag seine yielded about 1 bushel of variegated minnows. This minnow generally travels in schools, and on a rising tide swims about near the shore's edge in water 1 or 2 inches deep. At high tide the fish work their way up among the grass on overflowed banks, returning to open water with the receding tide.

The largest specimen taken during the investigation was 76 millimeters (3 inches) long, and it represents the maximum size attained by this species. It is too small to be of commercial importance, but on account of its abundance and wide distribution no doubt it is an important food for larger fishes.

Habitat.—Coastwise in brackish water from Cape Cod, Mass., to Mexico.

Chesapeake localities.—(a) Previously recorded from "Chesapeake Bay," "Lower Potomac," St. George Island, Tolchester, and Chestertown. (b) Specimens in collection: From many parts of the bay, from Love Point, Md., to Lynnhaven Roads, Va.

45. Genus LUCANIA Girard. Rain-water fishes

Body rather short, compressed; head small; mouth small, nearly terminal; teeth in the jaws pointed, in a single irregular series; scales rather large; dorsal and anal fins rather small, the dorsal above or in advance of the anal; the anal fin not modified; oviparous. A single species occurs in the waters of the Chesapeake.

58. Lucania parva (Baird and Girard). Rain-water fish.

Cyprinodon parvus Baird and Girard, Ninth Smithsonian Report, 1854 (1855), p. 345; Greenport, L. I.
Lucania parva Smith, 1892, p. 68; Jordan and Evermann, 1896–1900, p. 665, Pl. CIX, fig. 292; Evermann and Hildebrand, 1910, p. 159.

Head 3.4 to 3.7; depth 3 to 3.8; D. 11 or 12; A. 10 or 11; scales 25 or 26. Body rather short, compressed; caudal peduncle rather strongly compressed, its depth 1.6 to 2.4 in head; head small, about as deep as broad at eyes; snout blunt, its length 3.5 to 6 in head; eye 3 to 4; interorbital 1.3 to 1.8; mouth nearly terminal, the lower jaw projecting slightly; premaxillaries protractile; teeth small, pointed, in a single irregular series in each jaw; scales rather large, 6 or 7 oblique series between the upper angle of gill opening and origin of dorsal; dorsal fin moderate, higher in adult males than in females, inserted nearly an eye's diameter in advance of origin of anal; caudal fin straight or slightly rounded posteriorly; anal fin similar to dorsal, the rays proportionately longer in adult males than in females; ventral fins rather small, inserted about an eye's diameter nearer the tip of snout than base of caudal; pectoral fins moderate, 1.3 to 1.8 in head.

Color of female dark olive above, pale underneath; scales on sides with dusky punctulations and dark edged, those on anterior part of sides with bluish and silvery reflections; a dark vertebral streak present in advance of dorsal; dorsal, caudal, and pectorals more or less greenish or olivaceous; other fins colorless. General color of male similar to female but somewhat brighter, at least during the breeding season; the anterior rays of the dorsal black or sometimes with only a black spot at base, the black occasionally extending on the outer margin of the fin; anal fin also sometimes with a dark margin; anal and ventrals with more or less red during the breeding season. The color of the young is similar to that of the adult female, the markings differentiating the sexes appearing when the fish has reached a length of about 25 millimeters.

Many specimens of this species were taken, ranging in length from 24 to 58 millimeters (1 to $2\frac{3}{10}$ inches). This fish is recognized by its small size and plain greenish coloration, no bars or stripes being present.

The only food present in 28 stomachs examined consisted of small crustaceans. However, the species no doubt also feeds on other small animal life. In the aquarium it readily takes finely chopped fish and beef.

Ripe or nearly ripe fish were taken from early in April until near the end of July. It seems probable that the fish spawn more than once suring a season, as the gravid females taken during the early part of the saeason, in addition to the ripe eggs, contained another size of eggs easily visible with the unaided eye. This may also account for the long spawning season. The eggs, when mature, are about 1 millimeter in diameter, and the largest number found in one fish was 104. Sexual maturity appears to be attained very soon after the color differentiation between the sexes takes place, or when the fish are about 25 millimeters in length. The largest inidividual taken during the present investigation was 58 millimeters ($2\frac{3}{10}$ inches) long, and it appears to represent the maximum size attained by the species. The females reach a somewhat larger size than the males, the average difference in the length being about one-fourth inch.

This fish is very abundant in all brackish waters of Chesapeake Bay, but was not taken in strictly fresh water. It is especially plentiful in coves, bays, ponds, creeks, and open flats, where vegetation is present. Its abundance is indicated by the following catches made with a 30-foot collecting seine: Love Point, Md., May 12, 1922, brackish creek, bottom mud and vegetation, 7 hauls, 18,300 *Lucania parva*; Annapolis, Md., May 2, 1922, brackish pond, bottom mud, dense vegetation, 20 hauls, 14,600 *Lucania parva*. The fish travel in schools and are often found in association with Gambusia and Fundulus.

Several investigators have mentioned this species as being of value for mosquito control, but this appears to have been based upon the nature of its habitat rather than upon direct investigations, and no definite information is available.

This fish is too small to be of commercial importance, but, because of its abundance and wide distribution, it no doubt is an important food for larger fishes.

Habitat.—As given by Smith (1907, p. 151) and others, Cape Cod to Key West.

Chesapeake localities.—(a) Previous records: "Lower Potomac," St. George Island, and Cape Charles City. (b) Specimens in collection: From shore waters of all parts of the bay from Love Point, Md., to Lynnhaven Roads, Va.

46. Genus FUNDULUS Lecépède. Killifishes; Mummichogs

Body elongate, posteriorly compressed, back little or not elevated; head rather broad, usually depressed; mouth terminal or the lower jaw slightly projecting; teeth usually villiform and in narrow bands; dorsal and anal fins usually higher in males than in females; caudal fin with straight or rounded margin; the anal fin not modified; oviparous, the sexes differing in color and size, the females being the larger. Several species of this genus frequent salt and brackish water and others, not considered in this report, are confined to strictly fresh water.

KEY TO THE SPECIES

a. Dorsal fin inserted over or in advance of origin of anal, with 10 to 15 rays; its base equal to or longer than that of the anal.

  b. Scales rather large, 33 to 38 in a lateral series.

   c. Body robust, average depth in length about 3.2; 11 to 13 oblique series of scales between upper angle of gill opening and origin of dorsal; ventral fins usually inserted equidistant from tip of snout and base of caudal; size rather large, maximum length about 125 millimeters_____*heteroclitus*, p. 138

   cc. Body less robust, average depth in length about 4; 18 or 19 oblique series of scales between upper angle of gill opening and origin of dorsal; ventral fins usually inserted about an eye's diameter nearer base of caudal than tip of snout.

     e. Snout rather long, 2.6 to 3.5 in head; dorsal fin with 13 to 15 rays; young of both sexes with vertical black bars along sides; adult males with vertical side bars and a black ocellus on posterior rays of the dorsal; adult females with longitudinal black stripes and with variable oblique and vertical bars; size large, maximum length about 200 millimeters_____*majalis*, p. 140

     ee. Snout shorter, 3.7 to 5.5 in head; dorsal fin with 10 or 11 rays; both sexes usually with a black ocellus on the posterior rays of the dorsal; size rather small, maximum length about 65 millimeters_____*ocellaris*, p. 141

  bb. Scales rather small, 41 to 46 in a lateral series, 14 to 18 oblique rows between upper angle of gill opening and origin of dorsal; ventral fins inserted about an eye's diameter nearer tip of snout than base of caudal; size medium, maximum length about 111 millimeters_____

  _____*diaphanus* p. 143

aa. Dorsal fin inserted over or a little behind origin of anal, about equidistant from the tip of the tail and anterior half of eye, very small, with only 8 rays its base shorter than that of the anal; snout short, 4 to 4.8 in head; size small, maximum length about 50 millimeters_____*luciae*, p. 144

59. Fundulus heteroclitus (Linnæus).   Mummichog; Mud minnow; Killifish; Common killifish; "Pike minnow"; "Mud dabbler"; "Gudgeon."

Cobitis heteroclita Linnæus, Syst. Nat., ed. XII, 1766, p. 500; South Carolina.
Fundulus viridescens Uhler and Lugger, 1876, ed. I, p. 147; ed. II, p. 126.
Fundulus heteroclitus Uhler and Lugger, 1876, ed. I, p. 149, ed. II, p. 127; Bean, 1891, p. 92; Smith, 1892, p. 66, Pl. XIX; Jordan and Evermann, 1896–1900, p. 640, Pl. CII, fig. 273; Smith and Bean, 1899, p. 184; Evermann and Hildebrand, 1910, p. 159.
Fundulus pisculentus Uhler and Lugger, 1876, ed. I, p. 149; ed. II, p. 127.
Fundulus heteroclitus macrolepidotus Fowler, 1912, p. 54.

Head 3.2 to 3.7; depth 2.8 to 3.7; D. 11 or 12; A. 10 to 12; scales 35 to 38.   Body rather robust, compressed; caudal peduncle strongly compressed, its depth 1.65 to 2.1 in head; head depressed; snout short, broad, its length, 2.7 to 4.7 in head; eye 4 to 5.8; interorbital 2.1 to 2.3; mouth terminal, mostly transverse; premaxillaries protractile; teeth all pointed, in villiform bands, the outer ones somewhat enlarged; scales moderate, 11 to 13 oblique series between upper angle of opercle and origin of the dorsal; dorsal fin rather long, higher in adult males than in females, its origin slightly in advance of origin of anal, inserted nearer end of caudal than tip of snout in adult females, about equidistant from tip of snout and end of caudal in adult males; caudal fin broadly rounded; anal fin with a slightly shorter base than the dorsal but with longer rays, the oviduct attached to the first ray; ventral fins rather small, inserted about equidistant from tip of snout and base of caudal; pectoral fins rather broad, 1.3 to 1.6 in head.

Color of large female in life plain brownish green above, paler underneath.   Small females usually with 13 to 15 dark cross bars, narrower than the interspaces.   No dark vertebral line; a small dark area at origin of dorsal; fins all unmarked, the vertical ones often with a greenish tinge. Color of a 3-inch breeding male dark green or olive above, blending into silvery on lower part of sides; yellow underneath; sides with about 15 narrow silvery vertical bars and numerous irregular white or yellowish spots, pale spots extending on lower half of vertical fins; head brownish between the eyes; silvery-blue punctulations below eyes; operculum dusky above, golden below, with punctulations; chin olive; caudal peduncle the color of back; dorsal and caudal dusky with yellow margins, a dark spot on posterior 4 or 5 rays of dorsal (this spot may not always be present); anal and ventrals golden; pectorals dusky yellow. When not in breeding condition the colors of the male fade somewhat. The young are mostly grayish and bear dark cross bars, which vary greatly in number. The color markings differentiating the sexes usually become evident when the fish have attained a length of 1½ to 1¾ inches. A large variation in the intensity of color exists among specimens, depending upon the environment in which they are taken, color adaptation being developed to a considerable degree.

Numerous specimens of this species taken in many localities, ranging in length from 20 to 123 millimeters (⅘ to 4⅞ inches), were preserved. This species is recognized by its chubby form, short and broad snout, and by the coloration. The smaller, brightly colored males are sometimes difficult to separate from the adult males of *F. luciæ*. When the color can not be relied upon for identification, the length and position of the dorsal fin must be taken into consideration. The dorsal fin in the present species consists of 11 or 12 rays and is inserted over or a little in advance of the origin of the anal. In *F. luciæ* the dorsal fin has eight rays and is inserted a little posterior to the origin of the anal.

This fish feeds on a large variety of foods. Among the contents of 48 stomachs examined, the following foods were found: Small crustaceans, small mollusks, annelid worms, insects, small fish, and vegetable matter, such as blades of grass, bits of roots, algæ, and seeds. A considerable amount of sand also was present in some of the specimens examined, but this may have been taken incidently in the capture of foods.

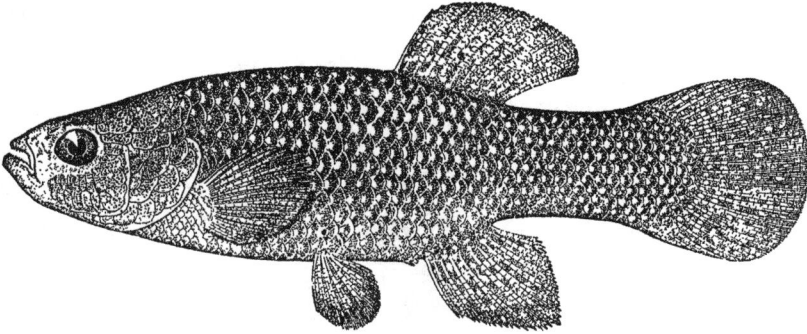

FIG. 72.—*Fundulus heteroclitus*, adult female

Spawning takes place from April to August, and it seems probable that one female may produce several broods of eggs during one summer. The ovary is single and the number of eggs produced varies greatly among specimens. The largest number of ova of one size found among several dozen specimens examined was taken from a female 98 millimeters in length, which contained 460. The eggs of one brood are of uniform size, rather large, and spherical when mature, measuring about 2 millimeters in diameter. Sexual maturity is attained by the female when it has reached a length of approximately 1½ inches, and the male may be sexually mature when 1¼ inches long. The largest fish taken during the present investigation was a female 4⅞ inches long, which represents the maximum size attained by the species, and the largest male was 4 inches long. The females reach a somewhat larger size than the males, the difference in the average length between the sexes among the adult fish being about one-half inch.

This fish is very common in the shallow, brackish-water coves and inlets of Chesapeake Bay, ascending streams to fresh water. It was rarely taken in strictly salt water, the species being more fresh than brackish water in its habits than *F. majalis*, athough the habitats of the two overlap. *F. heteroclitus* is found on many kinds of bottom, but it prefers mud, one of its common names being "mud dabbler," in allusion to its mud-frequenting habit. Chidester (1920, pp. 551-557), who made a special study of the habits of *F. heteroclitus* on the Bonhamptown Marshes, N. J., and at Woods Hole, Mass., states that a spring migration from the mouth of the Raritan River to the brackish and fresh waters takes place, and that when cold weather comes they again retire to the deeper waters. Those caught in pools burrowed in the mud upon the approach of cold weather. During the winter fish were found burrowed in the mud at a depth of 6 to 8 inches.

This killifish has been found to be of considerable value as an eradicator of mosquito larvæ on the brackish-water marshes of New Jersey and elsewhere. In addition to its value as a mosquito destroyer, it is of importance, no doubt, as food for larger predacious fishes. In some localities, notably New York and New Jersey, large quantities are sold as bait.

Habitat.—Coastwise, in brackish and fresh water, from Anticosti Island, Labrador, to Tampico, Mexico.

Chesapeake localities.—(*a*) Previously recorded from many sections of the bay and from tributary streams from Baltimore to Cape Charles. (*b*) Specimens in the collection: From all parts of the bay; taken almost daily during shore collecting, from April to November, in coves, inlets, and streams from Havre de Grace, Md., to Lynnhaven Roads, Va.

60. Fundulus majalis (Walbaum). Killifish; Striped killifish; "Gudgeon"; "Bull minnow."

Cobitis majalis Walbaum, Artedi Genera Piscium, III, 1792, p. 12; Long Island.
Hydrargyra majalis Uhler and Lugger, 1876, ed. I, p. 150; ed. II, p. 128.
Fundulus fasciatus Uhler and Lugger, 1876, ed. I, p. 148; ed. II, p. 126.
Fundulus majalis, Bean, 1891, p. 92; Smith, 1892, p. 64; Jordan and Evermann, 1896–1900, p. 639, Pl. CI, figs. 271, 271a, 271b; Evermann and Hildebrand, 1910, p. 159; Fowler, 1912, p. 54.

Head 3 to 3.6; depth 3.3 to 4.7; D. 13 to 15; A. 11 or 12; scales 33 to 36. Body rather slender, compressed posteriorly; caudal peduncle moderate, its depth 2 to 2.8 in head; head rather long, depressed; snout long, blunt, its length 2.6 to 3.5 in head; eye 6 to 8.6 in adults, 4 to 5.6 in young; interorbital 2.7 to 3.1; mouth horizontal, terminal, small; premaxillaries protractile; lower jaw

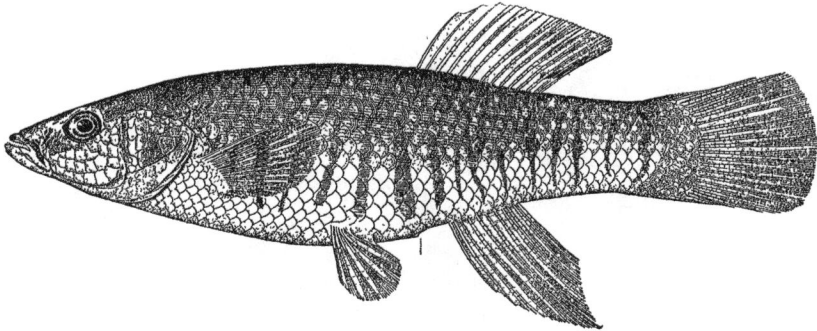

FIG. 73.—*Fundulus majalis*, adult male

slightly in advance of the upper; teeth all pointed, in villiform bands; scales moderate, 18 or 19 rows in advance of dorsal; dorsal fin rather long, notably higher in adult males than in the females. The rays of about equal length, its origin a little nearer upper anterior angle of gill opening than the base of caudal in the male, about equidistant from these two points in the female; caudal truncate or slightly rounded; anal rather short, much higher in adult males than in the females, the fourth or fifth rays longest, the origin of fin below anterior third of dorsal; ventrals inserted about equidistant from anterior margin of eye and base of caudal, usually reaching origin of anal in adult, shorter in young; pectoral rounded, the upper rays longest, 1.4 to 1.9 in head.

Color of adult male, back olive, sides salmon yellow, with 15 to 20 vertical black stripes; belly salmon yellow; cheeks and opercles diffused with black; dorsal dusky, a black ocellated spot on last rays; caudal somewhat dusky; anal, ventrals, and pectorals lemon. Female olivaceous above, white below, black markings on side, with considerable variation in adults of same size. The most common markings are two or three longitudinal stripes with several vertical crossbars near base of tail. Some fish marked with mixture of longitudinal, oblique, vertical, and complete or broken lines. The young of both sexes are marked with about 7 to 12 vertical bands along the sides. The markings differentiating the sexes usually occur when the fish are 1½ or 2 inches in length.

Many specimens of this species were preserved, ranging from 20 to 170 millimeters (⅘ to 6¾ inches) in length. This species is readily recognized by the large size that it attains, by the long head and snout, and by the distinct dark bars and stripes. The sexes, as shown by the color description, are readily recognized by the difference in the markings, the males having dark cross-bars only and a prominent dark spot on the last rays of the dorsal fin, while the females have two or three longitudinal dark lines along the sides and no black spot on the dorsal fin.

The food of this fish consists of small mollusks, small crustaceans, small fish, and insects and insect larvæ. Many of the stomachs examined contained a considerable amount of sand and some vegetable débris. In the field it was noticed that this killifish fed greedily on pieces of meat, bread, and on shrimp eggs.

Spawning occurs from April to September, one female probably producing several broods of eggs during a single season. The eggs are rather large and spherical, measuring about 2 millimeters (12 to 14 to an inch) in diameter. Those of one brood are of uniform size. The ovary is single, and the largest number of eggs of uniform size contained therein in specimens examined was 540. In general the large fish produce more eggs at one time than the smaller ones. Sexual maturity is attained by the female when it has reached a length of approximately 3 inches, and the male may be sexually mature when 2½ inches long. The largest fish caught during the present investigation was a female 8 inches long, which represents the maximum size of the species. The females reach a somewhat larger size than the males, the difference in the average length between the sexes among the adult fish being about one-half inch.

This fish is very common in the vicinity of Chesapeake Bay and is found in bays, coves, creeks, tide pools, and along the outer shores. It is most abundant in small protected bodies of water, preferring especially to hover near the entrances to such places. It travels in schools of a few individuals to several hundred or more. On an ebbing tide it may be found on shallow flats, where the water it but a few inches in depth; but on a flood tide it adheres to the very shore's edge, where it often is cast on the beach by the waves, from which it easily returns to the water. If placed on the beach some distance from the water the fish has the ability to reach its habitat by a series of jumps. Experiments of this kind were made by the authors. Fish were placed at various distances (5 to 20 feet) from the water's edge. In almost every instance they jumped unerringly toward the water, progressing from several inches to several feet at a time. A special article on this apparent "homing instinct" in this species, written by Prof. S. O. Mast, appears in the Journal of Animal Behavior, vol. 5, No. 5, September–October, 1915, pp. 341 to 350.

This fish is of no commercial value, but because of its general distribution and great abundance in Chesapeake Bay is of importance as food for other species.

Habitat.—Massachusetts to Florida in coastwise protected waters, brackish ponds, creeks, mouths of rivers, tide pools, etc.

Chesapeake localities.—(a) Previous records: Patapsco, Patuxent, and Potomac Rivers, Hampton Creek, Cape Charles, and other localities. (b) Specimens in collection: From all parts of Chesapeake Bay, along the shores from Baltimore, Md., to Lynnhaven Roads, Va., taken almost daily during the entire period of shore collection from April until November, 1921.

61. Fundulus ocellaris (Jordan and Gilbert). Killifish; Ocellated killifish.

Fundulus ocellaris Jordan and Gilbert, Proc., U. S. Nat. Mus., 1882, p. 255; Pensacola, Fla. Jordan and Evermann, 1896–1900, p. 642, Pl. CII, fig. 274.

Head 3.1 to 3.6; depth 3.7 to 4.3; D. 10 or 11; A. 9 or 10; scales 34 to 36. Body rather slender, compressed; caudal peduncle strongly compressed, its depth 1.8 to 2.3 in head; head depressed; snout moderate, 3.7 to 5.5 in head; eye 3.6 to 4.6; interorbital 2.2 to 2.7; mouth slightly superior, largely transverse; premaxillaries protractile; teeth all small, in a villiform band in each jaw; scales moderate, 18 or 19 oblique rows between upper angle of gill opening and origin of dorsal; dorsal fin moderate, much higher in the male than in the female, inserted over or slightly in advance of the anal, about equidistant from tip of tail and anterior margin of eye; caudal fin convex, somewhat more so than in *F. heteroclitus;* anal fin of about the same size as the dorsal, notably higher in the male than in the female; the oviduct ending at base of first anal ray; ventral fins rather small, inserted about equidistant from anterior margin of eye and base of caudal; pectoral fins moderate, 1.65 to 2 in head.

Color of female in life, brownish olive above, pale or slightly greenish below; lower part of sides from eye to anal yellowish; head, back, and sides of body irregularly sprinkled with black dots; sides with about 13 blackish cross bars; dorsal and caudal dusky golden, the base of these fins with small black dots similar to those on the body; a large black ocellus on posterior rays of the dorsal (elongate on some specimens and not completely surrounded by white); anal fins wine color; ventral fins plain translucent; pectoral fins dusky golden. Color of male in life dark green above, pale below, sides with pearly spots, these most numerous on posterior half of body, frequently forming indistinct vertical bars; dorsal mostly dusky, with pearly spots at base, a black ocellus near base of last ray usually present; caudal and anal fins slightly dusky, each with pearly spots on base; the margin of the anal pale or slightly pinkish; other fins plain translucent. In young individuals of 30 millimeters and less in length the color is uniformly that of the adult female. The color of the

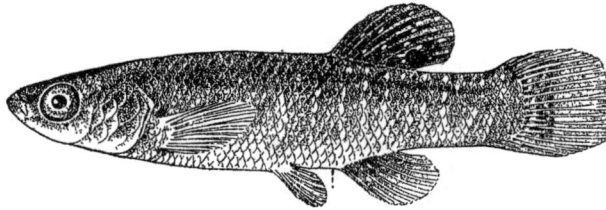

♂

FIG. 74.—*Fundulus ocellaris*, adult male

species appears to be rather variable. As no notes were taken on the life colors of the males taken in Chesapeake Bay, the color description of that sex offered herewith is based on specimens from Beaufort, N. C.

This species is represented by 25 specimens in the present collection, ranging from 30 to 60 millimeters (1¼ to 2⅜ inches) in length.

The females of this species may be recognized by the brownish olive color, the upper parts being sprinkled with black dots and the sides having short, black, vertical bars. The males may be recognized by the same general greenish color of the female, but instead of being sprinkled with black dots this sex is sprinkled with pearly dots, which sometimes form indistinct vertical bars on

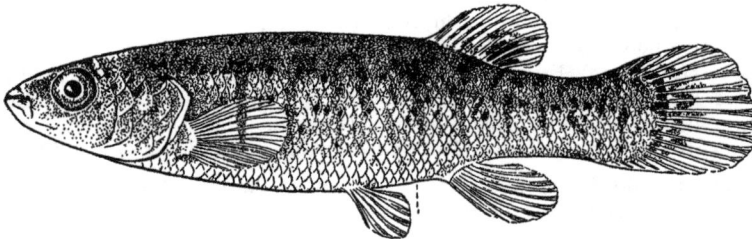

♀

FIG. 75.—*Fundulus ocellaris*, adult female

the sides. Both sexes usually have a large ocellus on the posterior rays of the dorsal fin in allusion to which the species received its specific name, *ocellaris*.

The food of this fish, as shown by the contents of 15 stomachs, consists of insects, insect larvæ, small crustaceans, and small mollusks. A small amount of vegetable matter, too, was removed from three of the stomachs examined.

Gravid specimens were taken only during April and May. At Beaufort, N. C., however, where the senior author made special observations of this and related species, spawning fish were found from April to October. Sexual maturity appears to be attained when the fish are about 40 millimeters long. The largest female taken during the present investigation measured 60 millimeters (2⅜ inches) in length. The largest specimen taken at Beaufort during extensive collecting was 65 millimeters (2⅝ inches) long. The size of the male averages somewhat smaller.

This fish was taken only on the muddy marshes of Lynnhaven Bay. Our imperfect knowledge of the life history of this fish would indicate that the species is limited in its habitat to shallow, muddy, brackish-water swamps, where it is taken in company with *F. heteroclitus*.

This fish is said to be rather common on the Gulf coast. On the Atlantic coast it has been recorded only from Beaufort, N. C. (Hildebrand, 1916, p. 306). The range is now extended northward to Lynnhaven Bay, Va.

Habitat.—Coastwise from Chesapeake Bay to Louisiana.

Chesapeake localities.—(*a*) Previous records: None. (*b*) Specimens in collection: Lynnhaven Marshes, May 9 to 16, June 10 to 17, September 26, 1921, and April 6, 1922.

62. Fundulus diaphanus (LeSueur). Killifish; Fresh-water killy.

Hydrargyra diaphanus LeSueur, Journ., Ac. Nat. Sci., Phila., 1817, p. 130; Saratoga Lake.
Fundulus multifasciatus Uhler and Lugger, 1876, ed. I, p. 150; ed. II, p. 127.
Fundulus diaphanus Bean, 1883, p. 366; Smith, 1892, p. 65, Pl. XIX; Jordan and Evermann, 1896-1900, p. 645, Pl. CIII, figs. 275 and 275a; Smith and Bean, 1899, p. 184; Evermann and Hildebrand, 1910, p. 159; Fowler, 1912, p. 54.

Head 3.3 to 4; depth 4.1 to 5.7; D. 13 or 14; A. 10 to 12; scales 41 to 46. Body rather elongate, compressed; depth of caudal peduncle 2.2 to 3.3 in head; head depressed; snout broad and rather long, 2.8 to 4 in head; eye 2.8 to 4; interorbital 2.4 to 3.4; mouth mostly transverse, slightly superior; teeth in villiform bands, rather fewer and stronger than in related species; scales rather small, 14 to 18 oblique rows between upper angle of gill opening and origin of dorsal; dorsal fin rather long, scarcely higher in males than in females; inserted nearly an eye's diameter in advance of origin of anal in both sexes; inserted about equidistant from tip of snout and tip of caudal in females; fully an eye's diameter farther forward in adult males; caudal fin straight or slightly concave; anal fin shorter than the dorsal; the oviduct extending slightly above the base of the first anal ray; ventral fins of moderate size, inserted about an eye's diameter nearer tip of snout than base of caudal; pectoral fins rounded, 1.3 to 1.7 in head.

Color in life, female, olive above, silvery white on lower part of side, abdomen white; sides with about 16 to 20 narrow greenish bars, becoming dark in spirits; dorsal, caudal, and pectorals yellow; anal and ventrals plain translucent. Male, greenish olivaceous above, abdomen white; caudal peduncle bluish white underneath; sides with about 20 to 22 silvery iridescent vertical bars; dorsal more or less dusky, sometimes with small dark dots on the base; caudal dusky; anal and pectorals more or less yellowish; ventrals mostly bluish white, tinged with yellow. The young, as usual in this group of fishes, are similar in color to the adult female, the differentiation in color between the sexes taking place when the fish reach a length of approximately 50 millimeters.

Numerous specimens, ranging from 32 to 111 millimeters (1¼ to 4⅜ inches) in length, were preserved. This fish is distinguished from related species by the elongate body, long depressed snout, small scales, and by the many narrow vertical bars on the sides, the latter being darker than the ground color in the female and silvery in the male.

The food of this species, as indicated by the contents of 15 stomachs examined, consists of small crustaceans, insects, mollusks, annelid worms, and of miscellaneous unidentified vegetable matter.

Gravid fish were taken from April until September. The eggs, when fully developed, are spherical and about 2 millimeters in diameter. The ovary is single, and it usually contains eggs of more than one size. The largest number of eggs of one size found in one ovary in a limited number of specimens examined was 252. The males in this fish, as well as in the other species of the genus, are somewhat smaller than the female, the average difference in size being about one-half inch. Sexual maturity appears to be attained when the fish is about 2¼ inches in length. The largest individual of this species taken during the present investigation was a female 111 millimeters (4⅜ inches) in length, which appears to represent the maximum size attained by the species.

This minnow is more fresh-water in its habits than are the other members of the genus discussed in the present report. It is common in bays, rivers, and coves where the water is only slightly brackish, and it runs up the streams tributary to Chesapeake Bay into fresh water.

This killifish no doubt is of some value as an eradicator of mosquito larvæ. It is said to be an excellent bait for the larger predatory fishes, and it also is valuable as a natural food for these larger fishes.

Habitat.—From Quebec to North Carolina, represented by the variety *menona* in the region of the Great Lakes and in the northern part of the Mississippi Valley.

Chesapeake localities.—(*a*) Previously recorded from many sections of the fresher arms of the bay and from the streams tributary to the bay. (*b*) Specimens in the collection from the lower Rappahannock River, Va., to Havre de Grace, Md., on the western shore of the bay and as far south as Cape Charles City, Va., on the eastern shore. Not taken in the lower York River, Buckroe Beach, and Lynnhaven Bay.

63. Fundulus luciæ (Baird).   Killifish; Baird's killifish.

Hydrargyra luciæ Baird, Ninth Smithsonian Report, 1854 (1855), p. 334; Beasley's Point, N. J.
Zygonectes luciæ Smith, 1892, p. 68, Pl. XVIII, fig. 3.
Fundulus luciæ Jordan and Evermann, 1896–1900, p. 654, Pl. CVII, fig. 286; Crawford, 1920, p. 75.

Head 3.1 to 3.6; depth 3 to 4.4; D. 8; A. 10; scales 34 to 36.   Body rather elongate, compressed; caudal peduncle strongly compressed, its depth 2 to 2.4 in head; head depressed, snout short, 4 to 4.8 in head; eye 4 to 4.8; interorbital 2.1 to 2.6; mouth slightly superior, largely transverse; premaxillaries protractile; teeth pointed, in villiform bands in each jaw, with the outer teeth in each jaw considerably enlarged; scales moderate, 15 or 16 oblique rows between upper anterior angle of gill opening and origin of dorsal; dorsal fin short, its origin in both sexes over or slightly behind origin of anal and inserted about equidistant from the tip of the tail and the anterior half of the eye; caudal fin convex; anal fin with a somewhat longer base than the dorsal; ventral fins very small, inserted a little nearer base of caudal than tip of snout; pectoral fins moderate, 1.4 to 1.7 in head.

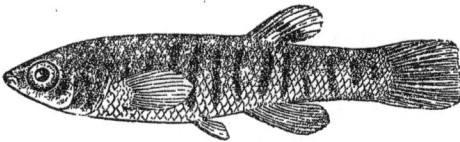

FIG. 76.—*Fundulus luciæ,* adult male

Color of female plain grayish green, pale below; eye dark, with narrow golden band; opercle with brownish peppery spots, forming a blotch; a dark vertebral line present; the fins all plain yellowish brown. Male olive-green above, lower sides golden, orange-white underneath; sides with 11 to 14 crossbars slightly darker in color than the back; the fins orange, pinkish, or light brown, the dorsal and anal usually bright orange to reddish; the dorsal with a black ocellus on the posterior rays. The young of about 26 millimeters and less in length all bear the modest color of the female, and the sexes are not distinguishable.

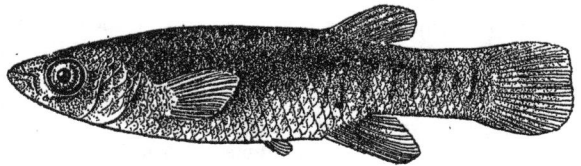

FIG. 77.—*Fundulus luciæ,* adult female

This species is represented by about 75 specimens in the present collection, ranging in length from 22 to 40 millimeters (⅞ to 1 9/16 inches). The females are readily recognized by the plain grayish green color, but the males resemble very closely the smaller brightly colored males of *F. heteroclitus.* For example, at Love Point, Md., both species were taken and placed in the same jar. Later it was found impossible in some instances to separate the males by the color. Usually the species may be distinguished by the slightly more posteriorly placed dorsal fin in *F. luciæ,* but the most reliable character, for purposes of identification, is the length of the dorsal fin, which in the present species constantly has two or three fewer rays, the usual number being eight.

The food of this species, as shown by the contents of nine stomachs, appears to be similar to that of *F. heteroclitus,* consisting largely of small crustaceans, small mollusks, and annelid worms.

Gravid specimens were taken only during April and May, but at Beaufort, N. C., the senior author found that in that vicinity, at least, the species spawns throughout the spring and summer, or from April until October. The eggs are rather large and spherical, measuring about 2 millimeters in diameter when mature. Since the fish reaches a much smaller size than *F. heteroclitus,* sexual maturity is attained when the fish is much smaller. The exact size at which the sexual organs begin to develop has not been determined, but fish 1¼ inches long are plainly adult fish.

This minnow is not generally common. It was taken at only five localities in Chesapeake Bay, and it was only fairly common in one of these places, namely, Love Point, Md. It frequents very shallow, brackish water, and is taken in company with *F. heteroclitus*. This species has rarely appeared in collections. It was first taken and described by Baird (1855, p. 334) from Beasely Point, N. J. Then it appears not to have been taken again until Smith (1892, p. 68) secured two specimens from the lower Potomac. Fowler records the species from Pecks Bay, N. J. (1912, p. 36), and from Cedar and Parramores Islands, Lotusville Branch, and Virginia Beach, Va. (1912, p. 57). The senior author found the species common in restricted areas near Beaufort, N. C. (1916, p. 306), and Crawford took some specimens at Lewisetta, Va. (1920, p. 75). The color of the female appears not to have been described previously.

The species reaches a small size. The largest specimen taken in Chesapeake Bay during the present investigation was a female measuring 40 millimeters (1$\frac{9}{16}$ inches) in length. This fish, because of its small size and general scarcity, is of little or no commercial importance, either directly or indirectly.

Habitat.—Coastwise in brackish water from New Jersey to North Carolina.

Chesapeake localities.—(a) Previously recorded: Only from the lower Potomac. (b) Specimens in collection: Love Point, Md., May 11 and 12, 1922; Annapolis, Md., September 9, 1921, and May 1 to 3, 1922; Solomons, Md., April 27, 1922; Crisfield, Md., November 21, 1921; brackish swamp opposite Lynnhaven Roads, Va., June 10 to 17, 1921.

Family XXXII.—PŒCILIIDÆ. The top minnows

Body elongate; compressed posteriorly; head depressed; mouth terminal, or nearly so; teeth pointed; no lateral line; dorsal fin small, composed of soft rays only; caudal fin usually round, never forked; anal fin in the male modified, some of the rays produced, others short and more or less coalesced, the fin forming an intromittent organ. Species viviparous.

47. Genus GAMBUSIA Poey. Top minnows

Body moderately robust; head rather short, depressed; mouth moderate, the lower jaw projecting; teeth all fixed, pointed, in bands in each jaw; scales rather large; fins small, the anal fin in the male modified, the third, fourth, and fifth rays much produced, forming an intromittent organ; color plain; intestinal canal short; species viviparous.

64. Gambusia holbrooki Girard. Top minnow.

Gambusia holbrooki Girard, Proc., Ac. Nat. Sci., Phila. 1859, p. 61; Palatka, Fla.
Gambusia affinis Smith, 1892, p. 69, Pl. XX (female); Jordan and Evermann, 1896–1900, p. 680, Pl. CXIII, figs. 299 and 299a; Evermann and Hildebrand, 1910, p. 160.

Head 3.5 to 4.4; depth 3.5 to 4.7; D. 7 or 8; A. 9 to 11; scales 26 to 30. Body rather robust moderately compressed, usually deeper in the female than in the male; caudal peduncle strongly compressed, its depth 1.75 to 2.3 in head; head depressed; snout short and broad, it length 2.85 to 4.65 in head; eye 2.5 to 3.5; interorbital 1.8 to 2.5; mouth slightly superior, the lower jaw projecting; premaxillaries protractile; teeth small, pointed, in a band in each jaw; scales moderate, 12 to 14 oblique series between the upper angle of gill opening and origin of dorsal; dorsal fin small, placed behind origin of anal; caudal fin rounded; anal fin similar to dorsal in female, modified into an intromittent organ in the male, the third, fourth, and fifth rays being much produced, placed proportionately farther forward in adult males than in females; ventral fins small, inserted much nearer tip of snout than base of caudal; pectoral fins moderate, 1.1 to 1.4 in head.

Color of male and female similar but with considerable variation according to the environment because of the development of color adaptation; usually olivaceous above, grayish on sides, and pale underneath; scales on upper parts with dusky punctulations, these often concentrated on the margins of certain scales, forming irregular dark dots; a dark vertebral streak present in front of dorsal; a dusky area usually present under the eye and at occiput; dorsal and caudal usually slightly greenish and with dark dots (in light colored specimens the dots are frequently wanting); other fins plain translucent. The female with a black blotch on each side of abdomen just above and in front of vent when gravid, increasing in size with the development of embryos and becoming very prominent before parturition takes place.

Many specimens were collected from various restricted areas, ranging in length from 19 to 42 millimeters (¾ to 1⅝ inches), and these form the basis for the above description. The species is recognized by its small size (although not unlike *Lucania parva* in size), the plain greenish or grayish coloration, and the depressed head. The males are readily distinguished by the peculiar prong-shaped development of the anal fin, which occurs in no other minnow in the vicinity of the Chesapeake. Geiser recently (1923, pp. 175 to 188) has shown that Gambusia of the eastern part of the

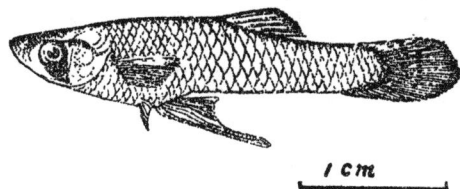

FIG. 78.—*Gambusia holbrooki*, adult male

United States differ from those of the Mississippi drainage in the detailed structure of the modified anal fin (intromittent organ; also called gonopod by Geiser). The eastern form has a larger number (11 to 14) of antrorse teeth on the distal ossicles of the third ray; and the larger ossicles, situated toward the base of the fin from the distal ones, are posteriorly denticulate, whereas in the Mississippi fish these segments are entire. The posterior branch of the fourth ray usually has seven ossicles, with teeth on the posterior margin in the eastern fish, while the Mississippi Valley fish has only about five ossicles with teeth. These differences, together with other minute differences, appear to be constant and therefore of specific value. Consequently the eastern Gambusia must be regarded as distinct from the central western one; and, as pointed out by Geiser, the Atlantic slope one, in accordance with rules of nomenclature, becomes *G. holbrooki* Girard and the Mississippi Valley fish should stand as *G. affinis* (Baird and Girard).

This fish feeds on a large variety of foods, which, however, are taken principally at the surface, the habit of surface swimming and feeding being more strongly developed in Gambusia than in any of the other so-called "top minnows" inhabiting the waters of the United States. In 15 stomachs examined from specimens in the present collection the principal food consisted of insects and insect larvæ, although a few small crustaceans and a few egg masses, too, were present.

The young of this species are born alive and well developed, being 8 to 10 millimeters long. The first young appear in May, and spawning continues until about September. One female may deliver several broods of young during a single season. The number of young produced at one time may vary from a few to a hundred or more. The largest brood on record (Hildebrand, 1921, p. 12) consisted of 211 young. As a rule a small female

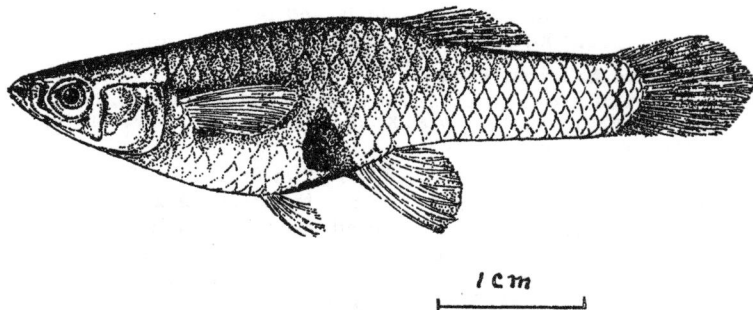

FIG. 79.—*Gambusia holbrooki*, gravid female

has fewer young than a large one, although there are many exceptions to this rule. The average number of young in a brood probably does not exceed 25. A female kept in the aquarium (Hildebrand, 1917, p. 6) once produced six broods of young during one summer. The young grow rapidly, and usually when about 20 millimeters long the sexual differentiation in the anal fin takes place, and the fish are sexually mature very soon thereafter. Young born in May and reared in the aquarium have been known to produce young by September of the same year.

The largest specimen taken during the present investigation was only 1⅝ inches in length. This would indicate that the fish run rather small in the vicinity of the Chesapeake, for the maximum size attained by the species is about 2⅜ inches. The males, however, are much smaller, rarely reaching a length of 1½ inches.

This fish is common only in restricted areas in the brackish and fresh-water arms and disconnected pools and marshes of Chesapeake Bay, where, as a rule, the water is quiet and more or less stagnant. Although a few specimens of this fish have been reported as far north as New Jersey, the species may be said to reach the northern limits of its distribution in the Chesapeake region, where it is much more particular in the selection of its habitat than in more southern localities. However, it is usually quite common within the restricted places where it is found.

This top minnow and closely related varieties or species are now widely employed in the South for the destruction of the aquatic stages of the mosquito. Gambusia are so effective for this purpose that it is doubtful if a more valuable fish swims in American waters. For accounts of the importance of this fish for the control of mosquito breeding see Hildebrand (1919, 1921, and 1925) and Howard (1920).

Habitat.—As here understood, the range of *G. holbrooki* extends from New Jersey to Florida. The top minnows of this genus that occur in the United States have been considered identical and as belonging to a single species by some writers while by others they are considered three distinct species. Geiser, from studies based upon the microscopic anatomy of the anal fin of the male, as already stated, has recently determined (1923, pp. 175 to 188) that the structure of the distal part of this fin ("gonopod") is different in the eastern Gambusia from those of the South Central States, and he also found slight differences in this structure between specimens from the Central and Southwestern States; but he regards this last difference as of varietal importance only, while the more pronounced difference between the eastern and central specimens he regards as of specific value. He also found the Mexican species, *G. senilis*, extending into Texas. According to Geiser, the species of Gambusia should now stand as follows: Eastern form, *G. holbrooki;* central form, *G. affinis;* southwestern form, *G. affinis* var. (unnamed, presumably *patruelis* Baird and Girard); and the Mexican form, *G. senilis,* from southern Texas. On the basis of these new divisions, the exact limits of distribution of the species and varieties remain to be established. (A further revision of the species of Gambusia and their distribution has appeared since the foregoing was written, in "Studies of the Fishes of the Order Cyprinodontes, VI," by Carl L. Hubbs, in Miscellaneous Publications No. 16, Museum of Zoology, University of Michigan, 1926, pp. 26–40, to which the reader is referred.)

Chesapeake localities.—(a) Previous records: "Lower Potomac" and St. Georges Island. (b) Specimens in collection: Annapolis, Love Point, Solomons, Oxford, and Crisfield, Md.; also from the marshes of Lynnhaven Bay, Va.

Order SYNENTOGNATHI

Family XXXIII.—BELONIDÆ. The needlefishes

Body very elongate, slender, compressed or not; both jaws produced, forming a beak, the lower one a little the longer; maxillary united with the premaxillaries; each jaw with a band of short, pointed teeth and a series of enlarged ones; lateral line low, running along the edge of belly; scales small; dorsal fin opposite the anal; no finlets; air bladder present.

KEY TO THE GENERA

a. Body only moderately or not compressed, the depth not greatly exceeding the width
-- Tylosurus, p. 147
aa. Body rather strongly compressed, the width less than half the depth_____ Ablennes, p. 150

48. Genus TYLOSURUS Cocco. The needlefishes; the garfishes

Body very elongate, little or not compressed; gill rakers obsolete; lateral line on sides of abdomen, becoming median on the caudal peduncle; dorsal and anal elevated anteriorly, falcate.

KEY TO THE SPECIES

a. Origin of dorsal behind origin of anal, about equidistant from base of ventral and base of caudal, with 13 to 17 rays; caudal fin scarcely forked, the margin merely concave; a dark lateral band present on sides and a median band on back_____ *marinus*, p. 148

aa. Origin of dorsal over the origin of the anal, much nearer the base of the ventral than base of caudal, with 23 rays; caudal fin well forked; no dark bands on sides or back____ *acus*, p. 149

65. Tylosurus marinus (Walbaum).   Garfish; Houndfish; Billfish; Needlefish.

Esox marinus Walbaum, Artedi Piscium, III, 1792, p. 88; Long Island.

Belone longirostris Uhler and Lugger, 1876, ed. I, p. 142; ed. II, p. 121.

Tylosurus marinus Bean, 1883, p. 366; Bean, 1891, p. 92; Smith, 1892, p. 69; Jordan and Evermann, 1896–1900, p. 714; Smith and Bean, 1899, p. 185; Evermann and Hildebrand, 1910, p. 160; Fowler, 1912, p. 54.

Head 2.45 to 3.25; depth 14.3 to 24; D. 13 to 17; A. 17 to 21; scales about 325. Body slender, not compressed, very slender in young; caudal peduncle depressed, broader than deep, with a slight keel on the sides, its depth 15 to 20 in head; head long, flat above and with a broad groove; cheeks and opercles straight and nearly vertical; snout produced into a long slender beak, its length 1.47 to 1.87 in head; eye 9.5 to 13; interorbital 9.5 to 13; mouth large, horizontal; lower jaw the longer; upper jaw scarcely arched; teeth in bands in each jaw, very sharply pointed, the inner ones enlarged, caninelike; lateral line complete, on lower edge of sides, curved upward on posterior part of caudal peduncle; scales small, cycloid, present on the preopercle but not on the opercle; dorsal fin inserted behind origin of anal, about equidistant from base of ventrals and base of caudal, the anterior 6 or 7 rays much longer than the posterior ones; caudal fin with concave margin; anal fin similar to the dorsal but with a somewhat longer base; ventral fins rather small, inserted about an eye's diameter nearer margin of opercle than base of caudal; pectoral fins moderate, the upper rays the longest, 3.4 to 4.7 in head.

Color greenish above, silvery on sides, white below; a dark greenish stripe on median line of back and a narrower bluish silvery stripe along sides, becoming broader and less distinct posteriorly and frequently disappearing on caudal peduncle in large specimens; snout dark green; cheeks and opercles silvery; a blackish blotch, deeper than long, on upper part of preopercle; dorsal plain translucent or somewhat dusky, the longest rays yellowish at tips; caudal bluish at base, the lobes yellowish; anal plain, the longest rays dusky in some specimens, yellowish in others; ventrals and pectorals plain.

Many specimens of this common species were preserved. The above description is based upon specimens ranging from 47 to 619 millimeters (1⅞ to 24⅜ inches) in length. This fish is distinguished from related species by the short dorsal and anal fins, which, however, show considerable variation among individuals in the number of rays present. In 130 specimens, 3 had 13 rays in the dorsal, 11 had 14, 35 had 15, 66 had 16, and 15 had 17. In the same number of fish, 7 had 17 rays in the anal fin, 40 had 18, 65 had 19, 17 had 20, and 1 had 21. The body is cylindrical at all ages, but it is much more slender in the young than in larger fish, which accounts for the great variation in the depth of the body shown in the description.

The food of this gar, according to the literature consulted, consists almost wholly of fish. We are able to substantiate this statement by the fact that of 18 stomachs examined, the contents, with a single exception, consisted of small fish, including the silver mullet (*Mugil curema*), the killifish (*Fundulus diaphanus*), and one or more species of silverside. One specimen had fed on shrimp. It is said to take small fish crosswise in its jaws, afterwards turning its prey around for the purpose of swallowing it.

Little is known of the spawning habits of this fish, but it is said that the eggs are deposited during the summer in bays and estuaries (Smith, 1907, p. 157). The ovary is single and a very large number of eggs are produced at one time. A specimen taken on May 21, 1921, has the ovary greatly distended with eggs that slightly exceed 1 millimeter in diameter.

The habit of surface swimming, which is correlated with surface feeding, makes this gar one of the most conspicuous fishes, and it is therefore well known to those living on the sea shores within the range of the species. Its movements are very swift and it is extremely difficult to catch with a dip net. It readily becomes entangled in the meshes of a seine or drag net, however, because of its large mouth and long teeth, sometimes doing a considerable amount of damage to the nets. At times it is a nuisance to anglers on account of its bait-stealing habits.

This garfish is said to attain a length of 4 feet (Smith, 1907, p. 157), but the largest fish observed in Chesapeake Bay did not exceed 2½ feet. Little is known concerning the rate of growth. Young fish taken during the late spring and summer were of the following sizes: June 10, 1921, Lynnhaven Roads, 47 to 50 millimeters (1⅞ to 2 inches); June 25–30, Buckroe Beach, 50 to 106 millimeters (2 to 4⅕ inches); July 10, lower York River, 54 to 70 millimeters (2⅛ to 2⅘ inches); July 25–30, lower Rappahannock River, 62 to 129 millimeters (2⅖ to 5 1/12 inches). Larger fish are difficult to group. Thus, while using a small haul seine at Ocean View, Va., on October 3, 15 fish were taken, measuring from 11 to 16 inches, the intervening sizes being well represented.

Several specimens were weighed in the field, giving the following results: Length 13.3 inches, weight 1.4 ounces; length 13.4 inches, weight 1.6 ounces; length 21.1 inches, weight 9.4 ounces; length 23 inches, weight 9.6 ounces; length 23.2 inches, weight 10 ounces.

This species is comparatively common in the Chesapeake region and is found in all parts of the bay from Havre de Grace, Md., to Cape Henry, Va. It ascends the various tributaries and has been recorded from fresh water (Smith and Bean, 1899, p. 185). Bean (1901, p. 405) states that "it ascends rivers far above the limits of the tide, feeding upon minnows and other small fishes." It is generally found in small schools of from a few to several dozen fish, and occasionally an individual is taken alone.

Although the flesh of the gar is palatable, the fish has no commercial importance in Chesapeake Bay. It was never observed in any of the fish markets, and it was noted that fishermen always culled this species from the other food fishes.

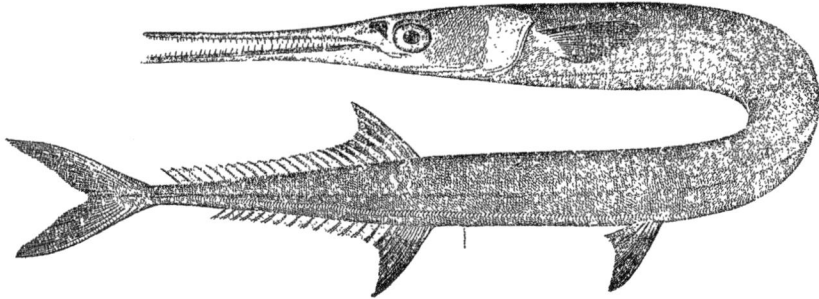

FIG. 80.—*Tylosurus acus.* From a specimen 43 inches long

Habitat.—Massachusetts to Texas, sometimes straying northward to Maine.

Chesapeake localities.—(*a*) Previous records: Nearly all brackish waters of Chesapeake Bay and its vicinity; Bryans Point and Aqueduct Bridge, Potomac River. (*b*) Specimens in collection: All parts of the bay from Havre de Grace, Md., to Lynnhaven Roads, Va.

66. Tylosurus acus (Lacépède). Houndfish; Garfish; Needlefish.

Sphyræna acus Lacépède, Hist., Nat. Poiss., V, 1803, p. 6, Pl. I, fig. 2; Martinique.
Tylosurus acus Jordan and Evermann, 1896–1900, p. 716, Pl. CXVI, fig. 309.

Head 2.6; depth 18.5; D. 23; A. 22; scales about 400. Body very elongate, not quite as broad as deep; caudal peduncle depressed, broader than deep, with a keel on the sides, its depth 24 in head; head long, flat above, and with a broad, shallow groove; cheeks and opercles straight and nearly vertical; snout very long and slender, its length 1.85 in head; eye 13.5; interorbital 11; mouth large, horizontal; the lower jaw slightly projecting; the upper jaw weakly arched, the mouth not quite capable of being closed; teeth in bands in each jaw, the inner ones enlarged, long and pointed, not compressed; lateral line complete, curved upward on caudal peduncle; scales quite small, cycloid, present on the preopercle but not on the opercle; dorsal fin inserted over the origin of the anal, much nearer the base of the ventral than base of caudal, the anterior rays elevated, much longer than the posterior ones; caudal fin forked, the lower lobe much the larger; anal fin similar to the dorsal, but with a shorter base; ventral fins moderate, inserted about equidistant from posterior margin of eye and base of caudal; pectoral fins moderate, the upper rays longest, 5.25 in head.

Color of preserved specimen brownish above; sides silvery; pale underneath; ventral surface of head white; no bands on back or sides; dorsal largely black, the anterior part yellowish at base; caudal, anal, and ventrals mostly plain translucent, with more or less dusky, at least on the outer or anterior rays; pectoral fins largely black, only the base and the lower rays yellowish. The general color of the back is greenish in life.

A single specimen, 830 millimeters (32¾ inches) in length, occurs in the collection. In addition, a fish 32 inches in length, in a poor state of preservation, and which was thrown away by fishermen, was found on the shores at Buckroe Beach. This fish was colored blue along the back. The species of this genus are not all well known and the identification is more or less tentative. This disposition of the species is based mainly upon the long snout, the round, sharply pointed teeth, the small scales, the length of the dorsal fin, and the position of the ventral fins. This species is readily separated from the common garfish of Chesapeake Bay by the longer dorsal fin, the more deeply forked tail, and the absence of dark lateral and vertebral bands.

This gar appears to be a West Indian species that occasionally strays northward. In Chesapeake Bay it occurs only as a straggler and has no commercial importance. It reaches a length of 4½ feet.

Habitat.—Massachusetts to the West Indies; also said to occur in the Mediterranean.

Chesapeake localities.—(*a*) Previous records: None. (*b*) Specimens in collection or observed in the field: Lynnhaven Roads, June 8, 1916, pound net; Buckroe Beach, Va., June 21, 1921, found on beach.

49. Genus ABLENNES Jordan and Fordice. The garfishes

This genus differs from Tylosurus chiefly in the compressed body.

67. Ablennes hians (Cuvier and Valenciennes). Garfish; Needlefish; "Silver gar"; "Silver fish."

Belone hians Cuvier and Valenciennes, Hist. Nat. Poiss., XVIII, 1846, p. 432; Havana, Bahia.
Athlennes hians Jordan and Evermann, 1896–1900, p. 718.

Head 3.1 to 3.8; depth 15.9 to 16.5; D. 24; A. 25 or 26. Body very elongate, strongly compressed, head compressed, more or less quadrate, flat above, and with a broad groove, much narrower underneath, the sides nearly straight; snout very long and slender, 1.4 to 1.5 in head; eye 9.65 to 11.6; interorbital 9.7 to 11.3; mouth large; teeth in narrow bands, sharply pointed, the inner ones enlarged; lateral line following the lower edge of the body throughout; scales minute, too small to enumerate accurately; dorsal fin inserted behind origin of anal, about an eye's diameter nearer the base of ventral than base of caudal, the anterior rays elevated; caudal fin broadly forked; anal fin similar to the dorsal; ventral fins rather small, inserted considerably nearer eye than base of caudal; pectoral fins moderate, the upper rays the longest, 3.5 to 3.9 in head.

Color greenish with bright bluish-green reflections above; lower part of sides and abdomen bright silvery; snout bright red at tip; dorsal fin mostly greenish, the tip of the produced rays as well as the posterior rays black; caudal greenish, with more or less dusky and a pale margin; other fins mostly pale green, the pectorals with dusky tips.

Only two specimens of this gar, 400 and 445 millimeters (15¾ and 17½ inches) in length, were secured. The species is readily recognized by the compressed body and the straight and nearly vertical sides.

A considerable variation in color exists among specimens, as shown by a series of fresh specimens examined at Beaufort, N. C. Some specimens have dark blotches on the sides, others have distinct black crossbars, and in still others, as in the Chesapeake Bay specimens at hand, the sides are plain silvery. Much variation with age in the depth of the body, too, was noticed in the Beaufort specimens. In nine specimens, ranging in length from 4⅝ to 26 inches, the depth in the length to base of caudal varied from 11 in the larger fish to 24 in the smaller ones.

This fish is carnivorous. The alimentary canal is a straight tube, without a definite differentiation between the stomach and intestine. The air bladder is very long and narrow and it has very thin walls. The peritoneum is silvery and it bears dusky punctulations. The bones are greenish.

Spawning apparently takes place in the spring. During May, 1915, we took several large fish, with ripe roe, at Beaufort, N. C. The ovary in this species, as in the other members of gar-fishes, is single, and when fully distended with eggs it is fully one-third the total length of the body. The ripe ova are large, measuring about 3 millimeters in diameter.

This gar attains a length of at least 3 feet. It is not common in Chesapeake Bay, but is occasionally seen in pound nets (in the meshes of which it becomes entangled by means of its long snout, large mouth, and sharp teeth) in the southern sections of the bay. This fish is of no commercial importance.

Habitat.—Massachusetts to Brazil.

Chesapeake localities.—(a) Previous records: None. (b) Specimens in collection or observed in field: Buckroe Beach, June 21, 1921; Cape Charles, September 23, 1921.

Family XXXIV.—SCOMBERESOCIDÆ. The sauries

Body long, slender, compressed; both jaws prolonged in the adult, forming a slender beak; maxillary and premaxillary united; teeth feeble; gill rakers numerous, long, and slender; scales small, thin, deciduous; dorsal and anal low, similar, each with 4 to 6 detached finlets, as in the mackerels. A single genus and species comes within the scope of the present work.

50. Genus SCOMBERESOX Lacépède. Sauries; Skippers

Both jaws produced, forming a slender beak, the lower jaw the longer, the jaws short in the young; air bladder large; lateral line near ventral edge of body; scales small, partly covering the opercle.

68. Scomberesox saurus (Walbaum). Skipper; "Northern billfish"; Saury.

Esox saurus Walbaum, Artedi Genera Piscium, III, 1792, p. 93; Cornwall.
Scomberesox scutellatus Uhler and Lugger, 1776, ed. I, p. 144; ed. II, p. 123.
Scomberesox saurus Jordan and Evermann, 1896–1900, p. 725, Pl. CXVII, fig. 314.

Head 3.5; depth 9 to 13; D. 10 or 11–V; A. 12 or 13–VI; scales about 115. Body elongate, compressed; head broad above, narrow below, tapering gradually to the very slender beak; snout longer than rest of head, proportionately shorter in young; lower jaw longer; eye about 3 in postorbital part of head; air bladder large; scales small, about eight rows on upper part of opercle; dorsal and anal fins similar, small, and mainly opposite each other, each followed by five or six detached finlets; caudal fin forked; ventral fins small, inserted about equidistant from eye and base of caudal; pectoral fins shorter than postorbital part of eye.

Color greenish brown above, sides and belly silvery; sides with a silvery lateral band about width of eye, bounding the darker color of the back.

This species does not occur in the present collection. It is included in the present work on the basis of a record by Uhler and Lugger (1876, ed. I, p. 144, and ed. II, p. 123), who state that this fish is found very rarely near the entrance to Chesapeake Bay. The foregoing description is compiled from published accounts.

The skipper is primarily a fish of the open sea, where it travels in large schools and is preyed upon by mackerel, pollock, tunny, and other fish. It is a warm-water fish, and in the western Atlantic it probably lives largely between the latitudes of 11° and 40° N., in which region the young are very numerous. (Bigelow and Welsh, 1925, p. 166.) Its appearance along our immediate shores during the summer is very erratic, and in places where large catches may be made one year none at all will be taken the succeeding year. Although the skipper evidently occurs in the subtropical part of the open Atlantic it has not been reported south of Beaufort, N. C. Its center of abundance along our coast appears to be around Provincetown, Cape Cod, north and south of which it is uncommon. It is strictly pelagic, living exclusively at the surface.

The skipper feeds on small pelagic Crustacea. Doctor Linton listed annelids, fragments of fish, vegetable débris, copepods, and crustacean larvæ as the food of one specimen examined at Woods Hole.

Spawning occurs in the open sea, probably at the surface. "The most interesting phase in the development of the skipper is that the jaws do not commence to elongate until the fry have attained a length of about 40 millimeters, and that the lower outstrips the upper at first, so that fry of 100 to 150 millimeters look more like halfbeaks (Hemiramphus stage) than like their own parents." (Bigelow and Welsh, 1925, p. 166.)

The skipper attains a length of 18 inches, the usual size being 12 to 16 inches.

Habitat.—Temperate parts of the Atlantic Ocean, on both the European and American coasts; Mediterranean Sea; New Zealand. On the western Atlantic coast, ranging from Nova Scotia to Beaufort, N. C.; rare south of Cape Cod.

Chesapeake localities.—(*a*) Previous records: Entrance of Chesapeake Bay (Uhler and Lugger, 1876). (*b*) Specimens in collection: None.

Family XXXV.—HEMIRAMPHIDÆ. The balaos (halfbeaks)

Body elongate, more or less compressed; upper jaw short; lower jaw various (much produced in the specimens included in the present work, the "beak" projecting beyond the upper jaw, being nearly or quite equal to the length of the rest of the head); lateral line placed low on side; scales in regular rows, cycloid; dorsal and anal fins small, placed posteriorly; caudal fin rounded or forked, if forked the lower lobe is the larger. This is a family of rather small, surface-swimming, warm-water shore fishes. A single species is common in Chesapeake Bay, where it has no commercial value.

KEY TO THE GENERA

a. Air bladder simple, not divided into compartments; sides of body not quite vertical, more or less convex; dorsal and anal fins opposite each other, the last ray of dorsal not produced; ventral fins placed about equidistant from gill opening and base of caudal_____Hyporhamphus, p. 152
aa. Air bladder divided into compartments, cellular; sides of body vertical; dorsal fin beginning in advance of anal, its last ray somewhat produced; ventral fins placed much nearer base of caudal than gill opening_____Hemiramphus, p. 153

51. Genus HYPORHAMPHUS Gill. Halfbeaks

Body long and slender; sides not quite vertical and usually more or less convex; lower jaw much produced, the produced portion never much shorter than the rest of head; air bladder simple, not divided into compartments; dorsal and anal fins similar and opposite each other, the last ray of dorsal not produced; ventral fins inserted about equidistant from gill opening and base of caudal.

69. Hyporhamphus unifasciatus (Ranzani). Halfbeak; "Skipjack."

Hemirhamphus unifasciatus Ranzani, Novi. Comment. Ac. Sci. Bonon, V, 1840, 326; Brazil. Lugger, 1877, p. 83.
Hemirhamphus roberti Uhler and Lugger, 1876, ed. I, p. 143; ed. II, p. 122; Bean, 1891, p. 92.
Hyporhamphus roberti Jordan and Evermann, 1896–1900, p. 721, Pl. CXVII, fig. 312.

Head [a] 4.1 to 4.8; depth 6 to 10; D. 13 to 15; A. 15 to 17. Body long, compressed, becoming deeper with age; head low, somewhat depressed above, the sides nearly straight; mandible much produced, except in very young (10 millimeters), developed in somewhat older specimens (25 millimeters), proportionately shorter in large examples than in smaller ones, its length from tip of upper jaw equal to rest of head in specimens about 150 millimeters in length; snout 2.5 to 4 in head; eye 3.2 to 4.8; interorbital space 4 to 4.5; teeth in the jaws in villiform bands; gill rakers rather short and blunt, 23 to 26 on the lower limb of the anterior arch; scales rather firm, cycloid; dorsal and anal similar, placed opposite each other, each scaled at base; caudal forked, the lower lobe much the longer; ventral fins small, inserted nearer base of caudal than eye; pectoral fins rather short, 1.25 to 2 in head.

Color largely silvery, with more or less greenish above; sides with a plumbeous band; middle of back with three dark lines; upper surface of head and mandible dark, the latter with a red tip in life; fins mostly plain translucent.

This species is represented by numerous specimens ranging from very small (larvæ) to 290 millimeters (11½ inches, without mandible) in length. This halfbeak is common in Chesapeake Bay but it is of no commercial importance.

---

[a] Head and body are measured from tip of upper jaw.

The alimentary canal is almost a straight tube, without a definite differentiation between the stomach and intestine. The food of the adult, according to the contents of eight stomachs, consists of small crustaceans, mollusks, and vegetable matter.

The ovary is single. Spawning takes place during summer, and the mature egg, when it is first laid, is approximately 2 millimeters in diameter, almost transparent, and semibuoyant. The very young have no "beak," but in specimens 15 millimeters in length its development has definitely begun. Specimens 100 to 200 millimeters in length have a proportionately longer beak than larger ones.

The youngest specimen (3 millimeters long), taken with a bottom net (July 8), had recently been hatched. Specimens taken with townets (June 10–11 and July 8–9) were 15 to 19 millimeters in length, measured from upper jaw. By the end of July many fish 24 to 49 millimeters long were taken with collecting seines.

The halfbeaks are commonly seen swimming near the surface. Their movements are often sudden and quick, making them rather difficult to capture, and because of their slenderness they pass through all except the smallest meshed nets. The greatest length attained is little in excess of 1 foot.

Habitat.—Cape Cod to Brazil, rarely straying to Maine; most common from Chesapeake Bay southward; on the Pacific coast from the Gulf of California to the Galapagos Islands.

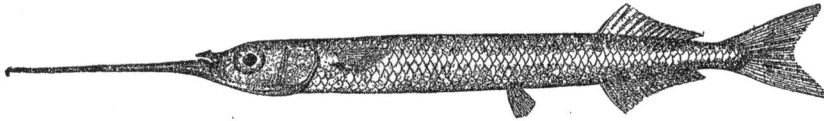

Fig. 81.—*Hyporhamphus unifasciatus*

Chesapeake localities.—(*a*) Previous records: St. Marys County, "southern part of Chesapeake Bay," and Cape Charles city. (*b*) Specimens in collection: From many points from Baltimore, Md., south to Cape Charles and Ocean View, Va., generally common; also taken in brackish water in the lower courses of streams.

52. Genus HEMIRAMPHUS Cuvier. Halfbeaks

Body more robust than in Hyporhamphus; sides nearly vertical and parallel; lower jaw much produced, usually longer than rest of head; air bladder divided into many compartments, cellular; dorsal fin a little longer than the anal and its origin a little farther forward, its last ray slightly produced; ventral fins inserted far backward, much nearer base of caudal than gill opening.

70. Hemiramphus brasiliensis (Linnæus).

Esox brasiliensis Linnæus, Syst. Nat., ed. X, 1758, 314; Jamaica.
Hemirhamphus pleei Bean, 1891, p. 92.
Hemiramphus brasiliensis Jordan and Evermann, 1896–1900, p. 722, Pl. CXVII, fig. 313; Fowler, 1912, p. 54.

Head 4.3 to 4.6; depth 5.4 to 6.3; D. 13 or 14; A. 11 to 13; scales 53 to 57. Body elongate, compressed, the sides vertical; head rather low; mandible much produced, its length from tip of upper jaw 3.3 to 3.9 in length of body; snout 2.8 to 3.5 in head; eye 3.6 to 4.1; teeth in jaws short, mostly in three series; gill rakers very short, 21 to 24 on lower limb of first arch; dorsal fin placed posteriorly, the last ray slightly produced; caudal fin forked, the lower lobe much the larger; anal fin small, beginning under middle of base of dorsal; ventral fins small, inserted about half as far from base of caudal as tip of upper jaw; pectorals 5.9 to 6.8 in body.

Color dusky brown above; sides and below bright silvery; median part of back with an indistinct dark streak, with a black line on each side; an inconspicuous dark streak extending from upper angle of gill opening to base of caudal; dorsal, caudal, and pectoral with more or less dusky; other fins pale. (The caudal fin is said to be orange in life.)

This fish does not occur in the present collection. It was reported from Cape Charles city by Bean (1891, p. 92), under the name *H. pleei*, as more common than *H. unifasciatus*. Jordan and Evermann (1896–1900, p. 722) record a specimen from Hungers Wharf, Va., and Fowler

(1912, p. 54) purchased one in the Baltimore fish market, "said to have been taken in Chesapeake Bay." It is probable that this fish occasionally enters Chesapeake Bay in some numbers, but it quite certainly is not a regular resident there, or it would have been secured during the extensive collecting done in connection with the present investigation. The foregoing description was compiled from published accounts.

Virtually nothing seems to be known of its life history and habits. The greatest length attained is about 15 inches.

Habitat.—Chesapeake Bay to Bahia, Brazil; also recorded from Angola, West Africa.

Chesapeake localities.—(*a*) Previous records: "Chesapeake Bay," Hungers Wharf, Va., and Cape Charles city, Va. (*b*) Specimens in collection: None.

Family XXXVI.—EXOCŒTIDÆ. The flying fishes

Body elongate; head with more or less vertical sides; mouth terminal, or the lower jaw projecting, the latter not produced in the adult; premaxillaries not protractile; maxillary short, slipping under preorbital; nostrils double, near the eye; teeth various, small or weak; lateral line running low, along edge of belly; scales cycloid, more or less deciduous, extending forward on head; dorsal fin without spines, placed on posterior part of body; caudal fin forked, the lower lobe the longer; anal fin opposite the dorsal and more or less similar to it; ventral fins abdominal, sometimes more or less enlarged; pectoral fins inserted high, usually greatly enlarged, serving as organs of flight.

53. Genus EXOCŒTUS Linnæus. Flying fishes

Body elongate; sides flattened; head rather short; snout blunt; eyes large; mouth small; jaws very short, about equal; scales large, deciduous; caudal fin broadly forked, the lower lobe the longer; pectoral fins very long and large, reaching nearly or sometimes quite to the base of the caudal. The species of this genus are inhabitants of the warm seas, many of them being largely cosmopolitan in their distribution, and they may at times work their way into Chesapeake Bay. A single species of flying fish, however, so far has been recorded from this body of water.

71. Exocœtus heterurus Rafinesque. Flying fish.

Exocœtus heterurus Rafinesque, Caratteri di Alauni Nouvi Generi, etc., 1810, p. 58; Palermo. Jordan and Evermann, 1896–1900, p. 735.

?*Exocœtus mesogaster* Uhler and Lugger, 1876, ed. I, p. 143; ed. II, p. 122.

Head 4.66; depth 5.33; D. 14; A. 9; scales 58. Body moderately robust; snout 3.75; eye 3.2; scales moderate, 33 before dorsal, 26 before ventrals, 7 rows between dorsal and lateral line; anal fin short, its origin behind that of the dorsal, base of anal 1.66 in base of dorsal; ventral fins inserted about equidistant from pupil and base of caudal, their length about 2.75 in body; pectorals about 1.45 in body, reaching last ray of dorsal. Dorsal and anal plain; ventrals white, their axils scarcely dusky; pectorals with an oblique white band across lower half of fin.

This species does not occur in the present collection and it is not certain that it definitely belongs to the Chesapeake Bay fauna. Uhler and Lugger (1876), however, record a species of flying fish under the name *Exocœtus mesogaster*, which may have been *Parexocœtus mesogaster*. The description is very inadequate, but it seems to suit *E. heterurus* rather better than *P. mesogaster*. The species is said to reach a length of about 15 inches.

Habitat.—Atlantic Ocean, common southward on both the European and American coasts, straying northward to Newfoundland and England.

Chesapeake localities.—(*a*) Previous records: Mouth of Potomac River and southern part of Chesapeake Bay (Uhler and Lugger, 1876). (*b*) Specimens in collection: None.

Order ANACANTHINI

Family XXXVII.—GADIDÆ. The codfishes

Body more or less elongate; the caudal region moderately long; mouth large, usually terminal; chin with a barbel more or less developed; gill openings very wide; gill membranes separate or somewhat united, usually free from the isthmus; gills 4, a slit behind the fourth; fins without spines; dorsal fins, 1, 2, or 3, extending almost over the length of the back; caudal fin separate or confluent with the dorsal and anal; anal fin long, single or divided; ventral fins jugular, with one to eight rays. Three genera of this rather large family come within the scope of the present work.

KEY TO THE GENERA

a. Dorsal fin divided into three separate parts; anal fin divided into two parts; ventral fins expanded, with about seven short rays.

  b. Lower jaw projecting; caudal fin forked; vent under first dorsal_____Pollachius, p. 155

  bb. Upper jaw projecting; caudal fin nearly square; vent under second dorsal_____Gadus, p. 156

aa. Dorsal fin divided into two separate parts; anal fin long, undivided; ventral fins with two or three filamentous rays_____Urophycis, p. 158

54. Genus POLLACHIUS Nilsson. Pollocks

Body rather elongate; mouth moderate or large; lower jaw projecting; teeth in the jaws equal or the outer ones slightly enlarged; pointed teeth on vomer; none on palatines; gill membranes more or less united; barbel at chin small or obsolete; scales numerous; dorsal fins three; anal fins two; caudal fin lunate or forked; vent under first dorsal.

72. Pollachius virens (Linnæus). Pollock.

Gadus virens Linnæus, Syst. Nat., ed. X, 1758, p. 253; seas of Europe.
Pollachius virens Jordan and Evermann, 1896–1900, p. 2534, Pl. CCCLIX, fig. 886.

Head 3.68 to 3.88; depth 4.53; D. 13 or 14–21–19 or 20; A. 24 to 28–20 or 21; scales 154 to 156; body rather elongate, somewhat compressed, tapering posteriorly; head conical; snout 2.84 to 3 in head; eye 5.73 to 6.35; interorbital convex 3.28 to 3.80; mouth oblique; lower jaw projecting; maxillary scarcely reaching anterior margin of eye, 2.90 to 3.19 in head; teeth small, pointed, cardiform, present on jaws and vomer, none of them notably enlarged; gill rakers rather long, slender, equal to diameter of pupil, 28 to 30 on lower limb of first arch; scales very small, cycloid; dorsal fins separate, first and second of about equal height, outer margin of first convex, margins of second and third nearly straight, the fins tapering posteriorly; caudal moderately forked; anal fins separate, outer margins gently rounded; ventrals small, inserted below posterior margin of gill cover, slightly in advance of pectorals, 3 to 3.37 in head; pectorals moderate, 2 to 2.31 in head.

Color dark green above, silvery to silvery gray below; lateral line pale; dorsals and caudal dark green, anals bluish white; pectorals pale, ventrals white.

This species was not observed by us in the field. The above description is based on three specimens, 450 to 460 millimeters (about 18 inches) in length, caught off Gay Head, Mass. Small pollock, below 15 inches in length are usually brownish green, while large pollock are dark green, with some dusky on the fins, particularly the caudal. Published accounts give the following range in the fin counts: First dorsal, 12 to 14; second dorsal 19 to 24; third dorsal 19 to 22; first anal 23 to 27; second anal 20 to 23.

This species was not secured during the present investigation. It is included here through the courtesy of Dr. William C. Kendall, who kindly permitted us to use his unpublished notes bearing upon certain investigations made in the vicinity of Hampton, Va., in 1894. We find a note, dated March 26, reporting the capture of a pollock 12 inches in length in a pound net. This fish is recognized by its projecting lower jaw, small ventrals, and forked caudal. It is also distinguished from all other species known from the Chesapeake (except the cod) by its three separate dorsal fins.

The food of the pollock is reported to consist chiefly of small fish and of pelagic crustaceans.

Spawning takes place late in the fall and early in the winter along the New England coast. The eggs are reported (Bigelow and Welsh, 1925, p. 405)[10] to be numerous, as many as 4,000,000 from a fish weighing 23½ pounds, about 1.15 millimeters in diameter, and bouyant, hatching in nine days at a temperature of 43° F. The newly hatched larvæ are about 3.4 to 3.8 millimeters in length. Young fish from 25 to 30 millimeters in length show most of the characters of the adult.

Young hatched in the winter in the Gulf of Maine attain a length of 1 to 2 inches by spring and 3 to 5 inches by late fall. The second spring, or when they are a little more than 1 year old, they are 5 to 6 inches long. At Provincetown, on June 26, 1925, we secured numerous young 125 to 140 millimeters (5 to 5½ inches) in length, apparently about 1½ years old.

On our side of the Atlantic pollock are most abundant from Woods Hole, Mass., to Cape Breton, and within this region about 40,000,000 pounds are caught and marketed annually. From eastern Long Island to New Jersey it occurs in small numbers, but below Cape May apparently it is only a straggler.

A pollock 44 inches in length and weighing 36 pounds, taken in the Gulf of Maine by the junior author, is the largest of which we have record. The average length, however, is from 2 to 3 feet, with a weight of 4 to 15 pounds, few exceeding a length of 40 inches or a weight of 30 pounds. The pollock is an active swimmer, occupying any or all levels between the surface and the bottom, and sometimes it schools. A small part of the catch is salted, and in that state the pollock is said to be as good as or better than cod.

Habitat.—Both sides of the North Atlantic; on the American coast from Hudson Bay and Davis Straits to Cape Lookout, N. C., chiefly between Narragansett Bay and the Gulf of St. Lawrence. Dr. Russell J. Coles (in Copeia, No. 151, Feb. 25, 1926) reports the capture of a 10-inch pollock at Cape Lookout, N. C., on February 13, 1925. This establishes the most southerly record for the species.

Chesapeake localities.—(*a*) Previous records: None. (*b*) Specimens in collection: None; a single 12-inch individual is mentioned in the unpublished notes of Dr. W. C. Kendall, this specimen having been taken in a pound on March 26, 1894, at Buckroe Beach, Va.

55. Genus GADUS Linnæus. Cods

Body moderately elongate, compressed posteriorly, and tapering to the rather slender peduncle; head large, becoming narrower anteriorly; mouth large; upper jaw projecting; teeth on jaws and vomer; a barbel on chin; lateral line pale; scales very small; dorsal fins 3; anal fins 2; ventral fins with seven rays.

73. Gadus callarias Linnæus. Cod.

Gadus collarias Linnæus, Syst. Nat., ed. X, 1758, p. 252; European seas. Jordan and Evermann, 1896-1900, p. 2541, Pl. CCCLXI, fig. 891.

Head 3.53 to 3.76; depth 4.74 to 5.14; D. 13 or 14—19 to 22—18 or 19; A. 20 to 22—17; scales 150 to 170. Body elongate, slightly compressed, tapering posteriorly; head conical; snout 2.70 to 2.90 in head; eye 5 to 5.76; interorbital convex, 3.67 to 4.28 in head; mouth horizontal; upper jaw and snout projecting; maxillary reaching anterior third of eye, 2.32 to 2.52 in head; teeth small, pointed, cardiform, in bands, present on jaws and vomer, those of outer row of upper jaw and inner row of lower jaw somewhat enlarged; chin barbel about equal to diameter of eye; gill rakers moderate, length less than diameter of pupil, 15 or 16 on lower limb of first arch (excluding rudiments); scales very small, cycloid; dorsal fins separate, the first the highest, outer margin convex; second and third tapering gradually posteriorly, outer margins nearly straight; caudal truncate or slightly emarginate; anal fins separate, tapering posteriorly, outer margins nearly straight; ventrals inserted below posterior margin of gill cover, slightly in advance of pectorals, the first ray slightly filamentous, the second ray more so, 2.40 to 2.95 in head, without filament; pectorals moderate, 1.86 to 2 in head.[11]

---

[10] For a detailed account of the pollock, see Bigelow and Welsh, 1925, pp. 396–406.

[11] For an account of the cod see Bigelow and Welsh (1925, pp. 409–430).

Color green to red brown above and on sides; pale below; sides and back with green, brown, or reddish irregular spots, usually of one color on any one fish; spots sometimes extending on dorsal fins; lateral line pale; fins usually the same as ground color.

This species was not seen by us in the field. The above description is based on five specimens, 350 to 497 millimeters (about 14 to 20 inches) in length, caught off Gay Head, Mass. The color of the cod is very variable. Out of 25,000 cod observed by the junior author during fish-tagging operations, the colors ranged from pale green to deep red brown and bright red. The greenish cod have dark green or reddish spots, while the reddish cod usually have dark brown spots. We have seen several specimens, as large as 30 inches, that were colored bright red everywhere on body, head, and fins, including the lower parts.

The fin rays vary in number. A large series of Gulf of Maine fish examined by Welsh gave the following variation: First dorsal, 13 to 16; second dorsal, 19 to 24; third dorsal, 18 to 21; first anal, 20 to 24; second anal, 17 to 22. (Bigelow and Welsh, 1925 p. 410.) The cod is distinguished by its three dorsal fins, two anal fins, barbel at the chin, projecting upper jaw, nearly square tail, vent under the beginning of the second dorsal, and by the pale (not black) lateral line.

The cod is typically a bottom feeder but at times rises to the surface in pursuit of schools of small fish and squid. The cod is omnivorous, including in its diet many species of fish, mollusks, crustaceans, worms, echinoderms (chiefly brittle stars), as well as hydroids and algæ. Many foreign

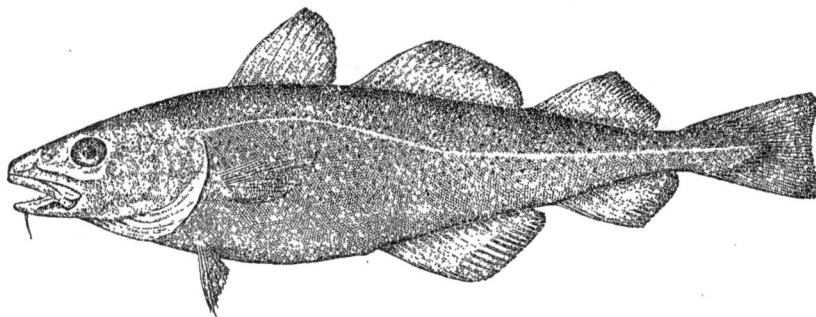

FIG. 82.—*Gadus callarias*. From a specimen 18¾ inches long

objects have been recorded from cod stomachs, including pieces of metal, gravel, wood, rope, and rubber, and fragments of clothing. Aboard the *Halcyon* in 1925, fishing 35 miles from shore, we were surprised to find an empty cigarette container in a cod stomach. It probably had been thrown from a vessel and was seized by the cod because of the bright tinfoil with which it was wrapped. However, the cod does have a preference for certain foods, among which are rock crabs (Cancer), hermit crabs, sea clams (Mactra), cockles (Lunatia), lobsters, brittle stars, blood worms (Nereis), sand launces (Ammodytes), and young herring (Clupea). The chief baits used in the western Atlantic for catching cod are fresh or frozen herring, squid, cockles, and clams (*Mya arenaria*).

The scales of cod form a good index for determining age, but this study is rendered complex because of the protracted spawning season (fall to spring) and because of the fish's wide distribution, wherein growth is more rapid in some localities than in others. European cod average 5, 8.3, and 12.2 inches in length for the first, second, and third years, respectively. Bay of Fundy cod average 5.7, 14.2, and 19.6 inches, and we find the size of Nantucket Shoals cod to be about 8, 15, and 23 inches for the first, second, and third years, respectively.

Spawning takes place along our coast from October to June, the height of the season, however, occurring during the winter. Spawning begins earlier in the southern part of the range than farther north and also ends correspondingly earlier. An inshore migration of spawning fish has been noticed. The eggs of the cod are buoyant, transparent, 1.16 to 1.82 millimeters in diameter (Bigelow and Welsh, 1925, p. 428), and the incubation period extends over about 10 to 11 days at 47° F. Lower temperatures lengthen this period and higher ones shorten it. The newly-hatched larvæ are about 4 millimeters in length. When the cod first hatch they float about helplessly on their backs. An upright position is acquired in a few days, however. The dorsal and anal fin rays begin to appear when the young have attained a length of 10 to 13 millimeters, and when 30 millimeters long many of the adult characters, including the spotted color pattern and scales, have developed.

Young cod live at the surface but descend to the bottom when about 2 or 3 months old, or at a length of about 1 to 1½ inches, and thereafter they are chiefly bottom dwellers, coming to the surface only occasionally, as indicated elsewhere.

The cod was not seen during the present investigation, nor do we find any published record of its occurrence in Chesapeake Bay. Some years, however, a few stragglers pass the capes and are taken in pound nets between Cape Henry and Ocean View. A set of two-pound nets in Lynnhaven Roads caught one cod on March 4, 1919, in the first day's fishing, but no more were caught during that year. The same nets were set on March 12 in 1923, and, on March 16, 16 cod, the only ones caught that year, were taken. The number of fish caught on this occasion illustrates the schooling habit of the cod.

The few cod caught in the lower Chesapeake are taken in March only. It is well known that cod appear yearly in November each year off the coast of New Jersey, from Seabright to a few miles southeast of Cape Henlopen, Del., and fair quantities are caught by hand lines until well into December. No doubt some cod continue down the coast late in the fall as far as North Carolina (where it has been reported by Smith (1907, p. 382)), but none are caught in November in the Chesapeake. A few fish are present on the New Jersey coast throughout the winter, but in March and early April fair-sized schools appear, and for a short period good hand-line fishing is had from Capes Henlopen and May to Atlantic City and Seabright. It is during the beginning of this run that a few stragglers are sometimes taken in the Chesapeake, and it seems probable that such fish belong to small schools that are migrating from North Carolina to the New England coast, their summer home. Recent cod-tagging experiments made by the Bureau of Fisheries have proven conclusively that cod migrate in the fall from Nantucket Shoals, Mass., at least to Cape Henlopen. No tagging has been done in southern waters, but the return of the cod from there to New England in the spring follows as a natural sequence. Virtually no fishing is done in the open Atlantic during the winter from Cape Henlopen to North Carolina, consequently the rarity of cod records within that region can be attributed largely to nonfishing.

At times cod attain an enormous size; the largest specimen recorded weighed 211¼ pounds and was more than 6 feet long. This size, however, is very exceptional, as individuals of more than 75 pounds are rare. Fish of 50 to 60 pounds are not unusual. The cod is too rare in Chesapeake Bay to be of economic importance, as apparently only occasionally a straggler passes between the capes. The cod is one of the most highly prized and among the most valuable commercial food fishes in the North Atlantic Ocean. It is found in depths as great as 250 fathoms, but most of the commercial fishing is done between 10 and 75 fathoms.

Habitat.—Both sides of the North Atlantic; on the American coast from Greenland and Hudson Straits southward to North Carolina; not taken in commercial numbers south of Delaware.

Chesapeake localities.—(a) Previous records: None. (b) Specimens in collection: None; listed in the invoice of the Buchanan Bros. fishery, in Lynnhaven Roads, Va., on March 4, 1919, and March 16, 1923.

56. Genus UROPHYCIS Gill. Codlings; Hakes

Body rather elongate; head subconic; mouth rather large; maxillary reaching below eye; lower jaw included; unequal teeth on jaws and vomer, none on palatines; chin with a small barbel; dorsal fins 2, the first one short, the second long and similar to the anal; ventral fins far apart, each consisting of 3 slender rays, closely joined, appearing like a bifed filament.

KEY TO THE SPECIES

a. First dorsal with a produced filamentous ray, its color uniform dusky; dorsal fins with 9 to 11— 56 to 61 rays; anal 52 to 56; scales about 104 to 112_____*chuss*, p. 159

aa. First dorsal low, without a produced ray, its color distally jet black margined with white; dorsal fins with 8 or 9—46 to 51 rays; anal 43 to 49; scales about 89 to 97_____*regius*, p. 160

74. Urophycis chuss (Walbaum).   Squirrel hake; "Ling."

Blennius chuss Walbaum, Artedi, Gen. Piscium, III, 1792, p. 186; New York.
Urophycis chuss Jordan and Evermann, 1896–1900, p. 2555, Pl. CCCLXV, fig. 902; Evermann and Hildebrand, 1910, p. 163.

Head: 4.25 to 4.5; depth, 4.8 to 5.05; D. 9 to 11—56 to 61; A. 52 to 56; scales about 104 to 112. Body elongate, compressed; head somewhat depressed; snout tapering, 2.95 to 3.6 in head; eye, 2.55 to 3.54; interorbital, 5.05 to 6.8; mouth horizontal; upper jaw and snout projecting; maxillary scarcely reaching opposite posterior margin of eye, 1.9 to 2.05 in head; teeth small, pointed, present on jaws and vomer, those in lower jaw in a very irregular series, those of the upper jaw more or less definitely in two series; gill rakers short, slender, 12 or 13 on lower limb of first arch; scales very small, cycloid; dorsal fins separate; the third ray of first dorsal produced, filamentous; second dorsal long, of nearly uniform height throughout; caudal fin round; anal fin long and low; ventral fins composed of two filamentous rays, inserted on margin of gill opening; pectoral fins rather narrow, 1.2 to 1.35 in head.

Color brownish above; lower part of sides more or less silvery; white, gray, or yellowish underneath; ventrals and pectorals pale; the other fins with dusky punctulations.

Many specimens of this species, ranging in length from 80 to 205 millimeters (3¼ to 8⅛ inches), were preserved. This ling is recognized by its produced, filamentous ray in the first dorsal fin, by the numerous rays in the dorsal and anal fins, and by the number of oblique rows of scales on the sides. It differs, with respect to the scales, from *U. regius* in having more numerous oblique rows and from *U. tenuis* (a northern species that has been recorded from the coast of Maryland

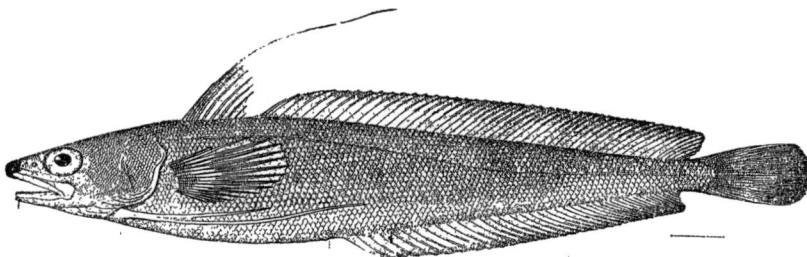

Fig. 83.—*Urophycis chuss*

and also reported from North Carolina, although not yet taken in Chesapeake Bay) in having fewer oblique rows of scales on the sides. *U. regius* has about 88 to 97 oblique rows, *U. chuss* has approximately 104 to 112, and *U. tenuis* has about 140.

The food in eight stomachs examined consisted wholly of crustaceans, principally shrimp. Squids, the smaller sizes of many species of fish, prawns, shrimps, and amphipods have been recorded by various investigators. However, it is said to seldom take mollusks of any kind (excepting squid) or the larger crustaceans, such as cod feed upon (Bigelow and Welsh, 1925, p. 450). As with many other species of fish, the sand launce (Ammodytes) is a favorite food of the ling. We have observed ling caught off Sandy Hook, N. J., gorged with launce, in some cases the tails extending into their mouths because their stomachs could hold no more.

With respect to the spawning of this ling, Bigelow and Welsh (1925, p. 452) state that they trawled fish with running spawn and milt in Ipswich Bay in July. The height of the spawning season falls in the early summer in the Massachusetts Bay region and begins in June south of Cape Cod. It is quite certain that the ling spawns at least as far south as New Jersey, for we have observed large schools of fish in April and May off Rockaway, N. Y., and Sandy Hook, which were distended with spawn. No ripe fish were observed in Chesapeake Bay. However, one individual, taken April 15, 1922, had the ovaries somewhat developed and contained eggs plainly visible under a low-power hand lens.

The eggs are buoyant, 0.72 to 0.76 millimeter in diameter. The fry attain a length of 27 to 70 millimeters in late summer and autumn off the New England coast. (Bigelow and Welsh, 1925, p. 452.) The fry are greenish on the back and silvery on the sides. According to Bigelow and Welsh (1925, p. 449), the young fish are pelagic until 2 to 4 inches long, individuals as small as 2 inches having been taken on the bottom and as large as 4 inches on the surface.

The ling is found on muddy or sandy bottom (rarely in rocky places), at all depths down to the deepest parts in the Gulf of Maine. During the summer it is taken in moderate and deep water, but in the fall it moves inshore. Whether there is a north and south migration is not known, but it is certain that along the New Jersey coast ling are almost entirely absent during the summer, appearing in vast numbers in November, however. Throughout New York Bay they are caught from docks and small boats anchored a few hundred feet from shore. They disappear from the immediate shores during the winter, but reappear in April, when they are distended with spawn. They remain close inshore only a short period, but are caught 2 to 6 miles off at least until late May. On the Cholera Bank, 12 miles south from Long Beach, Long Island, ling are sometimes abundant in July at a time when they have virtually forsaken the shore waters. The evidence seems to indicate that the migration is an inland-offshore one, particularly in the southern parts of its range, where the shore waters no doubt become uncomfortably warm during the summer.

It is of interest to note that all ling that we have seen in the Chesapeake have been taken either with offshore pound nets or with the beam trawls of the *Fish Hawk* and *Albatross* from Cape Henry to Bloody Point, off Annapolis, and not one specimen has been taken along the shores. A few fish 12 inches or less in length are mentioned in our field notes as occurring in catches of pound nets at Ocean View between November 15 and December 5, 1921. However, all fish taken in the beam trawl were caught between March 7 and May 23 of various years, and none of these fish exceeded 12 inches in length. It is reasonably certain, therefore, that the ling does not spawn in Chesapeake Bay, that adult fish are uncommon, and that young from 2 or 3 to 8 or 9 inches long enter the bay late in the fall or in the spring and leave for offshore waters by early June.

Assuming that the ling spawns off the coasts of Maryland and New Jersey as early or somewhat earlier than off Woods Hole, Mass., the larvæ living nearest Chesapeake Bay begin their existence in June or July. Young trawled in Chesapeake Bay were 97 to 134 millimeters (3.8 to 5.2 inches) in March, 81 to 126 millimeters (3.2 to 5 inches) in April, and 111 to 169 millimeters (4.3 to 6.6 inches) in May. At this rate of growth they would probably reach the length of 6 to 7 inches given for yearlings by Bigelow and Welsh (1925, p. 457).

Ling of marketable size are too scarce to be of commercial importance in the Chesapeake, and the very small catch is utilized by the fishermen themselves. From New Jersey northward this ling is an important market fish; and in the Gulf of Maine, where it is taken together and marketed with the white hake (both the ling and white hake are called "hake" in New England and are not separated), an annual catch of 20,000,000 to 35,000,000 pounds for the past 25 years is reported by Bigelow and Welsh (1925, p. 449).

The extreme length of the ling is given as 30 inches, with a weight of 8 pounds. The usual run of fish is from 1 to 4 pounds, and individuals over 5 pounds are uncommon.

Chesapeake localities.—(a) Previous record: Off Cape Henry, Va. (b) Specimens in the collection: From many deep-water stations from Bloody Point, Md., to the mouth of the bay.

75. Urophycis regius (Walbaum). Spotted hake; "Cod."

Blennius regius Walbaum, Artedi, Gen. Piscium, III, 1792, p. 186; New York.
Urophycis regius Lugger, 1877, p. 67; Jordan and Evermann, 1896–1900, p. 2553, Pl. CCCLXIV, fig. 898.

Head 3.85 to 4.4; depth 3.9 to 5.05; D. 8 or 9—46 to 51; A. 43 to 49; scales 89 to 97. Body elongate, compressed; head rather small, somewhat depressed; snout tapering, 3.4 to 4 in head; eye 2.9 to 4.6; interorbital 4.9 to 6.25; mouth large, horizontal; upper jaw and snout projecting; maxillary reaching well beyond eye, 1.75 to 2.08 in head; teeth present on jaws and vomer, those in the jaws pointed, irregular in size, those of upper jaw mostly in a single irregular series, those of lower jaw in two irregular series; gill rakers rather short, 13 or 14 on lower limb of first arch; scales rather small, thin, cycloid; dorsal fins 2, the first not elevated, scarcely higher than the second, none of the rays produced; second dorsal long and rather low, enveloped in scaly skin at base; caudal fin round; anal fin long and low, similar to second dorsal but not quite as long; ventral fins inserted on margin of gill opening, consisting of two long filaments; pectoral fins rather long, 1.1 to 1.35 in head.

Color in life of a specimen 153 millimeters (6 inches) in length, brownish above, darkest on back; white below; a row of white spots, connected by black lines, situated on lateral line; a vertical row of two to four small black spots on head back of eye; two similar spots about an eye's diameter behind the first row; first dorsal dusky, edged with white, with a prominent jet black spot; second dorsal and caudal uniformly dark; caudal dusky; anal white or pinkish at base, bluish along center, edged with black; ventrals white; pectorals pale dusky edged with light yellow. After death the body and fins become suddenly pale, the black spot on first dorsal remains, and the second dorsal and anal are edged with black; the caudal is dusky only at edges; the black spots on head become almost obscure.

Many specimens of this hake, ranging in length from 50 to 310 millimeters (2 to 12¼ inches), were preserved and were before us when the foregoing description was prepared. This hake differs from *U. chuss* chiefly in the low first dorsal fin, which has no produced ray, in the fewer rays in the dorsal and anal fins, and in having somewhat larger scales. A good field mark is the color of the first dorsal fin, which is distally black and margined with white, whereas in *U. chuss* this fin is uniform dusky. The row of white spots along lateral line in *U. regius* is absent in *U. chuss*.

Dr. Edwin Linton examined for us 141 specimens for food and found over 95 per cent of the stomach contents to consist of crustaceans. Nearly 82 per cent consisted of Mysis alone. Shrimp, crabs, amphipods, and isopods also were included. Negligible amounts of fish, annelids, leeches, sponges, and hydroids also had been eaten. Bigelow and Welsh (1925, p. 455), basing their con-

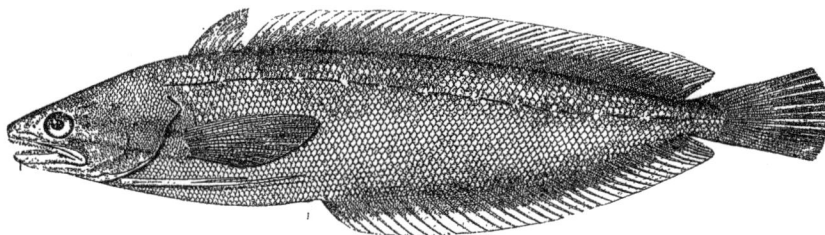

FIG. 84.—*Urophycis regius*

clusions on stomach examinations made at Woods Hole, Mass., by Vinal N. Edwards, state that this hake is more of a fish eater than the other hakes. Doctor Linton's examinations appear to show that in Chesapeake Bay, at least, this hake is quite as much of a crustacean eater as the other species.

Comparatively little is known of the spawning habits of this hake. Welsh, working on the *Albatross*, took spawning fish off the Carolinas in December, 1919 (Bigelow and Welsh, 1925, p. 455), which indicates that the species spawns during the winter. There is no indication that this hake spawns in Chesapeake Bay, for not only are adult fish scarce there, but no young are found until they have attained a length of about 3 inches and presumably are a year old. It is quite likely that this species, like *U. chuss*, spawns outside the bay, and that, like most of the hakes, the larvæ are pelagic for from one to several months. The following table gives the sizes and apparent rate of growth of spotted hake taken in Chesapeake Bay.

Date	Number of fish caught	Range in length	Average length
		Inches	*Inches*
Jan. 22	1	3	3
Mar. 30–31	2	2.7–3.7	3.2
Apr. 10–15	64	3.5–5.5	4.5
Apr. 19–29	170	3.3–6.0	4.7
May 11–20	28	3.7–7.2	5.5

The total catch (except for January 22, when one large fish 11.4 inches in length was taken) is shown for each date given in the foregoing table. Although only three small specimens were taken during the winter months, their size fits in so well with the subsequent collections that it seems safe to assume that all the fish listed in the foregoing table belong to the same year class. Assuming, then, that Welsh's discovery of spawning fish in December may be interpreted as restricting the spawning period to the winter, it seems reasonable to conclude that 1-year-old fish have a length of approximately 3 inches.

This hake, unlike *U. chuss*, was taken twice (between April 10 and 24, 1922) alongshore off Buckroe Beach and the lower York River in collecting seines. It was taken in the trawl at a depth of 38 to 144 feet, as follows: Two fish in January, two in March, many in April, and several in May. During other months it did not appear in the collections. This fish was present in collections made from Cape Henry to Bloody Point, near Annapolis. The majority of the fish, however, were taken south of the mouth of the Potomac River.

It appears, therefore, that most of the young (yearlings) enter the bay by the end of March or early in April, probably depending on the water temperature, and that they leave toward the end of May when the water begins to get warm. Larger fish over 8 inches in length are very scarce in the bay, and only occasionally one is taken either in the late fall or the spring in pound nets set in the lower part of the bay.

This fish is nowhere abundant (in comparison with other hakes), and this, together with its small size, makes it of very slight commercial importance. Certainly in Chesapeake Bay its value is negligible, although in some localities it serves as a source of food supply for other fishes.

The largest size attained by the spotted hake is about 16 inches, but the largest observed in the Chesapeake was only 12¼ inches long.

Habitat.—Nova Scotia to South Carolina; rare north of Cape Cod and south of Virginia.

Chesapeake localities.—(a) Previous record: Off Kent Island, Md. (b) Specimens in collection: From many localities, from Annapolis to the mouth of the bay, mostly taken at depths of 38 to 144 feet; occasional along the immediate shores.

Family XXXVIII.—MERLUCCIIDÆ. The hakes

Body moderately elongate; head elongate, depressed, shaped as in the pikes, its upper surface with a triangular excavated area; no barbels; suborbital bones moderate; dorsal fins 2, the first one short, the second long, consisting of soft rays only; tail isocercal; ventral fins subjugular, well developed. This family consists of a single genus with about four species, a single one of which comes within the scope of the present work.

57. Genus MERLUCCIUS Rafinesque. Hakes

Body elongate; head slender, its upper surface with well-defined ridges, converging backward into a low occipital crest; snout long, depressed; eye rather large; edge of opercle free; mouth large, oblique; maxillary reaching opposite eye; lower jaw projecting; sharp teeth present on the jaws and vomer; branchiostegals 7; gill membranes not united; scales small, deciduous; two well-separated dorsal fins, the first short, the second deeply emarginate; anal similar to the second dorsal; ventral fins normal, well developed.

76. Merluccius bilinearis (Mitchill). Silver hake; Whiting; "Winter trout."

Stomodon bilinearis Mitchill, Rept., Fishes, New York, 1814, p. 7; New York.
Merluccius bilinearis Jordan and Evermann, 1896–1900, p. 2530.

Head 3.55 to 3.65; depth 4.6 to 5.45; D. 12 to 14—40 or 41; scales about 105. Body elongate, compressed; caudal peduncle slender; head rather long and low, flat above, with rather prominent ridges; snout moderately broad, 2.65 to 3 in head; eye, 3.35 to 4.7; interorbital 3.85 to 4.2; mouth large, slightly oblique; maxillary reaching to or a little beyond middle of eye, 1.8 to 2 in head; teeth in the jaws sharp, recurved, in 2 or 3 irregular series, similar teeth present on the vomer; gill rakers slender, about 12 on lower limb of first arch; lateral line distinct; scales rather larger than in related species, deciduous; dorsal fins 2, well separated, composed of soft rays only, the base of the first contained about 3.5 times in the base of the second, its origin over base of pectorals; the second dorsal with longer rays anteriorly and posteriorly than in middle portion of its length; caudal fin nearly straight in the young, somewhat emarginate in the adult; anal fin similar to the second dorsal and placed opposite it; ventral fins well developed, inserted under and slightly in advance of pectorals, 1.45 to 1.6 in head; pectorals narrow, 1.25 to 1.4 in head.

Color in life, dark gray to brownish above, sides and belly silvery, highly iridescent. The iridescence fades soon after death and the color becomes dull silvery on sides and below.

This northern fish is represented in the present collection by three specimens, ranging in length from 105 to 460 millimeters (4⅛ to 18⅛ inches). The silver hake is readily distinguished from the true hakes (Urophycis) by the presence of well-developed ventrals, instead of feelerlike ventrals, by two well-developed dorsals, the second of which, and the anal, are emarginate instead of straight, and by the absence of a chin barbel.

The food of the silver hake, according to Bigelow and Welsh (1925, p. 389), consists principally of fish of suitable size, regardless of the species. Squids and occasionally crabs and other crustaceans also are eaten.

Spawning takes place along the New England coast from June to October, the principal months being July and August. Most of the ripe fish have been taken at depths of 50 fathoms or less, but others have been taken at 300 fathoms off southern New England (Goode and Bean, 1896, p. 387). Spawning, therefore, not only is protracted, but covers a wide range of depths. No ripe silver hakes have been observed in Chesapeake Bay, and it is so rare that we do not hesitate to eliminate this region as a probable spawning ground.

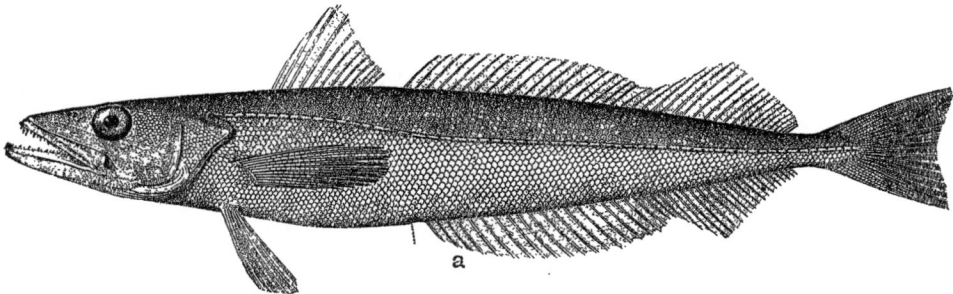

FIG. 85.—*Merluccius bilinearis.* From a specimen 18 inches long

The eggs float at the surface and are about 0.88 to 0.95 millimeters in diameter.[12] Incubation is fairly rapid, being about 48 hours at Woods Hole, Mass., but no doubt somewhat longer in the cooler waters of its natural breeding grounds. Bigelow and Welsh (1925, p. 394) believe that temperatures of 55° to 60° F. are the most suitable for normal incubation. Recently hatched larvæ are 2.8 millimeters in length, and many of the adult characters have been assumed at 20 to 25 millimeters. Newly hatched silver hake are pelagic but take to the bottom the first autumn at a length of 1 or 1½ inches. Little is known about the rate of growth.

In the Chesapeake the silver hake is taken only in the spring and only in the lower sections of the bay, not far from the capes. Its appearance from year to year is very erratic; in some years none are taken, in others only stragglers are caught, and occasionally, as in 1920, a fair catch is made. In a set of two pound nets, operated in Lynnhaven Roads from 1908 to 1923, catches of about 10 pounds or more on any one day were taken only in 1918 and 1920. In 1918 the silver hake was taken on only two days in these nets—100 pounds on May 3 and about 200 pounds on May 4. The year 1920 was exceptional, for this hake was caught from April 28 to May 17 in quantities of 10 to 150 pounds daily, the aggregate catch being about 1,000 pounds. The small local catch is easily disposed of in the Norfolk markets, where the fish is known as winter trout.

The silver hake is an important market fish in New England, but its value has only recently been realized. Bigelow and Welsh (1925, p. 396) point out that only 37,000 pounds were saved in Massachusetts and Maine in 1895, but that 14,000,000 pounds were marketed in 1919. It is exceedingly abundant at Provincetown, where it is taken in mackerel wiers in the spring and summer. There it is frozen, and a large market for it has been developed in the Middle West. Oddly enough,

---

[12] For an account of the embryology of this fish see Kuntz and Radcliffe, 1918, pp. 109–112.

large fish are discarded by the freezers, who state that the western demand is only for "pan size" fish of about 8 to 14 ounces. So abundant is the silver hake at Provincetown that the mackerel freezers will accept only 35 barrels a day from each fishing crew (who usually operate 3 to 6 wiers); and when cold-storage space becomes scarce no more are accepted. The surplus silver hake are allowed to escape through a hole cut in the trap near the surface, an operation we saw repeatedly in June, 1925. The flesh of the silver hake is very sweet when fresh but it softens quickly, which greatly lessens its value.

The maximum size is said to be 8 pounds, but fish over 4 pounds are rare. The usual length of adults is 14 to 24 inches.

Habitat.—Off Newfoundland to the Bahama Islands; most common between Cape Sable and Cape Cod; not recorded from the coast south of Virginia; taken off the New England coast from the shore line to a depth of about 300 fathoms, off Chincoteague, Va., in 90 and 190 fathoms, and only in deep water in the vicinity of the Bahama Islands.

Chesapeake localities.—(a) Previous records: None. Once recorded from Cedar Island, Va., which is on the eastern shore of the peninsula. (b) Specimens in the collection or observed in the field: Off Barren Island, Md., Ocean View, Lynnhaven Roads, and off Cape Henry Lighthouse, Va.

Order HETEROSOMATA. The flat fishes

Family XXXIX.—PLEURONECTIDÆ.[18] The flounders

Body much compressed, deep, and more or less oval in shape; eyes and color on one side, the skull being twisted, the fish swimming in the water horizontally, with the blind side down; premaxillary protractile; gills 4; pseudobranchiæ present; preopercular margin more or less distinct and not concealed by skin; air bladder wanting; vent close behind the head; lateral line rarely absent, extending on caudal fin when present; scales various, usually small; dorsal fin long, beginning on head, composed of soft rays only; anal similar but shorter; caudal fin sometimes continuous with dorsal and anal; ventral fins small, one of them sometimes wanting; pectoral fins rarely absent, placed rather high on the sides. The family is composed of many genera and numerous species. Some of them are important food fishes. Only two of the seven species from Chesapeake Bay, discussed in the following pages, are locally of economic importance.

KEY TO THE GENERA

a. Ventral fins similar in position and shape, or at least lateral; neither on the ridge of the abdomen.
  b. Mouth large; maxillary reaching beyond lower eye; eyes and color on the left side,
                                                         Paralichthys, p. 165
  bb. Mouth small; maxillary about reaching lower eye; eyes and color on the right side.
    c. Lateral line anteriorly with a distinct arch; body thin_____Limanda, p. 167
    cc. Lateral line not arched; body notably thicker_____Pseudopleuronectes, p. 168
aa. Ventral fins dissimilar in position and shape, the one of the eyed side being longer and inserted on the ridge of the abdomen.
    d. Lateral line strongly arched anteriorly; anterior rays of dorsal fin rather high and distally free and branched _____Lophopsetta, p. 171
    dd. Lateral line nearly straight; anterior rays of dorsal fin not elevated, not free and not evidently branched.
      e. Body oval, with the eyes and color on the left side_____Etropus, p. 172
      ee. Body quite elongate, with the eyes and color on the right side,
                                          Neoetropus gen. nov., p. 174

---

[18] We have examined the recent works of Regan (1910, pp. 484 to 496) and Jordan (1923, pp. 166 to 169), in which these authors subdivide the large family, Pleuronectidæ, as previously understood, into several smaller families. However, we prefer to leave the large family intact until more information relative to the merits of the characters upon which the separations are made is available and until the families can be described more adequately.

58. Genus PARALICHTHYS Girard. Summer flounders

Body oblong; eyes and color normally on left side; mouth large, oblique; jaws with a single row of sharp teeth; no teeth on vomer or palatines; gill rakers rather long; lateral line simple, with a strong curve anteriorly; scales small, cycloid or ctenoid; dorsal fin beginning before eye, its anterior rays produced; caudal fin double concave or double truncate; no anal spine; both ventral fins lateral. A single species of this large genus is included in the Chesapeake fauna.

77. Paralichthys dentatus (Linnæus). Flounder; Summer flounder; Fluke; Plaice.

Pleuronectes dentatus Linnæus, Syst. Nat., ed. XII, 1766, p. 458.
Chænopsetta ocellaris Uhler and Lugger, 1876, ed. I, p. 96; ed. II, p. 80.
Paralichthys dentatus Bean, 1891, p. 85; Smith, 1892, p. 72; Jordan and Evermann, 1896–1900, p. 2629, Pl. CCCLXXIII, fig. 922; Evermann and Hildebrand, 1910, p. 163; Fowler, 1918, p. 19.

Head 3 to 3.95; depth 2.15 to 2.45; D. 85 to 94; A. 60 to 73; scales 92 to 105. Body moderately elongate; dorsal and ventral outlines about evenly convex; head rather large; snout pointed, 4.05 to 5 in head; eye 3.35 to 5.7; interorbital varying greatly in width with age, narrower than pupil in very young, about three-fourths width of eye in large examples; mouth large, oblique; the jaws somewhat curved; maxillary reaching beyond eye in specimens ranging upward of 180 millimeters, not reaching posterior margin of eye in young, 2 to 2.6 in head; teeth rather prominent, pointed, in a

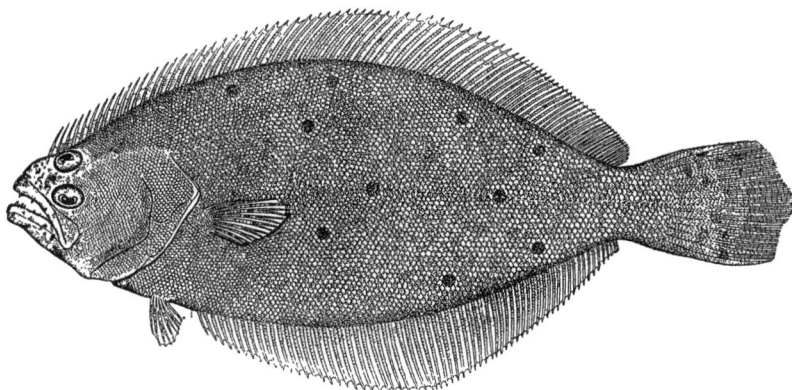

FIG. 86.—*Paralichthys dentatus.* From a specimen 6⅛ inches long

single series in each jaw; gill rakers rather long and slender, 14 to 18 on lower limb of first arch; lateral line anteriorly with a short, prominent arch; scales rather small, cycloid; origin of dorsal over or more usually slightly in advance of upper eye; caudal fin round in young, slightly double-concave in adult; origin of anal under base of pectorals; ventral fins symmetrically placed, inserted under and behind margin of preopercle; pectoral fins moderate, the one on eyed side somewhat more strongly developed, 1.8 to 2.1 in head.

Color brownish, variable, some specimens being much darker than others; most specimens marked with dark ocellated spots, the most prominent of these situated on posterior part of body, three of these forming a triangle with the apex directed forward, the anterior spot being situated on the lateral line, one of the posterior ones over posterior part of base of anal and the other under posterior portion of base of dorsal, these spots often extending on head; fins mostly uniform, sometimes more or less spotted with brown, pale, and dusky specks and bars.

Numerous specimens of this species, ranging in length from 20 to 445 millimeters (⅘ to 17¾ inches), were preserved. The smallest specimen at hand already has the eyes well on one side, and it has essentially the form of the adult. The summer flounder resembles the winter flounder in its outlines. It differs, however, in having the eyes and color on the left instead of the right side. It also has a much larger mouth than the winter flounder, and the lateral line has a distinct arch anteriorly. The rather distinct dark spots on the body of the summer flounder usually are convenient for identification in the field. The three most posterior ones form a triangle, the anterior spot situated on the lateral line being at the apex. This character, together with the more numerous gill rakers, readily separates the summer flounder from its relative and associate in more southern waters, *Paralichthys albiguttus*.

The food of the summer flounder, as shown by 41 specimens examined for us by Dr. Edwin Linton, consists mainly of fish. Squids, shrimp, crabs, and Mysis also were eaten. This is the same general diet reported in published accounts. One author, at least, adds to this list small shelled mollusks, worms, and sand dollars. This fish is frequently seen on sandy shores, partly buried in the sand. Its movements, however, are rapid when in pursuit of bait. Color adaptation is developed to a very high degree, as it is able to assimilate to a very remarkable extent the color of the bottom that it inhabits. For an account of its color adaptation see Mast (1916, pp. 177 to 238, pls. XIX to XXXII).

Comparatively little is known of the spawning habits of the summer flounder. The eggs quite certainly are deposited during the winter. Specimens taken in Chesapeake Bay during October had comparatively large gonads. The large individuals appeared to be more advanced in this respect than the smaller ones. The senior author, working at Beaufort, N. C., has given considerable attention to the spawning of this flounder and during the fall and early winter has found numerous individuals with partly developed gonads, but never any that were even nearly ready for spawning. By March and April the fish appear to be fully spent. It seems probable that this flounder may go to deep water to spawn. Further evidence that this flounder is a winter spawner is given in the following table enumerating the young caught in Chesapeake Bay in collecting seines:

Date	Locality	Number of specimens	Inches
May 21	Cape Charles (beach)	1	0. 9
May 20	Lynnhaven Bay (creek)	3	. 9–1. 1
June 24	Buckroe Beach (creek)	2	2. 0–2. 4
June 25	Back River (creek)	1	2. 4
July 1	Willoughby Point (beach)	3	1. 9–4. 1
July 10	Lower York River (sandy flats)	8	2. 5–4. 3
July 26	Lower Rappahannock River	2	3. 0–5. 0

A study of length frequencies of young fish indicates that by December and January following hatching, or at about the age of one year, the flounder has attained a length of 12 to 18 centimeters (4.7 to 7.1 inches); that by the next October, when about 1¾ years old, the length ranges from about 20 to 26 centimeters (7.9 to 10.2 inches); and that by May, or at a little over 2 years of age, the length centers around 27 and 28 centimeters (10.6 to 11 inches). During December and January no fish smaller than 12 centimeters or larger than 18 centimeters were caught, and there was a complete absence of fish below 20 centimeters in the seine catches from May to November.

The summer flounder is a valuable food fish in Chesapeake Bay. During 1920, among the food fishes of the bay this species ranked twelfth in quantity and eleventh in value, the catch being 285,100 pounds, valued at $13,763.

In Maryland it ranked twelfth in quantity and thirteenth in value, the catch being 26,746 pounds, worth $1,150. About 60 per cent of this amount was caught in pound nets, 38 per cent in fyke nets, and 2 per cent with haul seines. The counties leading with the largest catch were Somerset, 9,118 pounds; Dorchester, 5,520 pounds; and Talbot, 4,411 pounds.

In Virginia it ranked tenth in quantity and eleventh in value, the catch being 258,354 pounds, valued at $12,613. About 77 per cent of this amount was caught in pound nets; 9 per cent in fyke nets; 8 per cent with haul seines, and 6 per cent with lines. The counties having the largest catch were Norfolk, 51,200 pounds; Warwick, 36,230 pounds; and Accomac, 34,080 pounds.

The summer flounder is caught in the Chesapeake throughout the fishing season, from March until November. In the lower part of the bay, pound nets are set for shad early in March. Beginning about the middle of that month stray flounders are taken, generally one or two in the catch of one net or a set of nets. During the first week or two of April the daily catch per net generally varies from one to six flounders. For the last half of April, however, the daily catch may vary from 25 to 500 pounds per set of two or three nets. The largest catches of this flounder throughout the bay are made during May, October, and November. The bulk of the catch often is taken during November. The summer flounder can be taken during December, but nearly all fishing in the bay, except for small fyke nets, ceases the latter part of November. The most productive flounder fishing remembered by Chesapeake Bay fishermen occurred during 1921. In one set of two pound nets located at Lynnhaven Roads, Va., 25,605 pounds were caught during 22 fishing days in November. This amount was equivalent to the total catch of flounders made by these two nets during the entire fishing seasons of the years 1916 to 1919, inclusive. In a set of two pound nets at Ocean View, Va., a total of 4,000 pounds of flounders was taken on November 25 and 26, constituting the largest two-day catch ever known in this locality.

The greater part of the catch of summer flounders in the Chesapeake is taken below the Potomac River, particularly from Mobjack Bay to the Capes. However, it is taken in commercial quantities, at least, as far north as Love Point, Md. The summer flounder is rarely taken north of Baltimore. It takes the hook freely and offers good resistance. Sport fishermen catch many flounders from Annapolis, Md., southward, of which no record is made.

The summer flounder is a well-flavored fish, although the meat is rather dry. Its good flavor is best brought out when baked. A large part of the catch is shipped to markets located principally between Washington and New York. Comparatively good prices are frequently obtained, particularly during the heavy November run when fresh fish of other species are not plentiful in northern markets. During 1921 and 1922, the wholesale price usually ranged from 4 to 10 cents a pound.

The name "summer flounder," which is used in the northern part of the range of the present species, doubtless originated from its occurrence in those waters only during the summer. This name distinguished it from the winter flounder, which is most abundant during cold weather. The present species is nearly always simply called "flounder" on Chesapeake Bay. The winter flounder, *Pseudopleuronectes americanus*, is rarely referred to as flounder, for it is known there as "halibut" or "holibut." The summer flounder is the same fish that is called "fluke" or "plaice" in New York.

This species is reported to reach a maximum length of 3 feet and a weight of 10 to 25 pounds. The usual size of the fish seen in the markets of the Chesapeake, however, range from $\frac{1}{2}$ to 6 pounds in weight, and only occasionally one weighing 8 or 9 pounds is seen.

Habitat.—Shore waters from Maine to South Carolina and probably to Florida.

Chesapeake localities.—(a) Previous records: St. George Island, Md., Hampton Creek and Cape Charles City, Va., and entrance of Chesapeake Bay. (b) Specimens in collection: From many localities from Annapolis, Md., to the Capes; taken chiefly in shallow water during the summer; many small specimens also were taken by the *Fish Hawk* during the winter months in water measuring as much as 25 fathoms in depth.

Comparison of lengths and weights of P. dentatus

Length, in millimeters	Number of specimens	Average length, in inches	Average weight, in ounces	Length, in millimeters	Number of specimens	Average length, in inches	Average weight, in ounces
189	1	7.4	2.3	340–359	3	13.7	14.2
200–219	15	8.3	3.0	360–379	6	14.6	17.7
220–239	17	9.0	3.8	380–399	3	15.3	21.2
240–259	20	9.7	4.7	400–439	3	16.3	25.7
260–279	23	10.6	5.9	440–450	3	17.7	32.8
280–299	14	11.4	7.9	465	1	18.2	35.3
300–319	6	12.2	9.4	497	1	19.5	41.3
320–339	5	13.2	11.9				

59. Genus LIMANDA Gottsche. Mud dabs

Teeth chiefly uniserial; lateral line with a distinct arch in front and without an accessory dorsal branch; scales imbricate, strongly ctenoid. The single species of this genus is reported from Chesapeake Bay.

78. Limanda ferruginea (Storer).   Sand dab; Rusty dab.

Platessa ferruginea Storer, Rept., Fish., Mass., 1839, p. 141, Pl. II; Cape Ann, Mass.
Myzopsetta ferruginea Uhler and Lugger, 1876, ed. I, p. 95; ed. II, p. 79.
Limanda ferruginea Jordan and Evermann, 1896-1900, p. 2644, Pl. CCCLXXVII, fig. 929.

"Head 4 in length; depth 2⅖; D. 85; A. 62; scales 100.   Body ovate-elliptical, strongly compressed; teeth small, conical, close set, in a single series on each side in each jaw, about 11+30 in the lower jaw; snout projecting, forming a strong angle above upper eye, with the descending profile; gill rakers of moderate length, very weak, not toothed; eyes moderate, 4½ in head, the lower slightly in advance of upper, separated by a high, very narrow ridge, which is scaled posteriorly and is continued backward as an inconspicuous but rough ridge to the beginning of the lateral line; scales imbricate, nearly uniform, those on right side rough, ctenoid, those on left side nearly or quite smooth; scales on body rougher than on cheeks; caudal peduncle short, higher than long; dorsal inserted over middle of eye, its middle rays highest; pectoral less than two-fifths length of head; caudal fin rounded; anal spine present; lateral line simple, with a rather low arch in front, the depth of which is barely two-fifths the length; a concealed spine behind ventrals; ventral of colored side partly lateral, the other wholly so; anal spine strong.   Brownish olive, with numerous irregular reddish spots; fins similarly marked; left side with caudal fin, caudal peduncle, and margins of dorsal and anal fins lemon yellow."   (Jordan and Evermann.)   The fin counts given by other authors show a range for the dorsal of 76 to 85 rays and for the anal 57 to 63 rays.

This species was not seen during the present investigation.   It is known from Chesapeake Bay only from a record by Uhler and Lugger (1876), who state that it occurs occasionally in the southern part of the bay.   The rusty dab is distinguished from the other flounders of Chesapeake Bay by having its eyes and color on the right side, together with a small mouth and an arched lateral line.

The food of the rusty dab, according to Bigelow and Welsh (1925, p. 498), consists chiefly of the smaller crustaceans, such as amphipods, shrimps, schizopods, etc., and likewise of the smaller shellfish, both univalves and bivalves, and of worms.   It is also known to eat small fish.

Spawning is reported to take place on the New England coast all summer, a single female spawning over a considerable period of time, as only a small part of the eggs ripen simultaneously in any one fish.   The eggs are spherical, buoyant, transparent, and from 0.87 to 0.94 millimeter in diameter.   Incubation extends over a period of about five days at a temperature of 50° to 52° F.   Metamorphosis—i. e., the twisting of the skull and the migration of the left eye to the right side, which becomes the colored side—is reported to be completed at a length of about 14 millimeters.

The maximum size reported for this species is 21¾ inches.   This northern species is of commercial value from New York northward.   However, it is considered less valuable, because of its thin body, than the winter flounder.   It evidently is an extremely rare species in Chesapeake Bay if, in fact, it occurs there at all at the present time.

Habitat.—Northern shores of the Gulf of St. Lawrence; northern Newfoundland to Virginia; apparently rare south of New York.

Chesapeake localities.—(*a*) Previous record: "Occasional in the southern part of Chesapeake Bay."   (Uhler and Lugger, 1876.)   (*b*) Specimens in collection: None.

60. Genus PSEUDOPLEURONECTES Bleeker.   Winter flounders

Body oblong, with firm flesh; mouth small; teeth in a single series, close-set; lateral line nearly straight; scales strongly ctenoid on eyed side, firm and regularly imbricated; fin rays with scales.

79. Pseudopleuronectes americanus (Walbaum).   Winter flounder; "Halibut"; "Holibut."

Pleuronectes americanus Walbaum, Artedi, Piscium III, 1792, p. 113; New York.
Pseudopleuronectes americanus Uhler and Lugger, 1876, ed. I, p. 94; ed. II, p. 79; Jordan and Evermann, 1896-1900, p. 2647 Pl. CCCLXXIX, fig. 933.

Head 3.4 to 4.4; depth 1.75 to 2.55; D. 62 to 69; A. 46 to 53; scales 77 to 83.   Body eliptical, varying greatly in depth; dorsal and ventral outlines about evenly curved; head rather small; snout pointed, 4.6 to 5.2 in head; eye 3.05 to 4.9; mouth small; the jaws unsymmetrical; maxillary on the right side reaching anterior margin of the lower eye, 3.45 to 4.45 in head; teeth small, present only on the left side of each jaw; gill rakers rather short, about eight on lower limb of first arch; lateral line scarcely arched anteriorly; scales rather small, strongly ctenoid on right side, extending on the fin rays, less strongly ctenoid to nearly smooth on left side; origin of dorsal over anterior part of upper eye; caudal fin round; origin of anal about an eye's diameter behind base of pectorals; ventral fins small, inserted under base of pectorals, reaching to or somewhat beyond origin of anal; pectoral fins rather small, the one on right side slightly the longer, 1.45 to 2.6 in head.

Color in life of a Chesapeake Bay specimen 12 inches long, eyed side, olive green, reddish brown spots of various sizes, irregularly placed, everywhere on head and body; lips pale pinkish; fins reddish brown with darker blotches, dorsal and anal pink along edges; underneath white; dorsal and anal grayish blue, anteriorly pink; pectoral pink. Considerable color variation exists in this flounder, depending largely upon the locality where it is caught and somewhat upon size. The ground color may vary from light to dark; the spots may be prominent, obscure, or almost absent. Underneath it sometimes has areas of dark coloration. Young of 18 to about 50 millimeters (¾ to 2 inches) in length are paler than the adults and more distinctly spotted. Like the summer flounder, this species is able to bring about color changes adaptable to the bottom.

Many specimens of this species, ranging from 18 to 280 millimeters (¾ to 11 inches) in length, were preserved. The principal characters distinguishing this fish are the colored right side, the

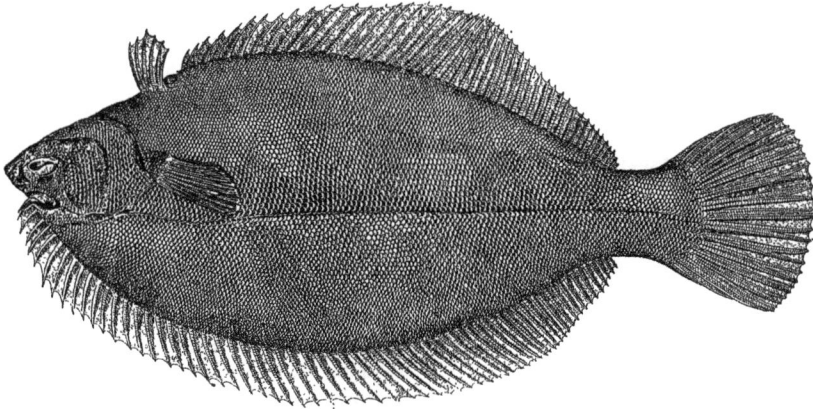

FIG. 87.—*Pseudopleuronectes americanus.* From a specimen 16⅛ inches long

nearly straight lateral line, the strongly ctenoid scales of the colored side, and the scaly fin rays. The young do not differ greatly from the adults except in color, as stated in the description.

The food of the winter flounder consists of many kinds of animal life, mostly of small size, for its small mouth restricts the size of the prey that can be taken. Small crustaceans appear to supply the principal food. Worms, small mollusks, and small fish also enter into the food. The first food taken by the larvæ is said to consist of diatoms.

Spawning takes place along sandy shores during the winter and early spring, the season beginning somewhat earlier in the southern part of the range of the species than farther north. The eggs are heavier than sea water and sink to the bottom in clusters. They are from 0.74 to 0.85 millimeter in diameter. The period of incubation occupies 15 to 18 days at a water temperature of 37° to 38° F. The newly hatched fish is about 3 to 3.5 millimeters in length. Metamorphosis—i. e., the twisting of the skull and the migration of the left eye to the right side, which becomes the colored side—is said to be completed when the fish is only 8 to 9 millimeters in length.

Many young were collected in Chesapeake Bay, as follows: April 24–26, Coan River, Lewisetta, length 18 to 28 millimeters (0.7 to 1.1 inches); Love Point, Oxford, and Crisfield, seined along beach and trawled in deep holes, May 11 to 16, length 24 to 44 millimeters (1 to 1.73 inches); mouth of Potomac River, June 11, 2 specimens 50 and 58 millimeters (2 to 2.3 inches) in length.

Flounders trawled in Chesapeake Bay during December and January, at about one year of age, ranged from 4¼ to 7 inches in length, agreeing very well with the 4 to 6 inch fish recorded by Bigelow and Welsh (1925, p. 504) for southern New England during January and February.

This flounder is a valuable food fish in Chesapeake Bay, and its importance is greatly increased because it occurs in the bay during the winter, when other fish are scarce. In 1920 it ranked sixteenth in quantity and seventeenth in value, the catch being 53,719 pounds, worth $5,372.

In Maryland it ranked ninth in quantity and eighth in value, the catch amounting to 40,119 pounds, worth $4,012. About 60 per cent of the quantity taken was caught in pound nets and 40 per cent in fyke nets. The counties having the largest catches were Kent, with 11,960 pounds; Somerset, with 8,117 pounds; and Dorchester, with 6,780 pounds.

In Virginia it ranked twenty-first in quantity and twentieth in value, the catch amounting to 13,600 pounds, worth $1,360. About 75 per cent of this amount was taken in pound nets and 25 per cent in fyke nets. The counties having the largest catches were Accomac, with 2,300 pounds; Gloucester, with 1,400 pounds; and Warwick, with 1,300 pounds.

The winter flounder is caught in the Chesapeake from November to April and a few stragglers in May. Nearly all the pound nets in the bay are taken up by December and reset in March. The winter flounders taken with this apparatus, therefore, are nearly all caught in November and April and a few in May. Fyke nets, being small and easily fished by one man, are used throughout the winter, and the winter flounder is one of the principal species caught. This flounder takes the baited hook. However, very little hook-and-line fishing is done in the Chesapeake during the winter months.

Unlike most of the salt-water species of the bay, the winter flounder appears to be more common in Maryland waters than in the lower sections of the bay. Around Norfolk the principal catch is taken with pound nets in November and early December. At Cape Charles it is taken in pound nets in late fall and early spring, and in Cherrystone Inlet it is taken during the winter with fyke nets. In the lower York River the winter flounder is one of the principal fish taken with fyke nets during the winter. At Lewisetta, Va., and Crisfield, Md., it is taken in November, December, March, and April in pound nets and fyke nets. At Solomons a few are taken in the spring, but the principal season is late fall. At Annapolis the winter flounder is one of the chief fish taken in the late fall and early spring. In the vicinity of Love Point it is considered an important winter fish. Few, however, are taken northward of this locality. The best catch of winter flounders made at Love Point by a set of two pound nets on one day during the spring of 1922 was 400 pounds.

As this flounder is caught during the colder months, when other species of fish are comparatively scarce in the bay, a market is always available and a good price generally is obtained. A large part of the catch is sold in Norfolk, Portsmouth, Washington, and Baltimore. The price received by the fishermen in 1921 and 1922 ranged around 10 cents a pound.

As a food fish this species is considered somewhat superior to the summer flounder. It is an important fish in the fall and early spring along the entire middle Atlantic and New England coasts, where in places it is quite abundant. In the vicinity of New York and elsewhere this is a favorite fish of anglers, who catch large numbers with hook and line. The name most used for this species in the Chesapeake is "holibut" or "halibut."

The maximum size recorded for the winter flounder is 21 inches. Individuals over 18 inches in length and over 3½ pounds in weight are unusual. The usual size of fish seen in the markets on the Chesapeake range from one-half to 2 pounds, and only occasionally are larger ones seen.

Habitat.—Northern Labrador to Georgia, not taken in commercial numbers south of Chesapeake Bay; entering brackish to nearly fresh water.

Chesapeake localities.—(*a*) Previous records: Southern part of Chesapeake Bay, Potomac River, and coast of St. Marys County, Md. (*b*) Specimens in collection: From many localities from Robins Point, on the Susquehanna River, where the water was fresh enough to drink, to the mouth of the bay; taken both in shallow and deep water, ranging from a foot or two to 29 fathoms in depth, and nearly all catches having been made between November 1 and June 1.

Comparison of lengths and weights of P. americanus

Length, in millimeters	Number of specimens	Average length, in inches	Average weight, in ounces	Length, in millimeters	Number of specimens	Average length, in inches	Average weight, in ounces
110–129	14	4.7	0.6	265	1	10.2	7.7
130–149	10	5.5	1.0	290	1	11.0	10.2
150–169	12	6.3	1.5	305	1	11.8	13.8
170–189	2	7.1	2.4	370	1	15.0	20.0
220	1	8.6	5.0				

61. Genus LOPHOPSETTA Gill.   Sand flounders

Body broad, much compressed, translucent; eyes and color on the left side; mouth large; maxillary reaching opposite the pupil; lateral line with a high arch anteriorly; scales small, cycloid; dorsal fin beginning in front of eye, the anterior rays long, distally free and branched; ventral fins dissimilar in shape and position, broad at base, the left one inserted on ridge of abdomen. This genus consists of a single species.

80. Lophopsetta maculata (Mitchill).   Sand flounder; Window-pane; Spotted flounder; Sand dab.

Pleuronectes maculatus Mitchill, Rept., Fish., New York, 1814, p. 9; New York.
Lophopsetta maculata Jordan and Evermann, 1896–1900, p. 2660, Pl. CCCLXXXII, fig. 938.
Bothus maculatus Bean, 1891, p. 85.

Head 2.9 to 4.05; depth 1.45 to 1.8; D. 63 to 69; A. 46 to 52; scales 92 to 102. Body rhomboid, very strongly compressed; head rather small; snout short, 3.7 to 4.45 in head; eye 2.7 to 3.95; interorbital narrower than eye, proportionately much broader in adult than in young, 1.8 to 3.5 in eye; mouth nearly vertical; lower jaw projecting, with a bony knob at chin; maxillary broad, reaching below middle of lower eye, 2 to 2.45 in head; teeth in jaws small, in a single series laterally, in a band anteriorly; gill rakers slender, 24 to 26 on lower limb of first arch; lateral line with a prominent arch anteriorly, shorter than head; scales small, scarcely imbricate; origin of dorsal nearer tip of snout than eye, the anterior rays distally free and branched; caudal fin round; origin of anal between base of ventrals; ventral fins small with broad bases, the left one nearly on ridge of abdomen; pectoral fins moderate, the left one somewhat larger than the right, 1.25 to 1.65 in head.

Color light brown mottled with numerous lighter brown and black spots, these spots extending on the vertical fins, where they sometimes become elongate; white below. Some specimens are much darker than others but the pattern is about the same.

Many specimens, ranging in length from 40 to 260 millimeters (1⅗ to 10¼ inches), were preserved. The young do not differ greatly from the adults. The sand flounder is recognized by the ventral fins, which are broad at the base and dissimilar in shape and position. These characters, in combination with the strongly arched lateral line and the free and branched rays forming the anterior part of the dorsal fin, distinguish this flounder from all others known from Chesapeake Bay.

We have nothing new to add concerning the food of this flounder. It is known to feed freely on fish of suitable size and on crustaceans, certain mollusks, annelids, and ascidians.

Spawning, according to Bigelow and Welsh (1925, p. 520), takes place in late spring and summer in the Gulf of Maine. It seems probable that spawning takes place much earlier in Chesapeake Bay, as fish with fairly well developed gonads were taken as early as the latter part of September. The larval development is said to be rapid. Tracy (1910, p. 166) gives a length of 2 to 3 inches for the young in July and 4 inches or more by December in Rhode Island waters.

The eggs are spherical, transparent, buoyant, and 1 to 1.08 millimeters in diameter. Incubation requires about eight days at 51° to 56° (Bigelow and Welsh, 1925, p. 520). The development of the larvæ is rapid, and at 10 millimeters the migration of the eye is completed and the fry are ready to take to the bottom (Williams, 1902, p. 2).

The following young fish were caught in Chesapeake Bay:

Date	Number of specimens	Total length, inches	Date	Number of specimens	Total length, inches
Feb. 19	6	1.6–2.3	Apr. 12	13	2.2–3.6
Mar. 7	1	2.6	Apr. 13	8	4.1
Mar. 31	1	2.1	Apr. 14	1	
Apr. 10	26	2.3–3.4	Apr. 15	4	2.7–3.0
Apr. 11	2	2.2–3.5	May 20	10	3.0–4.5

The sand flounder is fairly common in Chesapeake Bay, but, owing to the small size attained and the extreme thinness of its body, it has no commercial value. The common name "windowpane" has reference to the extremely thin body, which in some places is almost transparent.

A maximum length of 18 inches is reported for the sand flounder. Such a length, however, must be very exceptional. It would appear from published accounts that the species grows larger in the northern part of its range than farther south. A fish as much as 10 inches long is exceptional both in Chesapeake Bay and at Beaufort, N. C., whereas Bigelow and Welsh (1925, p. 517) report a usual length of 10 to 12 inches for adult fish. A fish 6¼ inches long weighs about 1.7 ounces; 7½ inches, 3 ounces; and 8½ inches, 4.7 ounces.

Habitat.—Shallow water from the Gulf of St. Lawrence to South Carolina.

Chesapeake localities.—(a) Previous record: Cape Charles City, Va. (b) Specimens in collection: From many localities from Bloody Point, Md., to the entrance of the bay, common chiefly in the southern sections of the bay; taken both in the shallow and deeper waters, ranging up to 25 fathoms, and principally from September to June.

62. Genus ETROPUS Jordan and Gilbert

Body oval; eyes and color on left side; head small; eyes small, separated by a narrow ridge; mouth very small; teeth small, pointed, in a single series, few or none in upper jaw of colored side; vomer toothless; scales thin, ctenoid on left side; smooth on right side; lateral line simple, nearly straight; origin of dorsal over upper eye; anal without a spine; left ventral on ridge of abdomen.

KEY TO THE SPECIES

a. Body somewhat elongate, the depth somewhat less than half the length in the adults; gill rakers about 13 on lower limb of first arch _____*microstomus*, p. 172

aa. Body more ovate, the depth equal to or greater than half the length in the adult; gill rakers about 8 on lower limb of first arch _____*crossotus*, p. 173

81. Etropus microstomus (Gill).

Citharichthys microstomus Gill, Proc., Ac. Nat. Sci., Phila., 1864, p. 223; Bean, 1891, p. 84.
Etropus microstomus Jordan and Evermann, 1896–1900, p. 2687.

"Body ovate. The depth of the body is contained two and one-tenth times in its length, which is three and one-half times the length of the head. Mouth small, very oblique, the gape curved; maxillary two and two-thirds times in length of head, reaching beyond middle of orbit; snout projecting; eyes small, even, shorter than snout, about six in head, separated by a narrow ridge, which is concave and scaleless anteriorly; teeth all small, front teeth of upper jaw wide set, much larger than posterior, which are close together and very small, teeth of lower jaw few, wide apart; gill rakers short and strong, 13 below angle; pectorals short, less than half length of head; scales large, those on middle of sides posteriorly largest. D. 80; A. 61; Lat. 1. 45. Individuals from Great Egg Harbor Bay have D. 74; A. 55; scales 41 to 42.

"Olive brownish, usually with large blotches of darker; a series of distinct, obscure, blackish blotches along the basal portions of the anal and dorsal fins. Size small. Tropical America, north to Long Island occasionally in summer." (Bean, 1903, p. 725.)

No specimens of this species were obtained during the present investigation. The species, however, has been recorded from Chesapeake Bay and we also find a reference to it in Dr. William

C. Kendall's field notes of 1894, based on investigations made in Chesapeake Bay, in which he lists the capture of a few specimens during March. This flounder is a rather obscure species and its relationship is not well known. It apparently differs from *E. crossotus* principally in having a more slender body, the depth being less than half the length. Nothing is known concerning its habits and life history. Its length probably seldom exceeds 4⅓ inches.

Habitat.—Long Island, N. Y., to Virginia, and probably southward.

Chesapeake localities.—(a) Previous records: Cape Charles City, Va. Also recorded in Dr. William C. Kendall's unpublished notes in 1894 from off Lynnhaven Roads and off Cape Henry, Va. (b) Specimens in collection: None.

82. Etropus crossotus Jordan and Gilbert.

Etropus crossotus Jordan and Gilbert, Proc., U. S. Nat. Mus., IV, 1881, p. 364; Mazatlan, Mexico. Jordan and Evermann, 1896–1900, p. 2689, Pl. CCCLXXXVI, fig. 946.

Head 4 to 4.85; depth 1.75 to 2.55; D. 75 to 81; A. 55 to 64; scales 41 to 44. Body elliptical, very strongly compressed; dorsal and ventral outlines about evenly convex; head short; snout very short, 5 to 7 in head; eye 3.05 to 3.8; interorbital very narrow, a mere ridge; mouth small, strongly oblique; maxillary reaching anterior margin of lower eye, 3.5 to 3.9 in head; teeth in jaws small, in a single series, greatly reduced or wanting in upper jaw on eyed side; gill rakers 7 to 9

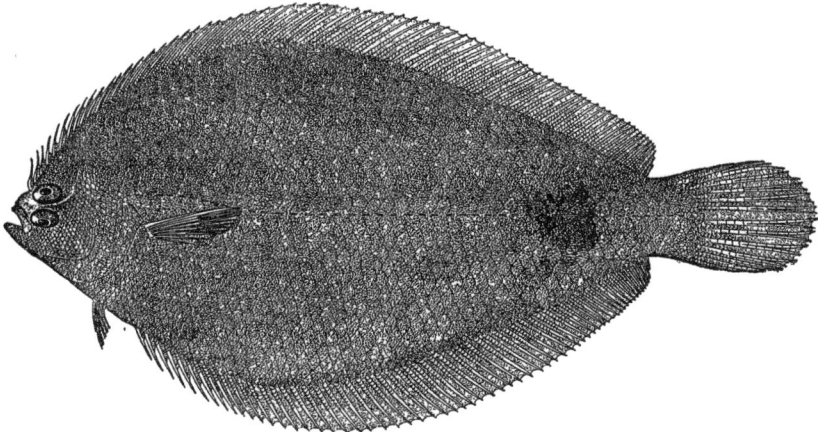

FIG. 88.—*Etropus crossotus.* From a specimen 4¾ inches long

on lower limb of first arch; lateral line nearly straight; scales rather large, ctenoid on eyed side, cycloid on blind side, with small accessory scales at base; origin of dorsal a little in advance of upper eye; caudal fin round; origin of anal a little behind vertical from base of pectorals; ventral fins small, the left one on the ridge of the abdomen; pectoral fins moderate, the one on the eyed side notably longer than the one on the blind side, 1.15 to 1.4 in head.

Color in alcohol brownish, more or less spotted with dusky markings, occasionally with a series of dark blotches along the side on the lateral line; fins pale, with dark specks.

This species is represented in the present collection by 20 specimens, ranging from 50 to 115 millimeters (2 to 4½ inches) in length. This small flounder is recognized by its extreme thinness and its deep body, which in the adult is quite half as deep as long, whereas in *E. microstomus* the depth is less than half the length. The small number of scales (41 to 44) in a lateral series also at once separates it from all other flounders of Chesapeake Bay.

The life history and habits of this little flounder are but little known. It is chiefly a fish of the American Tropics, occurring only as a straggler as far north as Chesapeake Bay. It is most frequently taken along sandy shores and is too small to be of commercial value, as it probably does not exceed a length of 6 inches.

Habitat.—Both coasts of tropical America; on the Pacific from Lower California to Panama, and on the Atlantic from Virginia to Rio de Janeiro, Brazil.

Chesapeake localities.—(*a*) Previous records: None. (*b*) Specimens in collection: All taken in the southern sections of the bay, from somewhat north of Cape Charles City, Va., to the mouth of the bay. Some were seined in shallow water; others were taken by the *Fish Hawk* in water 22 fathoms in depth. Catches were made during the months of July, September, October, and December.

63. NEOETROPUS gen. nov.

Type *Neoetropus macrops* sp. nov.

This genus has the eyes and color on the right side, as in the winter flounders, Pseudopleuronectes. However, it has the unsymmetrical ventral fins (the one of the right side being on the ridge of the abdomen), the small mouth, and very narrow interorbital of the small flounders, Etropus, the members of which have the eyes and color on the left side. It differs from both these genera in the elongate body (depth about 3 in length). The teeth are pointed, in a single series, present on both sides of the jaws, and apparently wanting on the vomer. The scales are rather large and deciduous.

83. Neoetropus macrops sp. nov.

Type No. 87653, U.S.N.M.; length 55 millimeters; off Smiths Point, Va.

Head 4.1; depth 3.2; D. 83; A. 67; scales about 40. Body elongate, strongly compressed; dorsal and ventral profiles evenly rounded; head small; snout short, 5.25 in head; eyes large,

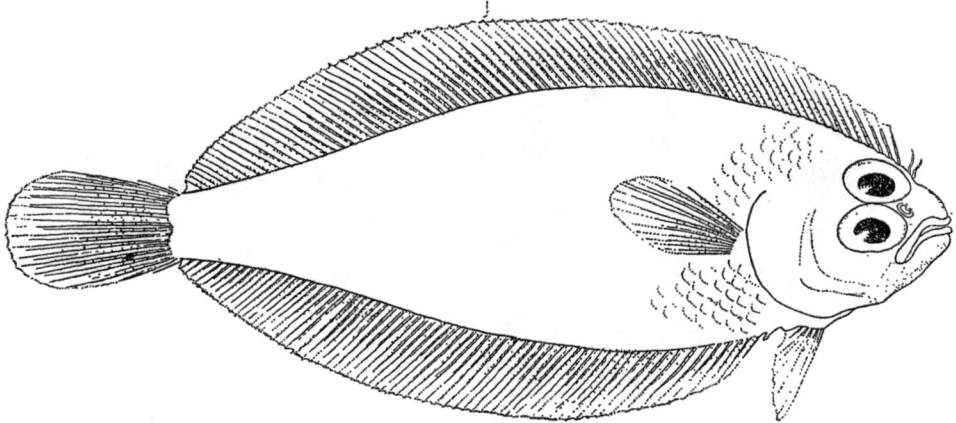

Fɪɢ. 89.—*Neoetropus macrops.* From the type 2⅙ inches long

the lower one scarcely in advance of the upper, interorbital space extremely narrow, 3.2 in head; mouth very small, oblique; maxillary reaching slightly past anterior margin of eye, 3.5 in head; teeth in the jaws pointed, in a single series, not quite as strongly developed on the eyed side as on the blind side; vomer apparently toothless; preopercular margin free; gill rakers rather short and blunt, placed far apart, six on lower limb of first arch; lateral line without an arch; scales deciduous (all lost from the specimen in hand), large (as shown by the marks on the body); origin of dorsal over snout, somewhat in advance of eye; dorsal and anal similar, both well separated from the caudal; caudal fin somewhat injured in the specimen at hand, apparently round or pointed; ventral fins unsymmetrical, the one of the blind side on ridge of abdomen in advance of origin of anal; pectoral fins moderately developed on both sides, about 1.9 in head.

Color of preserved specimen uniform pale, the eyed side being nearly as pale as the blind one.

A single specimen, 55 millimeters in length, is at hand. We are unable to place this specimen in any known genus or species. It has the eyes and color on the right side, as in the winter flounders, Pseudopleuronectes. On the other hand, it has the very small mouth, narrow interorbital, and the unsymmetrical ventral fins of the small flounders of the genus Etropus. Apparently it differs from all the known forms of both genera with which it has characters in common in the elongate body and the large eyes. The specimen might be considered an aberrant individual of Etropus, with the eyes and color on the reverse side, were it not for the fact that when so considered there is still no species known in that genus with which the specimen in hand may be identified. No other course seems open to us, therefore, than to set up a new genus and species, a matter that is nearly always highly unsatisfactory when only a single specimen happens to be at hand.

Chesapeake localities.—Off Smiths Point, Va., somewhat below the mouth of the Potomac River, taken by the *Fish Hawk* in an 8-foot beam trawl at a depth of 25 fathoms on February 21, 1914.

Family XL.—ACHIRIDÆ. The broad soles

Body oblong or ovate; eyes and color dextral (i. e., on the right side); eyes moderate or small, separated by a distinct bony ridge, upper eye usually more or less in advance of lower; mouth small, more or less twisted toward the blind side; teeth little developed, in villiform bands if present; edge of preopercle adnate, usually concealed by scales; gill openings more or less narrowed; gill membranes adnate to shoulder girdle above; scales usually ctenoid, rarely wanting; blind side of head usually with fringes; lateral line single, straight; caudal fin free from the dorsal and anal; right ventral on the ridge of the abdomen and continuous with the anal fin; one or both pectorals often absent.

64. Genus ACHIRUS Lacépède. Hog chokers; American soles

Body ovate, bluntly rounded anteriorly; head small; eyes small, separated by a bony ridge; mouth small, somewhat turned toward the colored side; gill openings narrow, not confluent below; teeth minute or wanting; color and eyes on the right side; scales very strongly ctenoid, similar on both sides; lateral line simple, nearly straight; origin of dorsal over snout; anal spine wanting; caudal fin free, the peduncle very short; ventral fins both present, the one of colored side often nearly continuous with the anal; pectoral fin of blind side wanting, the one on eyed side small or obsolete.

84. Achirus fasciatus Lacépède. Hog choker; Sole.

Achirus fasciatus Lacépède, Hist. Nat. Poiss., IV, 1803, pp. 659, 662; Jordan and Evermann, 1896–1900, p. 2700, Pl. CCCLXXXVII, fig. 948; Smith and Bean, 1899, p. 187; Evermann and Hildebrand, 1910, p. 163; Fowler, 1923, p. 14.
Achirus lineatus Uhler and Lugger, 1876, ed. I, p. 93; ed. II, p. 78; Bean, 1883, p. 365; Bean, 1891, p. 84.

Head 3.45 to 4; depth 1.6 to 1.95; D. 50 to 56; A. 36 to 42; scales 66 to 75. Body broad; dorsal and ventral outlines about evenly convex; head short; snout blunt, 2.85 to 3.7 in head; eye small, 5.2 to 7.5; interorbital never broader than eye, notably narrower than eye in young; mouth rather small, terminal; the jaws considerably curved; maxillary reaching under lower eye, 2.5 to 2.9 in head; teeth in the jaws in villiform bands, present only on blind side; lateral line indicated by a narrow stripe, but without pores; scales small, strongly serrate on both sides of fish; blind side of head with numerous tentacles, extending backward and somewhat on the dorsal and anal fins; origin of dorsal over snout; caudal fin round, separate from dorsal and anal; origin of anal under margin of opercle; ventral fins moderately developed, the one of the eyed side on ridge of abdomen, somewhat continuous with anal fin; pectoral fins wanting.

Color of eyed side brownish to dusky; some specimens much darker than others; usually with about seven or eight black vertical lines; sometimes much mottled with pale markings, these marks occasionally suggesting reticulations; blind side sometimes plain white, more usually with more or less brownish pigment and variously spotted with black, the spots varying in size from very small to larger than the eye, sometimes covering the entire side, sometimes present only posteriorly; fins with pale and dark streaks or spots.

Numerous specimens of this common sole, ranging from 38 to 202 millimeters (1½ to 8 inches) in length, are in the Chesapeake collection. The young appear to be somewhat more elongate than the adults, but in other respects they are similar. The hog choker is at once distinguished from all other flat fishes of Chesapeake Bay by the deep, round body, the fringed scales on the blind side of the head, and, when present, by the black crossbars extending across the body. Both species of soles known from Chesapeake Bay have no pectoral fins and, unlike in the flounders, the upper instead of the lower jaw is the longer.

The food of the hog choker, as shown by the stomach contents of 47 specimens examined, consists chiefly of annelids. A few specimens had fed on small crustaceans, also, and few others contained strands of algæ.

Spawning apparently takes places during late spring and summer. Specimens taken in April have the gonads somewhat developed, but no advanced development was observed until June. Ripe or nearly ripe fish were collected in June, July, and August. The eggs evidently are small and numerous. A female 165 millimeters long, taken June 14, 1921, with free eggs in the ovaries, contained approximately 54,000 eggs. The eggs were about 0.33 millimeter in diameter after preservation in alcohol. Sexual maturity at a small size is indicated by the well developed ovaries of a specimen only 4¼ inches in length taken on May 16, which undoubtedly would have spawned within a month or two. The following table is based on young that quite probably were in their second summer.

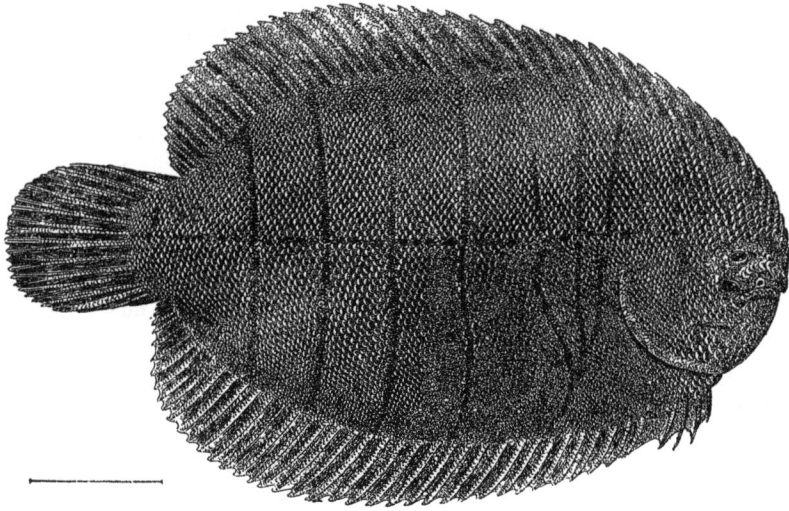

FIG. 90.—*Achirus fasciatus*

Date	Locality	Number of specimens	Inches
Apr. 17	Rappahannock River	31	1.3–3.0
Apr. 24	Potomac River	5	1.4–3.0
Apr. 29	Chesapeake Beach	7	2.7–4.0
May 8	Sassafras River	27	1.8–4.2
June 25	Back River	4	2.2–4.0
July 1	do	40	2.1–3.5
July 10	York River	2	2.3–4.1
Aug. 16	Chesapeake Beach	12	2.3–4.0
Oct. 29	Solomons	4	3.7–4.8
Nov. 22	Cape Charles	1	4.6

The maximum size attained by the hog choker, as given in published accounts, is 5 to 7 inches. However, we have a specimen at hand measuring 8 inches in length. The hog choker frequents shallow water during the summer and often ascends streams and is taken in fresh water. Smith and Bean, in their account of the fishes of the District of Columbia and vicinity (1899, p. 187), state: "Young specimens have been taken in Eastern Branch, Four-mile Run, and Little River. Adults are common in spring on the fishing shores below Washington but have not been observed in the immediate vicinity of the city." During the colder months of the year it is one of the commonest species, as shown by beam-trawl hauls made by the *Fish Hawk* in the deeper waters of the bay.

This sole, although common, has no commercial value and is discarded by the fishermen of Chesapeake Bay. The flesh is said to be well flavored. Apparently, however, because of the small size attained, this fish is seldom eaten. The name "hog choker" is reported to have originated from the fact that hogs, which in some sections feed on the fish discarded on the beaches, have great difficulty in swallowing this sole, because of the extremely hard, rough scales, and are said to choke on them.

Habitat.—Massachusetts to the Atlantic coast of Panama.

Chesapeake localities.—(*a*) Previous records: Havre de Grace, Md., Potomac River and several tributaries, Cape Charles city and Norfolk, Va. (*b*) Specimens in collection: From numerous localities from Havre de Grace, Md., to the capes; taken along the shores in the summer and in deeper waters during the winter.

Family XLI.—CYNOGLOSSIDÆ. The tongue fishes

Body elongate; eyes and color sinistral (i. e., on the left side); caudal fin joined to the dorsal and anal; ventral fins, if present, free from the anal; pectoral fins wanting.

65. Genus SYMPHURUS Rafinesque. Tongue fishes

Body quite elongate; eyes and color on the left side; eyes small, very close together, without a distinct interorbital ridge; mouth rather small, twisted toward the blind side; teeth minute, in

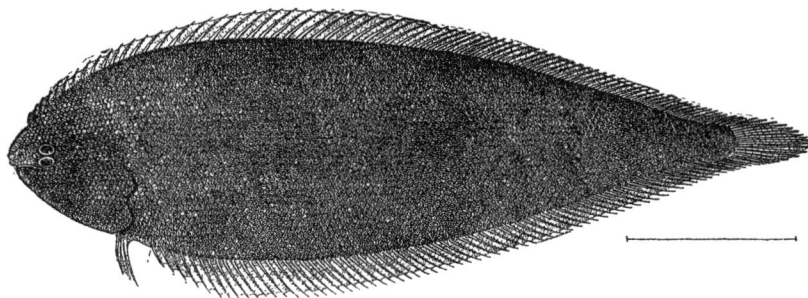

FIG. 91.—*Symphurus plagiusa*

villiform bands; gill openings rather small; the membranes joined below, free from the isthmus; scales ctenoid; lateral line wanting; vertical fins confluent; a single ventral fin present, situated on the ridge of the abdomen; pectoral fins wanting, at least in the adult. A single species of this genus comes within the scope of the present work.

85. Symphurus plagiusa (Linnæus). Sole; Tongue fish.

Pleuronectes plagiusa Linnæus, Syst. Nat., ed. XII, 1766, p. 455; probably Charleston, S. C.
Aphoristia plagiusa Bean, 1891, p. 84.
Symphurus plagiusa Jordan and Evermann, 1896-1900, p. 2710, Pl. CCCLXXXVIII, fig. 950.

Head 5.15 to 5.95; depth 2.9 to 3.3; D. 85 to 91; A. 69 to 75; scales 74 to 79. Body quite elongate; dorsal and ventral outlines about evenly convex; tail tapering; head short; snout blunt, 3.95 to 4.9 in head; eye small, 5.8 to 7.6; interorbital very narrow; mouth small, nearly horizontal; maxillary weakly developed, extending under lower eye, 3.25 to 4.6 in head; teeth in jaws small, in villiform bands, present principally on blind side; gill membranes broadly united; lateral line wanting; scales small, ctenoid; origin of dorsal slightly in advance of upper eye; dorsal and anal fin continuous with the pointed caudal; origin of anal slightly behind margin of gill opening; a single ventral fin present, situated on ventral ridge; pectoral fins wanting.

Color in life of a specimen 7¾ inches long: Brownish above, with 6 or 7 broad, dark crossbars extending from halfway to entirely across back; fins spotted with dusky markings. In some preserved specimens the crossbars have disappeared and the fins are plain.

This species is represented in the present collection by 8 specimens, ranging in length from 60 to 195 millimeters (2⅜ to 7¾ inches). The tongue fish is readily distinguished from the hog choker, the only other sole known from Chesapeake Bay, by the much more elongate body, the depth being contained about three times in the length. It also differs from its relative in having the eyes and color on the left side instead of the right; it has no fringed scales on the head, and the dorsal and anal fins are continuous with the caudal.

Little is known of the feeding habits of this species. Two examples examined had fed on annelids, small crustaceans, minute bivalve mollusks, and apparently on plants.

Nothing at all appears to be known concerning the spawning habits of the tongue fish.

This species reaches a length of only 7¾ inches, the size of our largest specimen, and it has no commercial value. It is a rare fish in Chesapeake Bay and unknown to most of the fishermen.

Habitat.—Chesapeake Bay to the northern part of the Gulf coast of Florida; rare north of Beaufort, N. C.

Chesapeake localities.—(*a*) Previous records: Cape Charles city, Va. (*b*) Specimens in collection: Off Hooper Island, off Point No Point, Md., and off Cape Charles city, Cape Charles, Old Point Comfort, and Ocean View, Va. One specimen (Ocean View) was taken in a haul seine on October 15, 1922. All the others were taken by the *Fish Hawk* in the deeper waters of the bay during January and March, 1914.

Superorder ACANTHOPTERYGII

Order THORACOSTEI

Family XLII.—GASTEROSTEIDÆ. The sticklebacks

Body elongate, somewhat compressed, tapering both anteriorly and posteriorly; caudal peduncle long and slender; mouth moderate, more or less oblique; premaxillaries protractile; skin naked or with vertically oblong plates on sides; middle or sides of abdomen shielded by the produced innominate bones; dorsal fin preceded by two or more free spines; caudal fin narrow, usually lunate; anal fin similar to soft dorsal, preceded by a single spine; ventral fins subthoracic, consisting of a strong spine and one or two rudimentary soft rays; pectorals rather short, inserted not far behind gill opening; air bladder simple; vertebræ 30 to 35.

KEY TO THE GENERA

a. Sides provided with vertically elongated bony shields; innominate bones united, forming a lanceolate plate on the middle of the abdomen_____Gasterosteus, p. 178
aa. Sides entirely naked; innominate bones not joined, forming a ridge on each side of abdomen
_____Apeltes, p. 180

66. Genus GASTEROSTEUS Linnæus. Sticklebacks

Body elongate, compressed; tail long and slender; sides with few or many bony plates, various; innominate bones coalesced, forming a triangular or lanceolate plate on median line of abdomen; gill membranes united to the isthmus; dorsal fin with two free, nondivergent spines and a third one partly connected with the soft dorsal. A single species of this genus comes within the scope of the present work.

86. Gasterosteus aculeatus Linnæus. Common eastern stickleback; Three-spined stickleback; "New York stickleback"; European stickleback.

Gaterosteus aculeatus Linnæus, Syst. Nat., ed. X, 1758, p. 489; Europe.
Gasterosteus bispinosus Jordan and Evermann, 1896–1900, p. 748.

Head 3.3 to 3.4; depth 4 to 4.2; D. II–I, 12 or 13, A. I, 8; lateral plates 32 or 33. Body elongate, notably compressed; caudal peduncle very slender, with a prominent keel on sides, its depth less than diameter of eye, about 6 in head; head rather long, compressed; snout pointed, 3.25 to 3.5 in head; eye 2.9 to 2.5; interorbital 3.7 to 4; mouth rather small, oblique, slightly superior; maxillary failing to reach eye, shorter than diameter of eye; sides with deep, well-developed, bony scutes; opercle finely striate; a large naked area in front of pectorals; innominate bones united, forming a narrow spinelike plate on the abdomen, somewhat shorter than the ventral spines; free dorsal spines strong, the first inserted over base of pectorals, the second the longest, nearly equal to length

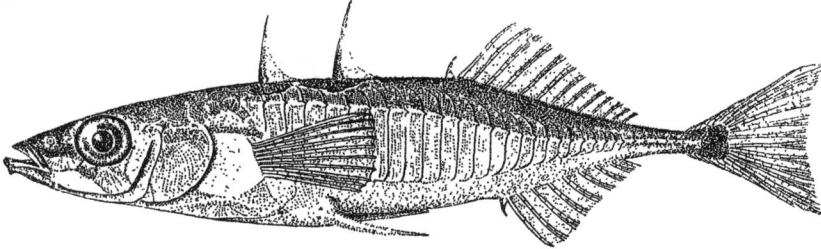

FIG. 92.—*Gasterosteus aculeatus.* From a specimen 2⅛ inches long

of snout and eye, the spine connected with soft dorsal short, the soft rays very low; caudal fin apparently slightly emarginate; anal fin similar to soft dorsal, the spine preceding it very short; ventral fins with large serrated spines, directed sidewise when set, reaching origin of anal when deflexed; pectoral fins moderate, inserted in advance of ventrals, 1.45 to 1.75 in head.

Color dark greenish above, lower parts silvery; back and upper parts of sides with indistinct dark bars, the last one of these on base of caudal.

Only two specimens, a male and female, respectively, 25 and 26 millimeters (1 inch) in length, were secured. This species is readily distinguished from the common four-spined stickleback of Chesapeake Bay by the bony plates on the sides. Published accounts give number of side plates as 28 to 33 and the number of dorsal spines as occasionally 3, rarely 4, in advance of the small spine at the base of the soft fin.

Much individual variation appears to exist among the group of mailed or partly mailed sticklebacks. Dr. W. C. Kendall, who has made an extensive study of these sticklebacks (unpublished), informs the writers that the American two-spined stickleback is doubtfully distinct from the European stickleback, and that, in any event, the name *bispinosus*, used by Jordan and Evermann (1896–1900, p. 748), is not available for this species. We therefore tentatively refer our specimens to the European species *aculeatus*.

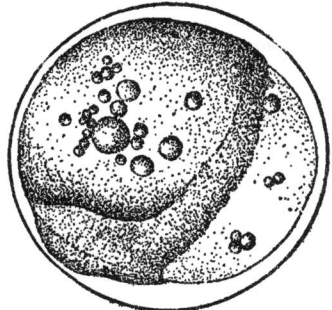

FIG. 93.—Egg with embryo

The stomachs of the two specimens at hand contained as food principally copepods; also small eggs (probably of insect origin) and fragments of alga.

Spawning apparently takes place in the spring at Woods Hole, the time being from June to July. This species, like the four-spined

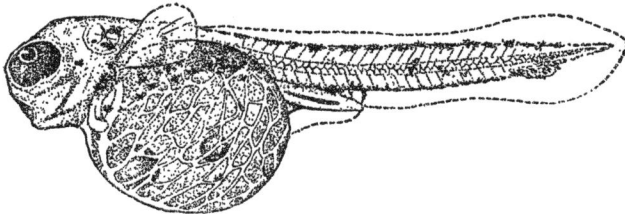

FIG. 94.—Newly hatched larva, 4.3 millimeters long

stickleback, is reported to build nests, in which the eggs are fanned and guarded by the male. For an account of the eggs, embryology, and larval development the reader is referred to Kuntz and Radcliffe (1918, pp. 130 to 132, figs. 113 to 121).

The species is said to reach a length of 4 inches. It is principally of northern distribution, occurring, according to Jordan and Evermann (1896–1900, p. 748), only as far south as New Jersey. Lugger (1878, p. 118), however, recorded it under the name *Gasterosteus noveboracensis*, from Sinepuxent Bay, on the Atlantic coast of Maryland. The present record appears to be the first from Chesapeake Bay, where it undoubtedly is rare.

Habitat.—Northern Europe and in America from Labrador to Virginia.

Chesapeake localities.—(*a*) Previous records: None. (*b*) Specimens in collection: Cape Charles city, Va., May 21, 1922. According to field notes by Dr. W. C. Kendall, two or three specimens were taken in the vicinity of Hampton, Va., on May 15, 1894.

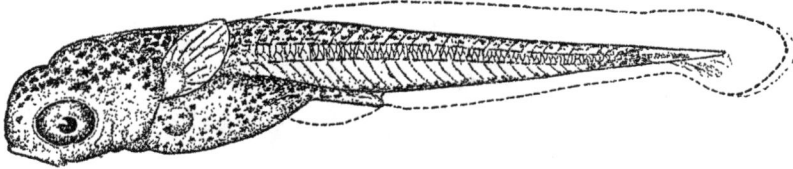

FIG. 95.—Larva 6.3 millimeters long

67. Genus APELTES De Kay. The four-spined stickleback

Body moderately elongate, somewhat compressed; tail very slender, not keeled; skin naked, no bony plates on sides; innominate bones not joined on the median line, forming a ridge on each side of abdomen; gill membranes attached to the isthmus, without free edge; two to four free spines in the dorsal; attached spine of dorsal and of anal strong; spines of ventrals strong, serrate; a bony ridge on each side of spinous dorsal. A single species is known.

87. Apeltes quadracus (Mitchill). Four-spined stickleback.

Gasterosteus quadracus Mitchill, Trans., Lit. and Philos. Soc., I, 1814, p. 430; New York.
Apeltes quadracus Uhler and Lugger, 1876, ed. I, p. 141; ed. II, p. 120; Jordan and Evermann, 1896–1900, p. 752, Pl. CXX, fig. 322; Smith and Kendall, 1898, p. 175; Evermann and Hildebrand, 1910, p. 160; Fowler, 1912, p. 55.

Head 3.6 to 4.2; depth 3.6 to 5.3; D. II to IV–I, 10 to 13 (commonly III or IV–I, 11 or 12); A. I, 8 or 9 (commonly I, 9). Body elongate, compressed, tapering anteriorly and posteriorly; caudal peduncle long and slender, not much deeper than broad, its depth less than diameter of eye, 6.6 to 10 in head; head rather long; snout pointed, its length 3 to 4.6 in head; eye 3.35 to 5.7; inter-orbital 4.7 to 6.8; mouth small, slightly oblique, nearly terminal; maxillary failing to reach the eye, scarcely as long as diameter of eye; teeth in the jaws small, pointed, in a single series; gill openings mostly restricted to the sides, the membranes united to the isthmus; body naked; innomi-nate bones extending back to the vent, bounding the lower lateral edges of the abdomen; dorsal

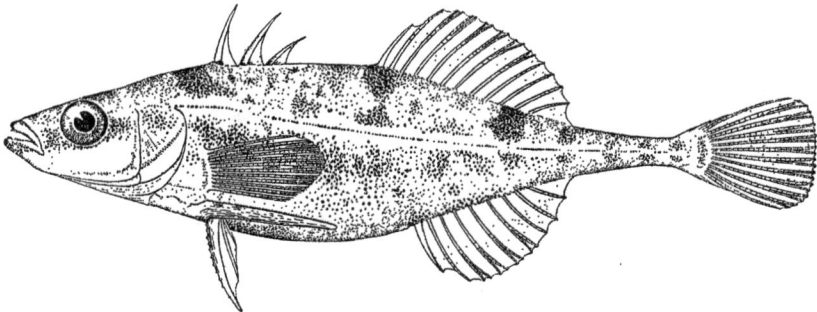

FIG. 96.—*Apeltes quadracus*

fin preceded by two to four free spines and another one largely free immediately in front of the soft rays, the free spines when deflexed fitting into a groove, strongly divergent when erect; the soft dorsal low, higher anteriorly than posteriorly; caudal fin broadly rounded; anal fin with a single nearly free spine, the soft part similar (although a little shorter) to that of dorsal; ventral fins with a strong serrated spine, pointing nearly sidewise when erect, lying on inside of innominate bone when deflexed; pectoral fins moderate, inserted almost exactly over the base of ventrals, 1.5 to 2.3 in head.

Color brownish above, mottled with darker; pale or silvery below; often an indefinite pale streak along side, with a broken dark band, extending through eye, below it; membrane of ventrals red, other fins mostly plain translucent.

This species is represented by many specimens, ranging from 25 to 65 millimeters in length. It is readily recognized by the naked body and the large stiff spines in the dorsal, anal, and ventral fins.

The food of this stickleback, according to the contents of 13 stomachs taken from specimens collected from early spring to late fall, consists almost wholly of small crustaceans, mainly amphipods.

Spawning in Chesapeake Bay takes place in the spring, apparently mainly during the last half of April and the early part of May, during which period we took many gravid fish. The sticklebacks build nests, which are guarded by the males after the eggs have been deposited.[14] The size attained is little in excess of 2½ inches.

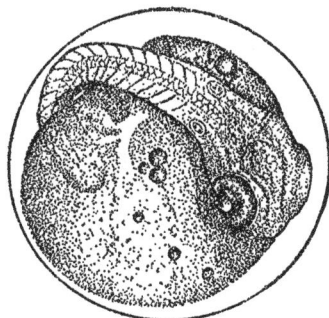

FIG. 97.—Egg with large embryo

This fish was taken in rivers and creeks and in nearly all parts of the bay along the immediate shores among vegetation. It commonly was found in company with pipefishes. It was especially abundant in quiet, brackish, grassy bays and was rarely taken along open sandy shores. On the flats of the lower Rappahannock River as many as 200 were secured in a single haul with a 30-foot seine. It is abundant as far north in the bay as Baltimore but not so common around Havre de Grace. On February 18, 1922, this stickleback was caught at three beam-trawl stations, 5 fish at a depth of 30 feet, 1 fish at 89 feet, and 1 fish at 102 feet, indicating that some of them, at least, spend the winter in the deeper waters of the bay.

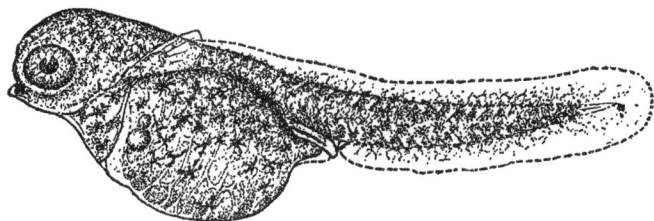

FIG. 98.—Newly hatched larva

This species is of importance only in the food that it furnishes for larger predatory fishes. It may not be easy to swallow by other fishes, however, because of its large, diverging, pungent spines.

Habitat.—Maine to Virginia, apparently reaching the southernmost limits of its distribution in Chesapeake Bay.

Chesapeake localities.—(a) Previous records: Fishing Creek, Big Bohemia River, Gunpowder River, Patapsco River, St. Georges Island, mouth of Windmill Creek, and Hampton. (b) Specimens in collection: From many points from Havre de Grace, Md., to Cape Charles and Lynnhaven Bay, Va.

Family XLIII.—SYNGNATHIDÆ. The pipefishes and seahorses

Body elongate, covered with bony rings; snout long, shaped like a tube, bearing a small mouth at the tip; jaws toothless; gill opening reduced to a small aperture near upper angle of opercle; tail long, sometimes prehensile; males with an egg pouch placed on the ventral side of the tail or under the abdomen, commonly formed by two folds of skin, meeting on the median line; dorsal fin simple, composed of soft rays only; caudal fin, if present, small; anal usually present, minute; ventrals wanting; pectorals small, occasionally missing.

---

[14] The eggs, embryology, and larval development of the four-spined stickleback are described by Kuntz and Radcliffe, 1918, pp. 132 to 134, figs. 122 to 126.

KEY TO THE GENERA

a. Tail not prehensile; head not shaped like that of a horse, usually in line with the axis of the body; egg pouch under the tail_____Syngnathus, p. 182

aa. Tail prehensile; head shaped like that of a horse, placed nearly at a right angle to the axis of the body_____Hippocampus, p. 185

68. Genus SYNGNATHUS Linnæus. Pipefishes

Body very elongate, 6 or 7 angled, not compressed, tapering into a long straight tail; snout long, tubelike, with a small toothless mouth at its tip; humeral bones firmly united to the "breast ring"; dorsal fin distinct; anal fin, if present, minute, placed close behind vent; pectorals present, short and rather broad. Male fishes with the egg pouch along the under side of tail.

KEY TO THE SPECIES

a. Snout rather short, 2 to 2.5 in head; dorsal fin with 35 to 41 rays, placed over 4 or 5 (rarely 3) body and 4 or 5 caudal rings; abdomen convex_____*fuscus*, p. 182

aa. Snout much longer, 1.6 to 1.85 in head.

   b. Dorsal fin short, with 28 to 30 rays, placed over 1 to 1.5 body and 5.5 to 6 caudal rings; body rings 16 to 18; abdomen more or less convex_____*floridæ*, p. 183

   bb. Dorsal fin rather long, with 32 to 37 rays, placed over 3 body and 5 caudal rings; body rings 20 or 21; abdomen flat_____*louisianæ*, p. 184

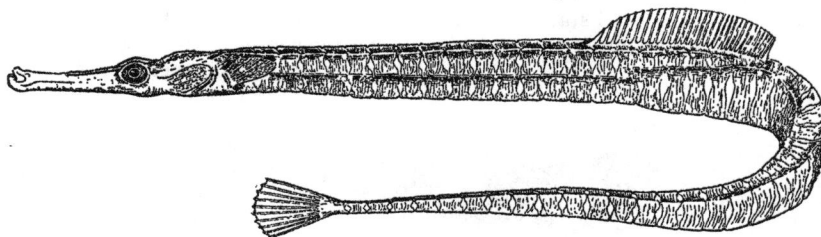

Fig. 99.—*Syngnathus fuscus*, adult

88. **Syngnathus fuscus** Storer. Common pipefish; "Banded pipefish."

   Syngnathus fuscus Storer, Report, Fish., Mass., 1839, p. 162; Nahant, Mass.

   Syngnathus peckianus Uhler and Lugger, 1876, ed. I, p. 91; ed. II, p. 76.

   Siphostoma fuscum Bean, 1891, p. 84; Jordan and Evermann, 1896–1900, p. 770; Smith and Bean, 1899, p. 185; Evermann and Hildebrand, 1910, p. 160.

Head 6.5 to 8.1; D. 35 to 41; body rings 17 to 19; caudal rings 35 to 40. Body somewhat broader below than above; caudal portion quadrangular, longer than rest of body, 1.6 to 1.7 in length; abdomen convex; snout rather short, 2 to 2.5 in head; eye 5.4 to 7.6; egg pouch on 13 to 16 rings; dorsal fin long, the end of its base equidistant from tip of snout and base of caudal or more usually somewhat nearer the former, normally occupying 4 or 5 body rings and 4 or 5 caudal rings (several specimens occur in the collection in which the dorsal occupies 3+5 rings); caudal fin rather long, rounded; pectoral fins short, broad, 4 to 6.4 in head.

Color in spirits brownish above, somewhat paler below; sides more or less mottled, variable; snout usually with a dark bar on sides, passing through eye; dorsal fin sometimes more or less blotched with black; caudal fin usually dark, with a pale margin.

Numerous specimens ranging from 15 to 205 millimeters in length were preserved. The species differs notably from *S. floridæ* (the only other common species in Chesapeake Bay) in the shorter snout, the much longer dorsal fin, and in its position with respect to the number of body and caudal rings which it occupies. It apparently is distinguished from *S. louisianæ* with some difficulty.

The food, according to the contents of 18 stomachs that were examined, consists largely of small crustaceans, including principally copepods and amphipods. One individual had fed on fish fry only, another had fed on an insect, and a few stomachs contained strands of alga in addition to small crustaceans. It seems probable that the plants were taken by accident when the animals in the food were captured.

The sexes, among adults, may be distinguished by the presence in the male of a membranous pouch—a marsupium—on the ventral surface of the tail just posterior to the vent. The eggs are deposited by the female in this pouch and are retained there until they are hatched. The young are carried for some time after hatching. One preserved specimen, for example, contained young 7 millimeters (about one-third inch) in length in the pouch. Young of this length still contained a yolk sac. Another specimen had young 10 millimeters (about three-eighths inch) in length within the pouch. In larvæ of this length the yolk sac was almost completely dissolved, and it is probable that at about this size an independent existence is begun.

Spawing takes place from April to October. The height of the spawning season, judging from the number of males having pouches filled with eggs, extends from April through July. Eggs in several stages of development may be present in the marsupium at one time, although occasionally they are all of uniform development. The largest number of eggs found in the pouch of a single specimen was 570. The fish carrying these eggs was 190 millimeters (about 7½ inches) in length. The smallest number of eggs found in a pouch was 104, carried by a specimen 120 millimeters (4¾ inches) in length. The largest number of ova of uniform size and apparently nearly ripe was 860, which were removed from the ovary of a specimen 190 millimeters (about 7½ inches) in length.

This fish is common in all sections of the bay from Baltimore southward, wherever vegetation occurs; it also ascends streams to fresh water. As many as 305 were taken in eight hauls of a 30-foot seine in the lower York River. It was seined in comparatively large numbers from the time collecting began in April until October. Even as late as November 23 we found it plentiful alongshore in a few feet of water at Cape Charles. None were trawled in deep water during the summer, but as early as October 22 several were taken in 24 feet, and on the 23rd one was taken in 54 feet of water, indicating that a migration from the shore already had begun. The next trawling record occurs on November 22, when one was taken at a depth of 30 feet and two at 125 feet. On a cruise in December a number were taken from the 6th to the 10th at depths of 84 to 126 feet; from January 17 to 21 at depths of 114 to 120 feet; from February 14 to 19, 23 specimens in 11 localities at depths of 48 to 162 feet, and on March 6 seven were trawled at 66 feet. These winter catches were made with the trawl was April 29. It is apparent from these records that the pipefish spends the winter in the deep waters of the bay and the remaining time along the immediate shores, most of the inshore migration occurring late in March and early in April and the offshore migration in November. This fish and the common four-spined stickleback, *Apeltes quadracus*, are common associates. This species is also common northward, where it is the only species of pipefish.

Habitat.—Nova Scotia to North Carolina.

Chesapeake localities.—(a) Previous records: St. Marys River and Riverside, Md.; Gunston, Hampton Creek, and Cape Charles city, Va. (b) Specimens in collection: From 96 localities lying between Baltimore, Md., and Cape Charles and Norfolk, Va.

89. Syngnathus floridæ (Jordan and Gilbert). Pipefish.

Siphostoma floridæ Jordan and Gilbert, Proc., U. S. Nat. Mus., 1884, p. 239; Key West, Fla.; Jordan and Evermann, 1896-1900, p. 766; Evermann and Hildebrand, 1910, p. 160.

Head 4.75 to 6.2; D. 28 to 30; body rings 16 to 18; caudal rings 32 or 33. Body slender, much more so in the young than in the adult; caudal portion quadrangular, usually somewhat longer than the rest of body, proportionately longer in males than in females, 1.7 to 1.95 in length; abdomen more or less convex; snout long, 1.65 to 1.85 in head; eye 6 to 11; egg pouch on 18 to 20 rings; dorsal fin rather short, the end of its base notably nearer the base of caudal than tip of snout, occupying 1 to 1.5 dorsal rings and 5.5 to 6 caudal rings; caudal fin moderate, rounded; pectoral fins short and broad, 8 to 10.2 in head.

Color in spirits dark brown above, lighter underneath; sides with gray specks; dorsal and pectorals plain translucent, the former sometimes with dark spots on the base; caudal fin usually dark brown. Some specimens are much darker than others, the color varying according to the environment in which they were taken, as "color protection" is measurably developed in the species.

This species is represented by 120 specimens, ranging from 30 to 190 millimeters (1¼ to 7½ inches) in length. This pipefish is readily recognized by the long snout, which is notably longer than the rest of the head, and by the short dorsal, which occupies only 1 or 1.5 body and 5.5 or 6 caudal rings.

A thorough and comprehensive study of the spawning habits, as well as the embryology of this species, was made at the United States Fisheries Biological Station at Beaufort, N. C., by Gudger (1905, pp. 447 to 500, Pls. V to XI). The act of spawning—that is, transferring the eggs from the female to the marsupium of the male—was observed by Gudger in fish confined in the aquarium. Spawning apparently took place only at night and a well lighted room did not seem to interfere with the process. The fish intertwined their bodies like two letter S's, the one reversed upon the other, the bodies coming in contact at three points, including the vicinity of the vent. Quoting Gudger directly from this point, he says:

> The anal papilla, or the protruding oviduct of the female, is, at the moment of contact of their bodies, thrust into the button-hole-shaped opening at the anterior end of the marsupium. Some eggs, in number a dozen or more, now pass into the pouch and are presumably fertilized at this moment.
>
> The eggs are now in the anterior end of the pouch, and no more can be received until these have been gotten into the posterior end. To bring this about, the male performs some very curious movements. He stands nearly vertically, and, resting his caudal fin and a small part of the tail on the floor of the aquarium, bends backward and forward and twists his body spirally from above downward. This is repeated until the eggs have been moved into the posterior end of the pouch. * * * Then a short period of rest was observed to take place, which was followed by a repetition of the described process. Four alternate periods of spawning and resting were observed in one pair of fish between 10.15 and 11.06 p. m.

Gudger further says that it is not likely that the eggs are all transferred at one time: First, because of the means used in moving the eggs backward in the pouch; second, because males are frequently found with the pouch only partially filled; and third, because males with eggs at two or three stages of development are not infrequent. The ripe egg at spawning, according to Gudger, is about 1 millimeter in diameter, and the incubation period is given as about 10 days. In Chesapeake Bay male fish with eggs in the pouch were taken from May to October, indicating a protracted spawning season.

The food, according to the contents of 13 stomachs examined, consists largely of small crustaceans, including schizopods, isopods, and copepods. Two specimens also had fed on ova of unknown origin.

This pipefish was found only in the southern part of Chesapeake Bay, and it was not taken during the winter, when it probably leaves the bay for warmer waters. This species reaches the northernmost range of its distribution in Chesapeake Bay, from which it has been recorded only once previously. The maximum length attained by this species is about 9 inches.

Habitat.—Chesapeake Bay to Texas.

Chesapeake localities.—(a) Previous record: Hampton Creek and Cape Charles city, Va. (b) Specimens in collection: Crisfield, Md.; and Lewisetta, lower Rappahannock River, lower York River, Cape Charles, creek tributary to Lynnhaven Bay, and Cape Henry, Va.

90. Syngnathus louisianæ Günther. Pipefish.

Syngnathus louisianæ Günther, Cat. Fish., Brit. Mus., VIII, 1870, p. 160; New Orleans.
Siphostoma louisianæ Bean, 1891, p. 84; Jordan and Evermann, 1896–1900, p. 770; Smith and Kendall, 1898, p. 176.

This species has twice been recorded from Chesapeake Bay, but it does not occur in the present collection unless we are in error in assigning to *S. fuscus* certain specimens in which the dorsal occupies 3 dorsal and 5 body rings. The position of the dorsal in these specimens is correct for *S. louisianæ*, but the dorsal rays and body rings come within the range of *fuscus*. The specimens, furthermore, have the short snout of *fuscus*. The distinguishing characters of these forms are given in the key to the species. This pipefish undoubtedly is very rare in Chesapeake Bay.

Habitat.—Virginia to Texas.

Chesapeake localities.—(a) Previous record: Cape Charles city, Va. (b) Specimens in collection: None.

69. Genus HIPPOCAMPUS Rafinesque. Seahorses

Body compressed, tapering abruptly into a long, quadrangular, prehensile tail; head placed nearly at a right angle to the body, shaped remarkably like that of a horse; top of head with a star-shaped coronet; egg pouch of males placed at base of tail, immediately posterior to vent; dorsal fin moderate, usually placed over vent; anal fin usually present, small; pectoral fins short and broad.

91. Hippocampus hudsonius De Kay. Common American seahorse.

Hippocampus hudsonius De Kay, Fauna of New York, Fishes, 1842, p. 322, Pl. LIII, fig. 171; New York. Uhler and Lugger, 1876, ed. I, p. 90; ed. II, p. 75; Jordan and Evermann, 1896–1900, p. 777, Pl. CXXI, fig. 327; Evermann and Hildebrand, 1910, p. 160.

Head in trunk, measured over back from gill opening to end of dorsal base, 1.6 to 2.3; D. 18 or 19; A. 4; body rings 12; caudal rings 33 to 36. Body with 7 angles; the tail with 4 angles; all angles provided with blunt spines; head also with spines; snout slender, 2.3 to 2.9 in head; eye 4 to 6.2; mouth very oblique; dorsal fin over 3.5 or 4 body rings, its base 1.7 to 2.5 in head; pectoral fins about as broad as long, 3.25 to 4.5 in head.

Color in preserved specimens uniform grayish brown or with dark lines and spots on sides, the lines most prominent on sides of head; dorsal fin spotted with black, the upper part of the anterior rays of the dorsal frequently black, forming a more or less definite black spot.

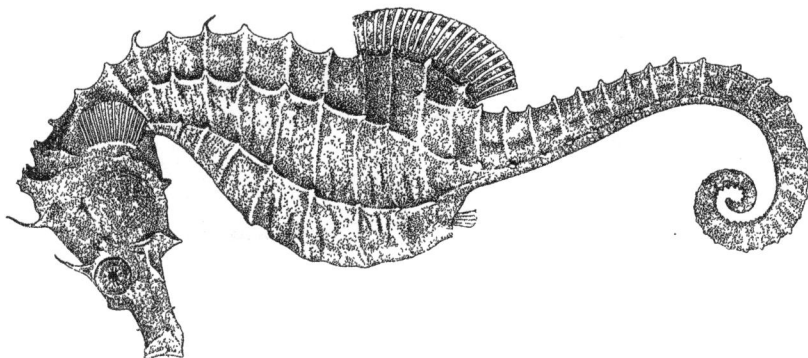

FIG. 100.—*Hippocampus hudsonius*, adult

This species is represented in the present collection by 11 specimens, varying in total length from about 40 to 150 millimeters. It is the only seahorse known from Chesapeake Bay. The southern allied species, *H. punctatus*, was once recorded from Ocean City and Somers Point, N. J., by Bean (1887, p. 134), which would indicate that stragglers may be expected, at least in the lower sections of the bay. *H. punctatus* does not have the dorsal placed wholly over body rings, the usual formula given for that species being 1½ or 2+1 or 2. The dorsal fin usually has somewhat fewer rays, although overlapping with *H. hudsonius*, the range being 16 to 18. *H. punctatus* is often profusely spotted with white or light blue, colors not occurring on *H. hudsonius*.

The food of this fish, as in the pipefishes discussed in this report, appears to consist mainly of small crustaceans. The egg pouch of the male is situated immediately posterior to the vent and it is rather short and slitlike when closed, but round when the young are about to be extruded. (Smith, 1907, p. 173.)[15]

This seahorse is not very common in Chesapeake Bay, and it was taken only from Cedar Point southward, about half of the specimens at hand having been taken in the vicinity of Cape Charles. A few specimens were taken in March with the beam trawl by the *Fish Hawk* at a depth of 150 feet. Other specimens were seined during September, October, and November. The usual length is about 6 inches; rarely a length of 7 inches is attained.

Habitat.—Massachusetts south to South Carolina, rarely straying northward to Nova Scotia.

Chesapeake localities.—(a) Previous records: St. Marys River, Md., and Cape Charles city, Va. (b) Specimens in collection: From the vicinity of Cedar Point, Md., and vicinity of Tangier Island, Yorktown, Cape Charles, and Lynnhaven Roads, Va.

---

[15] For an account of the early development of the seahorse see Ryder, 1882, pp. 191–199.

Order AULOSTOMI

Family XLIV.—FISTULARIIDÆ.　The cornet fishes

Body very elongate, much depressed, always broader than deep; head very long, the anterior bones much produced, forming a long tube, terminating in a small mouth; both jaws and usually the vomer and palatines with small teeth; branchiostegals 5 to 7; gills 4, a slit behind the fourth; scales wanting; bony plates on various parts of the body, mostly covered by skin; a single dorsal, placed posteriorly; caudal fin forked, the middle ray produced into a long filament; anal fin similar to dorsal and opposite it; ventral fins abdominal, far in advance of dorsal, with I, 4 rays; pectoral fins small, preceded by a smooth area.

70. Genus FISTULARIA Linnæus.　Trumpet fishes

The characters of the genus are included in the family description.

92. Fistularia tabacaria Linnæus.　Trumpet fish; "Tobacco trumpet fish."

Fistularia tabacaria Linnæus, Syst. Nat., ed. X, 1758, p. 312; "Tropical America." Jordan and Evermann, 1896–1900, p. 757.

Head 2.7 to 2.8; depth 28 to 37 (10 to 13 in head); D. 14 or 15; A. 13 to 15. Body very elongate, strongly compressed; head in the vicinity of the eye quadrate, slightly broader than deep; snout very long, depressed, its length 1.35 to 1.4 in head; eye 9.8 to 11.5; interorbital (bone) 4.7 to 5.5 in

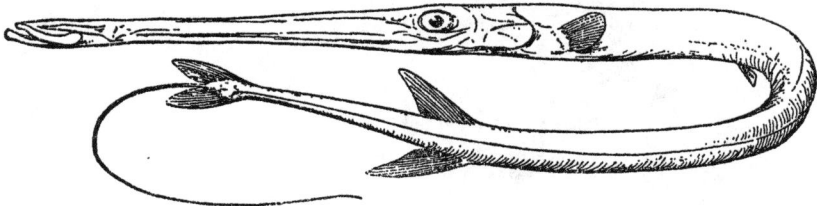

Fig. 101.—*Fistularia tabacaria*, adult

postorbital part of head; mouth oblique; lower jaw projecting; maxillary broad posteriorly, about 10 in head; skin slightly rough; lateral line posteriorly armed with bony scutes, these not evident in young; dorsal and anal fins similar, opposite each other, both somewhat elevated; caudal fin forked, the middle ray produced into a long filament; ventral fins small, inserted nearer base of caudal than tip of snout; pectoral fins rather small, 9 to 10 in head.

Color in life greenish brown above; pale below; sides with a row of blue spots close to vertebral line on back; sides and back with about 10 dark crossbars; caudal filament deep blue; the spots and bars disappearing in preserved specimens, leaving the back uniform brown.

This species is represented by four specimens ranging, without the caudal filament, from 190 to 285 millimeters (7½ to 11¼ inches) in length. This fish is peculiar in the greatly prolonged snout, which is somewhat similar to that of the pipefishes. The skin, however, is mostly naked, and the caudal fin is provided with a long filament.

The four specimens at hand had all fed on fish and one of them had also fed on shrimp and the ova probably of a fish. Its life history is virtually unknown.

The trumpet fish is interesting because of its peculiar structure, but it is without economic value. It is typically a tropical fish, said to be common in the West Indies and neighboring seas, but occasionally straying north in late summer as far as Woods Hole, Mass. Only one specimen has been recorded north of Nantucket, taken at Rockport, Mass., in September, 1865. (Goode and Bean, 1879, p. 4.) In the Chesapeake, where it is very rare, it probably enters the bay only late in the summer or early in the fall, when the water outside the capes is at its maximum temperature. The four specimens taken in this investigation were all caught on September 23, 1921, at the very end of Cape Charles in six hauls of a 250-foot bag seine. The maximum length attained is said to be about 6 feet.

Habitat.—Cape Cod to Rio Janeiro.

Chesapeake localities.—(*a*) Previous records: None. (*b*) Specimens in collection: From Cape Charles, Va. Another specimen was taken by Capt. L. G. Harron in Hampton Roads, Va., on September 18, 1899.

Order PERCOMORPHI

Family XLV.—ATHERINIDÆ. The silversides

Body rather elongate, more or less compressed; cleft of mouth moderate or rather small; teeth small, present on jaws, sometimes on vomer and palatines, rarely wanting; gill membranes separate, free from the isthmus; gills 4, a slit behind the fourth; branchiostegal 5 or 6; pseudobranchiæ present; scales moderate or small, cycloid or not; no pyloric cœca; air bladder present; dorsal fins two, the first with three to nine flexible spines, the second with one weak spine and with soft rays; anal fin similar to and usually longer than second dorsal; ventrals abdominal, with one small spine and five soft rays; sides with a silvery lateral stripe. The silversides are small fishes living in salt or fresh water.

KEY TO THE GENERA

a. Scales with smooth margins; base of dorsal and anal without scales_ _ _ _ _ _ _ _ _ _ _Menidia, p. 187
aa. Scales rough, with strongly lacinate margins; base of dorsal and anal each with a sheath of
    large decidious scales_ _Membras, p. 191

71. Genus MENIDIA Bonaparte. Smooth-scaled silversides

Margins of scales entire; no scales on base of dorsal and anal. Two species of this genus are common in Chesapeake Bay.

KEY TO THE SPECIES

a. Scales in lateral series 44 to 50 (15 to 18 oblique rows on side from upper angle of gill opening
    to origin of spinous dorsal); anal with I, 20 to 26 (usually 22 to 25) rays; peritoneum
    black_ *menidia,* p. 187
aa. Scales in lateral series 37 to 41 (12 to 14 oblique rows on sides between upper angle of gill open-
    ing and origin of spinous dorsal); anal with I, 14 to 20 (usually 15 to 18) rays; peritoneum
    silvery, with or without dark punctulations_ _*beryllina,* p. 189

93. Menidia menidia (Linnæus). Silverside; "Dotted silverside."

Atherina menidia Linnæus, Syst. Nat., ed. XII, 1766, p. 519; Charleston, S. C.
Chirostoma notata Uhler and Lugger, 1876, ed. I, p. 139; ed. II, p. 119. (Probably two or more species confused.)
Menidia notata, Bean, 1891, p. 92; Smith, 1892, p. 69; Jordan and Evermann, 1896–1900, p. 800; Evermann and Hildebrand,
1910, p. 160.
Menidia menidia Kendall, 1902, pp. 262 to 264; Jordan and Hubbs, 1919, p. 52.
Menidia menidia notata Kendall, 1902, pp. 262 to 264; Fowler, 1912, p. 54.

Head, 4.15 to 4.7; depth, 4.3 to 6.95; D. III to VII—I, 7 to 10 (usual formula IV to VI—I, 8 or 9); A. I, 20 to 26 (usual formula I, 22 to 25); scales—44 to 50 (15 to 18 oblique rows on sides between upper angle of gill opening and origin of spinous dorsal). Body variable, very slender to moderately deep and compressed; caudal peduncle rather long, its depth 2.2 to 3 in head; head depressed above, narrower below; snout moderately long, pointed, its length 2.7 to 3.75 in head; eye, 2.75 to 3.75; interorbital, 3.35 to 3.8; mouth small, moderately oblique, nearly terminal, the lower jaw being slightly included; teeth in the jaws pointed, in narrow bands, with the outer series somewhat enlarged; scales firm, with margins entire, extending somewhat on the base of caudal but not on the soft dorsal and anal; origin of spinous dorsal rather variable, sometimes about equidistant from tip of snout and base of caudal, more usually nearer the latter, the predorsal distance 1.75 to 2 in length to base of caudal; second dorsal situated over middle of anal base; caudal fin moderately forked; anal fin long, its base about an eye's diameter longer than head; ventral fins small, inserted equidistant from tip of snout and end of anal base, or more usually somewhat nearer the former; pectoral fins moderate, pointed, 1 to 1.3 in head.

Color greenish above; more or less silvery, with metallic luster in life below; sides with a bright silvery, well-defined band, about half diameter of eye, bounded above by a dark line; scales on upper part of sides and back with numerous brownish dots; fins plain translucent; peritoneum black.

Many specimens, ranging from very small (8 millimeters) to 130 millimeters (5⅛ inches) in length, are at hand. Two subspecies are recognized from Chesapeake Bay by Kendall (1902, pp. 262 to 267). We, too, find these forms, the extremes of which differ quite markedly. Intermediate specimens, however, are at hand, and the two varieties (subspecies) intergrade perfectly.

In general, the subspecies *notata*, which is greatly in the minority in the collection at hand, has a more slender body, rather more numerous scales in a lateral series, with a more forward position of the dorsal fins. The intergradations, however, are complete, and there are numerous specimens at hand that can not be said to belong to either typical form. Furthermore, the extremes as well as intermediates sometimes occur in a single lot collected in one locality within a few hours.

This species is recognized by the rather large size, long anal fin, rather small scales, and the black peritoneum. It is shown under *M. beryllina* that there is a slight overlapping with respect to the number of anal rays. In the present species, in 68 specimens, the following results were obtained: One specimen had 20 rays, 4 had 21, 7 had 22, 19 had 23, 17 had 24, 17 had 25, and 3 had 26 rays.

The food of this fish, according to the contents of 27 stomachs, consists largely of small crustaceans. Other foods are worms, insects, minute ova of unknown origin, and algæ.

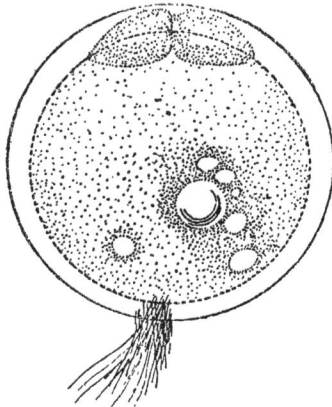

FIG. 102.—*Menidia menidia*, egg, two-cell stage

Spawning takes place from early spring to late summer. The largest number of ripe fish, however, were taken in April and May. In 1894, Kendall (unpublished notes) stated that many ripe fish were seined from March 15 to 20 in the vicinity of Hampton and Cape Charles, Va. The eggs (Hildebrand, 1922, p. 114) are deposited in shallow water, in "grassy" areas, where the fish collect in large schools. The eggs are provided with numerous gelatinous threads of considerable length, by means of which they become attached to vegetation and other objects in the water. In this species, as in *M. beryllina*, eggs of several sizes are present in the ovary at one time, and when one size is ripe and spawned the next is already large enough to be seen clearly with the unaided eye. The protracted spawning season, together with the fact noted relative to the various sizes of eggs in the ovary, suggests that the fish may spawn more than once during a season. The eggs are spherical in form, about 1.25 millimeters in diameter, and slightly heavier than sea water. The period of incubation was about 16 days in water varying in temperature from 40° to 60° F. The newly hatched larvæ are about 5 millimeters in length and highly transparent, only a few yellowish green pigment spots being present. The fish assumes virtually all the characters of the adult when it has reached a length of 13 millimeters, and it is then readily recognized.

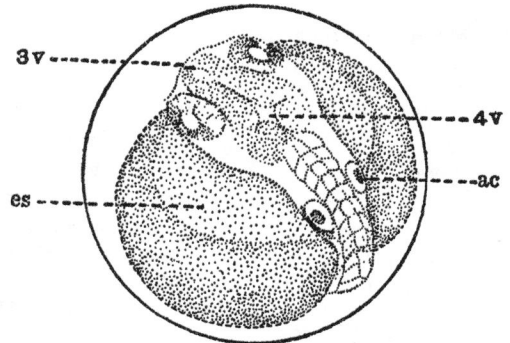

FIG. 103.—Surface view of egg of *Menidia menidia* 2 days after fertilization, water temperature 82° F. es, embryonic shield 3v, third ventricle of the brain; 4v, fourth ventricle; ac, auditory canal

This is the commonest and most abundant of the silversides, being found in all parts of the bay in both salt and brackish water. It is more salt-water in its habits than *beryllina*, rarely entering fresh water; and although the two species frequently associate in most sections of the bay, *M. menidia* is largely replaced by *M. beryllina* in the northern sections, and wholly so above the mouths of the rivers. It is among the most abundant of fishes in Chesapeake Bay and is present throughout the year. It was collected in large numbers along the shores from the time seining operations began, early in April, until late November. It probably remains near shore during most of the winter, but part of them at least retire to deeper water during the period of low temperatures, as shown by the following beam-trawl catches made in various parts of the bay: December 9 and 10, 1915, depths 108 to 126 feet; January 15 to 20, 1914, depths 33 to 162 feet; February 14 to 19, 1922, depths 46 to 162 feet; February 18 to 22, 1914, depths 33 to 150 feet; March 7 to 10, 1915, depths 50 to 63 feet; and March 21 to 23, 1914, depths 39 to 120 feet. Silversides were caught in many beam-trawl hauls during these winter months, but the aggregate catch was so small in comparison with the known abundance of the fish along shore during most of the year that it is doubtful if the deeper waters of the

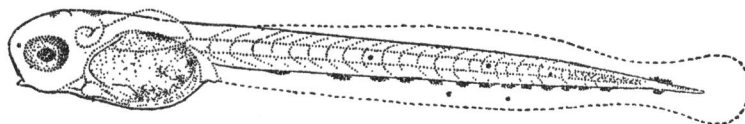

FIG. 104.—*Menidia menidia.* Recently hatched larva, 4 millimeters long

bay can be considered a wintering ground. The silverside is gregarious, usually traveling in schools of a few dozen to several hundred fish.

The largest silverside among many thousands taken in Chesapeake Bay was 5⅛ inches in length. However, it rarely exceeds a length of 4½ to 5 inches. Fish of this size could be utilized as food, but this is not done in the Chesapeake. It is of great economic importance as food for larger predatory fishes, however, notably of the striped bass.

Habitat.—Nova Scotia to the east coast of northern Florida. The variety *notata* predominates north of Chesapeake Bay and *menidia* from the Chesapeake Bay southward.

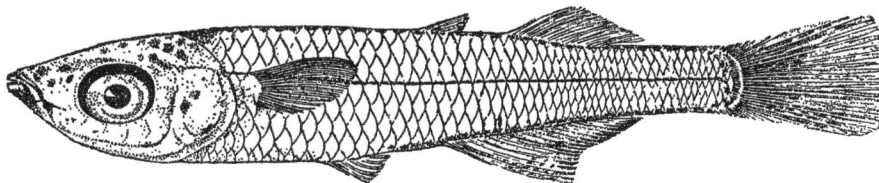

FIG. 105.—*Menidia menidia.* Young fish, 10 millimeters long

Chesapeake localities.—(a) Previous records: Havre de Grace, Baltimore, Riverside, lower Potomac, mouth of Rappahannock River, Fortress Monroe, Hampton, and Cape Charles city. (b) Specimens in collection: Many; from numerous localities from Havre de Grace, Md., to Cape Charles and Cape Henry, Va., throughout all months of the year; generally common, particularly southward.

94. Menidia beryllina (Cope). Silverside.

Chirostoma berrylinum Cope, Trans., Amer. Phil. Soc., 1866, p. 403; Potomac River, Washington, D. C
Menidia beryllina Smith, 1892, p. 70, Pl. XX; Smith and Bean, 1899, p. 185; Kendall, 1902, p. 260; Fowler, 1912, p. 54.
Menidia gracilis Jordan and Evermann, 1896–1900, p. 797; Evermann and Hildebrand, 1910, p. 160.
Menidia gracilis beryllina Jordan and Evermann, 1896, p. 797, Pl. CXXIV, fig. 338; Evermann and Hildebrand, 1910, p. 160.

Head 3.9 to 4.7; depth 5.4 to 6.6; D. IV or V—I, 8 to 11 (usual formula IV or V—I, 9 or 10); A. I, 14 to 20 (usual formula I, 15 to 18); scales 37 to 41 (12 to 14 oblique rows on sides between upper angle of gill opening and base of spinous dorsal). Body slender, moderately compressed; caudal peduncle rather long, its depth 2.2 to 3.1 in head; head somewhat depressed above, narrower below; snout moderately pointed, its length 3.2 to 4.6 in head; eye 2.4 to 3.1; interorbital 3 to 4.1; mouth rather small, terminal, strongly oblique, moderately protractile; teeth in the jaws small, pointed, in very narrow bands; scales firm, with margins entire, extending somewhat on the base of caudal but not on base of soft dorsal and anal; origin of spinous dorsal equidistant from tip of snout and base of caudal, or somewhat nearer the latter; soft dorsal over middle of base of anal; caudal fin moderately forked; anal fin rather short, its base equal to or slightly longer than head; ventral fins rather small, inserted about equidistant from tip of snout and end of anal base; pectoral fins moderate, 1.15 to 1.45 in head.

Color pale greenish; lower parts silvery; sides with a well-defined silvery band, narrower than half the eye, bounded above by a dark line; scales on the back with numerous brown dots; fins plain; peritoneum silvery, usually with dark dots.

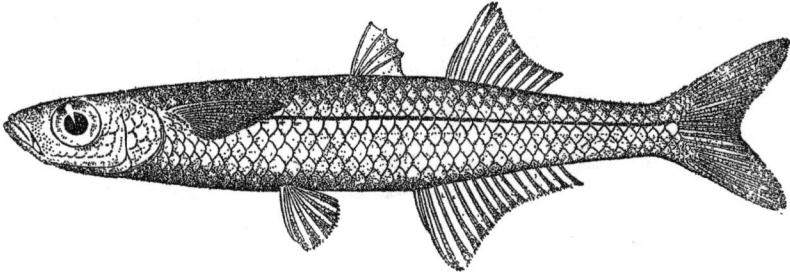

FIG. 106.—*Menidia beryllina*

Many specimens of this species are at hand, varying in length from 15 to 75 millimeters. Two varieties (subspecies) have been recognized by Kendall (1902, pp. 260 and 261), who regarded the Potomac River fish(the typical *beryllina*) as subspecifically distinct from the salt-water, coastwise form, which he named *cerea*. The salt-water form is said to have a somewhat blunter snout, less

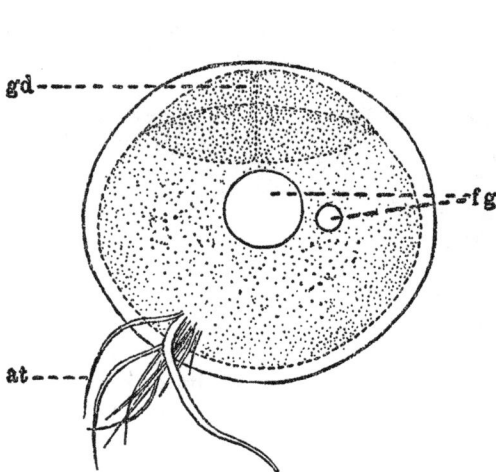

FIG. 107.—Egg, two-cell stage. *at*, adhesive threads; *gd*, germinal disk; *fg*, fat globules

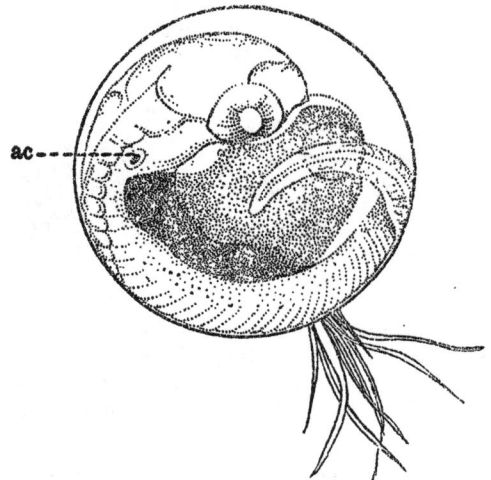

FIG. 108.—Egg with large embryo, 2½ days after fertilization; *ac*, auditory canal

compressed body, and usually a shorter caudal peduncle. The Chesapeake Bay specimens are partly from fresh, partly from brackish, and partly from salt water. The specimens, however, appear to be quite uniform, and they are here all regarded as representing the typical *beryllina*. This species is recognized by its small size, short anal fin, rather large scales, and the pale silvery

FIG. 109.—Recently-hatched larva, 3.5 millimeters long

peritoneum. The number of soft rays in the anal fin rarely overlaps with *M. menidia*. The average number, however, is quite distinct. In the present species in 156 specimens the anal fin has 14 soft rays in 2 specimens, 15 in 24, 16 in 54, 17 in 55, 18 in 11, 19 in 9, and 20 in 1 specimen. The number of scales in a lateral series, as well as the number in advance of the dorsal, appear to be quite distinct in the two species.

The food of this fish, as indicated by the contents of 20 stomachs, consists of the following, named in the order of their apparent importance: Small crustaceans, small mollusks, insects, and worms. A few strands of algæ also were found.

The spawning season of this fish is a protracted one, as ripe or nearly ripe fish were taken from April 10 to September 19, 1921. It is also quite probable that the fish spawns more than once during a season, since the female, for example, has ova of several sizes in the ovaries at one time; and when one lot of eggs is ripe those of the next largest size are big enough to be plainly visible to the unaided eye. The eggs, when spawned (Hildebrand, 1922, p. 120), are not quite spherical, and they are somewhat smaller than those of *M. menidia*, their greatest diameter being approximately 0.75 millimeters. The eggs, as in *M. menidia*, are provided with gelatinous threads, which in the present species are comparatively few in number, and one of them is always much enlarged. The eggs adhere to objects in the water by means of these adhesive threads. Hatching took place in 8 to 10 days in water varying from 78° to 82° F. The newly hatched larvæ are approximately 3.5 millimeters in length, very slender, and highly transparent.

The females appear to grow somewhat larger than the males. The maximum size, according to Chesapeake specimens, is 75 millimeters (about 3 inches) for the female and 70 millimeters (2¾ inches) for the male. The fish is abundant in the bay and it is found in association with *M. menidia*, being more fresh-water in its habits, however. It is more common in brackish than in salt water, and it ascends streams into strictly fresh water, the species having first been discovered in the Potomac River at Washington. It was not taken in deep water at beam-trawl stations. Because of its small size this silverside is of no direct commercial importance; its chief value is as food for larger predatory fish.

Habitat.—Cape Cod, Mass., to South Carolina, entering streams and fresh water.

Chesapeake localities.—(*a*) Previous records: Baltimore, Washington, Alexandria, Bryans Point, lower Potomac, Hampton, and Cape Charles city. (*b*) Specimens in collection: From many points from Havre de Grace, Md., to Cape Charles and Lynnhaven Roads, Va.

72. Genus MEMBRAS Bonaparte. Rough silversides

Margins of scales strongly lacinate; base of dorsal and anal with a sheath of large deciduous scales. The genus probably contains only two species, one from Martinique and one from the Atlantic and Gulf coasts of the United States.

95. Membras vagrans (Goode and Bean). Silverside; Sardine; Silverfish.

Chirostoma vagrans Goode and Bean, Proc., U. S. Nat. Mus., 1879, p. 148; Pensacola, Fla.

Kirtlandia vagrans Jordan and Evermann, 1896–1900, pp. 794 and 2840, Pl. CXXIV, fig. 336; Evermann and Hildebrand, 1910, p. 160.

Head 4 to 5.9; depth 4.9 to 6.7; D. IV to VI–I, 6 to 8; A. I., 17 to 22 (usual formula I, 18 to 21); scales 42 to 49. Body elongate, moderately compressed; caudal peduncle rather strongly compressed, its depth 1.1 to 1.4 in head; head rather flat above, narrower below; snout pointed, 2.85 to 3.15 in head; eye 2.5 to 3.55; interorbital 2.45 to 2.9; mouth small, oblique, strongly protractile; lower jaw included; teeth in the jaws small, pointed, in narrow bands, the outer series of teeth somewhat enlarged; scales firm, lacinate, distinctly rough to the touch in adults, the lacinations not evident in young of less than 30 millimeters; scales extending on the base of vertical fins; origin of spinous dorsal usually over origin of anal, about equidistant from posterior margin of opercle and base of caudal; soft dorsal placed over posterior part of anal base, the two fins being nearly coterminal; caudal fin moderately forked; anal fin rather long, its origin about equidistant from margin of opercle and base of caudal; ventral fins rather small, inserted about equidistant from tip of snout and base of caudal; pectoral fins pointed, the upper rays longest, 1.1 to 1.4 in head.

Color greenish on back, silvery on lower parts of side and belly; sides with a broad silvery band, bounded above by a dark line, width of band equal to about three-fourths diameter of eye; scales on the back with numerous dusky points; occipital region and tip of snout often bluish or dusky; caudal fin more or less dusky, yellowish in life; other fins mostly plain; peritoneum silvery with dusky punctulations.

49826—28——13

This species is represented by numerous specimens ranging in length from 22 to 115 millimeters (⅞ to 4½ inches). The young of less than 30 millimeters do not show the lacinated scales distinctly, although indications of projections on the margins of the scales may be detected under magnification. The young of this species and those of the genus Menidia, unlike the adults, can not be separated readily by the character of the scales. The present species, however, differs from *Menidia menidia* in having fewer anal rays, a character that is available in separating the young. This character, unfortunately, can not be used in *M. beryllina*, as the fin rays in that species are about the same in number as in *M. vagrans*. In these species, however, the position of the dorsal fin with reference to the anal, is helpful, for in *M. beryllina* the first dorsal is wholly in advance of the anal and the origin of the second dorsal is over about the middle of the anal. In *M. vagrans* the origin of the first dorsal is over the origin of the anal and that of the second dorsal is behind the middle of base of anal.

Two species of Membras have been described from the United States. One of these, *M. vagrans*, was supposed to represent the Gulf coast form and the other, *M. laciniatus*, the Atlantic coast form, and they were supposed to differ in the number of soft rays in the anal fin and in the number of scales in a lateral series. The Chesapeake Bay fish, according to the range assigned, therefore should be *M. laciniata*. In 119 specimens examined the number of anal rays varies from 17 to 22, 2 of this number having 17 rays, 15 having 18 rays, 30 having 19 rays, 48 having 20 rays, 19 having 21 rays, and 5 having 22 rays. The number of scales in a lateral series, in 38 specimens, varies from 42 to 50, as follows: Four specimens with 42 scales, 3 with 43, 5 with 44, 6 with 45, 4 with 46, 6 with 47, 8 with 48, 1 with 49, and 1 with 50. This range covers the extremes of both forms, as given in current works. The extremes are not covered with respect to the number of anal rays, since specimens from the Gulf coast with as few as 14 rays have been recorded; a pronounced intergradation nevertheless is evident. Smith (1907, p. 178) referred *laciniata* to the synonymy of *vagrans*. Jordan and Hubbs (1919, p. 57) show that Atlantic and Gulf specimens intergrade, but they retain the names, regarding them as representing subspecies. The data presented herein appear to show that the retention of the names as representing subspecies is scarcely tenable.

The food of this silverside, according to 13 stomachs taken from fish collected at various times and places, consists mainly of small crustaceans, and among them copepods constituted the main bulk of the material eaten. Other foods found consisted of fragments of insects and small ova of unknown origin. A few films of algæ also were found.

The spawning period in this species appears to be a protracted one, as specimens with well distended sexual organs, captured from May to August, occur in the collection. It seems probable that this silverside, like *Menidia menidia*, spawns among vegetation, to which the eggs become attached.

This fish reaches a maximum length in Chesapeake Bay of about 4½ inches. It is common in the southern part of the bay but rather rare in the northern sections. No specimens were secured during the winter months. It runs up streams to brackish water, being somewhat more salt-water in its habits, however, than *Menidia menidia*. The species is of no direct commercial importance; its main value is as food for larger predatory food fishes.

Habitat.—New York to Tampico, Mexico, if *M. vagrans* and *M. laciniata* are regarded as identical.

Chesapeake localities.—(a) Previous record: Cape Charles city. (b) Specimens in collection: From numerous localities from Havre de Grace, Md., to Cape Charles and Norfolk, Va.; common toward the mouth of the bay.

Family XLVI.—MUGILIDÆ. The mullets

Body elongate, somewhat compressed; mouth rather small, the jaws with small teeth or none; premaxillaries protractile; gills 4, a slit behind the fourth; branchiostegals 5 or 6; scales large, cycloid; no lateral line, the scales, however, with furrows forming lateral streaks; air bladder large; intestinal canal long; two short dorsal fins, well separated, the anterior one with four stiff spines; caudal fin forked; anal fin with two or three graduated spines; ventral fins abdominal with I, 5 rays. A single genus of this family of fresh-water and marine fishes, inhabiting the warmer regions of the world, occurs in Chesapeake Bay.

73. Genus MUGIL Linnæus. Mullets

Body robust, somewhat compressed; head moderate, usually about as broad as deep, scaled above and on sides; eye in adult with a strongly developed adipose membrane, small or wanting in young; mouth subinferior, oblique, the gape wide but not deep; lower jaw angulated; jaws with one or a few series of small, flexible, villiform teeth; no teeth on palatines and vomer; anal fin in very young with two spines, adults constantly with three spines, the first soft ray in the young transforming into a spine; stomach with very heavy, muscular walls, gizzardlike.

The species of this genus run in schools, frequently swimming at the surface, where their movements may be observed and whereby they betray their presence to the fishermen. Certain species, at least, have the habit of leaping from the water, sometimes clearing the water as much as 3 feet. It is from this habit that the common name "jumping mullet" has originated. Only two species of mullets are known from the Chesapeake.

KEY TO THE SPECIES

a. Anal fin with III, 8 (very young with II, 9) rays; second dorsal and anal fins with few or no scales; rows of scales on sides with dark longitudinal stripes (very young, bright silvery)_*cephalus*, p. 193

aa. Anal fin with III, 9 (very young with II, 10) rays; second dorsal and anal fins densely scaled in adults; rows of scales on sides without definite dark stripes (very young, bright silvery)_____*curema*, p. 196

96. Mugil cephalus Linnæus. Striped mullet; Jumping mullet; "Jumper"; Mullet; "Fatback."

Mugil cephalus Linnæus, Syst. Nat., ed. X, 1758, p. 316, Europe; Jordan and Evermann, 1896–1900, p. 811, Pl. CXXVI, fig. 343.
Mugil lineatus Uhler and Lugger, 1876, ed. I, p. 140; ed. II, p. 120.
Mugil albula Bean, 1891, p. 92.
(?) *Querimana gyrans* Jordan and Evermann, 1896–1900, p. 818; Evermann and Hildebrand, 1910, p. 160.

Head 3.3 to 4.5; depth 3.3 to 3.75; D. IV–I, 8; A. III, 8 (young of about 50 millimeters and less with II, 9 rays); scales 38 to 42. Body rather robust, somewhat compressed; caudal peduncle rather strongly compressed, its depth 3.1 to 3 in head; head at eyes about as broad as deep; snout

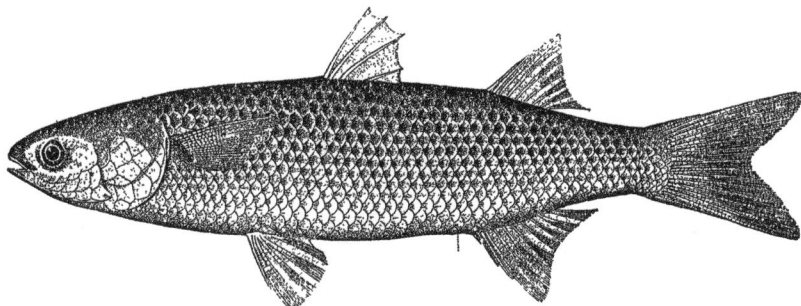

Fig. 110.—*Mugil cephalus.* From a specimen 7.1 inches long

short and broad, 4 to 6 in head; eye 3.3 to 4.2; interorbital 2.25 to 3.5; mouth moderate, oblique, the lower jaw included, the gape somewhat broader than deep; adipose eyelid strongly developed in adult, undeveloped in young; gill rakers numerous, slender, close-set; scales moderate, with crenate membranous borders, extending on caudal fin and a few on the anterior rays of dorsal and anal; origin of spinous dorsal nearer base of caudal than tip of snout in adults, the opposite being true of the young; origin of soft dorsal a little behind origin of anal; caudal fin forked, the lobes of about equal length; anal fin similar to second dorsal, but longer; ventral fins abdominal, inserted about equidistant from tip of snout and middle of anal base in adult, proportionately more posterior in young; pectoral fins not reaching opposite origin of first dorsal, 1.3 to 2 in head.

Color of adult bluish gray and greenish above, silvery below; scales on sides with dusky centers, forming dark longitudinal lines along the rows of scales; fins mostly plain, some of them more or less dusky; dorsals, caudal, and pectorals sometimes grayish green in life; axil of pectoral bluish. Young, bright silvery.

Many specimens, ranging from 28 to 255 millimeters (1⅛ to 10 inches) in length, are at hand. This species is distinguished from *M. curema* by the slightly shorter anal fin, fewer scales on the dorsal and anal fins, and by the dark stripes along the rows of scales. The young mullets of about 50 millimeters and less in length have two spines and nine soft rays in the anal. Later, however, the first soft ray is transformed into a spine. The anal count is quite constant, but Jacot (1920, p. 200) states that in rare cases the number of anal supports (spines and rays) may be either 10 or 12. The young mullets also differ notably in color from the adult, as they are bright silvery. For these reasons the young were considered as of a different genus for a long time.

The food of this mullet, according to the contents of 33 stomachs, consists of microscopic organisms, mainly of diatoms and Foraminifera, intermixed with considerable quantities of mud and vegetable débris.

Exact information as to where spawning takes place and a knowledge relating to the development of the eggs are wanting, notwithstanding that a number of investigators have made these matters a point of special study. It is known, however, that spawning takes place late in the fall and that the eggs are moderately large. In the Chesapeake region no roe mullet were observed by us and, so far as we know, none have been reported from the bay. At Beaufort, N. C., mullet with large roe (but not prime ripe) are taken mostly in October and early November. In northern Florida most of the spawning takes place during November and December. One of us (Schroeder) made

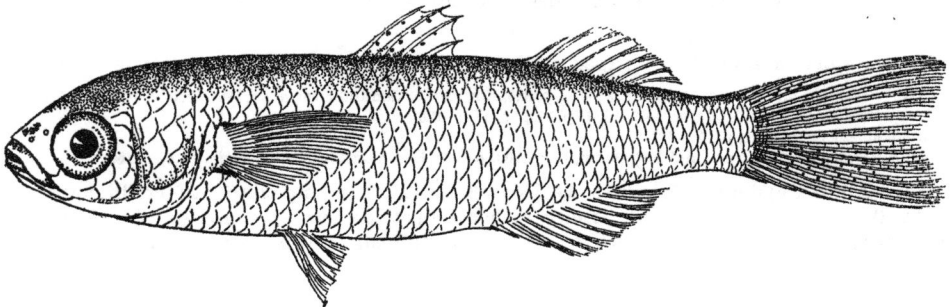

FIG. 111.—*Mugil cephalus.* Young, 25 millimeters long

a study of the spawning of the mullet in the region of Marco, on the southwest coast of Florida. Here it was found that the chief spawning period ranged from the middle of December until the end of January and that some fish spawned in February, but no fully ripe fish were seen nor was the locality of spawning found. It was evident from the appearance of the ovaries that all of the eggs were not spawned at one time.

Several fishermen in Florida have informed us that they have seen mullets spawning, and we include a description of their alleged observations in the hope that it will be an aid for future study. Capt. J. L. Sweat, of St. Petersburg, Fla., told us that since 1895 he had observed the mullet closely, and that off Indian Pass, near by, there is a locality where mullets spawn each year during the latter part of November and in December. This spawning ground is near the beach, where the water is about 24 feet deep and the bottom is of rock. The fish are so thick at the time of spawning "that a pole can scarcely be pushed through them." Captain Sweat stated that he had actually seen the spawn, and that the water was yellowish-white from the eggs and milt and "that the water smelled of fish" for some distance away. (The odor could have been caused by a flowering of diatoms.) He declared that the fish always spawn in the outside waters and not in the bays, rivers, or inlets. Some time after spawning he observed very small mullets near the spawning grounds. The fry swam so compactly that, looking down from above, they appeared like a large black ball. Several fishermen at Marco and Caxambas, Fla., agreed that mullets spawned in the outside waters and that at the time the water was "sticky and yellowish white from the eggs and milt."

We are satisfied that the mullet does not spawn in Chesapeake Bay, for the 10 to 12 inch fish that comprise the bulk of the catch in October proved to be immature. Mullets larger than these sizes seldom occur in the Chesapeake; yet spawning may take place not far from the mouth of the bay; otherwise it is difficult to account for the schools of fry slightly more than 1 inch long present in the lower bay beginning in April. These 1-inch fry probably are found along our entire coast, from Virginia to Texas, in the spring, for we have found them in North Carolina and along various parts of the Florida coast. It is not believed that these young migrate from southern waters to Chesapeake Bay.

The growth of the mullet fry is very slow during the winter, at least in the more northern waters. At Beaufort, Jacot (1920, p. 203), on December 22, collected fish 22 to 32 millimeters long. He found them the same size during January and early February and only 24 to 36 millimeters by February 24. These represented the full range of the sizes caught during each period. Many young mullets of the following sizes were collected in Chesapeake Bay: April, 25 to 36 millimeters; May, 30 to 49 millimeters; June, 43 to 60 millimeters. At the same time, and throughout the summer and fall, larger mullets were taken as follows: April, 175 to 194 millimeters; May, 118 to 160 millimeters; June, 102 to 132 millimeters; July, 113 to 167 millimeters; August, 150 to 155 millimeters; September, 107 to 200 millimeters; October, 230 to 301 millimeters.

The growth of the young during the spring appears to be quite regular, but we are unable to follow this growth after June. It is difficult to determine the age and growth of larger fish with the limited data at hand. In the spring, schools of mullets 4 to 6 inches long are rather common; in the summer, 5 to 8 inches; and in the fall 8 to 12 inches. However, mullets 4 to 12 inches long are taken throughout this period. Jacot (1920, p. 220) believes that, according to their scales, 5 to 8 inch jumping mullets arriving in April at Beaufort, N. C., are 14 to 17 months old.

The mullets in 1920 ranked thirteenth in quantity and fourteenth in value in Chesapeake Bay, the total catch amounting to 282,020 pounds, worth approximately $10,207.

In Maryland the mullets ranked tenth in quantity and twelfth in value, the catch being 35,337 pounds, value at $1,861. Approximately 36 per cent of this amount was caught in fyke nets, 23 per cent in haul seines, 21 per cent in gill nets, and 20 per cent in pound nets. The counties taking the largest quantities were Cecil, Kent, and Somerset.

In Virginia the mullets ranked eleventh in quantity and fourteenth in value, the catch being 246,683 pounds, valued at $8,346. Approximately 85 per cent of this amount was caught with gill nets, 14 per cent with haul seines, and 1 per cent in pound nets. The counties making the largest catches were Norfolk, Elizabeth City, Accomac, and Princess Anne. It is believed that a large part of the catch credited to Norfolk County was actually taken in North Carolina waters.[16]

The striped mullet is taken in the Chesapeake from June until November. It is found in all parts of the bay, particularly on the western shore, which is considerably broken up with islands, small bays, and creeks. It is especially common in Mobjack Bay, the lower York River, Back River, and Lynnhaven Bay.

The first run of fish usually appears in the lower part of the bay some time in June. The fish at this season are scarcely large enough to make fishing profitable, however, as they are only about 5 to 7 inches in length. As the season advances the size of the fish caught increases, and by September and October, when a definite run of fish occurs, the usual length is 10 to 12 inches. Fishermen and people residing along the water front catch a large number of mullets for their own use, of which no record is obtainable. During October, when the mullets are most abundant, of fair size, and in prime condition, many are caught with small seines and salted down for home consumption during the winter months. In the vicinity of Back River and Buckroe Beach as many as six small fishing crews are commonly seen at one time during October watching for schools of mullets. As the striped mullet usually betrays its presence by its habit of jumping out of the water, a practiced eye can discern an approaching school of fish at a considerable distance. The fish usually follow the shore and therefore are captured rather easily if the fishermen use precision and speed, for the mullet is an elusive fish. It is a fast swimmer and a good jumper, and when no other avenue of escape is available many individuals often obtain their freedom by jumping over the net. (An account of the methods used for catching mullet is given by Schroeder, 1924, p. 36.) A typical crew for catching mullets for home consumption is composed of the owner of the seine and several of his neighbors, who fish together and share the catch.

---

[16] The catch of mullets assigned to Virginia from Norfolk County, quite certainly is too great because of the shipment of mullets from North Carolina waters through the inland waterway to Norfolk City. However, it is impossible to separate or even to estimate with a degree of accuracy the fish taken locally and those coming from North Carolina.

The entire catch of mullets is consumed locally, as it is an esteemed food fish, and if a surplus exists it is salted for future use.   In the fall and winter the Baltimore and Norfolk markets receive mullets from North Carolina and Florida.   During 1921 and 1922 the fishermen received from 8 to 15 cents per pound and the retail price varied from 15 to 25 cents a pound.

"Mullet," "jumping mullet," and "jumper" are the names commonly used in the bay. "Fatback" is a term apparently applied only at Cape Charles.   The last-mentioned name probably alludes to the rather broad back and the layer of fat present there in the fish.   Jumping mullet and jumper, of course, refer to the characteristic jumping habit of this species, to which reference already has been made.

In North Carolina the mullet is one of the principal species taken, and in Florida it is the most valuable of the many species of fish taken within the waters of that State, the catch in 1918 amounting to 35,527,840 pounds, with a value of $1,565,843.

The size of the Chesapeake mullet averages from 8 to 12 inches in length, and the maximum is about 15 inches.   These sizes are considerably smaller than those attained farther south, where 20-inch fish are common and 30 inches is the maximum.   Following are weights of Chesapeake Bay striped mullets: Seven and one-half inches, 2.5 ounces; 8½ inches, 4.6 ounces; 9 inches, 5 ounces; 9½ inches, 5.6 ounces; 10 inches, 6.7 ounces 10½ inches, 7.5 ounces; 12 inches, 10 ounces.

Habitat.—Warm waters of both hemispheres; on the Atlantic coast of America from Cape Cod to Brazil.

Chesapeake localities.—(a) Previous records: "Salt water of Chesapeake Bay," Hampton and Cape Charles city, Va.   (b) Specimens in collection: From many localities from Love Point, Md., southward to Cape Charles and Norfolk, Va.

97. Mugil curema Cuvier and Valenciennes.   Silver mullet; White mullet; Mullet.

Mugil curema Cuvier and Valenciennes, Hist. Nat. Poiss., XI, 1836, p. 87; Brazil, Martinique, Cuba.   Bean, 1891, p. 92; Jordan and Evermann, 1896–1900, p. 813, Pl. CXXVI, fig. 344.
Mugil albula Uhler and Lugger, 1876, ed. I, p. 140; ed. II, p. 119.
(?) *Querimana gyrans* Evermann and Hildebrand, 1910, p. 160.

Head 3.6 to 4.1; depth 3.1 to 4.75; D. IV–I, 8; A. III, 9 (young of about 50 millimeters and less in length with II, 10 rays); scales 35 to 41.   Body moderately compressed; caudal peduncle rather strongly compressed, its depth 2.1 to 2.35 in head; head at eyes scarcely deeper than broad; snout rather short, its length 4.35 to 6 in head; eye 3.15 to 4.35; interorbital 2.4 to 2.9; mouth moderate, oblique, the lower jaw included, the gape somewhat broader than deep; adipose eyelid well developed in adult, undeveloped in young; gill rakers numerous, slender, close set; scales moderate, with crenate membranous edges, extending on the base of caudal and covering almost entirely the second dorsal and anal fins; origin of spinous dorsal about equidistant from tip of snout and base of caudal; origin of second dorsal a little posterior to origin of anal; caudal fin moderately forked; anal fin similar to but longer than second dorsal; ventral fins inserted about equidistant from tip of snout and middle of anal base; pectoral fins not quite reaching opposite origin of first dorsal, 1.25 to 1.6 in head.

Color in life dark greenish on back; silvery on sides; abdomen pale; opercle yellowish; dorsals, caudal, and pectorals more or less yellowish, sometimes with dusky tips; other fins plain; axil of pectoral bluish black.

Many small specimens, ranging in length from 30 to 150 millimeters (1⅓ to 6 inches), are at hand.   This species differs from *M. cephalus* in its plain coloration, having no evident dark stripes along the rows of scales; by the densely scaled, soft dorsal and anal fins; and by having one more ray in the anal fin.   The differences between the young and the adults of this species appear to be similar to those described for *M. cephalus*.   No doubt remains relative to the identity of the young mullets formerly recognized under the generic name "Querimana."   The young of this species and *M. cephalus* can be separated nearly as readily as the adults.   The number of spines and rays in the anal fins of these two mullets is quite constant but can not be relied upon wholly as a means of identification.   The young of this species usually has II, 10 rays and ordinarily is readily separable from the young of *M. cephalus*, which usually has II, 9 rays in the anal fin.   However, Jacot (1920, p. 200) points out that specimens of *M. curema* sometimes have 11 or 13 anal fin supports (rays or spines) and that *M. cephalus* occasionally has 10 (rarely 12) anal fin supports.

Earlier investigators appear not to have separated the young mullets of the Atlantic coast of the United States into species, but considered them all identical, calling them *Querimanna gyrans*. Bean (1903, p. 365) stated that Querimanna was nothing but the young of Mugil. He, however, placed *Querimana gyrans* in the synonymy of *Mugil trichodon*, a southern species, probably because the number of scales given by Jordan and Gilbert, who first described *Q. gyrans*, suited that species. Indeed, it is probable that *Q. gyrans*, the type of which we have not seen, was based upon the young of *M. trichodon*, as the type specimens came from Key West, where that species occurs. *Querimana gyrans* later was recorded from as far north as Cape Cod, where *M. trichodon* does not occur. Bean's reference led Smith (1907, p. 182) to say: " The present writer is inclined to accept Doctor Bean's general conclusion in the matter, but regards it as unfortunate that *Querimana gyrans* has not been shown to be the young of the striped mullet or the silverside mullet. The chief obstacle to such an identification is the difference in the number of scales in the lateral series, and until this is overcome the question must be considered unsettled." The scales in the young are quite easily lost, but in specimens from Chesapeake Bay in which they are present the usual number possessed by *M. curema* and *M. cephalus* may be counted.

The food of this mullet apparently is identical with that of *M. cephalus*, consisting almost wholly of minute organisms, which are found mixed with quantities of mud and vegetable débris.

The spawning habits of this species are as imperfectly known as in *M. cephalus*. Limited evidence has been produced (Jacot, 1920, p. 226) which would suggest that the species spawns in the spring of the year, whereas *M. cephalus* spawns in late autumn.

The growth of the silver mullet in Chesapeake Bay is difficult to follow with the limited data at hand. A single young fish taken in May measured 41 millimeters (1.6 inches), whereas the many young taken during June were 25 to 33 millimeters (1 to 1.3 inches) in length. No small fish were caught during the summer, and the next sizes at hand are 115 to 136 millimeters (4.5 to 5.3 inches), taken in September, and 95 to 148 millimeters (3.8 to 5.8 inches), caught in October. Jacot (1920, p. 203) seined young at Beaufort, N. C., 30 to 36 millimeters (1.2 to 1.4 inches) long, in May, took these sizes and somewhat larger fish throughout the summer, and young as small as 20 to 28 millimeters (0.8 to 1.1 inches) in September.

The silver mullet is taken in small quantities in Chesapeake Bay and is found under virtually the same conditions as the striped mullet. Much that has been said of the latter, therefore, applies also to this species. Neither the fishermen nor the markets distinguish the two kinds of mullets; *M. curema* forms but a very small part of the marketable catch.

This mullet seldom exceeds a length of 9 inches in the bay. Most of those seen are small unmarketable fish, 6 inches or less in length, and are taken incidentally with other species of fish. The small catch of silver mullets is included in the statistics given for the striped mullet. In southern waters the average length is about 10 inches and the maximum about 14 inches.

Habitat.—On the Atlantic coast, from Cape Cod to Brazil, and on the Pacific coast from the Gulf of California to Chile.

Chesapeake localities.—(*a*) Previous records: Cape Charles city, Hampton, and "southern part of Chesapeake Bay." (*b*) Specimens in the collection: From Annapolis, Md., lower York River, vicinity of Norfolk, and Cape Charles, Va.

Family XLVII.—SPHYRÆNIDÆ. The barracudas

Body very elongate, little compressed; head long, pointed, pikelike; mouth large, nearly horizontal; jaws elongate, the lower one strongly projecting; jaws and palatines with large teeth of uneven size; opercular bones without spines or serrations; gill membranes separate, free from the isthmus; gill rakers very short or obsolete; branchiostegals 7; gills 4, a slit behind the fourth; pseudobranchiæ well developed; air bladder large, bifurcate anteriorly; pyloric cæca numerous; scales small, cycloid, present on cheeks and opercles; lateral line well developed; first dorsal with five spines, second dorsal remote from the first, similar to anal and at least partly opposite it; caudal fin forked; pectoral fins short, placed in or below the axis of the body.

74. Genus SPHYRÆNA Röse. Barracudas

The characters of the genus are included in the family description.

KEY TO THE SPECIES

a. Scales moderate, about 120 in a lateral series; ventral fins inserted in advance of first dorsal;
  maxillary reaching eye in adults; pectorals reaching past base of ventrals__guachancho, p. 198
aa. Scales smaller, about 135 in a lateral series; ventral fins inserted under origin of first dorsal;
  maxillary failing to reach eye; pectorals failing to reach base of ventrals____borealis, p. 198

98. Sphyræna guachancho Cuvier and Valenciennes. Barracuda.

Sphyræna guachancho Cuvier and Valenciennes, Hist. Nat. Poiss., III, 1829, p. 342; Havana. Jordan and Evermann, 1896–1900, p. 824.

Head 3.4; depth 7.15; D. V–I, 9; A. I, 8; scales 120.   Body very elongate, nearly cylindrical, scarcely deeper than broad at base of first dorsal; caudal peduncle moderately compressed, its depth 4 in head; head low, quadrate, with prominent ridges above; snout long and pointed, 2.1 in head; eye 6.4; interorbital 4.8; mouth large, a little oblique; lower jaw sharply pointed and strongly projecting; maxillary broad, reaching anterior margin of eye, 2.15 in head; teeth in the jaws and on palatines large, lance-shaped, the lateral ones in lower jaw smaller; gill rakers obsolete; scales small, present on cheeks and opercles, also extending on the second dorsal, anal, and caudal fins; dorsal fins far apart, the first with five slender spines,.its origin behind base of ventrals; second dorsal short, its origin in advance of anal and about equidistant from origin of spinous dorsal and

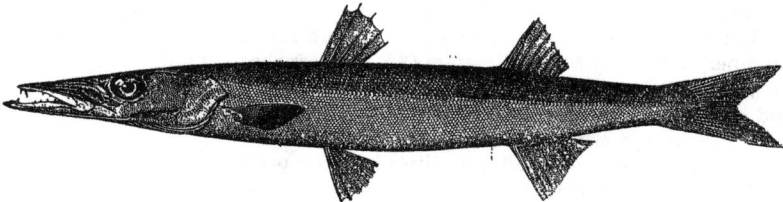

FIG. 112.—*Sphyræna borealis.* From a specimen 9 inches long

base of caudal; caudal fin forked, the lower lobe the longer; ventral fins rather small; pectoral fins small, reaching well beyond base of ventrals, 2.6 in head.

Color bluish gray above, silvery below; dorsals and caudal dusky, other fins pale.   The specimen in hand is of uniform color.   Some specimens, however, are irregularly blotched with black and the very young have black crossbars.

A single specimen of this species, 395 millimeters (15½ inches) in length, was secured.

This barracuda is reported from Woods Hole, Mass., but it does not appear to have been taken elsewhere north of Florida.   The specimen at hand was trapped in a pound net in Lynnhaven Roads on July 11, 1921.   The fish was unknown by the fishermen, which indicates that barracudas are very rare in Chesapeake Bay.   This species is said to reach a length of 2 feet.   In the Tropics it has some value as a food fish.

Habitat.—Woods Hole, Mass., to Panama; apparently rather rare north of Florida.

Chesapeake localities.—(a) Previous records: None.   (b) Specimen in collection: From Lynnhaven Roads, Va.

99. Sphyræna borealis DeKay. Barracuda; Northern barracuda.

Sphyræna borealis De Kay, Fauna, New York, Fishes, 1842, p, 39, Pl. LX, fig. 196; New York. Jordan and Evermann, 1896–1900, p. 825; Evermann and Hildebrand, 1910, p. 160.

This species is recorded from Chesapeake Bay by Evermann and Hildebrand (1910, p. 160), who had two small specimens, 1 and 3 inches in length.   Bean (1891, p. 83) states that William P. Seal observed a Sphyræna at Cape Charles city but secured no specimen.   The species that was observed, therefore, was undetermined.   No specimens of this barracuda occur in the present collection, and the fish quite certainly is very rare in Chesapeake Bay.

This species is distinguished from *S. guachancho* (the only other species of the genus recorded from waters north of Florida) principally by the smaller scales, the more posterior position of the ventral fins with reference to the spinous dorsal, and by the smaller mouth. A comparison of these characters is presented in the key to the species.

This is the smallest of the barracudas, rarely reaching a length of more than 1 foot. It has been called the Northern barracuda because it was thought to be entirely of northern distribution, its range having only recently been found to extend into the Tropics.

Habitat.—Cape Cod to Panama.

Chesapeake localities—(*a*) Previous records: Cape Charles city, Va. (*b*) Specimens in collection: None.

Family XLVIII.—POLYNEMIDÆ.   The threadfins

Body oblong, compressed; snout conical, projecting beyond mouth; eye anteriorly placed, with a well-developed adipose eyelid; mouth large, nearly horizontal; teeth in villiform bands on jaws, palatines, and sometimes on vomer; gills 4, a slit behind the fourth; branchiostegals 7; lateral line complete, continued on caudal fin; dorsal fins 2, rather remote from each other, the first with 7 or

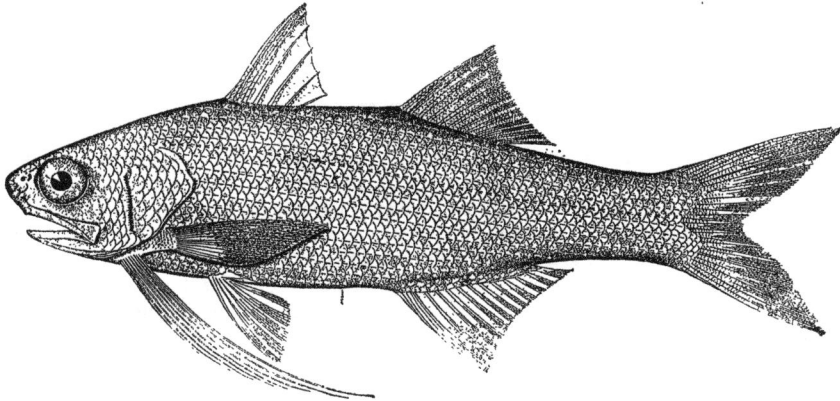

FIG. 113.—*Polynemus octonemus*

8 rather high, feeble spines; caudal fin deeply forked; anal fin either similar to second dorsal or much longer; ventral fins abdominal, with I, 5 rays; pectoral fins placed low, in two parts, the lower part consisting of free articulated filaments.

75. Genus POLYNEMUS Linnæus

Vomer with teeth; preopercle serrate, its lower posterior angle with a scaly flap; anal fin not much longer than second dorsal, consisting of about 13 or 14 rays; pectorals with 3 to 9 free filaments, all shorter than body. A single species was taken in Chesapeake Bay.

100. Polynemus octonemus Girard.   Threadfin.

Polynemus octonemus Girard, Proc., Ac. Nat. Sci., Phila., 1858, p. 167; Brazos, Santiago, and Galveston.
Ploydactylus octonemus Jordan and Evermann, 1896–1900, p. 830, Pl. CXXVIII, fig. 350.

Head 3.25 to 3.4; depth 3.1 to 3.35; D. VIII–I, 12; A. III, 13; scales about 58 (most of the scales are lost in specimens at hand). A fairly accurate count, however, is obtainable from one specimen, which gives the result stated when counting oblique series running upward and backward above lateral line. Jordan and Evermann (1906, p. 830) give 70 scales in a lateral series. It is probable that these authors started counting the series from the nape. Gill (1860, p. 280) states that the lateral line runs through 60 scales. Body compressed; caudal peduncle rather strongly compressed, deep, 2.2 in head; head moderate, compressed; snout conical, projecting far beyond mouth, its length 5.3 to 5.55 in head; eye 4.75; interorbital 4 to 4.3; mouth moderate, inferior, horizontal; maxillary broad, 2.35 to 2.45 in head; teeth small, in villiform bands on jaws, vomer, and palatines;

49826—28——14

gill rakers long, 21 or 22 on lower limb of first arch; scales moderate, ctenoid, moderately deciduous, extending forward on snout and on fins, the second dorsal and anal densely scaled; lateral line complete, forked at base of caudal, the branches extending on the fin; origin of spinous dorsal a little behind margin of opercle and about an eye's diameter nearer origin of second dorsal than tip of snout; the longest spine 1.4 in head; origin of second dorsal a little in advance of anal; caudal fin deeply forked; anal fin similar to second dorsal, its base only a little longer; ventral fins rather small, inserted about an eye's diameter nearer origin of anal than tip of lower jaw; pectoral fins rather long, 1.15 in head, the filaments well separated from the rest of the fin, eight in number, the longest ones reaching nearly to origin of anal.

Color in alcohol olivaceous; the fins dusky, the pectoral fins darkest. The fins are said to be mostly pale in the young.

Only three specimens of this species—respectively, 233, 233, and 235 milimeters (9⅛, 9⅛, and 9²/₅ inches) in length—are at hand. This fish, although recorded from as far north as New York, does not appear to have been taken previously in Chesapeake Bay, where it is scarcely known by the fishermen. The specimens at hand were the only ones seen during extensive collecting expeditions, and they were trapped in a pound net in Lynnhaven Roads, Va. The species of this genus are considered good food fishes on the Isthmus of Panama, where several species are abundant.

Habitat.—New York to the Rio Grande.

Chesapeake localities.—(*a*) Previous records: None. (*b*) Specimens in collection: From Lynnhaven Roads, Va.

Family XLIX.—SCOMBRIDÆ. The mackerels

Body fusiform, more or less compressed; head depressed above; snout pointed; caudal peduncle slender, with one or more keels; mouth large; premaxillary not protractile; maxillary without a supplemental bone; jaws with large or small sharp teeth; preopercle unarmed except in very young; opercle entire; gill openings large, the membranes separate, free from the isthmus; gills 4, a slit behind the fourth; pseudobranchiæ large; gill rakers long; dorsal fins 2, the first of weak spines, the second similar to anal, followed by detached rays, known as finlets; caudal fin large, forked.

Scombroid fishes of the Orient have recently been studied in great detail by Kishinouye,[17] who has grouped the fishes of that region into three families, placing (of the genera represented in the present report) Scomber in Scombridæ and Sarda in Cybiidæ. Scomberomorus, however, does not occur in the Orient, hence this genus is not included in Kishinouye's work. It seems probable that a detailed study would show this genus to be as distinct and equally as much deserving of family rank as some of the groups recognized as families by Kishinouye. While we do not question the arrangement of families in Kishinouye's very excellent work, we prefer to retain in this report the family Scombridæ as understood by Jordan and Evermann (1896–1900 p. 863) and others, principally because we do not know where to place *Scomberomorus* in the new arrangement and we have neither the time nor specimens to go into a study of this matter on this occasion.

KEY TO THE GENERA

a. Caudal peduncle without a median lateral keel but with two small keels, one above and one below the median line and placed more or less on base of caudal; first dorsal with 9 to 14 feeble spines.
　b. Air bladder wanting; first dorsal with 10 to 14 weak spines_____Scomber, p. 201
　bb. Air bladder present; first dorsal with 9 or 10 weak spines_____Pneumatophorus, p. 202
aa. Caudal peduncle with a median keel on each side and with a smaller one above and below this one; dorsal fin with 14 to 22 spines.
　c. Scales not forming a corselet on anterior part of body; pectorals inserted near level of the eyes; sides sometimes with yellowish spots and occasionally with one or more straight dark lines _____Scomberomorus, p. 203
　cc. Scales forming a corselet; pectoral fins placed lower than eye; sides without yellow spots.
　　d. Vomer toothless; sides with black longitudinal oblique bands_____Sarda, p. 205
　　dd. Vomer with villiform teeth; sides without black longitudinal bands_____Thunnus, p. 207

---

[17] Contributions to the comparative study of the so-called scombroid fishes. Journal, College of Agriculture, Imperial University, Tokyo, Vol. VIII, No. 3, (1923), pp. 293 to 475, Pls. XIII—XXXIV, fig. A–Z. Tokyo.

76. Genus SCOMBER Linnæus. Mackerels

Body elongate, fusiform; caudal peduncle slender, with two small keels on each side; mouth comparatively large; maxillary slipping under the preorbital; teeth small, in a single series on the jaws, paired oblique patches on vomer, and in one row on palatines; gill rakers long and slender; scales very small, not forming a corselet anteriorly; first dorsal with 10 to 14 feeble spines; second dorsal and anal similar, each followed by 5 to 9 finlets; caudal fin small, broadly forked; ventrals and pectorals small; air bladder wanting.

101. Scomber scombrus Linnæus. Common mackerel.

Scomber scombrus Linnæus, Syst. Nat., ed. X, 1758, p. 297; Atlantic Ocean. Uhler and Lugger, 1876, ed. I, p. 108; ed. II, p. 91; Jordan and Evermann, 1896–1900, p. 865, Pl. CXXXIII, fig. 363.

Head 3.6 to 3.8; depth 5.65 to 6; D. XI or XII—12—V; A. I, 11 or 12—V. Body fusiform, little compressed; caudal peduncle slender, broader than deep, its depth 10 to 11 in head; head long, slender; snout pointed, its length 2.85 to 3.25 in head; eye 4.75 to 5.55; interorbital 4.1 to 4.75; mouth moderate, terminal, oblique; maxillary reaching nearly to middle of eye, 2.5 in head; teeth small, in a single row on jaws, palatines, and vomer; gill rakers long, slender, about 30 on lower limb of first arch; scales very small; first dorsal with slender spines, its origin about an eye's diameter behind base of pectorals; second dorsal very small, followed by five finlets; caudal fin broadly forked; anal fin similar to and opposite second dorsal, also followed by five finlets; ventral fins small, inserted under or slightly in advance of vertical from origin of dorsal; pectoral fins short, 1.9 to 2.2 in head.

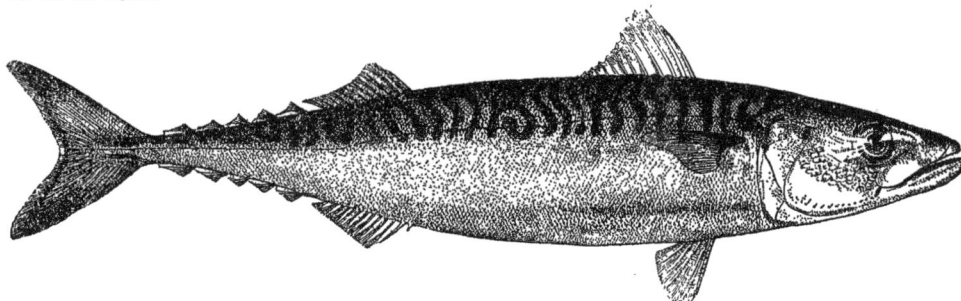

FIG. 114.—*Scomber scombrus*

Color bluish black above, with wavy, blackish transverse streaks; bright silvery below; dorsals, caudal, and pectorals largely dusky; axil of pectoral black, peritoneum black.

This species is represented by four small specimens, ranging from 220 to 240 millimeters (8¾ to 9½ inches) in length.

Garstang (1898, pp. 235–295) made some interesting comparisons among American and various groups of European mackerel (all *Scomber scombrus*) in order to determine what racial differences, if any, exist. In this study he utilized 100 fish from Newport, R. I., and 1,549 fish from Ireland, the English Channel, and the North Sea. As a result of these studies he indicated that a racial difference did exist between American and European mackerel but that this difference is so small that it can only be appreciated by an examination of many specimens.

The number of dorsal spines and rays of the mackerel shows rather wide variation when a large series of specimens is examined. The first dorsal usually contains 11, 12, or 13 spines, rarely 10 or 14. The second dorsal usually contains 12 rays, less frequently 9 to 11 or 13 to 15.

The mackerel is largely a plankton feeder, subsisting chiefly on pelagic crustaceans as well as on fish eggs and fish fry. For a comprehensive account of the feeding habits, spawning, migrations, etc., of the mackerel see Bigelow and Welsh (1925, pp. 188–208).

Spawning takes place during the last half of May and throughout the month of June in the Massachusetts Bay region and a few weeks later in the Gulf of St. Lawrence. Most of the spawning is done at night and when the water temperature ranges from 46° to 61° F. (Bigelow and Welsh, p. 208.) The egg is buoyant, from 0.97 to 1.38 millimeters in diameter, and hatches in about 96 hours at 60° to 62° F., and in about 120 hours at 55° F.

The mackerel is present off the New England coast from spring to fall. The first catches are made between Cape Hatteras and Chesapeake Bay between the end of March and middle of April, but the fish are not seen again in this region until the following year. Off the New England coast the first mackerel appear in May, remaining until November. Not much is known of the winter home of the mackerel in the western Atlantic, but stray fish have been taken on Georges Bank and in South Channel in February or March. It is suggested by Bigelow and Welsh that mackerel may winter on the continental shelf at a depth of 100 to 200 fathoms and not farther south than Cape Hatteras. No mackerel have ever been reported more than a few miles south of Cape Hatteras at any time.

The mackerel, although one of the most valuable food fishes of the north Atlantic, is of no commercial importance in Chesapeake Bay. Toward the end of April and early in May a few small mackerel sometimes stray inside the mouth of the bay and are caught in pound nets below Cape Charles city and at Lynnhaven Roads, Va. In a pound net operated in these localities usually only about five or six mackerel are caught during a season, and sometimes none at all are taken. As early as 1876, Uhler and Lugger state that at that time this fish was much less common in Chesapeake Bay than formerly. Whether the species ever was common enough within the bay to be of commercial value probably will remain unknown.

The mackerel attains a length of 22 inches and, when in prime condition in the fall, a weight of 4 pounds. Fish of the same school are usually all about the same size. We have observed that commonly 90 to 95 per cent of a catch at Provincetown consisted of fish that did not vary more than 1 inch in length (by actual measurement in the course of tagging the fish for the purpose of determining their migrations). At times large numbers of "tinkers" are caught; that is, fish about 8 to 10 inches in length. The usual size of market fish is 12 to 16 inches, but a length of 18 to 20 inches is not unusual.

Habitat.—North Atlantic, inhabiting both coasts; known on the American coast from Labrador to Cape Hatteras.

Chesapeake localities.—(*a*) Previous record: Chesapeake Bay (Uhler and Lugger, 1876). (*b*) Specimens in collection: From Lynnhaven Roads, Va.

77. Genus PNEUMATOPHORUS Jordan and Gilbert. Chub mackerels

This genus differs from Scomber in the possession of a well-developed air bladder. Externally it differs principally in having fewer (9 or 10) weak spines in the dorsal fin.

102. Pneumatophorus colias (Gmelin). Chub mackerel; Thimble-eye mackerel; Bull's-eye.

Scomber colias Gmelin, Syst. Nat., 1788, p. 1329; Sardina. Uhler and Lugger, 1876 ed. I, p. 109; ed. II, p. 91; Jordan and Evermann, 1896-1900, p. 866, Pl. CXXXIII, fig. 364.

This species was once recorded by Uhler and Lugger as entering Chesapeake Bay. It is not reported by other investigators, and it was not seen there by us. Uhler and Lugger (1876) do not say that they had specimens, and their record may have been based only upon an observation. We find no other record showing that this fish has been taken south of the New Jersey coast.

This species is readily distinguished from the common mackerel by the shorter first dorsal (which has only 9 or 10 spines), by the dusky spots extending well below the lateral line in the adult, and by the much larger eye (comparing fish of nearly the same size). Its feeding habits appear to be similar to those of the common mackerel. Nothing is known of its breeding habits.

Along our North Atlantic coast this mackerel occurs irregularly. In some years large catches are made off the New England coast, and again it appears to be entirely absent. In the spring of 1925 we observed that about 1 per cent of the mackerel catch at Provincetown consisted of this species.

As a food fish it is said to equal the common mackerel and is not culled from the catch. The maximum length is 14 inches, but fish 11 to 13 inches long are often caught.

Habitat.—Atlantic and Pacific Oceans; on the western Atlantic coast from the Gulf of St. Lawrence to New Jersey, and possibly very rarely to Virginia.

Chesapeake localities.—(*a*) Previous records: "Enters Chesapeake Bay from the ocean" (Uhler and Lugger, 1876). (*b*) Specimen in collection: None.

78. Genus SCOMBEROMORUS Lacépède. Spanish mackerels

Body elongate, more or less compressed; snout quite long, pointed; mouth large; maxillary not concealed by preorbital; teeth in the jaws strong, compressed; vomer and palatines with granular teeth; gill rakers rather short and few in number; scales small, rudimentary, not forming a corselet on anterior part of body; caudal peduncle with a keel in the lateral line and a supplemental one above and below it; first dorsal with 14 to 18 feeble spines; interval between dorsals slight; second dorsal and anal each followed by 7 to 10 finlets; ventrals small; pectorals moderate, inserted near level of eyes; alimentary canal short; air bladder present.

KEY TO THE SPECIES

a. Sides of body with roundish bronzy spots but without dark, longitudinal stripes; pectoral fins scaleless_____*maculatus*, p. 203
aa. Sides with elongate bronzy spots and with one or two dark, longitudinal stripes; pectoral fins mostly covered with scales_____*regalis*, p. 205

103. Scomberomorus maculatus (Mitchill). Spanish mackerel.

Scomber maculatus Mitchill, Trans., Lit. and Phil. Soc., N. Y., I, 1814, p. 426; New York.
Cybium maculatum Uhler and Lugger, 1876, ed. I, p. 110; ed. II, p. 92; McDonald, 1882, p. 12, fig. 1.
Scomberomorus maculatus Bean, 1891, p. 87; Smith, 1892, p. 71; Jordan and Evermann, 1896-1900, p. 874, Pl. CXXXIV, fig. 368.

Head 3.2 to 4.8; depth 4 to 5; D. XVIII-14 to 17-VIII or IX; A. II, 14 to 16-VIII or IX. Body elongate, compressed; dorsal and ventral outlines about evenly rounded; caudal peduncle

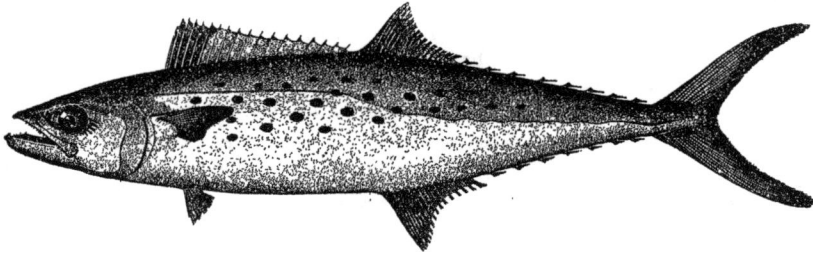

FIG. 115.—*Scomberomorus maculatus.* From a specimen 10¾ inches long

slender, with a median lateral keel and a small supplemental keel both above and below it; head compressed; snout long, pointed, its length 2.45 to 2.75 in head; eye 4.55 to 4.9; mouth large, oblique; lower jaw (at least in young) a little shorter than the upper; maxillary reaching opposite posterior margin of eye, 1.6 to 1.8 in head; teeth in the jaws compressed, variable in size and number; gill rakers about half the length of eye in adult, very short in young, 8 to 10 on lower limb of first arch; first dorsal with slender spines; second dorsal and anal similar, densely scaled, each fin followed by 8 or 9 finlets; the origin of second dorsal a little in advance of anal; caudal fin broadly forked; ventral fins small, shorter than snout; pectoral fins not scaly, short, 1.5 to 2.85 in head.

Color of a fresh specimen 260 millimeters (10¼ inches) in length, dark blue above, with sky-blue reflections; silvery below; sides with roundish yellow spots, forming three longitudinal rows, the lower one present only on anterior part of body; spinous dorsal mostly black, the base of the short spines white; soft dorsal greenish, with dusky tips; finlets pale green; caudal greenish dusky, the tips of lobes mostly black; anal and ventrals pale; pectorals greenish at base, dusky at tips.

No specimens of this common species from Chesapeake Bay were preserved. It is here described from specimens collected at Beaufort, N. C., ranging in length from 45 to 355 millimeters (1¾ to 14 inches).

The Spanish mackerels congregate in schools; they appear to be migratory in their habits, appearing off the Middle Atlantic States in spring and fall runs, the first presumably representing a northward migration and the latter a southward one. No such runs of Spanish mackerel appear to take place within Chesapeake Bay, however, as the fish usually arrive in May or June and are present continuously until about September. Relative to spawning, Smith (1907, p. 191) says:

> The lower part of Chesapeake Bay was formerly and is still a favorite spawning ground. The eggs are about 1 millimeter (0.04 inch) in diameter and float at the surface; they are laid mostly at night, and the hatching period is about 25 hours in a water temperature of 77° or 78° F. All the eggs of a given fish do not ripen at one time, and the spawning may thus extend over several weeks, during which several thousand eggs may be deposited.

The spawning period in Chesapeake Bay occurs during late spring and early summer.

During 1920 the Spanish mackerel ranked twenty-third in quantity and nineteenth in value in Chesapeake Bay, the catch amounting to 13,766 pounds, worth about $2,114. Only 337 pounds of the entire catch was taken in Maryland waters during 1920, all caught with pound nets. In Virginia the Spanish mackerel ranked twenty-second in quantity and eighteenth in value, the catch being 13,429 pounds, worth $2,052. The entire catch, exclusive of a few fish taken with haul seines, was caught in pound nets. The counties producing the largest quantities were Elizabeth City, 5,900 pounds; Northampton; 3,835 pounds; and Mathews, 2,479 pounds.

The Spanish mackerel is one of the most highly esteemed fishes occurring in the bay. It appears regularly each year, some time in May or June, but it is never taken in such large quantities as farther south. Along the Atlantic coast of Florida and in the Gulf of Mexico, millions of pounds of this fish are caught each year from November to March,[18] a period when it is entirely absent from the Chesapeake. Large quantities sometimes are taken along the coasts of Virginia and North Carolina also. In the South the fish are caught mainly with gill nets and purse seines and, to a lesser extent, with hook and line. In the Chesapeake at least 99 per cent of the catch is taken in pound nets.

The Spanish mackerel, upon their arrival in the spring, first enter the bay as stragglers. In certain pound nets, situated near Cape Charles, Lynnhaven Roads, and Ocean View, Va., one or two fish a day are caught, followed, perhaps, with a few days when none are caught. To illustrate this the following statements are taken from our field notes:

> The first three Spanish mackerel of the 1921 season were taken in Lynnhaven Roads May 12 to 17. A set of three pound nets at Ocean View caught the first Spanish mackerel on May 15, 1922, and until May 27 a total of six fish had been caught. A set of two pound nets at Lynnhaven Roads, the closest nets to the entrance of the bay in 1922, caught one mackerel on each of the following dates: May 16, 20, 24, 25, and 26; on May 30 these nets caught 15 pounds and on May 31, 85 pounds.

The first pound-net catches that were of commercial importance (that is, about 20 pounds of fish, or more, on one day by one net or set of nets) occurred in Lynnhaven Roads on the following dates: May 30, 1916; June 26, 1917; June 17, 1918; June 9, 1919; June 1, 1920; June 15, 1921; May 30, 1922; and June 19, 1923. Almost the entire catch is taken from June to September, and only a few stragglers are taken before and after these dates. A particularly good run of fish occurred in Lynnhaven Roads from June 27 to July 2, 1921, when two pound nets caught 150 to 350 pounds daily and the catch for the week amounted to 1,400 pounds. The catch is confined to the lower part of the bay, and it is seldom that the fish strays above the mouth of the Rappahannock River.

At one time the Spanish mackerel was considered abundant in Chesapeake Bay. The following excerpt is taken from the United States Fish Commission's report for 1880:

> Gill nets were introduced into the Spanish-mackerel fisheries of Chesapeake Bay in 1877, and, proving fairly successful, they soon came into general favor among the fishermen of the eastern shore, though they are even now seldom employed by those living on the opposite side. There are at present about 175 men engaged in "gilling" for mackerel between Crisfield, Md., and Occohannock Creek, which is 30 or 40 miles from the capes. The nets were at first set only in the night, but during 1880 the fishermen of Tangier Island obtained the best results by fishing from the middle of the afternoon until midnight. The nets range from 75 to 100 fathoms in length and have a mesh similar to those already mentioned (3½ to 4 inches). The catch varies considerably, as many as 500 mackerel having been taken at one set, though the average is only 20 to 40 daily to the net.

It is estimated that during the past 10 years the annual catch of Spanish mackerel taken in Chesapeake Bay has ranged from 10,000 to 25,000 pounds. Part of the catch is marketed locally, but when a good run of fish occurs shipments are made to other markets, located principally from Washington to New York. The wholesale price in 1921 and 1922 generally ranged from 18 to 25 cents a pound and the retail price varied from 25 to 40 cents a pound.

---

[18] For an account of the Spanish-mackerel fishery of southern Florida, see Schroeder (1924, p. 40).

The size of Chesapeake fish ranges mostly from 1 to 3 pounds. One of the largest Spanish mackerel observed from the bay was caught on September 21, 1922, at Ocean View, Va. This fish was 760 millimeters (30 inches) long and weighed 7 pounds 5 ounces. The largest fish ever recorded from anywhere, so far as known to us, weighed 25 pounds. This weight is very exceptional, however, the usual maximum weight for Atlantic coast fish being only about 10 pounds and the common range of market fish is 1½ to 4 pounds.

Habitat.—Both coasts of America; on the Pacific from Cortez Banks south to the Galapagos Islands, and on the Atlantic from Maine to Brazil; common as far north as New York, stragglers occurring as far north as Monhegan, Me.

Chesapeake localities.—(*a*) Previous records: Various localities, from the mouth of the Potomac River southward. (*b*) Specimens in collection: None. The species was observed in the lower York River, Cape Charles, Buckroe Beach, Lynnhaven Roads, and Ocean View, Va.

104. Scomberomorus regalis (Bloch). Cero.

Scomber regalis Bloch, Ichthyol., 1795, Pl. CCCXXXIII; Martinique.
Cybium regale Uhler and Lugger, 1876, ed. I, p. 111; ed. II, p. 93.
Scomberomorus regalis Jordan and Evermann, 1896–1900, p. 875, Pl. CXXXV, fig. 369.

This fish was once recorded from "Chesapeake Bay near the ocean" by Uhler and Lugger (1876). It has not been reported by other writers and it was not seen during the present investigation. This species is distinguished from the Spanish mackerel principally by the color. The present species has elliptical bronzy spots on the sides and one or two longitudinal dark streaks. The Spanish mackerel also has bronzy spots on the sides, which, however, are less elongate, and

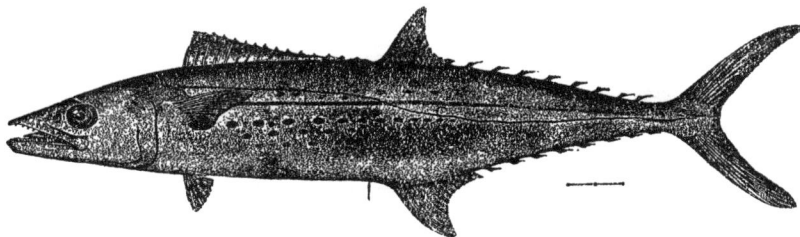

FIG. 116.—*Scomberomorus regalis*

the dark longitudinal stripe or stripes are missing. The cero has the pectoral fins mostly covered with scales, whereas in the Spanish mackerel these fins are naked.

The cero is a good fish, and in some localities where it is caught by trolling it is considered a good game fish. It is said to reach a weight of 35 pounds.

Habitat.—Cape Cod, Mass., to Brazil; not common north of Florida.

Chesapeake localities.—(*a*) Previous record: Chesapeake Bay near the ocean. (Uhler and Lugger, 1876.) (*b*) Specimens in the collection: None.

79. Genus SARDA Cuvier. Bonito

Body elongate, somewhat compressed; head large, pointed; mouth large; teeth on jaws rather strong, compressed, similar teeth on palatines, none on vomer or tongue; scales small, those of the pectoral region forming a corselet; a distinct lateral keel on caudal peduncle; first dorsal with 18 to 22 spines; second dorsal and anal similar, each followed by 6 to 9 finlets; caudal broadly forked; ventrals and pectorals small; upper parts with longitudinal more or less oblique stripes. A single species of this genus occurs in Chesapeake Bay.

105. Sarda sarda (Bloch).   Bonito; "Boston mackerel"; "Bloater."

Scomber sarda Bloch, Ichthyol., X, 1793, p. 35, Pl. CCCXXXIV; Europe.
Sarda pelamys Uhler and Lugger, 1876, ed. I, p. 109; ed. II, p. 92.
Sarda sarda Jordan and Evermann, 1896–1900, p. 872; Fowler, 1912, p. 58.

Head 3.6; depth 4.35 to 4.5; D. XXI—16–IX; A. II, 10 or 11–VIII.   Body elongate, compressed; caudal peduncle slender, broader than deep, with a membranous fold on the sides, its depth 10 or 11 in head; head rather long; snout pointed, its length 2.8 to 2.95 in head; eye 7.55 to 7.7; interorbital 3.7 to 3.8; mouth large, oblique; upper jaw slightly projecting; maxillary reaching past posterior margin of eye, 1.9 in head; teeth in the jaws rather strong, curved inward; a few teeth on palatines; gill rakers slender, 12 or 13 on lower limb of first arch; spinous dorsal long, the spines slender, highest anteriorly, its origin over or slightly in advance of base of pectorals; second dorsal small, wholly in front of anal, followed by 9 finlets; caudal fin rather small, broadly forked; anal fin similar to second dorsal but smaller, followed by 8 finlets; ventral fins small, inserted nearly under base of pectorals; pectoral fins short, with broad base, 2.6 in head.

Color bluish black above; lower parts silvery; dorsals and pectorals more or less dusky, other fins mostly plain translucent.   Young with black crossbars on upper part of sides; these bars replaced in the adult by 7 to 20 black, longitudinal, prominent stripes running backward and slightly upward.

This species was observed only in the southern part of the bay.   No specimens were preserved. The above description is based on two specimens, respectively 250 and 255 millimeters (9¾ and 10 inches) in length, from Buzzards Bay, Mass.

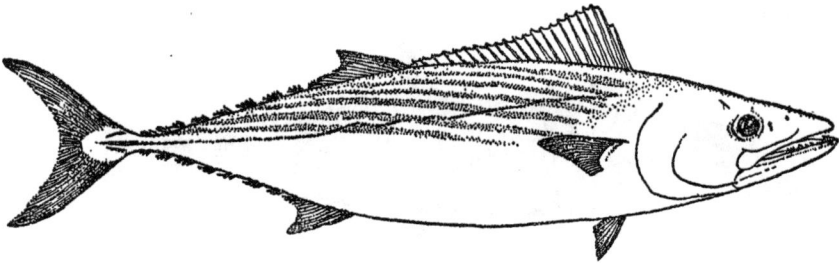

Fig. 117.—Sarda sarda

This fish, like most mackerels, travels in schools along the coast.   It is a rapid swimmer and feeds mainly at the surface.   When seen in the water, the adults are readily recognized by the dark stripes on the back.   Apparently the entire catch in Chesapeake Bay is taken in pound nets.   Along the coast, however, the bonito is often caught by trolling with tackle that is also used for catching bluefish.   Its spawning habits remain almost unknown.   Bean (1903, p. 395) says that this fish is believed to live in the open sea, coming to the shores only to feed or to deposit eggs.   Of its feeding habits this author says that it is predacious, feeding insatiably on mackerel and menhaden, and that it takes bait as freely as does the bluefish.   The catch of bonito in Chesapeake Bay during 1920 amounted to about 1,400 pounds, valued at $192.   The entire catch was made with pound nets set in the waters of Virginia.

The bonito is never taken in large quantities inside the bay.   Large schools of fish pass up and down the coast outside of the capes, but apparently only stragglers, and occasionally a small school, enter the Chesapeake.   This species is taken only in the southern part of the bay and seldom above the mouth of the York River or Cape Charles city, Va.

The first fish are caught some time in May, and small numbers are taken by the pound nets operated in the lower parts of the bay throughout the summer, or until September.   The first fish caught during the season of 1921 was taken at Lynnhaven Roads on May 20.   The following year (1922), in a set of three pound nets at Ocean View, bonitos were caught for the first time on May 17. The second catch, however, was not taken until May 26.   The records of fish taken in a set of two pound nets situated at Lynnhaven Roads show that the largest amount of bonito caught in any one month during a period of seven years (1916 to 1922) was 135 pounds.   This record catch was made in June, 1920.

The bonito, although a good food fish, is not generally regarded as highly as the common and the Spanish mackerels. However, it is more esteemed in the lower Chesapeake than in many other localities. The small catch is marketed locally and brings a good price, selling at retail (in 1921 and 1922) for 25 to 40 cents a pound. This species is known in the Norfolk markets as "Boston mackerel," a name generally given to the common mackerel, *Scomber scombrus*. *Rachycentron canadus* is known as bonito or black bonito in the lower Chesapeake. This fish is very different and must not be confused with *Sarda sarda*.

The bonito is said to reach a weight of 12 pounds along the Atlantic coast; the usual size of the fish from the Chesapeake, however, is only about 2 to 4 pounds.

Habitat.—Atlantic Ocean, on both coasts; common in the Mediterranean and on the coast of the United States from Cape Ann southward to Florida; rarely northward to Maine; no definite West Indian, Central, or South American records.

Chesapeake localities.—(*a*) Previous records: Entrance of Chesapeake Bay; Norfolk fish market. (*b*) Specimens in collection: None. Observed at Lynnhaven Roads and Ocean View, Va.

80. Genus THUNNUS South. Great tunnies or albacores

Body robust; mouth large; teeth in the jaws small, conical, in a single series; vomer and palatines with bands of villiform teeth; scales small, corselet well developed; dorsal fins close together, the first with 12 to 15 spines; dorsal and anal finlets about 9.

106. Thunnus thynnus (Linnæus). Tuna; Horse mackerel; Tunny; Albacore.

Scomber thynnus Linnæus, Syst. Nat. ed. X, 1758, p. 297; Europe.
Thunnus thynnus Jordan and Evermann, 1896–1900, p. 870.

"Head 3¾; depth 4. D. XIV–I, 13–IX; A. I, 12–VIII. Body oblong, very robust; corselet well developed, extending farther back than pectorals; caudal keel extending forward to second finlet from caudal. Mouth rather large; maxillary reaching pupil; posterior margin of preopercle somewhat shorter than inferior. Eye small. Dorsal and anal falcate, short, 2 in height of first dorsal; ventrals longer than anal; caudal very widely forked; pectorals short, reaching to about ninth dorsal spine. Dark blue above; below grayish, with silvery spots." (Jordan and Evermann, 1896–1900.)

This species was not seen during the present investigation. However, we find mention of "one horse mackerel" among the records of the Buchanan Brothers' pound-net fishery for August, 1909. We have only this record of the occurrence of this species in Chesapeake Bay.

The horse mackerel is generally readily recognized by its large size, and the presence of teeth on the vomer separate it from all the other mackerels of Chesapeake Bay.

Little is known of the life history of this large pelagic fish, which inhabits all warm seas. It feeds mainly on smaller fish, probably chiefly on other pelagic species, such as the menhaden, herring, and mackerels. Its eggs and fry are unknown. Even moderately small individuals seldom are taken along the Atlantic coast.

The horse mackerel, formerly considered of no value as a food fish, has gained rapidly in favor during recent years. In California and in the Mediterranean it is highly prized, and it now has some sale value on the Atlantic coast of America. It is also regarded with considerable favor by sport fishermen.

This fish is reported to attain a length of 14 feet or more and a weight of 1,600 pounds, whereas fish of 1,000 pounds are said to be not uncommon; the usual weight is from 100 to 400 pounds.

Habitat.—Warm parts of the Atlantic and Pacific Oceans; Mediterranean Sea.

Chesapeake localities.—(*a*) Previous records: None. (*b*) Specimens in collection: None. The species is here included because of a mention found in a record of fishes caught at the Buchanan Brothers' fishery located in Lynnhaven Roads, Va., who list "one horse mackerel" in their record of catches for August, 1909.

Family L.—TRICHIURIDÆ. The cutlass fishes

Body elongate, strongly compressed, band-shaped, the tail tapering to a point; head long, compressed; snout more or less beaklike; mouth large; lower jaw projecting; premaxillaries not protractile; pseudobranchiæ present; gills 4, a slit behind the fourth; gill membranes separate, free from the isthmus; teeth on the jaws strong, unequal; lateral line continuous; scales wanting; dorsal fin very long and low, beginning on head and extending over the entire length of body; anal long and very low, composed of separate spines; ventrals rudimentary and thoracic when present, sometimes wanting; air bladder present.

81. Genus TRICHIURUS Linnæus

This genus is distinguished from others of the family by the absence of ventral fins. A single species of wide distribution is known.

107. Trichiurus lepturus Linnæus. Silverfish; Cutlass fish; Hairtail.

Trichiurus lepturus Linnæus, Syst. Nat., ed. X, 1758, p. 246; America. Bean, 1891, p. 87; Jordan and Evermann, 1896–1900, p. 889, Pl. CXXXVII, fig. 375.

Head 7.2 to 8.2 in total length; depth 13 to 14.5; D. 133 to 140; A. XCVII to CVIII. Body extremely elongate, strongly compressed, bandlike; tail very slender, tapering to a point; head long, compressed; snout long, pointed, its length 2.75 to 2.9 in head; eye 6.1 to 7.45; interorbital 7.05 to 7.8; mouth large; lower jaw strongly projecting; maxillary concealed under preorbital, reaching about to anterior margin of pupil, 2.2 to 2.7 in head; teeth in the jaws strong and unequal, compressed, the largest ones with distinct barbs on posterior edges; gill rakers poorly developed

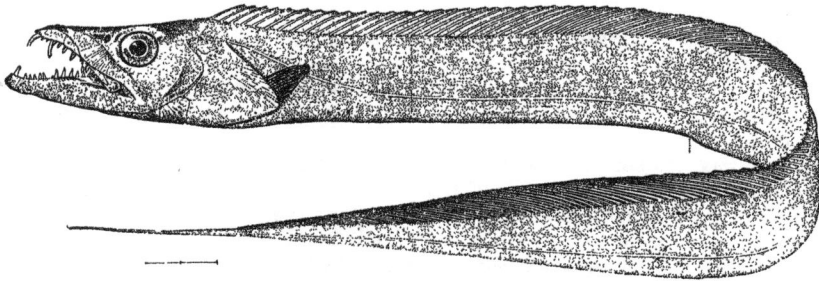

Fig. 118.—*Trichiurus lepturus*

and of unequal length, from 5 to 15 more or less developed on the lower limb of first arch; dorsal fin extremely long, beginning over the preopercular margin and occupying the whole length of the back; caudal and ventral fins wanting; anal fin consisting of very short spines, the anterior ones directed backward and the posterior ones forward; pectoral fins small, 3.3 in head.

Color plain silvery; tips of jaws blackish; dorsal plain, with dusky margin; pectorals plain, with dusky punctulations, at least on distal parts.

This species is represented by four specimens, ranging in length from 550 to 965 millimeters (21¾ to 38 inches). The fish is unique in the long, ribbon-shaped, silvery body, with the long, tapering, filamentous tail, large mouth, and very large, barbed teeth.

The cutlass fish, with its large mouth and formidable teeth, probably is a terror among small fish. Four stomachs examined contained only the remains of fish that gave the appearance of having been bitten into pieces before being swallowed. In one stomach the remains were recognized as those of anchovies; in the others maceration had proceeded too far to permit of identification. The ovary in this fish is single, and specimens taken during May have the sexual organs well distended with eggs and milt, showing that their spawning time was near at hand.

This fish is not common in Chesapeake Bay and it is not used there for food. In some other localities it is eaten, although it is not regarded as a choice food fish. It is said to attain a length of 5 feet; no individuals approaching that size, however, were seen in Chesapeake Bay. It was observed only during the month of May and from Mobjack Bay and Cape Charles city southward.

Habitat.—All warm seas; on the Atlantic coast of America from Massachusetts Bay southward.

Chesapeake localities.—(*a*) Previous record; Cape Charles city, Va. (*b*) Specimens in collection: Mobjack Bay and Lynnhaven Roads, Va.

Family LI.—XIPHIIDÆ. The swordfishes

Body elongate, compact; caudal peduncle slender, with strong median keel, upper jaw greatly produced, forming a sword, composed of the premaxillaries, ethmoid and vomer; lower jaw also prolonged in young; teeth wanting in adult, present in young; gills 4, the laminæ of each united into a single plate; gill membranes separate, free from the isthmus; scales wanting; dorsal fins 2 in adult (single, high, and continuous in young), the first beginning over gill opening, the second small, situated posteriorly; anal fins 2 in adult (single and continuous in young), the first rather large, the second small and opposite second dorsal; caudal fin large and broadly forked; pectorals long, narrow, pointed; ventrals absent; all the fins sharklike, the rays enveloped in skin; intestinal canal long; pyloric cæca numerous; air bladder present. A single genus and species, of large size and great power, is known. The young (as indicated in the foregoing description) differ very markedly from the adult.

82. Genus XIPHIAS Linnæus. Swordfishes

The characters of the genus are included in the family description.

108. Xiphias gladius Linnæus. Swordfish.

Xiphias gladius Linnæus, Syst. Nat., ed. X, 1758, p. 248; Europe. Uhler and Lugger, 1876, ed. I, p. 108; ed. II, p. 90; Jordan and Evermann, 1896–1900, p. 894.

Head about 2.25; depth about 5.5; D. 40–4; A. 18–4. Body fusiform, tapering uniformly from head to tail, deepest at anterior dorsal fin; head, with sword, longer than rest of body; snout long, flat, about 3 in total length, or about three times as long as rest of head; lower jaw in adult extending in advance of eye, a distance equal to half the length of postorbital part of head, much produced in young; first dorsal in adult very high, falcate, its height being as great as depth of body; second dorsal and second anal small, similar, placed nearly opposite each other; first anal similar to first dorsal, but smaller; caudal lobes long; pectoral fins inserted very low.

Color lustrous blue-black above, shading into whitish underneath; head and upper side of sword purplish blue; lower side of sword brownish purple; eye deep blue; fins mostly dark bluish.

This species was not observed in Chesapeake Bay during the present investigation. The above description is compiled from published accounts. The swordfish was once reported by Uhler and Lugger (1876) as "sometimes entering Chesapeake Bay," but it quite evidently is very rare there.

The swordfish is readily distinguished from all other fish by the greatly produced snout, high dorsal, and large size. The young differ greatly from the adult and have the skin covered with rudimentary scales, both jaws produced, and the dorsal and anal fins are high and not divided.

Swordfish feed upon many species of fish as well as upon squid. No ripe fish and very few bearing eggs have been found in the western Atlantic. The following records have been taken from the field notes of Marie Poland Fish, formerly of the Bureau of Fisheries: A 150-pound swordfish harpooned off Coxs Ledge, about 20 miles southeast of No Mans Land, Mass., on July 24, 1924, contained a pair of ovaries weighing 1,430 grams, full of partially matured eggs. The number of ova were estimated at 16,130,400 and measured from 0.1 to 0.55 millimeter in diameter. The eggs of another fish, taken at Provincetown on September 10, 1909, were further developed and averaged 1.2 millimeters in diameter. The smallest swordfish caught on our coast weighed 7⅜ pounds and was taken on August 9, 1922, on Georges Bank.[19] Individuals under 50 pounds are rare. In the Mediterranean, however, young of one-half pound and up are common.

The swordfish is one of the largest and most powerful fish sought by man. It is a fish of the high seas, appearing on our North Atlantic coast near the end of May or some time in June, leaving in late October. It is most abundant during July and August. The favorite fishing grounds extend

---

[19] U. S. Bureau of Fisheries Service Bulletin No. 88, Sept. 1, 1922, p. 3.

along the continental shelf from La Have Bank to Georges Bank, Nantucket Shoals, and Block Island. Swordfish are harpooned. Rarely it is caught on a halibut or cod trawl and in a few instances it has been taken on hand lines baited for other fish. The fishery is a valuable one, for in 1919 there were landed in Maine and Massachusetts 1,136,542 pounds, valued at $270,164.

The swordfish attains a length of 16 feet and a weight of about 800 pounds. Fish of this size are very rare, however, and usually only two or three weighing more than 500 pounds are taken each year. The usual size ranges from 200 to 350 pounds.

Habitat.—Both coasts of the Atlantic Ocean; also found in the Indian and Pacific Oceans. Known on the Atlantic coast of America, from Newfoundland southward beyond the Tropic of Capricorn.

Chesapeake localities.—(*a*) Previous record: Sometimes entering Chesapeake Bay (Uhler and Lugger, 1876). (*b*) Specimens in collection: None. The species was not seen or reported by fishermen from Chesapeake Bay during the present investigation.

Family LII.—STROMATEIDÆ. The butterfishes

Body compressed, moderately to extremely deep; head more or less blunt; mouth moderate or small; premaxillaries not protractile; teeth weak, usually present in the mouth only on jaws; œsophagus with lateral sacks provided with hooked or barbed teeth; pseudobranchiæ present; gills 4, a slit behind the fourth; scales small, cycloid; lateral line well developed; dorsal fin single, long, preceded by a few weak and often obsolete spines; anal similar, usually with three spines; caudal fin well forked; ventral fins thoracic, often wanting; pectoral fins usually rather long.

KEY TO THE GENERA

a. Body very deep, ovate, the depth about 1.2 to 1.4 in the length; dorsal and anal fins anteriorly prominently elevated, falcate; no conspicuous pores on back_____Peprilus, p. 210
aa. Body more elongate, the depth about 1.7 to 2.1 in the length; dorsal and anal fins not prominently elevated, never falcate; a row of conspicuous pores on back near base of dorsal
_____Poronotus, p. 212

83. Genus PEPRILUS Cuvier. Starfishes or harvestfishes

Body ovate or more or less elongate, strongly compressed; head short; snout very short and blunt; mouth small, terminal or nearly so; premaxillaries not protractile; opercles and preopercles entire; teeth small, in a single series on jaws; gill membranes separate, free from the isthmus; lateral line high, following the outline of the back; scales small, cycloid; rather loosely attached, small scales present on the dorsal and anal and sometimes on the caudal; dorsal and anal similar, elevated anteriorly; caudal deeply forked; ventrals represented by a single short spine attached to the pubic bone; pectorals long and narrow.

109. Peprilus alepidotus (Linnæus). Harvestfish; "Starfish"; "Star"; "Butterfish"; "Diamond."

Chætodon alepidotus Linnæus, Syst. Nat., ed. XII, 1766, p. 460; Charleston, S. C.
Peprilus gardenii Uhler and Lugger, 1876, ed. I, p. 115; ed. II, p. 97.
Stromateus paru Bean, 1891, p. 88.
Stromateus alepidotus Smith, 1892, p. 71.

Head 3 to 3.6; depth 1.2 to 1.4; D. III, 45 or 46; A. III, 42 to 44. Body very deep, oval, strongly compressed; dorsal profile anteriorly more strongly convex than the ventral; head short, deep; snout very blunt, 3.65 to 5.1 in head; eye 3 to 3.4; interorbital 2.15 to 3.1; mouth rather small, terminal or slightly inferior; maxillary scarcely reaching anterior margin of eye, 3.3 to 3.9 in head; teeth in the jaws minute; gill rakers rather short, 14 to 16 on lower limb of first arch; scales small, thin, deciduous; lateral line arched, following the curvature of the back; dorsal and anal similar, notably elevated anteriorly; caudal fin deeply forked; ventral fins wanting; pectoral fins long, 2.5 to 3.35 in length of body.

Color greenish-silvery above; lower parts of sides plain silvery or with a tinge of yellow; dorsal and anal dusky, slightly yellowish in some specimens; caudal and pectorals plain or slightly dusky, and sometimes slightly yellowish.

Many specimens of this species, ranging from 25 to 175 millimeters (1 to 7 inches) in length, were preserved. This fish is recognized by its deep body and the anteriorly elevated dorsal and anal fins. The young do not differ markedly from the adults. They appear to be proportionately deeper, however, and the dorsal and anal fins are less strongly elevated anteriorly. Specimens vary considerably in color, some being plain grayish-silvery, with no yellow; others are greenish above and the lower parts of the sides are yellowish or golden.

The food of this starfish appears to be identical with that of the butterfish. The stomach contents are always ground to pulp. Occasionally fish bones and scales are recognizable.

Spawning appears to occur simultaneously with the butterfish—that is, during June and July—and the eggs are similar, being spherical and approximately 1 millimeter in diameter.

The smallest starfish taken during the present investigation (which probably were the result of the same summer's spawning) consisted of a lot of four specimens, ranging from 25 to 28 millimeters (about 1 inch) in length. These fish were caught from July 25 to August 1, 1921, in the

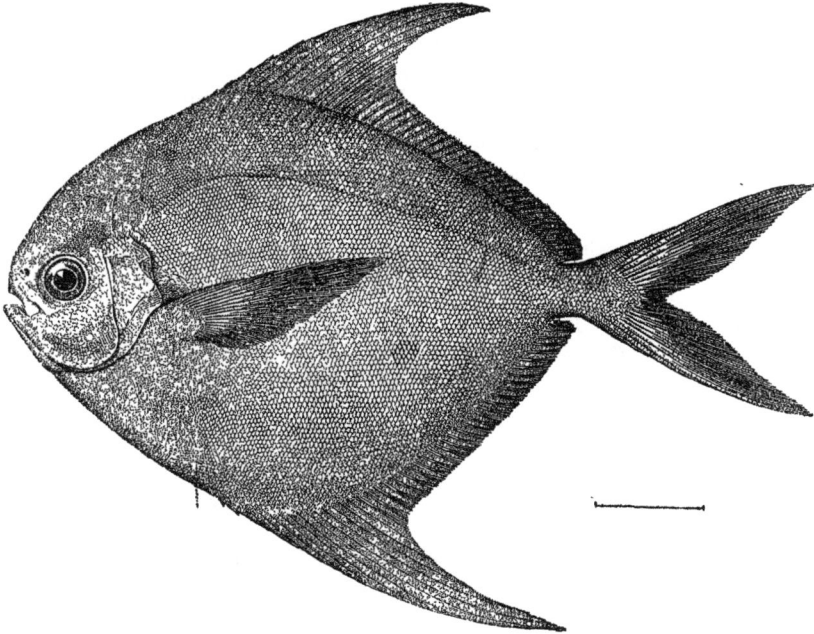

FIG. 119.—*Peprilus alepidatus*

lower Rappahannock, along the shore. A gap occurs in the collections, however, for no young fish were taken from the first of August until late in September, although intensive collecting was continued; those caught during the latter part of September and October ranged from 64 to 130 millimeters (2½ to 5 inches) in length. Only a few of the many that were collected exceeded a length of 4½ inches, and the average length was only 3½ inches. If the starfish spawns in Chesapeake Bay, as seems to be indicated by the small fish taken during the latter part of July and early in August, it seems rather strange that no young fish were taken from the first of August until late in September. Fish ranging in length from 2½ to 5 inches are abundant in October in the lower sections of the bay, and many of this size are often taken in pound nets. The fish within this range, of which numerous specimens were measured, may be separated into two definite size groups by means of plotting frequency curves, the break coming between 4 and 5 inches. The fish making up the group ranging around 5 inches in length are believed to be in their second year.

The starfish is one of the Chesapeake's valuable food fishes. During 1920 it ranked tenth in quantity and thirteenth in value, the catch being 319,681 pounds, worth $10,650.

In Maryland it ranked seventeenth both in quantity and value, the catch in 1920 being 3,765 pounds, worth $150. The entire catch was taken with pound nets. The counties having the largest catches were Somerset, Kent, and Dorchester.

In Virginia the starfish ranked ninth in quantity and thirteenth in value, the catch in 1920 being 315,916 pounds, worth $10,500. The entire catch was taken with pound nets. The counties taking the most fish were Elizabeth City, with 138,850 pounds; Mathews, with 72,399 pounds; and Warwick, with 38,880 pounds.

The starfish is caught in commercial quantities in the Chesapeake from May until October. The first catches, amounting to 10 pounds or more, taken by a set of two pound nets in Lynnhaven Roads, were made on the following dates: May 17, 1916; May 22, 1917; May 16, 1918; May 29, 1919; June 7, 1920; May 9, 1921; May 6, 1922; and May 21, 1923. The starfish appears in the bay about a month later and leaves the bay about a month earlier than the butterfish. Small numbers are taken during the latter half of April, and occasionally a fair catch is made near the mouth of the bay as late as early in November. Large catches are made from the end of May until the end of September. During the seven-year period, 1916 to 1922, the largest catch of starfish made by a set of two pound nets at Lynnhaven Roads, Va., on any one day of each successive month from May to October, was as follows: May 30, 1919, 3,400 pounds; June 4, 1918, 3,800 pounds; July 8, 1919, 2,100 pounds; August 23, 1922, 5,500 pounds; September 10, 1921, 4,000 pounds; October 4, 1919, 400 pounds.

The starfish is confined chiefly to the lower part of the bay, and, like the butterfish, only small quantities are caught above the Potomac River. The season in the upper parts of the bay is somewhat shorter. At Love Point, Md., the most northern locality where this species is taken in commercial numbers, the season commences in June and ends early in October.

This species is closely associated with the butterfish, *Poronotus triacanthus*, and in appearance, size, and edible qualities the two species are very much alike. Both species are frequently sold together under the name "butterfish."

Because of the great body depth, fish as small as 3 inches long can not escape through the meshes of the usual size used in the pound-net trap. Each year thousands of pounds of small, unmarketable starfish are trapped and destroyed in Chesapeake Bay. Frequently 1,000 or more undersized fish are taken in a single pound net on one day. The greater part of these waste fish could be returned to the water alive if fishermen would give the small amount of time and care that would be necessary.

Although large catches of starfish are made frequently, particularly in the lower parts of the bay, the total annual catch in the Chesapeake is only about one-fourth as great as that of the butterfish.

Large shipments of starfish are made to various points in Maryland and Virginia. Shipments are made to the large markets of the North, but there the species is not as well known as the butterfish. The retail price in 1921 and 1922 generally varied from 12 to 20 cents a pound, but large fish frequently brought 25 cents. The size of most of the market fish ranges from 7 to 9 inches in length, but fish 10 and 10½ inches long are not uncommon. The maximum length is about 11 inches and the weight about 1¼ pounds. Many fish were used in determining the following relationship between lengths and weights: Three inches, 0.4 ounce; 3½ inches, 0.7 ounce; 4 inches, 1 ounce; 5 inches, 1.6 ounces; 6 inches, 2.5 ounces.

Habitat.—Southern Massachusetts to Florida.

Chesapeake localities.—(*a*) Previous records: "Many parts of Chesapeake Bay, as far north as the Patapsco River" (Uhler and Lugger, 1876); lower Potomac River and Cape Charles city, Va. (*b*) Specimens in collection and observed: In many localities from Annapolis, Md., southward to Cape Charles and Cape Henry, Va.

84. Genus PORONOTUS Gill. Butterfishes

This genus is similar to Peprilus, differing, however, in having a more elongate body, a row of large conspicuous pores on the back near the base of the dorsal, and in having the dorsal and anal fins anteriorly much less strongly elevated.

110. Poronotus triacanthus (Peck). Butterfish; "Butter;" Harvest fish; Dollarfish.

Stromateus triacanthus Peck, Memoir., Amer. Ac., II, Part II, 1800, p. 48, Pl. II, fig. 2; Piscataqua River, N. H.
Poronotus triacanthus Uhler and Lugger, 1876, ed. I, p. 114; ed. II, p. 96; Jordan and Evermann, 1896–1900, p. 2849, Pl. CL, fig. 405.
Stromateus triacanthus Bean, 1891, p. 88.
Rhombus triacanthus Jordan and Evermann, 1896–1900, p. 967, Pl. CL, fig. 405.

Head 3.35 to 4.3; depth 1.7 to 2.1; D. III, 44 to 46; A. III, 40 to 42. Body moderately ovate, strongly compressed; dorsal and ventral outlines about evenly rounded; head short, deep; snout blunt, 3.25 to 4.5 in head; eye 3 to 4.05; interorbital 2.55 to 2.85; mouth moderate, oblique, slightly

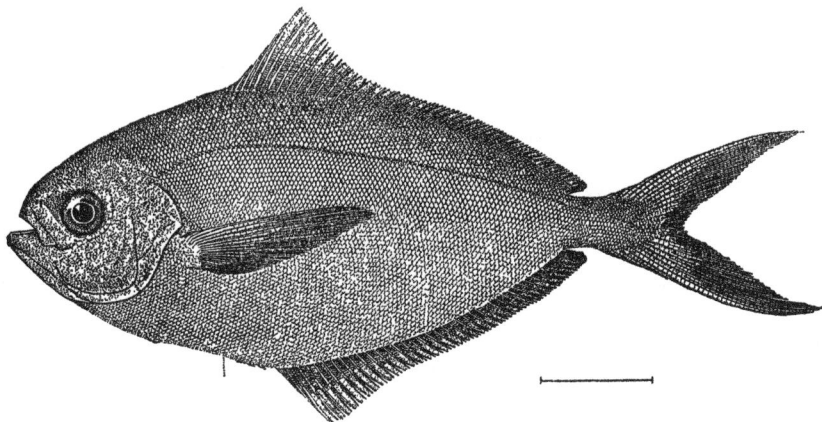

FIG. 120.—*Poronotus triacanthus*

superior; maxillary failing to reach eye, 2.15 to 2.45 in head; teeth present in jaws, feeble; gill rakers short, 15 to 17 on lower limb of first arch; scales small, thin, more or less deciduous; lateral line running high, following the curvature of the back; a row of conspicuous pores on back near base of dorsal; dorsal and anal similar, little elevated anteriorly, the spines small; caudal fin deeply forked; ventral fins wanting; pectoral fins long, 2.95 to 3.15 in length of body.

Color silvery blue or gray above; sides paler, with numerous irregular dark spots, prominent when the fish is seen swimming in the water, but fading completely after death; silvery below; pectoral fins plain, with dusky axil, other fins pale to dark gray, sometimes slightly dusky.

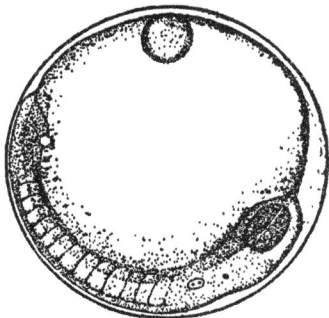

FIG. 121.—Egg with embryo

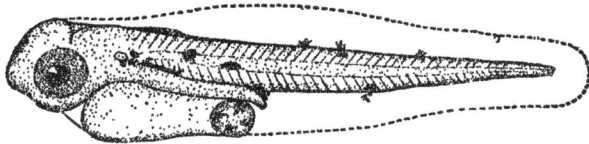

FIG. 122.—Larva 1 day old; 2.1 millimeters

This species is represented by many specimens, ranging from 20 to 235 millimeters (¾ to 9¼ inches) in length. This butterfish is readily distinguished from *Peprilus alepidotus* (the only other "butterfish" of Chesapeake Bay) by the more elongate body and the lower dorsal and anal fins, which are scarcely elevated anteriorly. The young of this species do not differ greatly from the adult. The pores, situated along the back near the base of the dorsal, which are very conspicuous in the adult, are not very noticeable until the fish reaches a length of about 50 millimeters.

It is of interest, as an illustration of the severe mutilations which fish sometimes overcome, to mention that we have at hand a specimen 205 millimeters in length, which met with an accident, during life in which it lost its snout from the nostrils forward. The injury became completely healed over; and, notwithstanding that it had not even rudiments of jaws, the mouth being represented by an oval-shaped opening, this fish not only survived but at the time of its capture was fatter and apparently in better condition than many normal individuals.

It has been difficult to determine upon what this fish feeds, for the stomachs examined contained mainly finely divided, flocculent substances, without definite shape or form. A few stomachs contained some cycloid scales, one contained fragments of shells of mollusks, and a few contained what appeared to be strands of algæ. It seems probable that the food is finely divided before it reaches the stomach. It is possible that grinding the food is a function of the teeth situated in

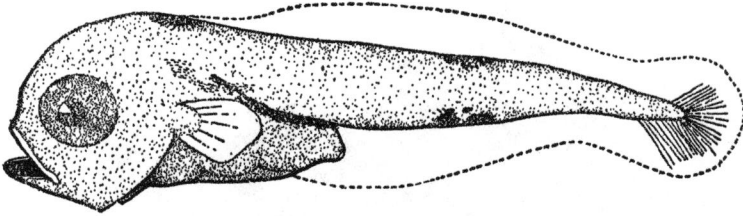

Fig. 123.—Larva 3.2 millimeters long

the œsophagus. Butterfish taken at Woods Hole were found feeding on small fish, squid, crustaceans, annelids, and ctenophores.

Spawning takes place chiefly during June and July in Chesapeake Bay. The earliest date when we observed ripe fish was May 26. Fish as small as 145 millimeters (5¾ inches) had well developed roe late in May, indicating that maturity is reached at a length of about 6 inches. Apparently, the butterfish spawns throughout most of its range, for Bigelow and Welsh (1925, p. 247) give the season for the Gulf of Maine from June to August. The eggs are transparent, round, and 0.7 to 0.8 millimeter in diameter, and hatch in less than 48 hours at a temperature of 65° F.[20] The larvæ are 2 millimeters long at the time of hatching. At 15 millimeters the tail is forked, the dorsal and anal fins are formed, and the fry can be identified readily.

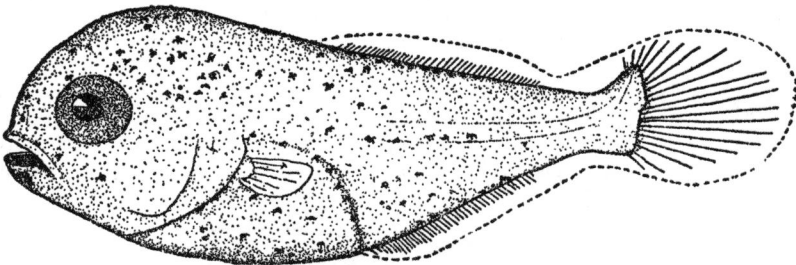

Fig. 124.—Larva 6 millimeters long

While butterfish spawn in Chesapeake Bay, no very young fish were collected, due in part, at least, to the fact that tow nets seldom were used during the summer. At the time when large catches of adults were being made with pound nets none were taken near-by along shore with our collecting seines. It was noted throughout the collecting that butterfish rarely were taken within a few hundred feet of the shore. Large numbers of unmarketable fish, 3 to 5 inches long, taken the middle of May with pound nets, probably were about 1 year old. In the lower York River, on October 13, 13 butterfish 128 to 155 millimeters (5 to 6 inches) long were seined, and in the evening of that same day the catch was 10 of 135 to 165 millimeters (5.3 to 6.5 inches) in length, indicating that the growth from May to October is from about 4 to 5¼ inches. It is probable that by the following spring, when 2 years old, the fish will have reached a length of about 7 inches and maturity.

---

[20] For an account of the embryology and larval development of the butterfish see Kuntz and Radcliffe, 1918, pp. 112 to 116, figs. 58 to 68.

The butterfish is one of the most valuable and abundant food fishes caught in Chesapeake Bay. During 1920 it ranked sixth in quantity and ninth in value, the catch being 1,278,628 pounds, worth $42,603.

In Maryland it ranked fourteenth in quantity and fifteenth in value, the catch for 1920 being 15,062 pounds, worth $603. The entire catch was taken in pound nets. Somerset County records the largest catch, having taken 6,550 pounds, followed by Kent with 6,440 pounds, Dorchester with 1,532 pounds, and Calvert with 540 pounds.

In Virginia it ranked fifth in quantity and seventh in value, the catch for 1920 being 1,263,566 pounds, worth $42,000. Virtually the entire catch was taken with pound nets. The remainder, amounting to less than 1 per cent, was taken with haul seines and fyke nets. The counties taking the largest quantities of fish were Elizabeth City with 555,550 pounds, Mathews with 289,596 pounds, and Warwick with 155,520 pounds.

The butterfish is caught in the Chesapeake from April until November. The first catches of the season, amounting to 10 pounds or more, were made by a set of two pound nets in Lynnhaven Roads, Va., on the following dates: April 19, 1916, April 24, 1917, April 3, 1918, April 15, 1919, March 30, 1920, April 15, 1921, April 15, 1922, and April 21, 1923. Generally no butterfish are taken above the Rappahannock River before May. Stray fish are taken in the lower part of the Chesapeake as early as the last week in March. Sometimes a set of two pound nets catches small quantities (from about 10 to 50 pounds) daily throughout the last half of April, but usually the first

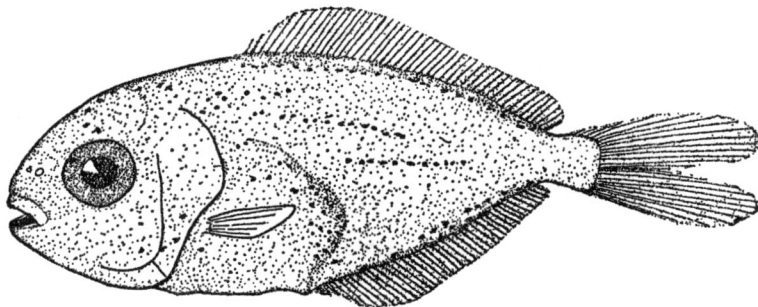

FIG. 125.—Young fish 15 millimeters long

large catches are not made until May. The fish is caught throughout the summer and fall and often well into November. During the seven-year period, 1916 to 1922, the largest catch of butterfish, made by a set of two pound nets at Lynnhaven Roads, Va., on any one day of each month from April to November, was as follows: April 21, 1919, 1,700 pounds; May 14, 1918, 5,225 pounds; June 12, 1918, 4,600 pounds; July 30, 1917, 8,100 pounds; August 8, 1919, 5,200 pounds; September 21, 1919, 3,000 pounds; October 6, 1919, 850 pounds; and November 14, 1918, 900 pounds. A run of fish occurring about the middle of November is not unusual, and pound nets in the lower sections of the bay make small catches until the nets are taken up for the winter at the end of November.

The greater part of the catch of butterfish in the Chesapeake is taken below the Potomac River, particularly from Mobjack Bay to the capes. A large part of the catch is shipped to markets principally between Washington and Boston, and good prices, especially early and late in the season, often are obtained. During a part of May, 1921, and in 1922 the wholesale price of butterfish in New York was around 20 cents a pound. Later in the season, however, the price, especially for the small sizes, dropped to 3 and 4 cents a pound.

Many small, unsalable butterfish, 3 to 5 inches in length, are caught in pound nets. As a rule these small fish are not culled from the catch until the pound-net boats are en route to or have reached shore, with the result that many thousands of fish are wasted annually in Chesapeake Bay. Most of the marketable butterfish are from 7 to 9 inches long, but fish of 10 or 11 inches are not uncommon. A butterfish 6 inches long weighs about 1¾ ounces; one 7 inches long, 2¾ ounces; and one 8 inches long, from 4 to 4½ ounces. A fish 11 inches long and in prime condition weighs about 1 pound. The maximum length is 12 inches and the maximum weight is 1¼ pounds.

The butterfish is a good pan fish and, particularly along the Atlantic coast, finds a ready sale. In the Chesapeake region it is frequently not separated from the starfish or harvest fish (*Peprilus alepidotus*), and the two are sold together.

Habitat.—Nova Scotia to Florida.

Chesapeake localities.—(*a*) Previous records: "* * * many parts of Chesapeake Bay, as far north as the Patapsco River." (Uhler and Lugger, 1876); Cape Charles city, Va. (*b*) Specimens in collection and observed: At many localities from Annapolis southward to the capes.

Family LIII.—CARANGIDÆ.   The crevallies, pompanoes, etc.

Body deep or elongate, usually more or less compressed; head compressed; mouth variable in size; premaxillaries usually protractile; maxillary with or without a supplemental bone; teeth variable, usually small, occasionally wanting in adult; gills, 4, a slit behind the fourth; pseudobranchiæ large, sometimes lost with age; branchiostegals commonly 7; scales small, cycloid, sometimes embedded, occasionally obsolete; lateral line complete, usually with a prominent arch anteriorly, sometimes wholly or in part armed with bony scutes; dorsal fins 2; spinous dorsal rather weak, usually preceded by a procumbent spine; second dorsal long, usually more or less elevated anteriorly; caudal fin broadly forked; anal fin similar in form to second dorsal, sometimes much shorter, preceded by two strong spines, these sometimes disappearing with age; ventrals thoracic, I, 5; pyloric cæca generally numerous.

KEY TO THE GENERA

a. Shoulder girdle with a deep furrow near its juncture with the isthmus and a fleshy knob above it; eye large_____Selar, p. 217
aa. Shoulder girdle normal, not as above; eye moderate.
  b. Anal fin much shorter than the second dorsal; body elongate, not strongly compressed; lateral line entirely unarmed_____Seriola, p. 217
  bb. Anal fin little, if any, shorter than second dorsal; body deep or elongate, rather strongly to very strongly compressed.
    c. Body elongate; scales linear, embedded, giving the skin a leathery appearance; premaxillaries not protractile, except in very young_____Oligoplites, p. 219
    cc. Body deeper, moderately to very deep; scales small, round, not as above; premaxillaries protractile.
      d. Back not much elevated; chest and abdomen deep; ventral outline much more strongly convex than the dorsal; lateral line unarmed; anterior rays of second dorsal and anal not produced_____Chloroscombrus, p. 220
      dd. Back notably elevated; ventral outline not more strongly convex than the dorsal.
        e. Pectoral fins rather long and pointed (except in very young); lateral line with a high arch anteriorly, armed with bony scutes posteriorly (obsolete in Selene).
          f. Body moderately elongate and not excessively compressed, the depth less than half the length of body; lateral line posteriorly strongly armed_____Caranx, p. 220
          ff. Body deep, ovate, very strongly compressed; the depth greater than half the length of body.
            g. Dorsal and ventral outlines both strongly convex; anterior rays of second dorsal and anal bearing long, threadlike filaments; bony scutes in straight part of lateral line well developed_____Alectis, p. 224
            gg. Dorsal outline much more strongly convex than the ventral; anterior rays of the second dorsal and anal produced or not, never bearing long, threadlike filaments.
              h. Anterior profile of head straight, oblique; anterior rays of the second dorsal and anal notably produced (except in very young); lateral line without definite bony scutes_____Selene, p. 224
              hh. Anterior profile of head nearly vertical, more or less concave in advance of eyes; none of the rays of the dorsal or anal produced; straight part of lateral line with small bony scutes_____Vomer, p. 226
        ee. Pectoral fins always short, rarely exceeding length of postorbital part of head; lateral line not definitely arched anteriorly, unarmed_____Trachinotus, p. 227

85. Genus SELAR Bleeker

Body elongate, little compressed; the back not elevated; eye very large; shoulder girdle with a deep furrow at its juncture with the isthmus and a fleshy projection above the furrow.

111. Selar crumenophthalmus (Bloch).   Big-eyed scad; Goggle-eye jack; Goggler.

Scomber crumenophthalmus Bloch, Naturg. Ausl. Fische, 1793, VII, p. 77, Pl. CCCLXIII; Guinea.
Trachurops crumenophthalmus Bean, 1891, p. 87; Jordan and Evermann, 1896–1900, p. 911, Pl. CXLI, fig. 385.

Head 3.35; depth 3.25; D. VIII–I, 26; A. II–I, 21.   Body elongate, little compressed, the back little elevated; head long and low; snout rather pointed, 3.4 in head; eye very large, 3.35, with a well-developed adipose membrane in adult; mouth large, oblique; lower jaw projecting; maxillary reaching anterior margin of pupil, 2.32 in head; teeth small, villiform, present on jaws, vomer, palatines, and tongue; gill rakers scarcely half as long as eye, 26 on lower limb of first arch; scales very small; lateral line without definite arch, armed with scutes, these increasing in size posteriorly; first dorsal with rather high, slender spines; second dorsal and anal similar, moderately elevated

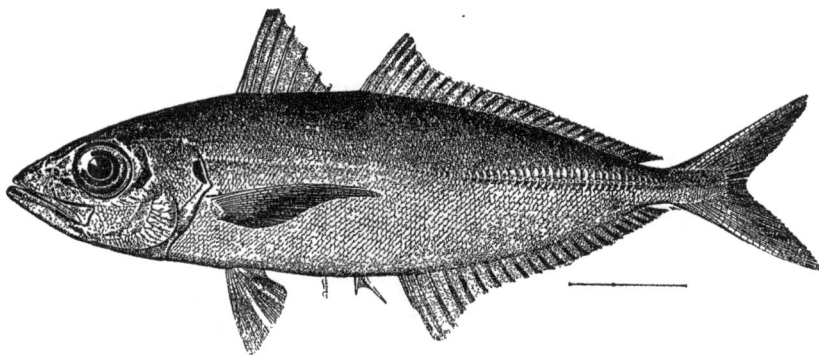

FIG. 126.—Selar crumenophthalmus

anteriorly, each with a low sheath of scales at base; caudal forked; ventrals rather long, reaching to or slightly beyond vent; pectorals long, falcate, about equal to length of head, 4 in length of body.

Color bluish above; silvery below; snout and tip of lower jaw dusky; fins mostly more or less dusky; second dorsal and caudal edged with black.

This species is represented in the present collection by a single specimen, 180 millimeters (7 inches) in length.

This fish has been recorded only once from Chesapeake Bay, and it was seen only once during the present investigation.   It is probable that only stragglers enter the bay.   The species is reported to reach a length of 2 feet.   One of us (Hildebrand) observed numerous examples on the coasts of Panama, where this fish is common and of some commercial value.   The maximum length in that vicinity appeared to be only about 15 inches.

Habitat.—Both coasts of tropical America; stragglers ranging northward on the Atlantic to Massachusetts.

Chesapeake localities.—(a) Previous record: Cape Charles city, Va.   (b) Specimen in collection: Lynnhaven Roads, Va., taken in a pound net, September 27, 1921.

86. Genus SERIOLA Cuvier.   Amber fish

Body elongate, moderately compressed; back not greatly elevated; head rather long; snout more or less pointed; mouth moderately large; premaxillaries protractile; maxillary very broad, with a wide supplemental bone; teeth in villiform bands on jaws, vomer, palatines and usually on tongue; lateral line with a long, low arch, unarmed, in a slight keel on caudal peduncle in adult; first dorsal with six to eight slender spines, connected by membrane; second dorsal long, more or less elevated anteriorly; anal similar to second dorsal, only much shorter; no finlets; ventral fins long; pectoral fins broad, shorter than ventrals.

112. Seriola dumerili (Risso). Amber fish; Rubber jack; Rudder fish; Shark pilot.

Caranx dumerili Risso, Ichthyol., Nice, 1810, p. 175, Pl. VI, fig. 20; Nice.
Scomber zonatus Mitchill, Trans., Lit. and Phil. Soc., N. Y., 1815, p. 427; New York.
Seriola zonata Jordan and Evermann, 1896–1900, p. 902, Pl. CXXXIX, fig. 381.
Seriola lalandi Cuvier and Valenciennes, Hist. Nat. Poiss., IX, 1833, p. 208; Jordan and Evermann, 1896–1900, p. 903, Pl. CXL, fig. 382.
Seriola dumerili Jordan and Evermann, 1896–1900, p. 903.

Head, 3.25 to 3.65; depth, 3.15 to 3.85; D. VI to VIII–I, 34 to 39; A. II, 18 to 20; scales about 150 to 180. Body elongate, not greatly compressed; head rather large; snout long, tapering, 2.45 to 2.85 in head; eye, 4.4 to 5.8; mouth large, terminal, a little oblique; maxillary broad, reaching about middle of eye, 1.95 to 2.25 in head; teeth small, in broad villiform bands on jaws, vomer, palatines, and tongue; gill rakers about the length of eye, 10 to 12 on lower limb of first arch; lateral line anteriorly scarcely arched, unarmed, in a keel on caudal peduncle; first dorsal composed of low, weak spines; second dorsal very long, elevated anteriorly; caudal fin broadly forked, the lobes of about equal length; anal fin shaped like the second dorsal, but much shorter; ventral fins large, longer than the pectorals, inserted nearly under base of pectorals; pectoral fins short, 1.9 to 2.25 in head.

Color grayish or purplish, with golden reflections, above; a bronze stripe along sides from snout to caudal; pale or white below; fins pale or dusky, pectoral, dorsal, and caudal yellowish. Young

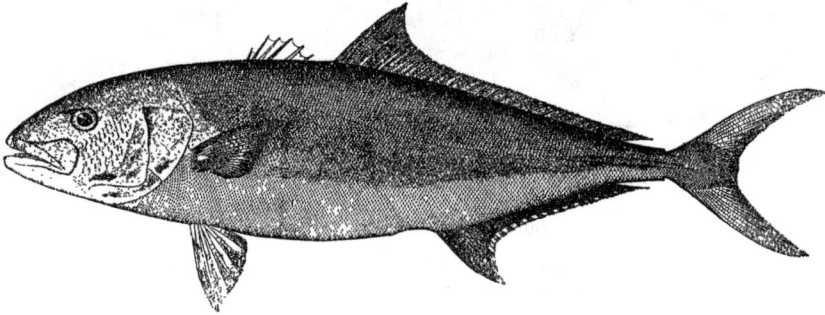

FIG. 127.—*Seriola dumerili*

with dark bars along sides, these bars disappearing with age. Color in alcohol bluish gray above, pale underneath; dorsal fins dark; other fins mostly pale or dusky. The longitudinal band, when present, is dark.

This fish was observed only twice during the present investigation and no specimens were preserved, because of their large size. It is here described from specimens collected at Beaufort, N. C., ranging in length from 205 to 440 millimeters (8 to 17¼ inches). One of us (Hildebrand) has examined a large number of specimens of this genus from various localities and has found great difficulty in separating species. It seems probable that too many species have been recognized. One of the specimens observed in the Chesapeake was recognized as *S. zonata*, the banded form, and two larger specimens, each weighing 16 pounds, were thought to be *S. lalandi*. It seems highly probable that *S. zonata* merely represents the young of *S. dumerili* and *S. lalandi*. The three are considered identical in this work.

This fish apparently seldom enters Chesapeake Bay and is known only from the lower sections of the bay, where only a few fishermen appear to have seen it. A 16-pound fish taken at Lynnhaven Roads with a pound net and displayed in the Norfolk fish market caused much comment because of its size and rarity. Various names were suggested for it, including "salmon," but no one seemed to recognize the fish. On the same day (June 16, 1921) another 16-pound fish was taken in a pound net off Back River. Evidently a small school, composed of fish of about equal size, entered the bay at this time. No others were seen during 1921 and 1922.

Some of the rudder fishes reach a large size and are valued as food. This species reaches a maximum weight of 100 pounds. At Key West amber fish weighing 20 to 70 pounds occur irregularly during the winter and are sold in the local markets (Schroeder, 1924, p. 7). The amber fish is considered a fine game fish.

Habitat.—Both coasts of the Atlantic, ranging from Massachusetts southward to Brazil on the American coast.

Chesapeake localities.—(a) Previous records: None. (b) Specimens in collection: None. Observed at Lynnhaven Roads, Va., July 15, 1916, and June 16, 1921, and at Back River, Va. June 16, 1921.

87. Genus OLIGOPLITES Gill. Leatherjackets

Body oblong, compressed, dorsal profile anteriorly with a keel; head short, compressed; snout pointed; mouth large, oblique; lower jaw usually projecting slightly; premaxillaries not protractile except in very young; maxillary long and very narrow; teeth in bands on jaws, vomer, palatines, and tongue; scales small, linear, embedded in the skin and placed at different angles to each other; lateral line anteriorly with a low arch or broad angle, unarmed; dorsal spines 3 to 5, connected at base by low membranes; second dorsal and anal similar, somewhat elevated anteriorly, with deep notches between their posterior rays; anal preceded by two strong spines; pectoral fins short. A single species of the genus ranges northward on the Atlantic coast of the United States.

113. Oligoplites saurus (Bloch and Schneider). Leatherjacket.

Scomber saurus Bloch and Schneider, Syst. Ichthy., 1801, p. 321; Jamaica.
Oligoplites saurus Jordan and Evermann, 1896–1900, p. 898, Pl. CXXXVIII, fig. 378.

Head 4.75 to 4.9; depth 3.4 to 3.6; D. V–I, 19 to 21; A. II–I, 20 or 21. Body moderately elongate, strongly compressed; caudal peduncle rather slender, its depth 4.3 to 4.5 in head; head

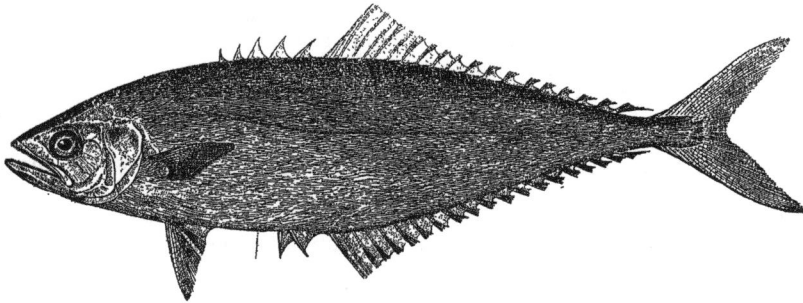

Fig. 128.—*Oligoplites saurus*

rather short, its upper surface without evident pores; snout pointed, its length 3.7 in head; eye 4.25 to 4.35; interorbital 3.6 to 3.7; mouth large, oblique; lower jaw projecting very slightly; maxillary very narrow, reaching to or a little beyond posterior margin of pupil, 1.7 to 1.8 in head; teeth small, villiform, in bands on jaws, vomer, palatines, and tongue; gill rakers rather short, 12 on lower limb of first arch; branchiostegal rays not connected across isthmus by a membrane; origin of dorsal over tips of pectorals, the spines short, pungent; soft dorsal and anal similar, somewhat elevated anteriorly, the short rays posteriorly deeply notched, scarcely forming separate finlets; caudal broadly forked, the lobes about equal; ventral fins moderate, inserted about equidistant from tip of snout and origin of second anal; pectoral fins short, 1.55 to 1.6 in head.

Color bluish above; sides bright silvery; fins all plain in spirits, mostly yellowish in life.

Only two specimens, 230 and 260 millimeters (9 and 10¼ inches) in length, were taken. This fish is the only one of the genus that ranges northward on the coast of the United States. It is recognized by its leathery jacket, sharply pointed snout, and short, pungent dorsal and anal spines. The feeding and spawning habits are unknown. The stomachs of the two specimens at hand, both taken in a pound net, contained fish remains and plant tissue. One specimen, taken in May, had the ovary well distended with eggs. In the other one, taken in September, the ovary was collapsed. Apparently it is only a straggler in Chesapeake Bay and so rare that it was unknown to the fishermen who saw the specimens. We find no record of its previous occurrence in the bay.

This fish reaches a length of about 12 inches. In the tropics, where the species is common, it has limited value as a food fish.

Habitat.—Both coasts of tropical America, ranging northward on our coast to Woods Hole, Mass.

Chesapeake localities.—(*a*) Previous records: None. (*b*) Specimens in collection: Lynnhaven Roads, Va., taken in pound nets, September 8, 1921, and May 25, 1922.

88. Genus CHLOROSCOMBRUS Girard. Bumpers

Body ovate, strongly compressed, the ventral outline much more strongly curved than the dorsal, both carinate; mouth strongly oblique, slightly superior; maxillary broad, emarginate behind, with a large supplemental bone; teeth small, present on jaws, vomer, palatines, and tongue; gill rakers long and slender; head mostly naked, the rest of the body covered with small, smooth scales; lateral line with a strong arch anteriorly and with or without bony scutes posteriorly; first dorsal composed of feeble spines; second dorsal and anal low, longer than the abdomen; caudal deeply forked; ventrals small; pectorals long and falcate.

114. Chloroscombrus chrysurus (Linnæus). Bumper.

Scomber chrysurus Linnæus, Syst. Nat., ed. XII, 1766, p. 494; Charleston, S. C.
Chloroscombrus chrysurus Bean, 1891, p. 87; Jordan and Evermann, 1896-1900, p. 938, Pl. CXLV, fig. 394.

Head 3.6 to 4.3; depth 2.1 to 2.4; D. VIII–I, 26 or 27; A. II–I, 26 to 28. Body very deep, ovate, strongly compressed; outline of abdomen extremely convex, much more strongly curved than the dorsal outline; head short and deep; snout blunt, shorter than eye, 3.34 to 4 in head; eye 2.65 to 3.2; mouth nearly vertical; maxillary reaching anterior margin of eye, 2.45 to 2.75 in head; gill rakers about two-thirds length of eye, 28 to 33 on lower limb of first arch; lateral line anteriorly with a prominent arch, shorter than straight part, without developed bony scutes posteriorly; second dorsal and anal about equal in length, the first somewhat more strongly elevated anteriorly, each with a sheath of scales at base; pectorals long and falcate in adult, proportionately shorter in young, 2.9 to 3.6 in length of body.

Color bluish gray above; sides silvery; a small opercular spot present; a prominent, quadrate, black blotch on upper part of caudal peduncle; fins mostly yellowish in life; vertical fins edged with dusky.

This species was not seen during the present investigation and only one specimen has been recorded from Chesapeake Bay. The foregoing description is compiled from published accounts of the species.

This fish reaches a small size, rarely exceeding 8 inches in length. It is not valued as food, for it is said to be dry and bony. In Colon, Panama, however, it is not infrequently seen in the market. Although straying northward to Massachusetts, it is not common north of the coast of South Carolina. It appears to be extremely rare in Chesapeake Bay, as already indicated.

Habitat.—Massachusetts to Brazil.

Chesapeake localities.—(*a*) Previous record: Cape Charles city, Va. (*b*) Specimens in collection: None.

89. Genus CARANX Lacépède. Crevallies

Body oblong or ovate, compressed; dorsal profile anteriorly sometimes strongly convex; head rather large, compressed; snout usually blunt; mouth moderate or large, oblique, usually terminal; maxillary with a supplemental bone; premaxillary protractile; teeth in the jaws more or less unequal, in one or a few series; villiform teeth usually present on vomer, palatines, and tongue; deciduous or wanting in some species; gill rakers slender; scales small; lateral line usually arched anteriorly, posteriorly armed with bony plates; first dorsal with slender spines, second dorsal and anal similar, usually more or less elevated anteriorly; anal preceded by two rather strong, short spines; caudal broadly forked; pectorals long and falcate in adult.

KEY TO THE SPECIES

a. Breast naked, except for a small triangular patch of scales immediately in front of ventrals; opercular spot prominent_____*hippos*, p. 221

aa. Breast fully scaled; opercular spot present or wanting.

   b. Body rather slender, the depth 3.45 to 4 in length; gill rakers numerous, 24 or 25 on lower limb of first arch_____*crysos*, p. 222

   bb. Body moderately deep, the depth 2.25 in length; gill rakers fewer, 14 on lower limb of first arch_____*latus*, p. 223

115. Caranx hippos (Linnæus).　Jack; Crevalle; Runner; "Jenny Lind;" "Rudder fish."

    Scomber hippos Linnæus, Syst. Nat., ed. XII, 1766, p. 494; Charleston, S. C.
    Carangus hippos Uhler and Lugger, 1876, ed. I, p. 113; ed. II, p. 94.
    Caranx hippos Jordan and Evermann, 1896–1900, p. 920, Pl. CXLI, fig. 387.

Head 3.3 to 3.7; depth 2.2 to 2.55; D. VIII–I, 20 or 21; A. II–I, 16 or 17; lateral scutes 26 to 30. Body rather robust, compressed; upper profile anteriorly very strongly convex; head short and deep; snout very blunt, 3.65 to 4.5 in head; eye 3.3 to 4.25; interorbital 2.95 to 3.65; mouth slightly oblique, terminal; maxillary reaching about middle of eye in young, nearly or quite to posterior margin of eye in adult, 2.1 to 2.45 in head; teeth present on jaws, vomer, palatines, and tongue, some of the teeth on anterior part of lower jaw enlarged; gill rakers about half the length of eye,

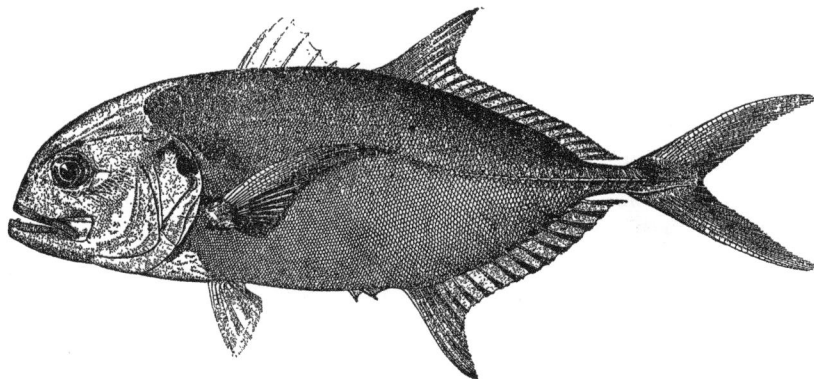

FIG. 129.—*Caranx hippos*

13 to 15, besides rudiments, on lower limb of first arch; scales small, cycloid, not present on breast, except for a small patch immediately in front of ventrals, not present on second dorsal and anal, except on elevated portions; lateral line with a long low arch anteriorly, usually slightly shorter than straight portion; lateral scutes strong posteriorly, forming a strong keel on caudal peduncle; first dorsal with slender spines; second dorsal and anal similar, moderately elevated anteriorly; caudal fin very broadly forked; ventrals moderate, inserted a little behind base of pectorals; pectorals long, falcate, 2.9 to 3.4 in length.

Color bluish green or greenish bronze above; lower parts pale silvery, sometimes with yellow blotches; a distinct black spot on opercle; fins usually more or less yellowish; spinous dorsal and elevated portion of soft dorsal distally dusky; ventral fins partly white; axil of pectoral dusky; the lower rays with a black blotch in adults, indistinct or wanting in young. Very young with five or six dark cross bars.

This species is represented by 15 specimens in the present collection, ranging from 115 to 195 millimeters (4½ to 7¾ inches) in length. The species differs from others of the genus in having the breast naked, except for a small triangular patch of scales just in advance of base of ventrals.

This fish is carnivorous and apparently highly predatory of other fish, fish remains only occurring in the contents of six stomachs examined. The spawning habits are entirely unknown.

Nothing is known of the rate of growth of the crevalle. Individuals taken during October at Ocean View, Va., with a single exception (203 millimeters), ranged in length from 129 to 175 millimeters (about 5 to 7 inches).

Small numbers of this species appear in the markets of Norfolk, Va., and vicinity in the late summer and during the fall. It is probable that the annual catch of marketable fish does not exceed 1,000 pounds, and nearly all are caught with pound nets.

The crevalle is confined to the lower sections of the bay and seldom is found above the mouth of the York River and Cape Charles city. It generally appears in July or August; during September and October small fish, 5 to 8 inches in length, are sometimes rather common at Ocean View and Lynnhaven Roads, Va. In September, 1921, about 400 small individuals were taken from a pound net on one day. While seining for spots and spotted squeteagues at Ocean View from September 27 to October 27, 1922, 40 small crevalles were caught in 30 hauls of an 1,800-foot net. This fish was taken in 11 of the 30 hauls, but it did not appear in the catch after October 15.

The rudder fish is not well known in the Norfolk markets, where it is sold with the related species, *C. crysos*, under the names "Jenny Lind" or "Rudder fish." The flesh of this fish is considered of medium quality.

The usual size of market fish ranges from one-third to 1 pound, but occasionally an individual of 2 or 3 pounds is taken. Along the southern coast and the Gulf of Mexico, where the species is common, fish weighing 10 pounds are not unusual, and the maximum size is said to be 20 pounds. The species is principally of southern distribution, the southern part of Chesapeake Bay being as far north as it is taken in considerable numbers. This species is of considerable importance as a food fish on the Gulf coast and southward. It is a common fish in the markets on both coasts of Panama, where it is esteemed as a food fish and brings a good price.

Small fish taken in Chesapeake Bay were of the following weights: 5½ inches, 1.3 ounces; 6 inches, 1.8 ounces; 6½ inches, 2.3 ounces; 7 inches, 2.9 ounces.

Habitat.—Widely distributed in warm seas; common on both coasts of tropical America, ranging northward on the Atlantic to Lynn, Mass.

Chesapeake localities.—(*a*) Previous record: "Enters Chesapeake Bay" (Uhler and Lugger, 1876). (*b*) Specimens in collection: Yorktown, Lynnhaven Roads, and Ocean View, Va.

116. Caranx crysos (Mitchill). Crevalle; Hard tail; Runner; "Jenny Lind"; "Rudder fish."

Scomber crysos Mitchell, Trans., Lit. and Phil. Soc., N. Y., I, 1814, p. 424; New York.
Caranx chrysus Bean, 1891, p. 87.
Caranx crysos Jordan and Evermann, 1896–1900, p. 921, Pl. CXLII, fig. 388.

Head 3.45 to 4; depth 2.65 to 3.1; D. VIII–I, 23 to 25; A. II–I, 19 or 20; lateral scutes 38 to 45. Body elongate, compressed; dorsal profile anteriorly rather strongly convex; head rather short; snout somewhat pointed, 3.1 to 3.95 in head; eye 3.65 to 4.45; interorbital 2.65 to 3.1; mouth oblique, terminal; maxillary reaching to or scarcely to middle of eye, 2.3 to 2.5 in head; teeth present on jaws, vomer, palatines, and tongue, some of the outer teeth in the jaws enlarged, those on anterior part of lower jaw not especially larger than the others; gill rakers somewhat longer than half the eye, 24 or 25 on lower limb of first arch; scales small, cycloid, fully covering breast, also present on soft dorsal and anal; lateral line with an arch anteriorly, equal to about two-thirds the length of the straight part; lateral scutes very strong posteriorly; first dorsal with slender spines; second dorsal and anal little elevated anteriorly, very low posteriorly, each with a wide sheath of scales at base; ventral fins rather small, inserted slightly behind base of pectorals; pectoral fins long, falcate, 2.75 to 4.15 in length of body.

Color greenish bronze above, shading into bronze silvery below; a more or less distinct opercular spot usually present; spinous dorsal dusky; second dorsal, caudal, and soft part of anal yellowish and more or less dusky on distal parts; ventrals mostly white with tinge of yellow; pectorals plain or slightly yellowish.

Fifteen specimens of this species, ranging in length from 145 to 360 millimeters (5¾ to 14¼ inches), were preserved. This fish has the breast fully scaled and it has a more slender body than the other species of the genus known from Chesapeake Bay. This species, like *C. hippos*, is carnivorous and preys on other fish. Four stomachs examined contained remains of fish only. The spawning habits of this fish are unknown.

This species is taken in the lower part of the bay under almost the same conditions as *C. hippos*, with which it associates. It appears, however, that small fish, 5 to 8 inches in length, are less common than *C. hippos* of similar size, while marketable sizes (⅓ to 1 pound) are taken in slightly greater numbers. As many as 40 fish of various sizes, all weighing less than 1 pound, were observed during September, 1921, among the day's catch of one pound net at Lynnhaven Roads, Va.

The small quantity taken is marketed in the vicinity of Norfolk, where this species and *C. hippos* were retailed together in 1921 and 1922 at about 20 cents a pound. The local names are "Jenny Lind" and "rudder fish."

This fish, like *C. hippos*, is principally of southern distribution. It is the common "runner" of the Gulf coast, where it is taken in large quantities and is esteemed as food. It does not reach as large a size as *C. hippos*, the maximum weight recorded being 3 pounds.

Habitat.—Cape Cod, Mass., to Brazil; rarely to Nova Scotia.

Chesapeake localities.—(*a*) Previous record: Cape Charles city, Va. (*b*) Specimens in collection: From Lynnhaven Roads and Ocean View, Va.

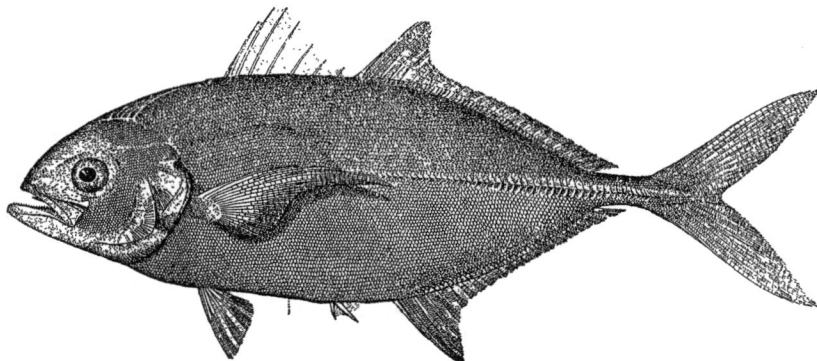

Fig. 130.—*Caranx crysos*

117. Caranx latus Agassiz. Jurel; Horse-eye jack; "Jenny Lind."

Caranx latus Agassiz, Pisc., Brasil., 1829, p. 105, Brazil; Jordan and Evermann, 1896–1900, p. 923, Pl. CXLII, fig. 389.

Head 3.65; depth 2.25; D. VIII–I, 21; A. II–I, 17; lateral scutes 35. Body rather deep, compressed; upper anterior profile strongly convex; head short, deep; snout rather blunt, 3.4 in head; eye 3.8; interorbital 3.3; mouth oblique, terminal; maxillary reaching a little beyond middle of eye, 2.1 in head; teeth present on jaws, vomer, palatines, and tongue, some of the outer ones in the jaws enlarged; gill rakers, a little longer than half the eye, 14 on lower limb of first arch; scales small, cycloid, covering the entire breast, not especially numerous on soft dorsal and anal; lateral line anteriorly with a prominent arch, about three-fourths as long as the straight part; lateral scutes strong posteriorly, forming a prominent keel on caudal peduncle; first dorsal low, with slender spines; second dorsal and anal moderately elevated anteriorly, each with a low sheath of scales at base; caudal fin broadly forked; ventral fins small, inserted under base of pectorals; pectoral fins long, falcate, three in length of body.

Color greenish blue above, silvery below; no opercular spot (this spot, although wanting in the specimen at hand, is sometimes present); fins more or less yellowish in life; the distal parts of dorsals and caudal dusky.

A single specimen, 195 millimeters (7¾ inches) in length, was seen and preserved. This species has a deep body like *C. hippos*, but less robust. It is most readily distinguished from that species by the fully scaled breast and usually by the absence of a dark spot on the opercle.

This fish, like the others of the genus, is of southern distribution. It appears to be the rarest of the three species herein recorded from Chesapeake Bay, as only a single individual was seen. On the Atlantic coast of Panama it is the most common species of the genus, occurring in the markets almost daily. The size attained is rather small, probably not exceeding 1 pound.

Habitat.—Virginia to Brazil.

Chesapeake localities.—(*a*) Previous records: None. (*b*) Specimen in collection: From Ocean View, Va., taken in a 1,800-foot seine on October 23, 1922.

90. Genus ALECTIS Rafinesque.   The threadfishes

Body strongly ovate in young, becoming much more elongate in adult, strongly compressed; head short and deep, its anterior profile convex; mouth rather large, maxillary reaching well past front of eye; teeth small, in bands on jaws, vomer, palatines, and tongue; scales minute, embedded; lateral line with bony scutes on straight part; first dorsal with six or seven short spines, becoming obsolete with age; second dorsal and anal similar, the anterior rays of each bearing filaments. The changes due to age are very marked in this genus. The body in large examples is much more elongate, the anterior profile is less steep, the outlines of the body are scarcely angulate at origin of second dorsal and anal, the filaments on these fins are much shorter, the ventral fins are much shorter and the pectoral fins are much longer.

118. Alectis ciliaris (Bloch).   Threadfish; Hair fish.

Zeus ciliaris Bloch, Naturg. Ausl. Fische, III, 1787, p. 36, Pl. CXCI; East Indies.
Blepharichthys crinitus Lugger, 1877, p. 76.
Alectis ciliaris Jordan and Evermann, 1896-1900, p. 931.

Head 2.95; depth, 1.3; D. VI-I, 18; A. II-I, 16. Body ovate (proportionately deeper in young than in adult); profile rather steep anteriorly, slightly concave over snout, strongly convex over the head, angulated at origin of soft dorsal and anal; head deep; snout projecting but little in advance of forehead, 3.35 in head; eye 3.55; interorbital 2.95; mouth slightly oblique, terminal; maxillary broad, reaching nearly opposite anterior margin of pupil, 2.5 in head; teeth in villiform bands on jaws, vomer, palatines, and tongue; gill rakers scarcely half the length of eye, 13 on lower limb of first arch; lateral line with a high arch anteriorly, arched portion a little longer than straight part, posteriorly armed with bony scutes; spinous dorsal very low, some of the spines almost obsolete; second dorsal and anal similar, the anterior rays of each fin greatly produced, forming long filaments; caudal fin broadly forked; ventral fins long (varying greatly with age); pectoral fins rather long, falcate in adult (shorter in young), 2.85 in length of body.

Color bluish above; sides silvery, with traces of darker bars and blotches (disappearing in large individuals); upper margin of eye dark; produced portion of the dorsal and anal bluish black; ventrals mostly black; the fins otherwise plain or slightly yellowish.

A single specimen 185 millimeters (7¼ inches) long was secured and it forms the basis for the foregoing description. This species is the only one of the genus. It is readily recognized by the extremely long dorsal and anal filaments. Virtually nothing is known of the feeding and spawning habits. The single stomach examined contained a few fragments of bones and plant tissue.

The threadfish is of southern distribution, being reported as common in southern Florida and Cuba. On the coast of Panama it apparently is rather uncommon. Only stragglers stray northward on our coast. According to fishermen in the southern part of the bay the species is quite rare, only an occasional example being taken. It is not known from the more northern sections of the bay.

Although used for food in the localities where it is common, the species has only limited commercial value. The largest example seen by us was from Key West, Fla., and measured 22 inches in length, which quite probably is the maximum size attained.

Habitat.—Both coasts of tropical America, straying northward on the Atlantic to Massachusetts.

Chesapeake localities.—(*a*) Previous record: Southern part of Chesapeake Bay (Lugger, 1876). (*b*) Specimen in collection: Lynnhaven Roads, Va., October, 1921. Fishermen reported one fish off Back River, Va., in June, 1921.

91. Genus SELENE Lacèpéde.   Moonfishes

Body ovate, very strongly compressed; head short and deep, its anterior profile steep but never vertical, forming a rather abrupt angle with dorsal outline; snout moderately projecting; teeth small, present on jaws, vomer, and tongue; palatines with a few teeth or none; lateral line anteriorly with a prominent arch, without definite bony scutes posteriorly; spines of first dorsal slender, bearing filaments in young; soft dorsal and anal elevated anteriorly, some of the rays much produced in adult.

119. Selene vomer (Linnæus). Moonfish; Lookdown; Horsehead.

Zeus vomer Linnæus, Syst. Nat., ed. X, 1758, p. 266; America.
Argyreiosus vomer Uhler and Lugger, 1876, ed. I, p. 112; ed. II, p. 94.
Selene vomer Bean, 1891, p. 87; Jordan and Evermann, 1896-1900, p. 936, Pl. CXLIV, fig. 393, and Pl. CXLV, fig. 393a.

Head 2.15 to 2.4; depth 1.15 to 1.3; D. VIII-I, 21 or 22; A. II-I, 18 or 19. Body ovate, very strongly compressed, the outlines trenchant; back much elevated; anterior profile steep, oblique, nearly straight, forming an angle with dorsal outline; head short, very deep; snout moderately projecting, 1.55 to 1.9 in head; eye 4 to 5.85; interorbital 4.2 to 5.45; mouth moderate, oblique, slightly superior; maxillary broad, 2.7 to 3.4 in head; teeth small, present on jaws, vomer, and tongue; gill rakers slender, about two-thirds length of eye, 24 or 25 on lower limb of first arch; lateral line with a prominent arch anteriorly, equal to or a little shorter than straight part, without developed bony scutes posteriorly; first dorsal with eight spines, the anterior ones slender, bearing

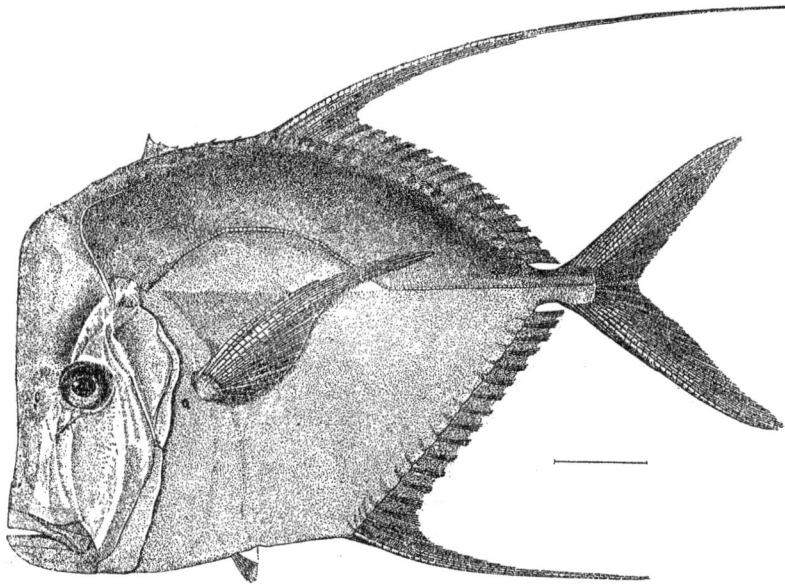

FIG. 131.—*Selene vomer*

long filaments in young; second dorsal and anal similar, the anterior rays much produced; caudal deeply forked; ventral fins very long in young, often reaching to or beyond base of caudal, very short in adult, about equal to length of eye in specimens 200 millimeters in length; pectoral fins long and falcate in adult, moderate in young, 2.05 to 3.25 in length of body.

Color in life of a specimen 175 millimeters long, bluish green, shading into bright silvery on sides; spinous and soft dorsals plain, the produced rays dusky; caudal yellowish, the lobes slightly dusky; produced rays of anal dusky white, anterior short rays yellowish, rest of fin plain; ventrals dusky brown; pectorals plain. A fish 144 millimeters long differed from the preceding as follows: Spinous dorsal dusky; no dusky markings on caudal; produced rays of anal dusky yellow. Young 90 millimeters long with a dusky yellow bar from spinous dorsal through eye, followed by four or five less distinct yellow bars; produced rays of dorsal black, anterior shorter rays dusky yellow, posterior rays plain; caudal tinged with yellow; produced rays of anal dusky yellow; ventrals black; pectorals plain.

This fish is represented in the present collection by 21 specimens, ranging in length from 60 to 235 millimeters (2⅜ to 9¼ inches). The moonfish is recognized by the very deep body, straight oblique forehead, and by the produced anterior rays of the second dorsal and anal. The young of this fish are proportionately shorter and deeper than the adults; the dorsal spines bear long filaments, which disappear with age; the ventral fins are greatly produced, becoming short in the adult; and the sides have four or more dusky cross bars, which later disappear.

The moonfish is carnivorous. Six specimens examined had fed on small crustaceans and on fish. Each alimentary tract, also, contained a considerable amount of smooth sand. Its spawning habits are unknown.

This species was observed in the southern part of the bay and only during September and October, when it was common and many were taken in pound nets and small numbers in large haul seines. During the period from September 23 to October 5, 1921, as many as 300 moonfish per day were caught in a set of two pound nets in Lynnhaven Roads. At Ocean View it was equally abundant. Our field records show that this fish was taken (1 to 15 individuals at a time) in 14 of a total of 32 hauls made at Ocean View, Va., with an 1,800-foot seine from September 23 to October 27, 1922. One fish was only 90 millimeters long. All the others ranged from 117 to 187 millimeters (4½ to 7¼ inches) in length. A number of individuals also were taken in small collecting seines along the beaches at Cape Henry, Cape Charles, Ocean View, and Buckroe Beach. The following weights were secured: 3½ inches, 0.5 ounce (1 fish); 4½ inches, 0.9 ounce (2 fish); 5 inches, 1.1 ounces (1 fish); 5½ inches, 1.5 ounces (5 fish); 6 inches, 1.9 ounces (14 fish); 6½ inches, 2.6 ounces (3 fish); 7 inches, 3 ounces (5 fish); and 7½ inches, 3.5 ounces (2 fish).

The moonfish is reported to reach a maximum weight of 2 pounds. In the Chesapeake fish weighing more than one-half pound are unusual, and virtually none are utilized in the markets. The moonfish is considered a good food fish in some parts of its range, notably at Key West, Fla., where one-half pound is a common size.

Habitat.—Southern Massachusetts (rarely to Cape Cod and Casco Bay, Me.) to Uruguay.

Chesapeake localities.—(a) Previous records: Mouth of Potomac River, "southern part of Chesapeake Bay," and Cape Charles city. (b) Specimens in collection: Cape Charles, Buckroe Beach, Lynnhaven Roads, and Ocean View, Va.

92. Genus VOMER Cuvier. Horsefishes

Body broad, ovate, very strongly compressed; head short and deep, its anterior profile nearly vertical; snout projecting only slightly; teeth small, present on jaws, vomer, and tongue; palatines with weak teeth or none; scales small, rudimentary; lateral line anteriorly with a prominent arch, the straight part with small bony scutes, at least in adult; spines of first dorsal very short; soft dorsal and anal very low, never bearing produced rays.

120. Vomer setipinnis (Mitchill). Moonfish; Sunfish; Horsefish.

Zeus setipinnis Mitchill, Trans., Lit. and Phil. Soc., New York, I, 1814, p. 384; New York.
Vomer setipinnis Uhler and Lugger, 1876, ed. I, p. 111; ed. II, p. 93; Jordan and Evermann, 1896-1900, p. 934, Pl. CXLIV, fig. 392.

Head 2.7 to 3.5; depth 1.2 to 1.8; D. VIII-I, 21 or 22; A. II-I, 17 or 18. Body rhombic, proportionately deeper in young than in adult, very strongly compressed; back much elevated; anterior profile very steep, concave in advance of eyes; head short and very deep; snout little projecting, 2 to 2.4 in head; eye 3.5 to 3.95; interorbital 3.4 to 4.6; mouth rather large, oblique, slightly superior; maxillary broad, 2.2 to 2.9 in head; teeth small, present on jaws, vomer, and tongue; gill rakers usually somewhat longer than half the eye, 26 or 27 on lower limb of first arch; lateral line anteriorly with a prominent arch, somewhat shorter than straight part, posteriorly with small but distinct bony scutes; first dorsal with eight short, pungent spines; second dorsal and anal similar, very low, scarcely elevated anteriorly; caudal fin broadly forked; ventral fins very small, shorter than eye; pectoral fins rather long, falcate in adult, 2.55 to 3.7 in length of body.

Color bluish green above, shading into bright silvery along sides; larger examples with an obscure dark bar, extending from tip of snout through eye to upper angle of gill opening; upper surface of caudal peduncle bluish black; some of the small specimens with a dusky spot on sides over the beginning of the straight part of lateral line; dorsal, anal, and ventrals plain; caudal greenish yellow; pectorals light yellow.

This species is represented by 35 specimens, ranging in length from 55 to 215 millimeters (2¼ to 8½ inches). The steep, concave forehead and the low dorsal and anal fins, bearing no produced rays, readily separate this fish from related forms occurring in Chesapeake Bay. The young, as in related species, are proportionately shorter and deeper than the adults, and the bony scutes on the sides of the tail are not evident.

The horsefish appears to be carnivorous. Four stomachs were examined and contained only the remains of fish. In one stomach an anchovy was recognizable among the contents. The spawning habits of this fish are unknown.

The size of a catch of 16 fish, taken one day in October with a haul seine at Ocean View, ranged from 49 to 94 millimeters (2 to 3¾ inches). Most of the fish taken on May 25, mentioned below, were 121 to 152 millimeters (4¾ to 6 inches) in length, which probably represents the growth from the preceding October.

The horsefish is not uncommon in the southern part of Chesapeake Bay during the fishing season. However, it was not found far from the entrance of the bay, and probably does not occur above the mouth of the Rappahannock River. Examples taken in the spring included more large fish than those taken during autumn. The fish is caught both in pound nets and in seines. The largest single

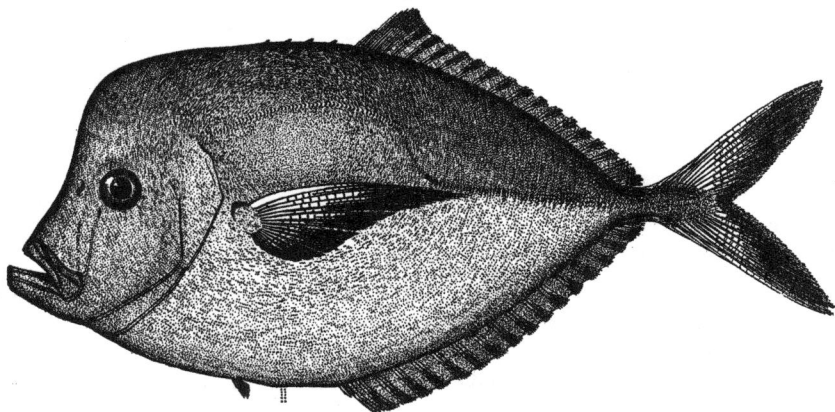

Fig. 132.—*Vomer setipinnis.* From a specimen 9¼ inches long

catch observed was taken from a pound net in Lynnhaven Roads, Va., on May 25, 1922, when about 2,000 horsefish, ranging from 4¾ to 8½ inches in length, occurred in the trap among other fish. It also was taken in small numbers during the fall of 1922, in haul seines at Ocean View, Va. This species, although esteemed as a food fish in some localities, is not utilized in the Chesapeake. The maximum weight attained is said to be about 1 pound. The fish in Chesapeake Bay, however, run rather small, and none approaching 1 pound in weight was seen.

Habitat.—Nova Scotia to Uruguay; rare north of Cape Cod.

Chesapeake localities.—(a) Previous record: "Not rare in * * * the southern part of Chesapeake Bay" (Uhler and Lugger, 1876). (b) Specimens in collections: Lynnhaven Roads and Ocean View, Va.

93. Genus TRACHINOTUS Lacépède. The pompanoes

Body, short, compressed, more or less ovate; abdomen shorter than anal fin, never trenchant; head short; snout blunt; mouth rather small, terminal or slightly inferior; premaxillaries protractile; maxillary without a distinct supplemental bone; teeth in young in villiform bands on jaws, vomer, palatines, and tongue, almost completely disappearing with age; preopercle serrate in very young, becoming entire with age; gill membranes somewhat united across the isthmus; gill rakers short and rather few; scales small, smooth; lateral line scarcely arched, unarmed; first dorsal with six short, strong spines in addition to a procumbent spine; second dorsal and anal similar, anteriorly more or less elevated in adult, long and falcate in some species; caudal broadly forked, the lobes sometimes produced in adults; pectorals never falcate, always shorter than head.

KEY TO THE SPECIES

a. Soft dorsal rays 19 or 20; soft anal rays 17 to 20; anterior rays of soft dorsal and anal notably produced in adult.

    b. Body deep, ovate, the depth 1.4 to 1.7 in length; no black cross bars on sides__*falcatus*, p. 228

    bb. Body more elongate, the depth 2 to 2.6 in length; sides with 4 or 5 black cross bars_*glaucus*, p. 229

aa. Soft dorsal rays 23 or 24; soft anal rays 20 to 22; anterior rays of soft dorsal and anal never notably produced_____*carolinus*, p. 229

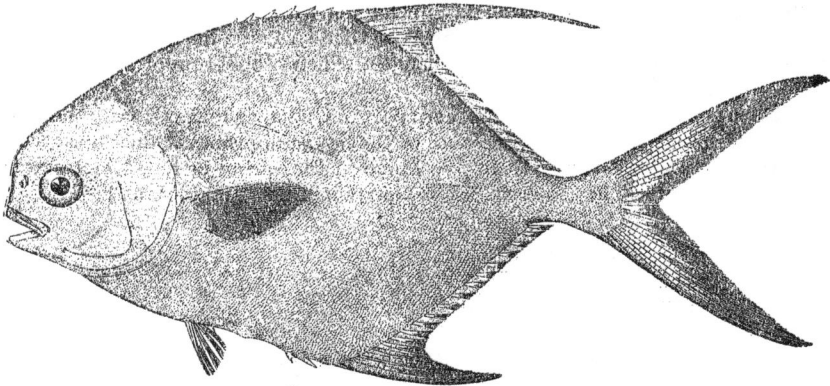

FIG. 133.—*Trachinotus falcatus*

121. Trachinotus falcatus (Linnæus). Round pompano.

Labrus falcatus Linnæus, Syst. Nat. ed. X, 1758, p. 284; America.
Trachynotus ovatus Uhler and Lugger, 1876, ed. II, p. 96.
Trachinotus falcatus Jordan and Evermann, 1896–1900, p. 941, Pl. CXLVI, fig. 396; Evermann and Hildebrand, 1910, p. 161.

Head 2.8 to 3.3; depth 1.4 to 1.7; D. VI–I, 19 or 20; A. II–I, 17 or 18. Body ovate, strongly compressed; dorsal and ventral outlines forming angles at origin of second dorsal and anal; head short; snout very short and blunt, 3.6 to 5.65 in head; eye 3 to 4; interorbital 2.3 to 2.8; mouth

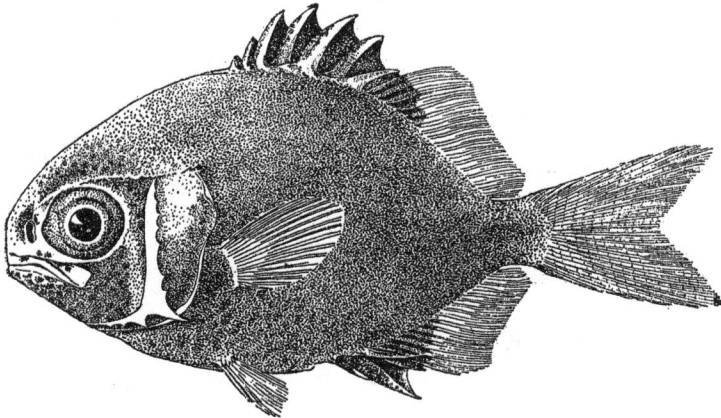

FIG. 134.—*Trachinotus falcatus.* Young, from a specimen 18 millimeters long

rather small, slightly inferior, little oblique; maxillary scarcely reaching middle of eye, 2.8 to 3.1 in head; gill rakers very short, scarcely a third diameter of eye, 9 on lower limb of first arch; lateral line with a slight arch anteriorly, more or less wavy; first dorsal composed of six short stiff spines, preceded by a strong procumbent spine; second dorsal and anal produced anteriorly in adult (the longest rays sometimes reaching base of caudal), scarcely elevated in very young; caudal fin forked, the lobes produced in adult; ventral fins short, scarcely as long as postorbital part of head; pectorals short, 1.35 to 1.9 in head.

Color bluish above; lower parts of sides silvery; dorsal and anal yellowish, with dusky punctulations, the produced part of dorsal sometimes black; caudal and pectorals plain yellowish; ventrals white. Very young (40 millimeters and less in length) densely punctulate with rusty dots, giving them the color of a dead leaf; dorsals and anal very dark; caudal pale.

This species is represented in the present collection by 28 small specimens, ranging from 20 to 95 millimeters (⅘ to 3¾ inches) in length. This species differs from the others of the genus occurring in Chesapeake Bay in the deep, ovate body and the long soft dorsal and anal fins, the anterior rays of which are much produced in adults.

This pompano is carnivorous, the small specimens at hand having fed on worms, crustaceans, mollusks, and fish. The spawning habits are unknown.

This fish is too rare and the individuals obtained are too small to make it of commercial importance in Chesapeake Bay. The species was seen only in the southern sections of the bay, and although it strays northward to Woods Hole, Mass., it is not abundant anywhere along our shores. Its chief habitat probably extends from the West Indies southward. On the Atlantic coast of Panama it is a food fish of some importance. The species is reported to reach a maximum weight of 3 pounds. Its flesh is of excellent quality.

Habitat.—Massuchusetts to Brazil.

Chesapeake localities.—(*a*) Previous record: Potomac River, St. Marys River, and Ocean View, Va. (*b*) Specimens in collection: Cape Charles, Buckroe Beach, Lynnhaven Roads, and Ocean View, Va., taken during September and October, 1921 and 1922.

122. Trachinotus glaucus (Bloch). Gaff-topsail pompano.

Chætodon glaucus Bloch, Naturg. Ausl. Fische, III, 1787, p. 112, Pl. CCX; Martinique.
Trachinotus glaucus Jordan and Evermann, 1896–1900, p. 940, Pl. CXLVI, fig. 395; Smith, 1907, p. 212, fig. 90.

Head 3 to 3.9; depth 2 to 2.6; D. VI–I, 19 or 20; A. II–I, 17 or 18.

This pompano was not seen during the present investigation and we find no definite record of its capture in Chesapeake Bay. Jordan and Evermann give Virginia as the northernmost limit of its range, and Smith says that it is found from Chesapeake Bay southward. It is on the basis of these records that we include these notes.

This species has the anterior rays of the soft dorsal and anal much produced in the adult, as in *T. falcatus.* The body is not ovate, however. Its depth is contained in the length about 2 to 2.6 times, and in this respect it is more like *T. carolinus.* It differs from both species in the presence of dark vertical bars on the sides.

This pompano is reported to reach a weight of about 2 pounds. Apparently nowhere along our coasts is it abundant enough to be of much commercial importance, and it is less highly regarded as a food fish than most pompanos.

Habitat.—Virginia to Panama. Once recorded from Uruguay.

Chesapeake localities.—(*a*) Previous records: No definite localities. (*b*) Specimens in collection: None.

123. Trachinotus carolinus (Linnæus). Pompano; "Sunfish."

Gasterosteus carolinus Linnæus, Syst. Nat., ed. XII, 1766, p. 490; Carolina.
Trachynotus carolinus Uhler and Lugger, 1876, ed. I, p. 113; ed. II, p. 95.
Trachinotus carolinus Bean, 1891, p. 87; Jordan and Evermann, 1896–1900, p. 944, Pl. CXLVII, fig. 398.

Head 3 to 3.6; depth 1.8 to 2.1; D. VI–I, 23 or 24; A. II–I, 20 to 22. Body moderately elongate, strongly compressed; dorsal and ventral outlines not forming pronounced angles at origin of soft dorsal and anal; head moderate; snout short and blunt, 3.8 to 4.5 in head; eye 3 to 4.1; interorbital 2.5 to 2.95; mouth moderate, slightly inferior, a little oblique; maxillary reaching opposite middle of eye, 2.8 to 3 in head; gill rakers very short, about one-fifth diameter of eye, 7 or 8 on lower limb of first arch; lateral line nearly straight; first dorsal composed of six short, stiff spines, preceded by a sharp procumbent spine; second dorsal and anal anteriorly not greatly elevated, none of the rays especially produced, the longest rays reaching about middle of base of fins in adult when deflexed; caudal fin deeply forked; ventral fins small, scarcely as long as postorbital part of head; pectoral fins short, 1.2 to 1.35 in head.

Color bluish green on back, shading into silvery on sides; fins mostly more or less yellowish, the elevated portion of the dorsal dusky; ventrals white.

This species is represented by 24 specimens, ranging in length from 85 to 200 millimeters (3⅜ to 7⅞ inches). This well-known pompano is recognized from other Chesapeake Bay pompanos by a body less deep, by the more numerous rays of the second dorsal (23 or 24, compared to 19 or 20 for *T. falcatus* and *T. glaucus*), and by the fact that none of the dorsal or anal rays are notably produced.

The contents of seven stomachs examined consisted of parts of mollusks, crustaceans, fish, and ova of unknown origin. The spawning habits of this fish are still unknown. No fish with developed gonads were seen.

The pompano is caught in small numbers in the lower parts of Chesapeake Bay. During 1920 the catch amounted to 1,650 pounds, valued at $330 to the fishermen. Almost the entire catch was taken in pound nets, the rest being caught with haul seines.

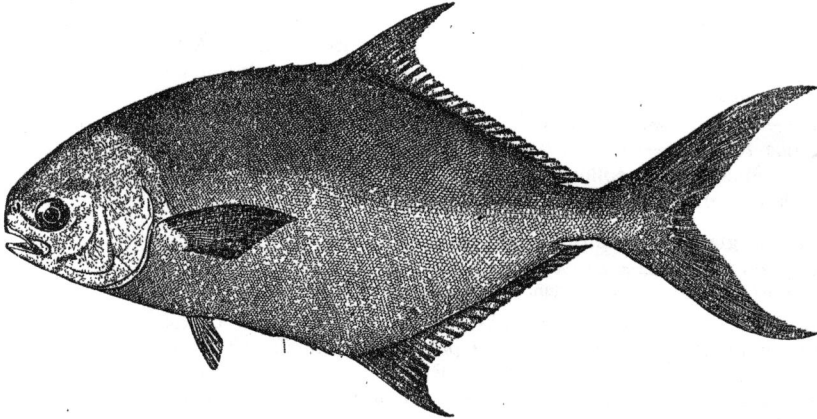

FIG. 135.—*Trachinotus carolinus*

A large catch of pompanos seldom is made in the Chesapeake. Although the quantity taken is small, the fish is valuable, nevertheless, and appears regularly each year. In the report of the United States Commissioner of Fisheries for 1893 (p. 67) the following statement occurs:

The pompano (*Trachinotus carolinus*) is of constant occurrence in the lower Chesapeake, but rarely appears in great abundance. The bay represents the northern limit of commercial fishing for this fish. In 1891 there was a remarkably numerous run of pompanoes in that part of the bay adjacent to its mouth. According to Mr. J. E. N. Sterling, of Cape Charles City, Va., the catch with pound nets and seines on the shores of Northampton County alone was between 20,000 and 25,000 pounds. The inquiries of the agents of the office disclosed a yield of 93,700 pounds in the Chesapeake, with a value to the fishermen of $9,520. In the following year the catch was much less, the Northampton County fishermen taking less than 5,000 pounds, according to Mr. Sterling, although there was said to be a large quantity in the bay which kept offshore out of reach of the nets.

Very few adult pompanos are caught above Cape Charles city, Va., or the mouth of the York River. Fish of commercial size—that is, fish of about 8 inches or more in length—first appear sometime in May. In 1921, in a pound net in Lynnhaven Roads, the first pompano of the season was taken on May 12. In 1922 the fish first appeared on May 23 at Ocean View, Va., followed by others on the 24th at Lynnhaven Roads and several more on the 25th and 26th at both localities. A set of two pound nets as a rule does not catch as much as 25 pounds of pompanos in one day before the middle of June. The greater part of the fish are taken during July and August. September appears to be a poor month, but in October both the pound-net and haul-seine fishermen sometimes make comparatively good catches. The last pompanos are caught about November 1.

While reviewing the records of the daily catch of pompanos taken by a set of two pound nets at Lynnhaven Roads for the years 1916 to 1922, it was found that the best catch for consecutive days' fishing occurred from July 24 to August 2, 1916, when 540 pounds were taken. The largest amount caught on any one day during this seven-year period was 150 pounds, taken on July 25,

1916. The nets to which the foregoing records apply are two of the largest and most favorably located of any in Chesapeake Bay.

Small pompanos from 3 to 8 inches in length are common along the shores of lower Chesapeake Bay from late summer to fall. Late in the season these small fish are found as far north as Solomons, Md., where a number were taken in October, 1922, with collecting seines and pound nets. During haul-seine fishing for spots and spotted squeteague in the lower part of the bay in the fall, many small pompanos are drawn on the beach, where they become smothered with sand and perish. Pompanos less than 7 inches in length are not marketed.

This species of pompano is one of the choicest of all salt-water fishes and everywhere commands a high price. The retail price in the Norfolk market during 1921 and 1922 ranged from 40 to 50 cents a pound. Most of the catch is marketed locally, but when a good run of fish occurs shipments are made to other markets.

This pompano reaches a weight of about 5 pounds. The Chesapeake fish, however, seldom exceed a weight of 3 pounds, the average size of the market fish ranging between 1 and 2 pounds.

Habitat.—Massachusetts to Brazil.

Chesapeake localities.—(a) Previous records: "Chesapeake Bay," Cape Charles city, Va. (b) Specimens in collection: Solomons, Md., Cape Charles, Buckroe Beach, Ocean View, and Lynnhaven Roads, Va.

Comparison of lengths and weights of pompanos

Number of fish measured and weighed	Length	Weight	Number of fish measured and weighed	Length	Weight
	Inches	*Ounces*		*Inches*	*Ounces*
1	3¼	0.3	7	6	1.6
1	3¾	.4	5	6¼	1.7
10	4	.5	3	6½	2.1
9	4¼	.6	2	6¾	2.4
27	4½	.7	3	7	2.6
24	4¾	.8	1	7¼	2.8
13	5	.9	4	7½	3.2
11	5¼	1.0	1	7¾	3.5
14	5½	1.1	1	8¼	4.6
9	5¾	1.4	2	15½	30.5

Family LIV.—POMATOMIDÆ. The bluefishes

Body oblong, compressed; head large; mouth large, oblique; premaxillaries protractile; maxillary not slipping under preorbital, provided with a large supplemental bone; lower jaw projecting; jaws each with a series of strong, compressed, unequal teeth, upper jaw with an inner series of small depressed teeth; villiform teeth present on vomer, palatines, and tongue; gill membranes separate, free from the isthmus; gills 4, a slit behind the fourth; branchiostegals 7; opercle ending in a flat point; preopercular margin serrate; scales rather small, weakly ctenoid; lateral line complete, unarmed; first dorsal composed of about eight weak spines; second dorsal and anal similar, the latter preceded by two small free spines; caudal forked; ventrals thoracic, with I, 5 rays; pectorals rather short. A single widely distributed genus and species is known.

94. Genus POMATOMUS Lacépède. Bluefishes

The characters of the genus are included in the family description.

124. Pomatomus saltatrix (Linnæus). Bluefish; Tailor; "Greenfish;" "Snapping mackerel."

Perca saltatrix Linnæus, Syst. Nat., ed. X, 1758, p. 293; Carolina.
Pomatomus saltatrix Uhler and Lugger, 1876, ed. I, p. 136; ed. II, p. 116; McDonald, 1882, pp. 12 and 13; Bean, 1883, p. 366; Bean, 1891, p. 91; Smith, 1892, p. 71; Jordan and Evermann, 1896–1900, p. 946, Pl. CXLVIII, fig. 400.

Head 3.1 to 3.4; depth 3.1 to 3.55 (in large fish about 4); D. VIII–I, 23 to 26; A. II–I, 25 to 27; scales about 95 to 105. Body elongate, moderately compressed; head rather long; snout pointed 3.35 to 3.95 in head; eye 4.4 to 5.6; interorbital 4.05 to 4.6; mouth large, moderately oblique; lower jaw projecting; maxillary reaching nearly or quite to posterior margin of eye, 2.05 to 2.3 in head; gill rakers short, 11 to 14 on lower limb of first arch; scales small, thin, more or less deciduous, densely

49826—28——16

covering the soft dorsal and anal; lateral line nearly straight, extending on base of caudal; first dorsal composed of very slender spines; second dorsal and anal similar, slightly elevated anteriorly; caudal fin forked; ventral fins inserted a little behind base of pectorals, notably shorter than postorbital part of head; pectoral fins short, 1.65 to 1.9 in head.

Color in life, of a fish 13 inches long, greenish above; silvery below; first dorsal and caudal dusky; second dorsal olive green; anal white along base, distal half translucent, with punctulations; ventrals white; pectorals yellowish green, with dark bases.

This fish is represented in the collection by many small specimens ranging in length from 70 to 265 millimeters (2¾ to 10½ inches). The bluefish is related to the family of crevallies and pompanos, from which it is distinguished, however, by the serrate preopercle, the rather large unequal teeth, and the stouter caudal peduncle.

The bluefish is a voracious feeder, being highly predatory on other fishes. Schools of bluefish are known to follow schools of menhaden and other fish, and after they have fed to the fullest extent of their capacity they still continue in the destruction of the fish, killing many more than they require for their own support. Nine small specimens, of which the stomach contents were examined in the laboratory, had fed almost exclusively on silversides. Others examined in the field had eaten, in addition to silversides, young gizzard shad (Dorosoma).

Nothing definite regarding the spawning habits of the bluefish has been published, and we are able to add but little. On June 8, Mr. Radcliffe saw a ripe male, which had been caught in a pound

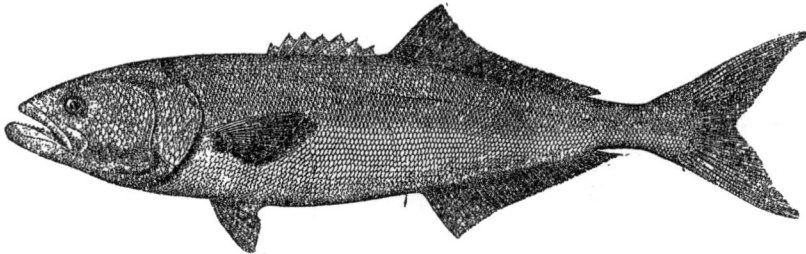

FIG. 136.—*Pomatomus saltatrix*

net at Lynnhaven Roads, Va., and the same investigator saw another there on July 15. It is probable, therefore, that spawning takes place in summer. Smith (1907, p. 216) states that the bluefish probably spawns offshore in summer.

Little is known of the rate of growth of the bluefish. Many young, 4 to 8 inches long, were taken daily throughout October. Fish 9 to 11 inches long were less common, but many ranging from 12 to 14 inches were caught and marketed in the lower bay later in the fall. The last-mentioned group may be one year older than the smallest group, or if fish 4 to 12 inches in length are the product of the same year it must be assumed that spawning is protracted or growth irregular.

The bluefish, among the fishes of Chesapeake Bay, ranked seventeenth in quantity in 1920 and sixteenth in value, the catch being 51,968 pounds, valued at $7,037.

In Maryland it ranked fifteenth in quantity and tenth in value, the catch amounting to 14,989 pounds, worth $2,112. Of this amount, 70 per cent were caught in gill nets, 24 per cent in pound nets, 3 per cent in purse seines, 2 per cent in haul seines, and 1 per cent with lines. The counties having the largest catches were Talbot, 7,200 pounds; Kent, 3,350 pounds; and Somerset, 1,769 pounds.

In Virginia it ranked sixteenth both in quantity and in value, the catch amounting to 36,979 pounds, worth $4,925. Of this amount, 52 per cent were taken in pound nets, 40 per cent in gill nets, 6 per cent in haul seines, and 2 per cent with lines. The counties catching the largest quantities were Northampton, 17,493 pounds; Elizabeth City, 6,870 pounds; and Mathews, 3,556 pounds.

The bluefish is one of the most valuable fish taken along the Atlantic coast and is everywhere highly esteemed. It enters Chesapeake Bay each season, its movements and abundance, however, varying considerably from year to year.

It is stated by many of the fishermen that large bluefish were plentiful 20 years ago, whereas now the fish is scarce and only small ones enter the bay. According to these fishermen, years ago the New England and New York fishing smacks followed the "blues" up the coast in the spring and entered Chesapeake Bay, where they caught large quantities. A law was enacted, however, excluding the northern bluefish fisherman from the bay, and since the enforcement of that law fishermen firmly believe that the bluefish has diminished in numbers in the bay, because the smacks bait the fish past the capes and prevent them from entering. The fact is, however, that vessels fishing for bluefish remain only a short while off the entrance of Chesapeake Bay, and during much the greater part of the fishing season no lures that prevent the fish from entering the bay are present.

The abundance of menhaden (a favorite food) to some extent governs the movements of the bluefish. During 1921 and 1922 young menhaden were plentiful throughout the lower half of the bay, and in 1922 the catch of bluefish was greater than it had been for many years.

Bluefish migrate up and down the Atlantic coast, following schools of menhaden and other fish, upon which they feed voraciously. In midwinter they are caught off southern Florida with purse seines and gill nets, together with Spanish mackerel. Large schools pass the Carolinas during March and April, and the first catches are made off New Jersey and Long Island during April and May. In southern Massachusetts the first catch is usually made late in May. From this vicinity northward the fishery is small, and only a few stragglers are taken along the coast of Maine during the summer.

Bluefish enter the Chesapeake as stragglers late in March or early in April. In 1922, in a set of pound nets at Lynnhaven Roads, Va., the first fish of the season was taken on April 3. The first capture was made at Ocean View on April 1. On April 7 a set of three pound nets at Ocean View caught 12 bluefish, weighing 1½ to 2 pounds each. The combined catch for the season of 75 small pound nets near Buckroe Beach, prior to April 11, was about 20 bluefish. The first fair catches are made some time in May. During June, July, and August the pound-net catches are small, although the fish are present in the bay and are taken in small numbers with hook and line and with gill nets. In September the pound-net catches increase, and during October the largest catches of the year are made. At Ocean View and Lynnhaven Roads small quantities often continue to be taken by the pound nets until about November 15.

Bluefish do not appear to ascend the bay to the northern sections, as the catch decreases rapidly above the mouth of the York River. All fishermen questioned unanimously stated that at one time bluefish, especially of large size, were abundant in Chesapeake Bay and were caught in comparatively large numbers as far north as Annapolis. The following statements pertaining to specific localities were obtained from conversations with fishermen:

Mouth of Rappahannock River, July 20, 1921: The fish are scarce this year, as in the past few years. The species is seldom seen now, but at one time large schools entered the lower part of the river. Last spring (1921) one pound net caught five bluefish and another caught four.

Lower Potomac River, August 6, 1921: Twenty years ago bluefish were plentiful opposite St. George Island and many large ones were caught in nets and with hand line. At the present time and during the past 10 years they have been very scarce, and only occasionally one is caught, usually a small one. April 21, 1922: Fifteen years ago bluefish were plentiful. They used to come in the spring and meet the glut herring passing out of the river and feed upon them. These fish were often large, weighing 6 to 12 pounds. Now only a few small ones are taken, generally weighing ½ to 3 pounds.

Crisfield, Md., November 21, 1921: Very few are caught and the size seldom exceeds 2 pounds. Most of the very small annual catch is taken in the fall with pound nets. A few are taken during the summer and fall with gill nets.

Oxford, Md., September 13, 1921: Bluefish have not been plentiful in this locality for 15 years. They have been decreasing steadily in numbers until now they are very scarce. During the present year fishing parties have been catching an odd one now and then with hook and line. One fisherman, who fishes regularly with hook and line, caught four this year. Very few are caught in pound nets. A school of these fish had not been seen in this locality for years, but on September 7, 1921, a small school was sighted off Cooks Point. The usual size of the fish caught at the present time is about 1 pound.

Solomons, Md., August 9, 1921: Twenty years ago large schools of bluefish entered the lower Patuxent and were caught about Solomons Island. The size of these fish averaged from 2 to 7 pounds. During the past eight years the fish have disappeared. This year hook-and-line fishermen have taken, to date, only two to five fish to a man. The size is generally below 2 pounds. October 26, 1921: A few large bluefish were taken in pound nets during September, and occasionally a small one was taken during October. No large run of these fish occurs here now, and not enough are caught at any time during the year to be of much commercial importance.

Annapolis, Md., August 19, 1921: Bluefish have been scarce for five or six years. They were once very plentiful and were caught near by with purse nets along with other species of fish. At the present time a fish is taken only occasionally in a pound net or with hook and line.

Love Point, Md., September 5, 1921: Some years ago bluefish were plentiful in the Chester River and about Love Point. Now they are scarce and are taken only occasionally with hook and line, haul seines, and pound nets. To-day one pound net caught two small ones.

Baltimore, Md.: Bluefish are not taken in commercial quantities in this vicinity. It seems probable that the species rarely strays north of this locality in Chesapeake Bay.

The majority of the bluefish are caught in the bay with pound nets and gill nets. From Annapolis to Ocean View quite a few are caught by sport fishermen with hook and line, but the total yearly catch by this method is not known.

Most of the annual catch is marketed in Norfolk and Baltimore, but when a good run of fish strikes in shipments are made to other points. The catch fluctuates so widely from week to week in the bay, and especially along the Atlantic coast, that wholesale prices are variable. The usual range during 1922 was from 10 to 20 cents a pound.

In the Chesapeake region small fish that weigh 2 pounds and less are known as "tailors," while the larger sizes are called "bluefish." The names "snapping mackerel" and "greenfish" are heard occasionally. The usual size of the fish taken in the bay ranges from ½ to 4 pounds. Fish weighing more than 6 pounds were comparatively rare in 1921 and 1922. Along the Atlantic coast a size of 5 to 10 pounds is common and the maximum authentic weight is given as 27 pounds.

Habitat.—Atlantic Ocean, Mediterranean Sea, Indian Ocean, straying northward on the coast of the United States to Maine.

Chesapeake localities.—(a) Previous records: Various localities from Havre de Grace, Md., southward. (b) Specimens in collection: Many localities from Love Point (Chester River), Md., southward to the mouth of the bay.

Comparison of lengths and weights of small bluefish

Number of fish weighed and measured	Length	Weight	Number of fish weighed and measured	Length	Weight
	Inches	*Ounces*		*Inches*	*Ounces*
1	4½	0.5	2	8¼	3.2
3	6½	1.5	3	8½	3.6
3	6¾	1.9	3	9	4.3
7	7	2.0	1	9¼	5.0
5	7¼	2.1	1	10	5.3
9	7½	2.3	2	10¼	6.8
3	7¾	2.7	1	12¼	11.7
5	8	3.0			

Family LV.—RACHYCENTRIDÆ. The crab-eaters

Body elongate, somewhat fusiform; head broad, strongly depressed; mouth wide, nearly horizontal; lower jaw projecting; premaxillaries not protractile; maxillaries reaching about to eye; teeth small, pointed, in bands on jaws, vomer, palatines, and tongue; opercle and preopercle unarmed; branchiostegals 7; gill rakers strong; first dorsal composed of eight or nine short, stiff, free spines; second dorsal and anal similar, somewhat elevated anteriorly but not falcate; caudal strongly rounded in very young, forked in adult; ventrals thoracic, with I, 5 rays; pectorals moderate, placed below level of lower margin of eye.

95. Genus RACHYCENTRON Kaup. Crab-eaters

The characters of the genus are included in the description of the family. A single species of wide distribution is known.

125. Rachycentron canadus (Linnæus). Crab-eater; Cabio; Sergeant fish; "Bonito;" "Black bonito."

Gasterosteus canadus Linnæus, Syst. Nat., ed. XII, 1766, p. 491; Carolina.
Elecate canadus Uhler and Lugger, 1876, ed. I, p. 137; ed. II, p. 117.
Elacate canada McDonald, 1882, p. 12; Bean, 1891, p. 91.
Rachycentron canadus Jordan and Evermann, 1896–1900, p. 948, Pl. CXLVIII, fig. 401.

Head 4; depth 5.45; D. VIII–I, 30; A. I, 23. Body elongate, anteriorly nearly as broad as deep, posteriorly compressed; caudal peduncle nearly round, its depth 4 in head; head long and broad, depressed; snout broad, 2.95 in head; eye 8.8; interorbital 2.45; mouth large; lower jaw projecting; maxillary reaching anterior margin of eye, 2.65 in head; teeth small, pointed, in bands on jaws, vomer, palatines, and tongue; gill rakers strong, seven on lower limb of first arch; scales minute; lateral line complete; first dorsal composed of very short, stiff spines; second dorsal long, elevated anteriorly; caudal fin forked (truncate in young); anal fin similar in shape to the second dorsal, but shorter; ventral fins moderate, inserted under base of pectorals, about as long as post-orbital part of head; pectoral fins large, 1.15 in head.

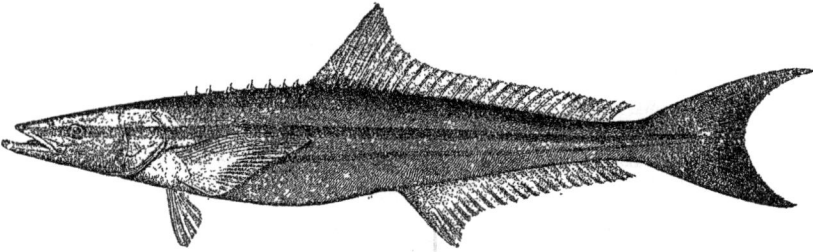

FIG. 137.—*Rachycentron canadus.* From a specimen 29¾ inches long

Color of a fresh specimen black above, grayish white to plain below; dorsal and caudal dusky; anal and ventrals white with gray or dusky markings; pectorals black. Some specimens observed in the Norfolk market were dark brown on back instead of black.

A single specimen, 810 millimeters (32 inches) in length, weighing 8 pounds when fresh, is contained in the present collection, and it forms the basis for the foregoing description. The young differ markedly from the adults in the more elongate body, less strongly depressed head, in having the caudal fin truncate instead of forked, and in being somewhat lighter in color and having a black lateral band, which extends from the snout, through the eye, to the base of the caudal.

This fish is carnivorous, feeding on fish and crustaceans. Relative to the spawning, Smith (1907, p. 221) says: "The fish is known to spawn in Chesapeake Bay in summer, and its eggs have been artificially hatched." Our specimen is a female (captured on May 27, 1922) with ova so small that they are not visible to the unaided eye.

The black bonito is confined to the lower part of Chesapeake Bay, being taken from the York River to Cape Charles and Lynnhaven Roads, Va. There are no published statistics of the annual catch of this fish. During 1921 about 3,000 pounds, worth $300 to the fishermen, were taken.

This fish is present in the bay from May until late summer, but it is most common in June. Most of the catch is taken with hook and line between the York River and Buckroe Beach, Va. On June 21, 1921, at Buckroe Beach, two hand-line fishermen caught four black bonito, weighing 40, 62, 75, and 84 pounds, respectively; while another boat had three fish, weighing 52, 62, and 82 pounds, respectively. Hook-and-line fishing appears to be done only in June, but a few fish are taken in pound nets throughout the summer.

The catch is readily disposed of in the markets of Norfolk, Portsmouth, and Phœbus, the large fish being cut into steaks. The usual size of the market fish is 10 to 50 pounds, and 84 pounds appears to be the largest fish recorded from the bay and the maximum size attained by the species.

Habitat.—New Jersey to Brazil; East Indies.

Chesapeake localities.—(*a*) Previous records: "Southern part of Chesapeake Bay" (Uhler and Lugger, 1876); Norfolk and Cape Charles city, Va. (*b*) Specimens in collection: Ocean View, Va.; also observed at Buckroe Beach and Lynnhaven Roads, and reported by fishermen from the lower York River, Va.

Family LVI.—PERCIDÆ. The perches

Body elongate, compressed or not; head moderate; mouth small or large, terminal or slightly inferior; maxillary without a distinct supplemental bone; teeth pointed (some species with a few canines), present on jaws and usually on vomer and palatines; pharyngeal bones with sharp teeth; branchiostegals 7; gills 4; preopercle serrate; opercle ending in a flat spine; pyloric cœca few; lateral line present; scales firm, ctenoid; dorsal fins 2, with 8 to 16 spines; anal similar to second dorsal or smaller, with two spines; ventrals well developed, situated below or a little behind base of pectorals, with one spine and five soft rays; air bladder present. A single genus and species comes within the scope of the present work.

96. Genus PERCA Linnæus. Yellow perches

Body elongate, only moderately compressed; mouth moderate, terminal; premaxillaries protractile; teeth small, pointed, in bands on jaws, vomer, and palatines; opercle ending in a spine; preopercle serrate; gill membranes separate and free from the isthmus; scales rather small, ctenoid; lateral line complete; dorsal fins 2, well separated, the first with 12 to 16 rather high and slender spines; anal with two weak spines; ventral fins close together, with a strong spine. A single widely distributed species is known from American waters.

126. Perca flavescens (Mitchill). Yellow perch; Red-fin.

Morone flavescens Mitchill, Rept., Fish., New York, 1814, p. 18; near New York City.

Perca flavescens Uhler and Lugger, 1876, ed. I, p. 128; ed. II, p. 109; Jordan and Evermann, 1896–1900, p. 1023, Pl. CLXV, fig. 435; Smith and Bean, 1899, p. 186; Fowler, 1912, p. 55; Snyder, 1917, pp. 18 and 28.

Perca americana Bean, 1883, p. 365.

Head 3.1 to 3.6; depth 3.35 to 3.85; D. XIII or XIV–II, 14 or 15; A. II, 7 or 8; scales 7–67 to 83. Body elongate, moderately compressed; head long and low; snout pointed, 3.3 to 4.1 in head; eye 3.65 to 5.35; interorbital 4.2 to 4.55; mouth rather large, a little oblique, terminal; maxillary broad, reaching to or a little beyond anterior margin of pupil, 2.4 to 2.9 in head; gill rakers short and stout, 15 or 16 on lower limb of first arch; scales small, firm, ctenoid, present on cheeks and opercles and extending on base of caudal; lateral line complete; dorsal fins well separated, the origin of the first over base of pectorals, the spines pungent, the longest about half the length of head; caudal fin rather deeply emarginate; anal fin small, with two rather long, slender spines; ventral fins close together, inserted about an eye's diameter behind base of pectorals; pectoral fins rather short, rounded, 1.7 to 1.9 in head.

Color dark olive green above; sides yellow; abdomen and chest pale; back and sides with six to eight black crossbars (these obsolete in a few of the adult specimens at hand); dorsal and caudal dusky green; spinous dorsal sometimes with a black blotch posteriorly; anal, ventrals, and pectorals red or orange, brightest in males during the breeding season. Young usually with indefinite dark spots on sides in addition to black crossbars.

Numerous specimens, ranging in length from 55 to 250 millimeters (2¼ to 9¾ inches) were collected. This fish is readily recognized by the black crossbars and the yellow coloration on the lower part of the sides. The young are not very different from the adults. They are somewhat more blotched, however, and rather more slender.

The yellow perch is carnivorous, feeding on a large variety of animal life, ranging (according to published accounts) from microscopic organisms to sizeable animals like crawfish, minnows, and young fish. The stomach contents of 20 specimens taken in brackish water contained the following foods, named in the order of their apparent importance: Isopods, amphipods, fish, crabs, shrimp, insect larvæ, and snails. Spawning takes place early in the spring and is described by Smith (1907, p. 251) as follows:

The spawn is very peculiar, in that the eggs are cemented together in a single layer in the form of long, hollow strings, which, when extruded, are several inches wide and folded or plaited like the bellows of an accordian, but are capable of being drawn out to the length of 3 to 7 feet. One fish in an aquarium at the Bureau of Fisheries, Washington, D. C., deposited a string 88 inches long. The weight after fertilization was 41 ounces, while the weight of the fish before the escape of the eggs was only 24 ounces. The egg masses are not attached to stones, vegetation, or other submerged objects, but are deposited loosely in the water. Spawning takes place at night in water having a temperature of 44° to 50° F., and the hatching period lasts from two to four weeks.

Vast numbers of eggs of this species are hatched annually by fish-cultural stations situated on streams tributary to Chesapeake Bay.

This fish averages less than 1 foot in length and 1 pound in weight. The maximum size recorded is 4½ pounds. The perch is of wide distribution and an important food and game fish in many parts of its range, including the streams tributary to the northern part of Chesapeake Bay. It is common enough in the brackish waters in certain sections of the Chesapeake to be of some commercial value, and ranks fairly high as to the quality of its flesh. It is a ready biter, taking a large variety of baits. It is also caught with fyke nets, pound nets, and seines.

Habitat.—North Carolina to Nova Scotia, the Great Lakes region, northern part of the Mississippi Valley, northward to the Red River Basin.

Chesapeake localities.—(a) Previous records: None definitely from brackish water. (b) Specimens in collection: From Havre de Grace, Baltimore, Annapolis, Love Point, Solomons, and Oxford, Md., and Lewisetta, Va. Highest salinity, at Annapolis, November 1, 1921, 12.94 per mille.

Comparison of lengths and weights of yellow perch

Number of fish weighed and measured	Length	Weight	Number of fish weighed and measured	Length	Weight
	Inches	*Ounces*		*Inches*	*Ounces*
1	3½	0.3	6	7¼	2.4
3	4½	.6	3	7½	2.6
4	4¾	.7	2	7¾	2.8
4	5	.8	8	8	3.6
9	5¼	1.0	5	8¼	3.9
5	5½	1.1	6	8½	4.2
11	5¾	1.3	1	8¾	4.6
9	6	1.4	3	9	5.0
10	6¼	1.5	1	9¼	5.2
8	6½	1.8	2	9¾	5.7
5	6¾	2.0	1	10½	7.3
4	7	2.3			

Family LVII.—ETHEOSTOMIDÆ. The darters

The darters usually have been considered dwarf or diminutive perches and are often all placed under the perch family, Percidæ. Jordan, in his "A Classification of Fishes" (1923, p. 187), however, gives the darter family rank under the name Etheostomidæ. The Etheostomidæ, as understood by Jordan, differ from the Percidæ in having six branchiostegals instead of seven; head (preopercle) unarmed; air bladder obsolete or nearly so; anal with one or two spines. A single genus and species of this large family of fresh-water fishes comes within the scope of the present work.

97. Genus BOLEOSOMA DeKay. Tesselated darters

Body elongate, fusiform; head small; snout dorsally strongly decurved; parietal region slightly convex; mouth small, horizontal; premaxillaries protractile; small, pointed teeth present on jaws and vomer; gill membranes narrowly to broadly connected; scales rather large, those on median line of abdomen not enlarged or deciduous; lateral line complete or interrupted; dorsal with 8 to 10 spines; soft dorsal notably longer than the anal; anal with a single short spine; ventral fins well separated; coloration rather plain, without red or blue.

127. Boleosoma olmstedi (Storer). Darter; Johnny darter.

Etheostoma olmstedi Storer, Jour., Bost. Soc. Nat. Hist., 1839, p. 61, Pl. V, fig. 2; Hartford, Conn.
Estrella atromaculata Girard, Proc., Ac. Nat. Sci., Phila., 1859, p. 66; Potomac River.
Boleosoma olmstedi Uhler and Lugger, 1876, ed. I, p. 134; ed. II, p. 115; Smith and Bean, 1899, p. 186.
Boleosoma nigrum olmstedi Jordan and Evermann, 1896–1900, p. 1057, Pl. CLXXI, fig. 451; Evermann and Hildebrand, 1910, p. 161; Fowler, 1912, p. 55.

Head 3.85 to 4.15; depth 5.45 to 6.1; D. VIII to X–14 or 15; A. I, 9 to 11; scales 48 to 55. Body slender, scarcely compressed anteriorly; caudal peduncle compressed, 2.4 to 2.8 in head;

head low, as broad as deep at eyes; snout rather short, decurved, 3.45 to 3.85; eye 3.4 to 3.75; interorbital very narrow, 11.15 to 13.6; mouth placed low, horizontal, terminal; maxillary reaching to or a little beyond anterior margin of eye, 2.9 to 3.25 in head; opercle ending in a strong spine; preopercle unarmed; gill membranes somewhat united, free from the isthmus; scales strongly ctenoid, covering entire body, including the chest, also present on cheeks and opercles; lateral line complete; dorsal fins well separated, the spines of the first slender, the longest one equal to distance from tip of snout to preopercular margin; soft dorsal somewhat higher; caudal fin straight to slightly rounded; anal fin small, its origin falling slightly behind that of soft dorsal; ventral fins moderate, inserted a little behind base of pectorals; pectoral fins rather long, pointed, 0.86 to 1 in head.

Color olivaceous above; pale underneath; back with six to eight black, saddlelike blotches; sides with irregular dark markings; a dark bar below the eye and one in front of it; dorsal fins, caudal, and pectorals spotted and barred with black; anal and ventrals plain.

Many specimens, ranging from 35 to 80 millimeters (1½ to 3⅛ inches) in length, were preserved. This is the only darter taken in brackish water. The species appears to be variable and varieties have been recognized. The form herein described, as a matter of fact, is often considered only a subspecies or variety of *B. nigrum*, the typical form of which occurs in the upper Mississippi and the Great Lakes region. *B. olmstedi* may be recognized by the complete lateral line, the completely scaled chest, cheeks, and opercles, the high fins, and the more profuse and distinct dark markings on the body and fins.

The stomach contents of 12 specimens taken in brackish water consisted wholly of small crustaceans (principally Gammarus) and insect larvæ. Specimens taken in May, 1922, are in spawning condition. The ovary of a female 62 millimeters in length contained 340 eggs of uniform size, approximately 1 millimeter in diameter after preserving in alcohol. The smallest sexually mature female found in the collection is 43 millimeters long, and the smallest sexually mature male has a length of only 40 millimeters.

The largest specimens in the collection (slightly exceeding 3 inches in length) appear to represent the maximum size attained by the Johnny darter. This fish is generally common in clear, running streams, occurring also in quiet and standing waters. The writers find no mention in literature of its occurrence in brackish water. It was taken only at the head of Chesapeake Bay, where it was common in water that was slightly saline.

Habitat.—Massachusetts to North Carolina in coastwise streams.

Chesapeake localities.—(a) Previous records: None from brackish water. (b) Specimens in collection: All from the vicinity of Havre de Grace, Md. Highest salinity, Elk River, opposite Turkey Point, November 11, 1921, 2.23 per mille.

Family LVIII.—CENTRARCHIDÆ. The fresh-water basses and sunfishes

Body usually rather short and compressed; mouth large or small; premaxillaries protractile; maxillary usually with a supplemental bone, obsolete in small-mouthed species; teeth pointed, in bands on jaws, vomer, and palatines, rarely on tongue; opercle ending in two points, or with a single long flap; preopercle entire or slightly serrate; pseudobranchiæ small; branchiostegals usually 6; gill membranes separate and free from the isthmus; gill rakers usually short; scales present on body, opercles, and cheeks; lateral line present, usually complete; dorsal fin continuous, sometimes deeply notched, with 6 to 13 spines; anal fin large or small, with 3 to 8 spines. This family includes some of the best known and most important fishes of American fresh waters. Nearly all the species are carnivorous, and they build nests in which the eggs and young are carefully guarded.

KEY TO THE GENERA

a. Anal fin somewhat longer than the dorsal, with six spines; body and fins profusely and irregularly spotted with black_____Pomoxis, p. 239

aa. Anal fin much shorter than the dorsal, with three spines.

  b. Body comparatively short and deep, strongly compressed, the depth frequently half the length, always exceeding one-third of the length; dorsal fin continuous, scarcely or not at all notched at the beginning of the soft rays.

c. Caudal fin rounded; opercle without a flap, ending in two flat points; dorsal with nine spines; size small, rarely exceeding a length of 3 inches_____Enneacanthus, p. 240

cc. Caudal fin emarginate, never rounded; opercle with a large black flap in the adult; dorsal with 10 spines; size larger_____Lepomis, p. 241

bb. Body elongate, not strongly compressed, the depth rarely as great as one-third the length; dorsal fin deeply notched; mouth large; maxillary reaching opposite middle of eye to beyond eye_____Micropterus, p. 242

98. Genus POMOXIS Rafinesque.   Crappies

Body rather deep, strongly compressed; upper anterior profile more or less concave over eyes; mouth large, oblique; lower jaw projecting; maxillary broad, with a large supplemental bone; teeth pointed, present on jaws, vomer, palatines, and tongue; gill rakers long and slender; scales moderate, feebly ctenoid; anal fin larger than the dorsal, with about 6 spines and 18 soft rays; dorsal with about 6 spines and 14 soft rays; caudal fin emarginate; ventral fins close together; with a strong spine and 5 branched rays.   One species was taken in slightly brackish water.

128. Pomoxis annularis Rafinesque.   Crappie; Speckled perch; Strawberry bass.

Pomoxis annularis Rafinesque, Amer. Month. Mag., 1818, p. 41; falls of the Ohio River. Jordan and Evermann, 1896–1900, p. 987, Pl. CLIV, fig. 415; Smith and Bean, 1899, p. 185.

Head 2.8; depth 2.75; D. VI, 14; A. VI, 18; scales 48.   Body elongate; the back not greatly elevated; a line at right angles to the posterior margin of the maxillary passing notably in front of origin of dorsal; upper profile moderately concave over eyes; head rather long; snout moderate, 3.75 in head; eye 4.1; interorbital 5.4; mouth large, oblique; lower jaw projecting; maxillary reaching nearly opposite posterior margin of pupil, 2.25 in head; preopercular margin with small serrations at lower posterior angle; gill rakers slender, 21 on lower limb of first arch; scales moderate, weakly ctenoid; dorsal and anal fins similar, but with the spines in the anal rather stronger; caudal fin rather deeply emarginate; ventral fin moderate, inserted a little behind base of pectorals; pectoral fins rather pointed, 1.7 in head.

Color olive silvery, mottled with darker, irregular spots and blotches, forming more or less definite bars on upper part of sides; the dark markings extending on dorsal and caudal and to a smaller extent on the anal; ventrals and pectorals plain translucent.

A single specimen of this species, 160 millimeters (6¼ inches) in length, was taken in slightly brackish water. *P. sparoides*, a closely related species, was not seen in brackish water.   Smith and Bean (1899, p. 185) offer the following remarks relative to crappies:

Both of these species were introduced into the Potomac River and the Chesapeake and Ohio Canal by the Fish Commission in 1884, and have become very common in places, noticeably Little River, Four-mile Run, Eastern Branch, and in the river near Seven Locks; also through the canal as far as Harpers Ferry. *P. annularis* is the more abundant here. These are excellent game and food fishes and many are now caught by anglers. The two species are much alike and not usually distinguished by local fishermen, who apply the names crappie, strawberry bass, strawberry perch, and speckled perch indiscriminately.

The species are distinguished principally by the difference in the depth of the body, the shape of the upper anterior profile, and the color; *P. annularis* being more slender, with a more concave profile, and having more or less definite dark bars on the sides; whereas, the dark markings in *P. sparoides* are irregular, more or less elongate, never forming vertical bars.   A straight line placed at the posterior margin of maxillary and at right angles to the anterior margin of the maxillary passes in front of the dorsal in *P. annularis*, while a line similarly placed in *P. sparoides* passes through the dorsal.

This crappie, according to published accounts, is strictly carnivorous, living mainly on insects, crustaceans, and fish.   Spawning takes place during the spring.

This fish lives under a large variety of conditions, adapting itself particularly well to artificial ponds.   It has been widely distributed by fish culturists and is an important food fish in many localities.   It is only a straggler in brackish water and is of no commercial importance among the fishes of Chesapeake Bay.

The maximum weight attained by this crappie is recorded as 3 pounds.   (Bean, 1903, p. 461.) The average weight of market fish, however, is less than a pound.

Habitat.—Great Lakes, southward to Alabama and Texas and westward to Kansas and Nebraska. Through canals and through the efforts of fish culturists, the species has become established in various places on the Atlantic slope, from New York to Georgia.

Chesapeake localities.—(a) Previous records: None known to be brackish-water. (b) Specimen in collection: Near the mouth of the Susquehanna River, Havre de Grace, Md., November 9, 1921, salinity, 1.53 per mille.

99. Genus ENNEACANTHUS Gill. Little sunfishes

Body short, deep; mouth small; teeth present on jaws, vomer, and palatines; maxillary with a well developed supplemental bone; margin of opercle entire; opercle ending in two flat points; scales large; lateral line usually complete; gill rakers short and rather few; dorsal with about nine spines; anal with three spines; caudal fin with round margin. One species was taken in brackish water in the northern sections of Chesapeake Bay.

129. Enneacanthus gloriosus (Holbrook). Speckled perch; Blue-spotted sunfish; Little sunfish.

Bryttus gloriosus Holbrook, Journ., Ac. Nat. Sci., Phila., 1855, p. 51; Cooper River, S. C.
Enneacanthus gloriosus Uhler and Lugger, 1876, ed. I, p. 131; ed. II, p. 112; Jordan and Evermann, 1896–1900, p. 993, Pl. CLVIII, fig. 422; Smith and Bean, 1899, p. 185; Evermann and Hildebrand, 1910, p. 161; Fowler, 1912, p. 55.

Head 2.75 to 3.05; depth 2 to 2.65; D. IX, 10 or 11; A. III, 9 to 11; scales 30 to 32. Body moderately deep, compressed; head rather short; snout blunt, 4 to 5.3 in head; eye 2.85 to 3.75; interorbital 3.4 to 4.6; mouth small, oblique, terminal or slightly superior; maxillary reaching opposite anterior margin of pupil, 2.8 to 3.1 in head; gill rakers short, 9 or 10 on lower limb of first arch; scales large, firm, ctenoid; lateral line usually wanting on several scales posteriorly; dorsal fin long, its origin over base of pectorals; caudal fin round; anal fin short, with three short, stout spines; ventral fins reaching somewhat beyond origin of anal, inserted a little behind base of pectorals; pectoral fins rather long, 1.25 to 1.55 in head.

Color of male dark green above, abdomen golden; sides with sky-blue spots, about half as large as scales, extending on dorsal, caudal, and anal; opercle with jet-black spots; ground color of dorsal greenish (dusky in spirits); caudal and anal reddish; ventrals plain, with red on longest ray; pectorals slightly greenish. Females olive green with purplish luster; bluish spots wanting. Some preserved specimens have indications of dusky crossbars on sides and pale lines along the rows of scales.

Numerous specimens of this little fish, ranging from 20 to 85 millimeters (¾ to 3⅜ inches) in length, were preserved. This species is separated from other sunfishes by the small mouth and round tail. It inhabits sluggish water, especially frequenting places with aquatic growths of vegetation. It is not uncommon in the brackish waters near mouths of streams in the northern sections of Chesapeake Bay, frequenting water with a specific gravity as great as 1.0095. It was particularly abundant in a brackish pond near Annapolis, where more than 1,000 were caught in 20 hauls of a 30-foot collecting seine.

The food of this fish, according to the contents of 13 stomachs examined, consists mainly of small crustaceans—that is, copepods, amphipods, and isopods. Insects and worms, too, were present in a few stomachs; also fragments of plants. Spawning apparently takes place in May and June, as specimens taken at about this time contained well-developed roe.

The blue-spotted sunfish is one of the most beautiful of our local fishes. It is hardy and an attractive aquarium fish. It is of no commercial importance because of the small size attained. The largest specimen at hand, having a length of 3⅜ inches, represents the maximum size for the species.

Habitat.—New York to Georgia.

Chesapeake localities.—(a) Previous records: None known to be from brackish water. (b) Specimens in collection: Havre de Grace, Baltimore, Annapolis, Love Point, and Oxford, Md., and Lewisetta, Va. Highest salinity, entrance to Lake Ogleton, Annapolis, Md., November 3, 1921, 12.88 per mille.

100. Genus LEPOMIS Rafinesque. Common sunfishes

Body ovate, quite strongly compressed; mouth moderate or small, terminal; maxillary narrow, not extending beyond pupil, with or without a small supplemental bone; no teeth on pterygoids or tongue, short or blunt teeth on the narrow pharyngeal bones; preopercular margin entire; opercle ending in a more or less elongated, conspicuously colored flap; gill rakers usually short and feeble; dorsal with 10 spines; anal with 3 spines and much shorter than the dorsal; caudal fin emarginate.

A single species of this genus was taken in the brackish waters of the Chesapeake. In the fresh waters the fishes of this genus are numerous, both as to species and individuals. Some of the species, especially the young, are difficult to distinguish.

130. Lepomis gibbosus (Linnæus). "Tobacco box;" Pumpkin seed; Sand perch; "Sunfish;" Bream.

Perca gibbosus Linnæus, Syst. Nat., ed. X, 1758, p. 292; Carolina.
Lipomis aureus Uhler and Luggar, 1876, ed. I, p. 132.
Pomotis aureus Uhler and Luggar, 1876, ed. II, p. 113.
Lepomis gibbosus Bean, 1883, p. 365; Smith, 1892, p. 71.
Eupomotis gibbosus Jordan and Evermann, 1896, p. 1009, Pl. CLXI, fig. 429; Smith and Bean, 1899, p. 186; Evermann and Hildebrand, 1920, p. 161; Fowler, 1912, p. 55.

Head 2.75 to 3.35; depth 1.85 to 2.9; D. X, 10 to 12; A. III, 9 or 10, scales 34 to 40. Body deep, ovate, strongly compressed; head rather short; snout broad, its length 3.05 to 4.8 in head; eye 3 to 4.3; interorbital 2.9 to 3.9; mouth small, oblique, terminal; maxillary scarcely reaching anterior margin of eye in some specimens, to anterior margin of pupil in others, 3.55 to 4 in head; preopercular margin entire; opercular flap very broad; gill rakers very short, 9 to 11 more or less developed on lower limb of first arch; scales of moderate size, firm, ctenoid; lateral line complete; dorsal fin long, its origin over or a little behind base of pectorals, the spinous portion longer than the soft part; caudal fin rather deeply emarginate; anal fin short, with three strong spines, the soft part similar to that of dorsal and coterminal with it; ventral fins moderate, reaching to vent, and in some specimens to or slightly beyond origin of anal, inserted somewhat behind base of pectorals, pectoral fins rather long, pointed, 1.05 to 1.4 in head.

Color above greenish, variously spotted with brown and with bluish reflections; head mostly brassy, with irregular sky-blue lines under eye; opercular flap jet black, margined with scarlet; abdomen golden; dorsal and caudal mainly dusky; anal and ventrals more or less dusky yellowish; pectorals light yellowish.

Many specimens of this common sunfish, ranging in length from 22 to 180 millimeters (7⁄8 to 7⅛ inches), were preserved.

This fish is not uncommon in the brackish waters in the northern sections of Chesapeake Bay, occurring at times in strongly brackish water. On September 6, 1921, at Thighmans Creek near Love Point, for example, 115 adults and 580 small fish were taken in eight hauls of a 30-foot collecting seine in brackish water. In the streams tributary to the bay it is among the commonest of species; and it is well known for its gameness and beauty, making it the joy of youthful anglers.

The food of the tobacco box in the brackish waters of Chesapeake Bay, according to the contents of 65 stomachs, consists of the following, named in the order of their apparent importance: Isopods, annelids, amphipods, mollusks, and insect larvæ. A comparatively large amount of vegetable débris also was present. However, it is uncertain whether this was eaten as food or obtained more or less by accident in the capture of small animal life.

Spawning takes place in the spring and early summer. The breeding habits are described by Smith (1907, p. 243) as follows:

The nest is a slight depression on the bottom, made by the fins, and after the eggs are laid and attached to stones or weeds, the male stands guard and repels intruding fishes or other animals; the care of the young also devolves on the male, which at this season is in his brightest colors and even in the water can be readily distinguished from his mate.

Many of the specimens at hand have isopods (Livoneca ovalis) attached to the gills. In the field hundreds of fish were examined, and at least 80 per cent of those found in brackish water bore this parasitic isopod. One or two, and sometimes three or four, were found on the gills on one or both sides. A number of sunfish found dead along the shore probably were killed by this large isopod, and the destruction from this cause may be large.

Habitat.—Northern part of the Mississippi Basin, the Great Lakes, and along the Atlantic seaboard from Maine to Florida (not found by Hildebrand (1923) in the Savannah River Basin in the vicinity of Augusta, Ga.).

Chesapeake localities.—(*a*) Previous records: "In the region of Baltimore they prefer the brackish water * * *" (Uhler and Lugger, 1876); St. George Island (other localities not definitely brackish water). (*b*) Specimens in collection: Many localities, from Havre de Grace, Md., southward to Lewisetta, Va. Highest salinity, Blackistone Island, Md., October 20, 1921, 18.17 per mille.

Comparison of lengths and weights of the tobacco box

Number of fish weighed and measured	Length	Weight	Number of fish weighed and measured	Length	Weight
	Inches	*Ounces*		*Inches*	*Ounces*
5	3½	0.5	5	5¼	1.7
5	4	.7	5	5½	2.0
4	4¼	1.0	4	5¾	2.4
11	4½	1.1	2	6	2.7
9	4¾	1.3	1	6¼	3.0
8	5	1.5	1	6½	4.1

101. Genus MICROPTERUS Lacépède. Black basses

Body elongate, compressed; back not much elevated; head long, rather low; snout conic; mouth large, oblique; lower jaw projecting; maxillary broad, with a well-developed supplemental bone, reaching opposite middle to beyond eye; teeth pointed, in bands on jaws, vomer, and palatines; opercle ending in two broad points; preopercle entire; gill rakers moderate; scales rather small, ctenoid; lateral line complete; dorsal fin continuous, rather deeply notched, with 10 rather low spines; caudal fin emarginate; anal fin with three spines, the soft part similar to that of the dorsal. This genus has three species; two are represented in the fresh and slightly brackish waters of the Chesapeake region.

KEY TO THE SPECIES

a. Mouth moderate; maxillary reaching to or a little beyond middle of eye; scales rather small, 69 to 72 in lateral series, 10 between lateral line and beginning of soft part of dorsal; color nearly uniform (young more or less barred and spotted)_____*dolomieu*, p. 242

aa. Mouth very large; maxillary reaching to or beyond posterior margin of eye; scales larger, 61 to 64 in lateral series, 7 or 8 between lateral line and beginning of soft part of dorsal; a distinct black lateral band usually present, occasionally obsolete in adults_____*salmoides*, p. 243

131. Micropterus dolomieu Lacépède. Black bass; Smallmouth black bass.

Micropterus dolomieu Lacépède, Hist. Nat. Poiss., IV, 1803, p. 325; locality doubtful. Bean, 1883, p. 365; Jordan and Evermann, 1896–1900, p. 1011, Pl. CLXII, figs. 430, 430a; Smith and Bean, 1899, p. 186; Fowler, 1912, p. 55.

Head 2.8 to 2.85; depth 3.35 to 3.6; D. X, 14; A. III, 11; scales 10–69 to 72. Body elongate, compressed; head rather long and low; snout pointed, 3.2 to 3.8 in head; eye 3.35 to 4.15; interorbital 5.25 to 5.65; mouth moderate, slightly oblique; lower jaw projecting; maxillary reaching about middle of eye, 2.15 to 2.3 in head; gill rakers moderate, about seven more or less developed on lower limb of first arch; scales rather small, firm, feebly ctenoid, scales on cheeks much smaller than on body; lateral line complete, arched a little anteriorly; dorsal fin long and low, the spinous part lower than the soft portion; caudal fin emarginate; anal fin short, preceded by three short spines, the soft part similar to that of the dorsal and coterminal with it; ventral fin inserted under or slightly behind base of pectorals, failing to reach vent; pectoral fins short, round, 2.2 to 2.3 in head.

Color in life, of young 2½ to 4 inches long, greenish blue or greenish brown, mottled with brown, the mottlings most conspicuous above lateral line; five brownish stripes on sides of head back of and below eye; underneath head and parts of opercle sky-blue; abdomen grayish white; dorsal dusky and light brown, slightly mottled; caudal dusky at base, median parts yellowish brown, distally with black, margined with white or salmon red; anal dusky yellow, slightly mottled, margined with white; ventrals and pectorals yellowish, the latter slightly dusky on one specimen. No adults were on the head in the young persist in the adult.

Five small specimens of this bass, ranging in length from 65 to 100 millimeters (2½ to 4 inches), were secured in slightly brackish water, and they form the basis for the foregoing description. This species differs from the large-mouthed black bass principally in the smaller mouth, slightly smaller scales on body, the much smaller scales on cheeks, and in the more uniform color.

The food of the adult smallmouth bass, according to published accounts, consists mainly of fish and crawfish. Three small specimens, taken in slightly brackish water, had fed on isopods, insects, and fish. Spawning takes place during the spring. The species is cultivated through pond-cultural methods, as it does not lend itself well to stripping. In this species, as in others of this family, the male builds a nest, in which the female deposits the eggs, which are fertilized when laid. The male then stands guard over the nest, fanning the eggs with its fins and fighting off intruders.

This species generally lives in higher altitudes than the large-mouthed species, preferring cooler and swifter water. It is not a native of the Chesapeake, having reached that vicinity through artificial means. Its occurrence in still, brackish water seems quite unusual.

The maximum weight recorded for this fish is 6 pounds. It is an important game fish in many sections, taking live minnows and other live bait, and it also rises to the artificial fly. It is only a straggler in the brackish waters of the Chesapeake, where it is of no importance either as a food or a game fish.

Habitat.—Vermont to the Great Lakes and Manitoba; southward to South Carolina, Mississippi, and Arkansas. (Its range has been considerably extended by artificial means beyond its natural habitat as given here.)

Chesapeake localities.—(a) Previous records: None from brackish water. (b) Specimens in collection: Havre de Grace and Love Point, Md. Highest salinity, Love Point, May 11, 1922, 7.39 per mille.

132. Micropterus salmoides (Lacépède).   Largemouth black bass; Black bass; "Chub;" "Trout."

Labrus salmoides Lacépède, Hist. Nat. Poiss., IV, 1803, p. 716; South Carolina.
Micropterus salmoides Uhler and Lugger, 1876, ed. I, p. 129; ed. II, p. 111; Jordan and Evermann, 1896-1900, p. 1012, Pl. CLXIII, fig. 431; Smith and Bean, 1899, p. 186; Evermann and Hildebrand, 1910, p. 161.

Head 2.8 to 3.05; depth 2.95 to 3.3; D. X, 12 or 13; A. III, 11; scales 7 or 8-61 to 64. Body elongate, compressed; head rather long and low; snout pointed, 3.55 to 3.85; eye 4.15 to 5; interorbital 4.1 to 4.45; mouth large, slightly oblique; lower jaw projecting; maxillary reaching to or a little beyond posterior margin of eye, 2 to 2.15 in head; gill rakers rather short, about seven more or less developed on lower limb of first arch; scales moderate, firm, rather weakly ctenoid scales on cheeks nearly as large as on body; lateral line complete, slightly arched anteriorly; dorsal fin long and rather low, deeply notched, the spinous part lower than the soft portion; caudal fin emarginate; anal fin preceded by three rather short, graduated spines, the soft part similar to that of the dorsal and coterminal with it; ventral fins short, failing to reach vent, inserted slightly behind base of pectorals; pectoral fins short, round, 2.15 to 2.3 in head.

Color dull green above, with brassy luster; sides silvery; abdomen white; three dark bars on sides of head back of and below eye; sides with a dark lateral band, usually more distinct in the young than in the adult, sometimes broken up into blotches and indistinct or occasionally wanting, sides usually also with irregularly placed dark blotches; dorsal fin slightly spotted, the caudal distally dusky with pale margin; fins otherwise plain.

This bass is represented by numerous specimens, ranging from 90 to 323 millimeters (3½ to 12⅝ inches) in length, which were taken in brackish water. This fish has a larger mouth than its congener, larger scales, particularly on the cheeks, and usually it has a dark lateral band, which also distinguishes it from the related species.

The food of the largemouth black bass in fresh water is reported to consist mainly of fish and crustaceans. The contents of 22 stomachs taken from specimens caught in brackish water of the Chesapeake consisted exclusively of fish remains. This fish is highly predatory, and where it is common the destruction of minows and smaller fish is great. What has been said relative to the spawning habits and artificial culture of the smallmouth black bass applies equally as well to the present species.

This bass is of wide distribution, apparently thriving best in the warmer waters of the United States, where it reaches the largest size. It is at home in sluggish streams, ponds, and lakes, and it enters brackish water freely. It is not a native of the Chesapeake region, having been first introduced in the Potomac River, according to Uhler and Lugger, prior to 1876, the date of publication of their "List of Fishes of Maryland." These authors say: "Introduced into the Potomac River from the Youghioghany, and now abounds in some of the upper parts of that stream; likewise in the Chesapeake and Ohio Canal, and has become naturalized in Lake Roland near Baltimore."

While "black bass" is the most universally used name throughout the fish's range, many local terms are used. In the Chesapeake region it is frequently called "chub" and in the South, "trout." The largemouth black bass is one of the most sought-after game fish in the United States, particularly throughout the Mississippi Valley, to the Atlantic seaboard, and even down to the lower reaches of streams flowing into the Gulf of Mexico, where very large fish frequently are taken. In St. Andrews Bay, northwestern Florida, we have taken the largemouth black bass in brackish water along with spotted squeteage (*Cynoscion nebulosus*) and blue crabs (*Callinectes*). In the Potomac and other tributary streams of the Chesapeake, as elsewhere, it is an important food and game fish, but it is not abundant enough in the brackish waters to be considered of commercial value in connection with the fisheries of the bay.

The following weights were obtained from Chesapeake Bay fish: Length 6¼ inches, 1.9 ounces (3 fish); 6½ inches, 2.3 ounces (6 fish); 7 inches, 2.7 ounces (1 fish); 7½ inches, 3.2 ounces (1 fish); 8 inches, 4.2 ounces (1 fish); 12½ inches, 17.3 ounces (1 fish).

Habitat.—Great Lakes to southern Florida and Mexico. Through the activities of the Federal and State fish commissions it has been introduced into nearly every State in the Union and also in Europe.

Chesapeake localities.—(*a*) Previous records: None definitely from brackish water. (*b*) Specimens in collection: Havre de Grace and Howells Point, Md., and Lewisetta, Va. Highest salinity, Lewisetta, August 6, 1921, 12.87 per mille.

Family LIX.—MORONIDÆ. The white basses

Body elongate, compressed; mouth large; teeth well developed, pointed, fixed; *maxillary without a supplemental bone;* scales of moderate size; dorsal fins separate. Two genera of important food fishes of this family are included in the fauna of Chesapeake Bay.

KEY TO THE GENERA

a. Body rather short and deep; jaws nearly equal; anal fin with III, 8 to 10 rays____Morone, p. 244
aa. Body more elongate; lower jaw projecting; anal fin with III, 10 to 12 rays_____Roccus, p. 247

102. Genus MORONE Mitchill. White perches

Body rather short and deep; jaws nearly equal; edge of tongue with linear patches of teeth; lower margin of opercle finely serrate; scales rather large; dorsal fins more or less connected by membranes, the spines stout; anal fin short, with three strong spines and with 8 to 10 soft rays.

133. Morone americana (Gmelin). White perch; "Blue-nosed perch"; "Gray perch"; Black perch.

Perca americana Gmelin, Linnæus's Syst. Nat., ed. XIII, vol. I, pt. 3, 1788, p. 1205; New York.
Morone americana Uhler and Lugger, 1876, ed. I, p. 127; ed. II, p. 108; Jordan and Evermann, 1896–1900, p. 1134, Pl. CLXXXI, fig. 479; Smith and Bean, 1899, p. 186; Evermann and Hildebrand, 1910, p. 161; Fowler, 1912, p. 55; Snyder 1917, pp. 18 and 28.
Roccus americanus Bean, 1883, p. 366.

Head 2.6 to 2.95; depth 2.6 to 3.2; D. IX–I, 12; A. III, 8 to 10; scales 7–48 or 49. Body rather deep, compressed; back elevated; head rather low; snout pointed, 3.6 to 4.2; eye 3.2 to 4.85; interorbital 4.6 to 5.65; mouth rather large, oblique, terminal or with the lower jaw slightly projecting; maxillary reaching about opposite anterior margin of pupil, 2.65 to 3.1 in head; teeth small, pointed, in bands on jaws, vomer, and palatines; opercle ending in two flat points; preopercular margin serrate; gill rakers rather slender, 13 or 14 on lower limb of first arch; scales strongly ctenoid, reduced scales extending on base of vertical fins, and forward on head to nostrils; dorsal fins separate, the spines large and strong; caudal fin slightly forked; anal fin with three strong spines, the soft part similar to that of the dorsal; ventral fins rather large, inserted a little behind base of pectorals; pectoral fins moderate, not reaching tips of ventrals, 1.4 to 1.7 in head.

Color variable, mostly silvery, often greenish to bluish and blackish above; sides sometimes brassy, frequently with irregular, dark longitudinal lines; fins often all more or less dusky; ventrals sometimes white; pectorals often plain. Large individuals often with bluish luster on head; young less than 4 or 5 inches in length usually silvery gray, never with blue on head.

Many specimens of this species, ranging from 55 to 280 millimeters (2⅛ to 11 inches) in length, were preserved. No individuals smaller than the ones preserved were seen in salt and brackish water. The plain coloration and the strong spines in the fins are among the distinguishing characters of this common and well-known fish. The young do not differ greatly from the adult. The white perch is essentially an anadromus species, but it is not infrequently landlocked in fresh water.

The food of this fish according to Smith (1907, p. 275) consists of minnows, shrimps, and other animals, and Bigelow and Welsh (1925, p. 258) mention small fish fry of all kinds, young squid, shrimps, crabs, and various other invertebrates, as well as the spawn of other fish. The contents of 130 stomachs of this fish taken in Chesapeake Bay consisted of fish, crustaceans of various kinds and sizes, annelids, and insect larvæ. A small amount of vegetable débris, too, frequently is present. The larger individuals had fed mainly on fish. Shrimp, Mysis, and annelids also are eaten by the larger fish. Young 4 inches and less in length had fed mainly on annelids, amphipods, isopods, copepods, and insect larvæ.

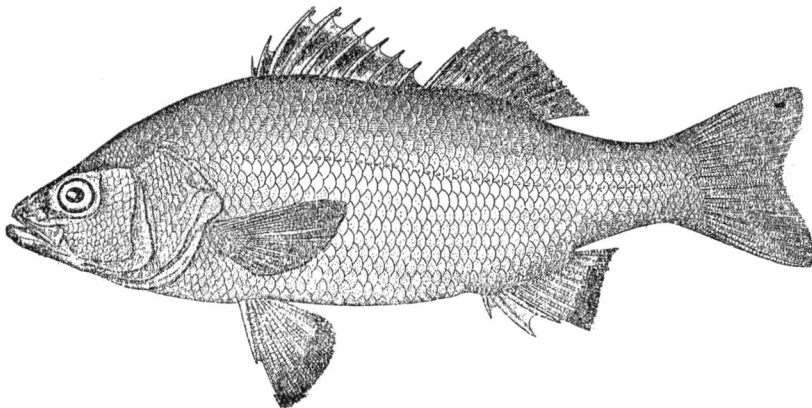

FIG. 138.—*Morone americana.* From a specimen 12¼ inches long

The white perch spawns from April to June, according to published accounts. Smith (1907, p. 275) gives a short season for the Albermarle, N. C., region, beginning between April 1 and 10 and lasting 10 days. Bigelow and Welsh (1925, p. 259) state that in southern New England the white perch breeds in April, May, and June. It has been learned through the present investigation that in Chesapeake Bay spawning occurs not only during April and May but that it may occur under certain conditions during December, for the *Fish Hawk* took 10 ripe males and 3 ripe females on December 9 and 10, 1915, at the following localities: Thomas Point Light, Sandy Point, and Sharps Inlet Light, in depths varying from 9½ to 21 fathoms. The ripe fish ranged in length from 5½ to 7¼ inches. It can not be concluded, however, that spawning takes place regularly at this season of the year, as no other ripe fish were taken in the winter months during subsequent collecting in the deeper holes of the bay. Further investigation relative to the winter spawning of this fish is highly desirable. Fish in spawning condition occur regularly in the shallow shore waters in April. The earliest date upon which ripe fish (two males, taken at Buckroe Beach, Va.) were taken is April 10, 1922.

According to Welsh (field notes) the spawning season was at its height at Havre de Grace on April 29, 1912. In this same locality many spawning fish were seined on May 8 to 10, 1922, in 3 or 4 feet of water (fresh) along the immediate shores. In one haul of a 300-foot seine, 600 males and 196 females 3 to 7 inches in length were taken. Nearly all the larger fish were ripe. In three other hauls in the same locality a preponderance of males also occurred, as a total of 91 males and 26 females was taken. The smallest ripe male in these lots of fish measured 109 millimeters (4⅛ inches), and the smallest ripe female measured 121 millimeters (4¾ inches) in length.

The eggs are small (0.73 millimeter in diameter) and hatch in six days at a water temperature of 51° to 53° F. At the time of hatching the fry measure 2.3 millimeters in length, reaching 3 millimeters at the end of 24 hours. (Ryder, 1887, pp. 518–519.) The eggs may be artificially stripped and fertilized; millions are hatched annually by fish-cultural stations situated on the streams tributary to Chesapeake Bay.

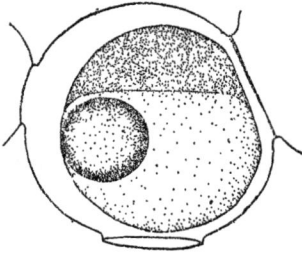

FIG. 139.—Egg recently fertilized

Virtually nothing is known of the rate of growth of the white perch, and our collections are inadequate and too erratic to permit the addition of any tangible information. Many fish taken in April ranged in length from 74 to 140 millimeters and may have been about 1 year old. The individuals of a collection made in the lower Rappahannock on July 25 ranged in length from 55 to 65 millimeters. These fish probably were the product of the last previous spawning season.

The white perch is one of the important food fishes of the Chesapeake. During 1920 it ranked eighth in quantity and seventh in value, the catch being 535,080 pounds, worth $51,914.

In Maryland the white perch ranked sixth in quantity and fifth in value, the catch being 316,915 pounds, worth $32,026. Of this amount, 42 per cent was caught in pound nets, 28 per cent in fyke nets, 21 per cent in haul seines, 6 per cent in gill nets, 3 per cent with lines, and 3 per cent with other apparatus. Dorchester, Cecil, Kent, and Baltimore Counties take first place in Maryland, with catches ranging from 53,400 to 41,000 pounds.

In Virginia it ranked twelfth in quantity and ninth in value, the catch being 218,165 pounds, worth $19,888. Of this amount, 32 per cent was taken in fyke nets, 28 per cent in pound nets, 24 per cent in haul seines, 9 per cent with gill nets, and 7 per cent with other apparatus. The largest catches are attributed to Norfolk and King George Counties, with 28,350 and 23,440 pounds, respectively. A catch of 420,000 pounds of white perch was made in Back Bay, Princess Anne County. This body of water, however, is removed from Chesapeake Bay and connects with the Atlantic Ocean, and therefore the catch is excluded from the statistics for the bay.

This fish is caught in all parts of Chesapeake Bay and its tributaries. It is commonest in brackish water, and the largest numbers are taken in the lower sections of tributary streams. Good catches, however, are also made in the spring, during the spawning season, far up various rivers where the water is always fresh. It is taken chiefly in the spring and the fall, from March until May and from September until November. A small number are caught in the winter with otter trawls or drift nets while fishing for striped

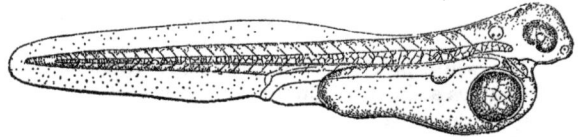

FIG. 140.—Larva, 6 days old, 8 millimeters long

bass. In the lower part of the bay, chiefly near the mouth of the Rappahannock River, a few pound nets fish all winter and report fair catches of white perch. In the upper sections of the bay good catches are made with pound nets during April and May. It was not infrequently taken by the *Fish Hawk* during the winter with the beam trawl in comparatively deep water. The greatest depth in which it was secured was 138 feet.

Various names are given to this fish in Chesapeake Bay, but the one in most general use, especially north of the York River, is "white perch." In the vicinity of Norfolk "blue-nosed perch" is the most common name, whereas in various other parts of the bay "gray perch" and "black perch" are used. These various common names lead to confusion, especially as the name "white perch" is used in the vicinity of Norfolk for *Bairdiella chrysura*, a species belonging to the croaker family. The name "white perch" is the most widely used of all the various names. It would seem advantageous, in order to prevent confusion, if this name alone were to be adopted throughout the Chesapeake region.

The white perch is a rather small fish, those seen in the markets usually weighing less than 1 pound. However, it is always in demand and is one of the favorite food fishes of the bay. The maximum weight attained by the species is 2 pounds, but in the Chesapeake it seldom exceeds 1½ pounds.

Habitat.—Nova Scotia to South Carolina.

Chesapeake localities.—(a) Previous records: From many localities, principally from the upper sections of the bay. (b) Specimens in collections: From many localities from Havre de Grace, Md., to Cape Charles and Cape Henry, Va.

Comparison of lengths and weights of white perch

Number of fish weighed and measured	Length	Weight	Number of fish weighed and measured	Length	Weight
	Inches	*Ounces*		*Inches*	*Ounces*
2	3¾	0.4	31	7½	3.6
4	4	.5	16	7¾	3.8
5	4¼	.6	25	8	4.4
10	4½	.7	17	8¼	4.7
8	4¾	.9	11	8½	5.2
14	5	1.0	8	8¾	5.7
25	5¼	1.1	5	9	6.2
26	5½	1.3	9	9¼	6.7
23	5¾	1.5	10	9½	7.9
32	6	1.7	3	9¾	8.8
36	6¼	2.0	3	10	9.1
35	6½	2.1	1	10¼	9.9
34	6¾	2.5	2	10¾	11.4
30	7	2.8	1	12¼	18.0
40	7¼	3.1			

103. Genus ROCCUS Mitchill. Striped basses or rockfishes

Body elongate, moderately compressed; head long, pointed; lower jaw projecting; base of tongue with patches of teeth; dorsal fins well separated, the first with about 9 or 10 spines; anal with 3 spines and 10 to 12 soft rays. Two species, one in the Great Lakes and Mississippi Basin and the other coastwise from New Brunswick to Alabama (and introduced on the Pacific), are known.

134. Roccus lineatus (Bloch). Striped bass; Rock; Rockfish.

Sciæna lineata Bloch, Ichthyol., IX, 1792, p. 53, Pl. CCCV; "Mediterranean Sea" (?).

Roccus lineatus Uhler and Lugger, 1876, ed. I, p. 126; ed. II, p. 107; Smith, 1892, p. 71; Jordan and Evermann, 1896–1900, p. 1132, Pl. CLXXX, fig. 478; Evermann and Hildebrand, 1910, p. 161; Fowler, 1912, p. 55; Snyder, 1919, p. 55.

Roccus saxatilis Bean, 1883, p. 365.

Head, 3.1 to 3.25; depth, 3.45 to 4.2; D. IX or X–I, 11 or 12; A. III, 10 or 11; scales, 7 or 8–60 to 67. Body elongate, compressed; head rather low and long; snout pointed, 3.3 to 4.15; eye, 3 to 4.9; interorbital, 3.75 to 5.4; mouth large, oblique; lower jaw projecting; maxillary broad, reaching middle of eye, 2.4 to 2.7 in head; teeth small, present in bands on the jaws, vomer, palatines, and in two parallel patches on tongue; preopercle serrate; gill rakers long, slender, 14 or 15 on lower limb of first arch; scales rather small, ctenoid, extending on the base of the vertical fins; dorsal fins well separated, the first with rather long stiff spines; caudal fin forked; anal fin with three rather strong graduated spines, the soft part similar to that of the dorsal, each with concave outer margin; ventral fins moderate, inserted a little behind base of pectorals; pectoral fins rather short, 1.8 to 2.1 in head.

Color in life of a 17-inch specimen, greenish above; silvery on sides and below, with a brassy luster, except on belly; sides with 7 or 8 prominent longitudinal black stripes, several above and below lateral line, one running along lateral line, those below decreasing in length; the longest stripes reach base of caudal, but none extend on head; dorsal, caudal, and anal dusky or black; ventrals white, slightly dusky; pectorals greenish. Some of the side stripes are sometimes interrupted, but in number they remain fairly constant. Young of less than about 60 millimeters in length usually without dark longitudinal stripes, sometimes with indications of dusky crossbars.

Numerous specimens of this common species, ranging in length from 30 to 360 millimeters (1¼ to 14¼ inches), were preserved, and larger ones were examined in the field. The dark longitudinal stripes on the sides from which it receives its scientific name, *lineatus*, at once distinguish this fish from all others in the Chesapeake. The young do not differ greatly from the adults, except in color, as stated in the description.

The rockfish is carnivorous, feeding on various kinds of animal life of suitable size. The contents of 48 stomachs taken from specimens caught in salt and brackish water of the bay consisted of fish, crustaceans, annelid worms, and insects. The larger fish had fed principally on fish, whereas the smaller ones had eaten mainly crustaceans. The young had fed on Mysis, Gammarus, annelids, and insects.

The striped bass is anadromous, coming in from the sea to brackish and fresh waters to spawn. Its chief spawning grounds perhaps are located in swift-running fresh-water streams. It ascends

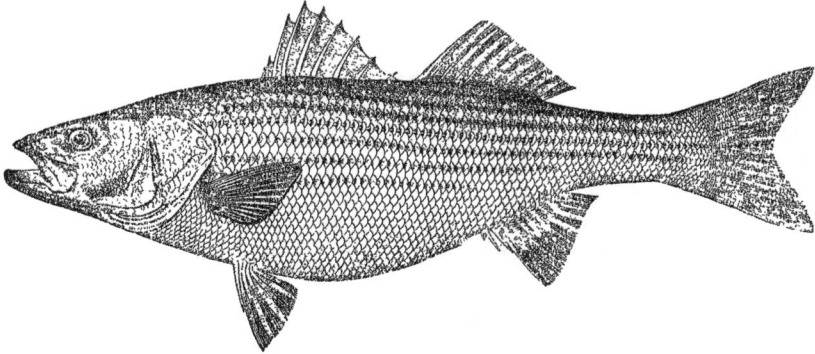

FIG. 141.—*Roccus lineatus*. From a specimen 21¼ inches long

the Potomac River to above Washington, where each spring it is taken among the rapids and bowlders. Smith (1907, p. 272) states that in the Roanoke River, N. C., the most important striped-bass spawning ground is in the vicinity of Weldon, where the river falls 50 feet in about 6 miles. "In these rapids, where the muddy current is exceedingly strong and rendered very erratic by islands, bowlders, and rocks, the fish spawns * * *." Bean (1903, p. 527) states that spawning takes place either in the rivers or in the brackish waters of bays and sounds. In North Carolina spawning takes place from late April to early May; in the Chesapeake region most of the spawning occurs in May; while in the Gulf of Maine (Bigelow and Welsh, 1925, p. 256) the chief spawning season apparently occurs in June.

The eggs of the striped bass may be incubated in shad jars, the young emerging in 48 hours in a water temperature of 67°, or in 36 hours at 70° F. After fertilization the eggs increase greatly in

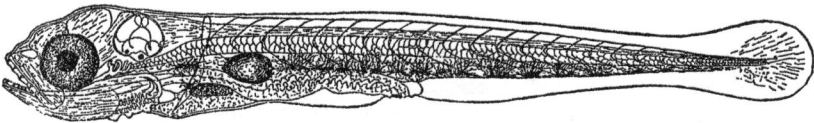

FIG. 142.—*Roccus lineatus*. Larva 5 millimeters long

size. Smith (1907, p. 272) says that in 1903 S. G. Worth, while conducting hatching operations on the Roanoke River, N. C., stripped from a 20-pound fish a mass of eggs that after fertilization and immersion in water measured 60 quarts and contained 1,500,000 eggs, on the basis of 25,000 eggs to a quart. According to the same author, who quotes Mr. Worth, a single female may be surrounded by many small males on the spawning grounds, and severe fights among the males take place.

The growth of the striped bass appears to be variable, and it is difficult to determine age by length frequencies alone. Bean (1903, p. 527) records young 1 inch in length taken the second week of June in the Delaware, and he states that some of these had grown to a length of 4½ inches by mid-October. In the Chesapeake the smallest fish, 30 millimeters (1⅕ inches) in length, were taken on

June 23. In July fish ranging from 45 to 53 millimeters (1⅘ to 2 1/12 inches) were taken, and in August large numbers, measuring 50 to 70 millimeters (2 to 2⅘ inches), were collected. After this date the sizes vary so greatly that in some cases it is difficult to determine whether a fish is in its first or second year. Specimens taken by the *Fish Hawk* during the winter months appear to give no clue to the rate of growth. Eleven fish taken in April and May range quite gradually from 3¾ to 8½ inches in length, and it seems probable that they are fish about 1-year old.

The striped bass, or rockfish, is one of the most valuable and esteemed fish caught in Chesapeake Bay. During 1920 it ranked fifth, both in quantity and value, the catch being 1,410,630 pounds, worth $261,918. In Maryland it ranked fourth in quantity and second in value, the catch being 1,040,274 pounds, worth $193,295. Of this amount, 45 per cent was caught in pound nets, 27 per cent in gill nets, 21 per cent in haul seines, 4 per cent in purse seines, and 3 per cent with other apparatus. The counties taking the largest quantities were Kent, 459,475 pounds; St. Marys, 101,645 pounds; Cecil, 85,105 pounds; and Dorchester, 83,151 pounds.

In Virginia it ranked eighth in quantity and fifth in value, the catch being 370,356 pounds, worth $68,623. Of this amount, 55 per cent was caught in pound nets, 18 per cent in fyke nets, 18 per cent with haul seines, 7 per cent with gill nets, and 2 per cent with lines. The counties with the largest quantities were Northumberland, 49,330 pounds; Westmoreland, 45,355, pounds; Richmond, 35,200 pounds, Warwick, 33,500 pounds; and Lancaster, 33,025 pounds.

This fish is caught in all parts of Chesapeake Bay and its larger tributaries during the entire year, but it is taken in greatest numbers in the spring and fall. During the winter a special fishery is pursued by means of drift nets, which are gill nets that fish the bottom and drift with the tide. This winter fishery is confined to the region between Swan Point and Bloody Point, Md., where large numbers of striped bass apparently are present in the deeper channels during the winter months.

In certain localities this fish assumes first rank, and because of the prolonged season during which it may be caught it affords various fishermen a constant source of income. In many parts of the bay, notably the lower York, Rappahannock, and Potomac Rivers, this fish is taken in small pound nets, which may be operated by one man. A small catch taken in this manner often is profitable.

Haul seining is carried on during the late summer and fall and is confined principally to the lower parts of the larger rivers. The haul-seine fishermen usually operate at night and select localities where the bottom is free of débris and that have a proper depth and a good landing place for the seine. Such places are known as "hauls." The striped bass is a very elusive fish and great care must be exercised in selecting the "haul" and in operating and landing the net in order to make a profitable catch. The quantity caught fluctuates greatly; it is not unusual for a fishing crew to catch only 2 or 3 fish during one night and as many as 500 or even more during the following night.

The striped bass commands a good price in the markets, the fishermen receiving from 16 to 24 cents a pound during 1922. Four trade names are used in the Baltimore wholesale market. "Shinie rock" are small fish; "hank rock" are fish weighing 3 pounds; "boilers," 3 to 6 pounds; "big rock" 6 pounds or more. "Big rock" command a slightly lower price than the other sizes, and very large fish, 20 pounds or more in weight, are worth considerably less. The great majority of the fish sent to market weigh less than 15 pounds, but large ones, weighing 50 pounds or more, are sometimes seen. The maximum weight recorded for this species appears to be 125 pounds.

The striped bass is always in demand. It bears shipment well and is considered one of the best of all the salt-water fishes. The names most used in the Chesapeake for this species are "rock" and "rockfish."

Habitat.—Atlantic and Gulf seaboards, ascending streams from the Gulf of St. Lawrence to Alabama; most numerous between Massachusetts and North Carolina. Introduced on the Pacific; now common and an important food fish in California.

Chesapeake localities.—(a) Previous records: Many localities, from Chesapeake Bay and streams tributary to it. (b) Specimens in collection: From many localities in fresh, brackish, and salt water, from Havre de Grace, Md., to Cape Charles and Cape Henry, Va. Many small fish were taken during the winter by the *Fish Hawk* with the beam trawl in water having a depth as great as 138 feet.

Comparison of weights and lengths of striped bass

Number of fish weighed and measured	Length	Weight	Number of fish weighed and measured	Length	Weight	
	Inches	*Ounces*		*Inches*	*Pounds*	*Ounces*
1	4½	0.5	2	10¾		8.3
1	4¾	.6	4	11		10.0
1	5	.7	4	11¼		8.2
1	6	1.3	1	11½		8.3
1	6¼	1.4	2	12¼		12.0
1	6½	1.6	3	13		14.2
2	7	1.8	1	15	1	7.0
2	7¼	2.0	1	17	1	14.2
5	7½	2.5	1	17½	1	15.2
1	7¾	2.8	1	18¼	2	9.2
8	8	2.9	1	19	2	10.2
2	8¼	3.7	1	19½	2	14.1
4	8½	3.8	1	20	4	----
2	9	4.7	1	21	3	7.4
4	9½	5.2	1	22¼	4	11.3
2	9¾	4.7	2	23	4	11.8
3	10	6.1	1	23¾	4	----
2	10¼	7.2	1	24½	5	2.1
			1	28	8	8.0

Family LX.—EPINEPHELIDÆ. The groupers

Body oblong, more or less compressed; teeth in jaws usually depressible; canine teeth more or less distinct; *maxillary with a supplemental bone;* scales small, firm, commonly extending on top of head; dorsal rays about VIII to XIV, 12 to 20; anal III, 7 to 12; ventral fins inserted slightly behind base of pectorals.

104. Genus MYCTEROPERCA Gill. Groupers

Head broad and transversely concave between the eyes; lateral crests of cranium strong, nearly parallel with the supraoccipital crest and extending farther forward, joining the supraoccipital crest above the eye; supraoccipital crest not extending on the frontals; lower jaw strongly projecting; scales small, mostly cycloid; anal fin rather long, rarely with 9 or 10, usually with 11 or 12, soft rays; spines of fins slender, none of them much elevated; caudal fin lunate.

135. Mycteroperca microlepis (Goode and Bean). Gag.

Trisotropis microlepis Goode and Bean, Proc., U. S. Nat. Mus., 1879, p. 141; west coast of Florida.
Mycteroperca microlepis Jordan and Evermann, 1896-1900, p. 1177, Pl. CLXXXVIII, fig. 494; Evermann and Hildebrand, 1910, p. 161.

Head 2.6; depth 3.5; D. XI, 16 to 19; A. III, 11; scales 140 to 145. Body elongate, compressed; head pointed; mouth large; maxillary reaching beyond eye; teeth in narrow bands, two canines in front in each jaw, the lower ones smaller; dorsal spines slender, the third and fourth spines longest; caudal fin with concave margin, pectoral fins reaching beyond ventrals, 2 in head.

Color variable; usually brownish gray above, paler below, with faint traces of darker spots; black mustache; dorsal dark green, edge of soft dorsal black; caudal black with blue shades, edge white; anal indigo blue with white edge; ventrals black, first ray white-tipped; pectorals green.

No specimens of this species were taken. The foregoing description was compiled from published accounts.

The gag is a common food fish in Florida. A maximum weight of 50 pounds has been reported, but market fish seldom exceed 10 pounds. It is known from Chesapeake Bay only from a single small specimen 140 millimeters (5½ inches) in length. (Evermann and Hildebrand, 1910, p. 161.)

Habitat.—Both coasts of Florida, northward to Chesapeake Bay.

Chesapeake localities.—(a) Previous record: Old Point Comfort, Va. (b) Specimens in collection: None.

Family LXI.—SERRANIDÆ. The sea basses

Body rather robust, compressed; teeth in jaws usually well developed, not depressible; *maxillary without a supplemental bone;* scales moderate or rather large. A single genus of this family is included in the fauna of Chesapeake Bay.

105. Genus CENTROPRISTES Cuvier. Sea basses

Body robust, slightly compressed; mouth large; maxillary without a supplemental bone; canines small; no teeth on tongue; preopercular margin serrate, the lower teeth somewhat antrorse; gill rakers rather long and slender; dorsal fin short, the spines with fleshy filaments at tips; caudal fin round or slightly double concave; anal rays III, 7; ventrals close together, inserted under or slightly in advance of pectorals. A single species is included in the fauna of Chesapeake Bay.

136. Centropristes striatus (Linnæus). Blackfish; Sea bass; "Black Will."

Labrus striatus Linnæus, Syst. Nat., ed. X, 1758, p. 285; "America."
Centropristis striatus Bean, 1891, p. 91; Jordan and Evermann, 1896–1900, p. 1199, Pl. CXC, fig. 500; Evermann and Hildebrand, 1910, p. 161.

Head 2.5 to 2.65; depth 2.4 to 2.95; D. X, 11; A. III, 7; scales 48 to 50. Body elongate, moderately compressed; back elevated; head rather thick; snout moderately pointed 3.35 to

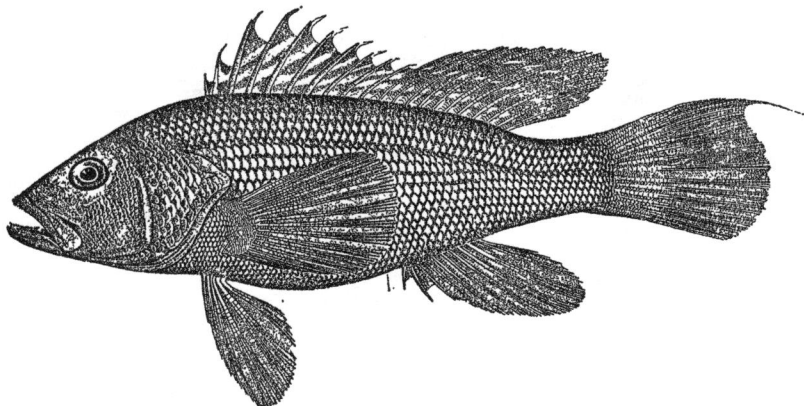

Fig 143.—*Centropristes striatus.* From a specimen 11¼ inches long

4.35 in head; eye 3 to 4.9; interorbital 6.65 to 9.75; mouth large, oblique; maxillary reaching about opposite middle of eye, 2.3 to 2.45 in head; teeth pointed, in bands on jaws, vomer, and palatines, no distinct canines; preopercular margin finely serrate; gill rakers scarcely longer than pupil, 17 or 18 on lower limb of first arch; scales moderate, ctenoid, reduced on head and cheeks, extending somewhat on the base of fins; lateral line complete, following the curvature of the back; dorsal fin continuous, the spines strong, the soft part elevated and notably higher than the spines in the adult; caudal fin round, large individuals with one of the upper rays produced; anal fin with three strong graduated spines, the soft rays very long in the adult; ventral fins moderate, inserted under base of pectorals; pectoral fins long, reaching beyond tips of ventrals, 1.35 to 1.45 in head.

Color of fish in the aquarium at Woods Hole, Mass., weighing from ¾ to 3 pounds, blue-black; centers of scales pale blue or white, forming longitudinal stripes along back and sides; several bluish streaks in front of or below eye present or absent; dorsal blue-black, with several pale stripes extending along both spinous and soft part, the stripes more numerous on the latter; caudal dusky or dark blue, streaked or mottled with pale markings; anal almost entirely pale or streaked with dark blue; ventrals bluish gray, the rays pale; pectorals grayish. Females are lighter than males, usually gray-blue instead of blue-black. The male develops an adipose hump on the nape, which in the breeding season is often bright blue; this hump sometimes evident in males only 12 inches

long. Large males frequently are colored bright blue between the eyes. Color of Chesapeake
Bay fish about 6 inches long: Blackish; centers of scales pale blue or white, often forming light
longitudinal streaks; a blue streak on border of lower outline of orbit; other blue streaks branching
from the first one; dorsal reddish or bronze, with white spots forming 3 longitudinal stripes on
spinous dorsal and 4 or 5 on soft dorsal; caudal with reddish bars at base, distally dusky, the lower
rays pale or reddish; anal and ventrals bluish white or dusky; pectorals plain, sometimes tinged
with yellowish brown. Young 2 to 3 inches in length brownish above, a dark brown or black
lateral stripe from eye to base of caudal; frequently with indefinite dark crossbars on sides; brick-
red markings below eye; spinous dorsal dusky, sometimes with a dark spot at base of posterior
spines; soft dorsal with brick-red spots forming three or four rows; caudal and anal with reddish
markings; ventrals plain; pectorals salmon.

Numerous specimens of this species, ranging from 60 to 225 millimeters (2⅜ to 9 inches) in
length, were preserved. This fish is recognized by its dark color, nonfilamentous spines of the
dorsal, and the round caudal with only one of the upper rays somewhat produced in the adult.
The young, as indicated in the description, differ rather prominently from the adults in color.
They also differ in having the soft parts of the dorsal and anal fins proportionately much lower,
and the caudal fin round, without a produced upper ray.

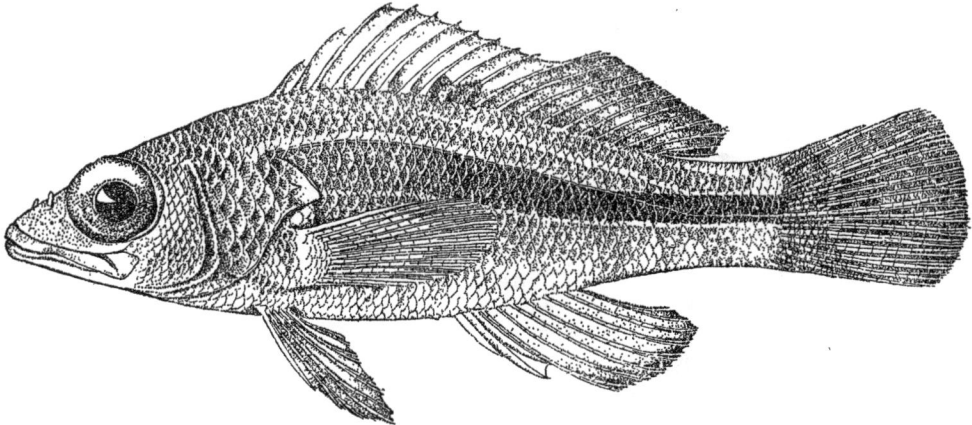

FIG. 144.—*Centropristes striatus.* Young, 58 millimeters long

The food of this fish in Chesapeake Bay, according to the contents of 19 stomachs, consists of
crustaceans, fish, mollusks, and plants, named in the order of their apparent importance. Adults
had fed chiefly on crabs and fish, and the young on shrimp, isopods, and amphipods.

Spawning occurs in May on the North Carolina coast, probably late in May near the mouth
of the Chesapeake, and from the middle of May until the end of June off the New Jersey, Long
Island, and southern New England coasts. The eggs are pelagic, about 1 millimeter in diameter,
and hatch in about 75 hours at a temperature of 60° F. (Wilson, 1891, p. 210.)

The sea bass is usually most common on rocky and coral bottom and around the piling of
wharves, etc. In several States it supports special fisheries. It is a voracious feeder and takes the
hook readily, being taken in commercial quantities chiefly with hook and line. In Chesapeake
Bay, however, it is of small commercial importance. During 1920 it ranked twenty-fifth in quantity and twenty-third in value, the catch being 5,100 pounds, worth $492. The entire catch is
credited to Virginia, and in this State the sea bass ranked the same as for the entire Chesapeake
region, namely, twenty-fifth in quantity and twenty-third in value. Eighty per cent of the fish
were caught with hand lines and 20 per cent in pound nets.

This species is taken only in the southern parts of the bay, the northernmost pound net for which
records were secured being located at Solomons, Md., where small numbers are taken throughout
the summer. A few fish are caught with hook and line in upper Tangier Sound, but the annual
catch for Maryland is probably not more than a few hundred pounds. It is most common near the

entrance of the bay, and the principal catches are made in the vicinity of Cape Charles, Ocean View, and Buckroe Beach, Va. The season extends from May until October, with no definite period when they are especially abundant.

The chief sea-bass grounds are off the coasts of North Carolina, Delaware, New Jersey, and western Long Island. To the eastward the fish are more scattered, but they are of some commercial importance in the following localities: Amagansett and Montauk, Long Island, off Block Island, and in the vicinity of Buzzards Bay. The sea bass is of chief importance along the New Jersey coast, where it is taken from May to November, not only by commercial fishermen but by a large number of anglers from the vicinity of New York. The magnitude of this sport fishing during the summer is remarkable, for no less than 100 seagoing power boats of various sizes carry as many as 4,000 or more persons daily to the fishing banks.

The sea bass is a well-flavored fish and finds ready sale in the Norfolk, Va., markets, where most of the Chesapeake catch is sold. The largest fish of which we have record weighed 7½ pounds and was caught on the Cholera Banks, off Long Island, on July 4, 1913. Examples weighing more than 6 pounds are rare, but 3 to 5 pound fish are rather common along the New Jersey coast during the summer. Its size in Chesapeake Bay, however, seldom exceeds one-half pound.

Habitat.—Massachusetts to northern Florida; rarely northward to Maine.

Chesapeake localities.—(*a*) Previous records: Cape Charles city, Norfolk, and Cape Henry. (*b*) Specimens in collection: From many localities from the southern part of the bay from Solomons, Md., to Cape Charles and Lynnhaven Roads, Va.

Family LXII.—PRIACANTHIDÆ. The big-eyes

Body oblong, compressed; head deep; snout short; eye very large; mouth rather large, very oblique to nearly vertical; teeth pointed, in bands on jaws, vomer, and palatines; preopercular margin serrate; pseudobranchiæ large; branchiostegals 6; lateral line continuous, not extending on caudal; scales small, firm, ctenoid, extending forward on head; dorsal fin continuous, with about 10 spines; anal fin with 3 spines; ventral fins thoracic, with I, 5 rays. Two genera of this family of tropical fishes occur as stragglers in Chesapeake Bay.

KEY TO THE GENERA

a. Scales small, 80 to 100 in lateral series; body elongate, the depth less than half the length; soft dorsal and anal, each with 12 to 15 rays_____Priacanthus, p. 253

aa. Scales larger, 35 to 50 in a lateral series; body deep, the depth about half the length; soft dorsal and anal, each with 9 to 11 rays_____Pseudopriacanthus, p. 254

106. Genus PRIACANTHUS Cuvier. Big-eyes

Body oblong, the depth less than half the length; preopercle with a well-developed, flat spine at angle; lateral line extending strongly upward and backward from upper angle of gill opening to anterior dorsal spine, then following curvature of back; scales small, 80 to 100 in a lateral series; dorsal with X, 13 or 14 rays; anal III, 13 to 15.

137. Priacanthus arenatus Cuvier and Valenciennes. Big-eye

Priacanthus arenatus Cuvier and Valenciennes, Hist. Nat. Poiss., III, 1829, p. 97; Brazil. Jordan and Evermann, 1896–1900, p. 1237, Pl. CXCV, fig. 511.

Head 3.2; depth 2.6; D. X, 14; A. III, 15; scales 98. Body elongate, rather strongly compressed; ventral outline anteriorly much more strongly convex than the dorsal; head deep; snout short, 3.85 in head; eye very large, 2.1; interorbital 5.1; mouth moderate, nearly vertical; lower jaw projecting; maxillary broad, reaching only a little past anterior margin of eye, 1.85 in head; teeth small, pointed, in narrow bands on jaws, vomer, and palatines; preopercular margin finely serrate, the angle produced into a short, flat, serrated spine; opercle with an indentation slightly above and behind preopercular spine; gill rakers long and slender, 21 on lower limb of first arch; scales small, ctenoid; dorsal fin continuous, the spines slender, pungent, the soft part not much higher than the spines; caudal fin with slightly concave margin; anal fin with three slender graduated spines, the soft part similar to that of dorsal; ventral fins very long, reaching beyond origin of anal, inserted under base of pectorals, the inner ray attached to the abdomen by membrane; pectoral fins very short, 1.7 in head.

Color in life bright red; brownish silvery in spirits; ventrals and anterior rays of anal whitish, tipped with black; fins otherwise mostly plain translucent.

A single specimen, 185 millimeters (7⅜ inches) in length, was secured. This specimen was saved by fishermen at Buckroe Beach as a curiosity, although they stated that the species had been seen previously by them. It is readily recognized by the bright red color, very large eye, nearly vertical mouth, and small scales.

Habitat.—Massachusetts southward to Brazil; chiefly from tropical waters, probably drifting northward during the summer in the Gulf Stream.

Chesapeake localities.—(*a*) Previous records: None. (*b*) Specimen in collection: From Buckroe Beach, Va., captured October 5, 1921.

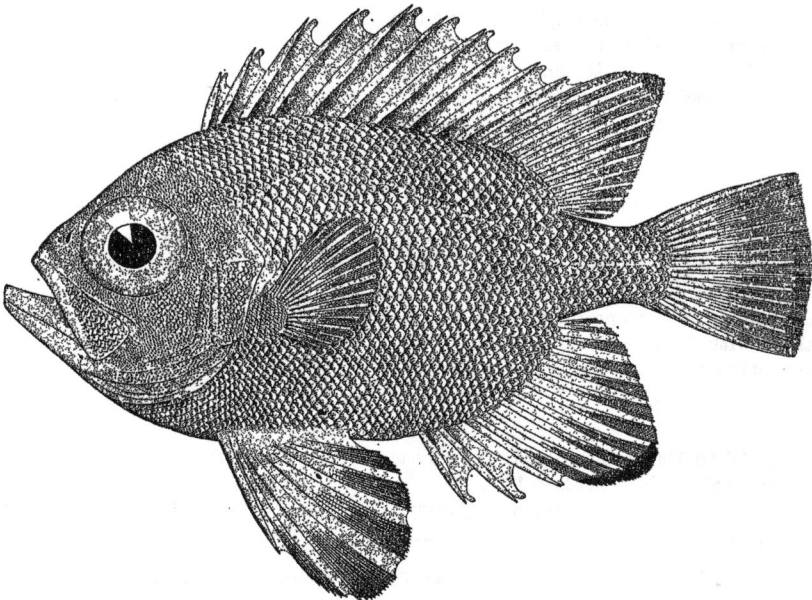

FIG. 145.—*Pseudopriacanthus altus.* From a specimen 5⅓ inches long

107. Genus PSEUDOPRIACANTHUS Bleeker. Short big-eye

This genus differs from Priacanthus principally in the larger scales, 25 to 50 in lateral series; deeper body, the depth about half the length; and the shorter dorsal and anal fins, dorsal rays X, 11, anal rays III, 9 to 11.

138. Pseudopriacanthus altus (Gill). Short big-eye.

Priacanthus altus Gill, Proc., Ac. Nat. Sci., Phila., 1862, p. 132; Narragansett Bay, R. I.
Pseudopriacanthus altus Jordan and Evermann, 1896–1900, p. 1239, Pl. CXCV, fig. 512.

Head 2.45; depth 1.7; D. X, 11; A. III, 10; scales 47. Body short and deep; head short; snout very short, 3.85 in head; eye large, 2.1; interorbital 5.1; mouth rather large, nearly vertical; maxillary reaching anterior margin of pupil, 1.85 in head; teeth pointed, in narrow bands on jaws, vomer, and palatines, the outer series in jaws slightly enlarged; preorbital very narrow, serrate; preopercular margin serrate, with two slightly enlarged spines at angle; gill rakers slender, about 20 on lower limb of first arch; scales strongly ctenoid, reduced on head; present on cheeks and maxillary; dorsal fin long, continuous, its origin slightly posterior to eye, the spines strong, the soft part somewhat higher than the spines; caudal fin round; anal fin with three graduated spines, the soft part similar to that of dorsal; ventral fins long, reaching beyond origin of anal, inserted under base of pectorals; pectoral fins short, 2 in head.

Color in life red; dorsal red, the spinous part edged with yellow, a few blackish dots on the soft rays; caudal fin pale, with blackish reticulations; anal red, edged with black; ventrals red at base, the rest of fins dusky or black; pectorals plain red. Color in spirits light brownish, with the dark marking on fins remaining as in the fresh specimen.

A single specimen 70 millimeters (2¾ inches) in length was taken, and it forms the basis for the foregoing description. The species is readily recognized by the deep body, very short snout, vertical mouth, large eye, moderately large scales, and the red color. The small specimen at hand differs somewhat from described specimens in having two enlarged spines at angle of preopercular margin, instead of having no spines at this place. It also differs in the much longer ventral fins and in the very deep body. These differences, however, may all be due to age.

This fish is principally a West Indian species that sometimes strays northward and occasionally is taken in considerable numbers on the coast of Massachusetts. Apparently it is very rare in Chesapeake Bay, and it was unknown to the fishermen who captured the specimen in hand. The largest individual of this species recorded was only 11 inches long.

Habitat.—West Indies to Massachusetts; occurring northward only as a straggler.

Chesapeake localities.—(a) Previous records: None. (b) Specimen in collection: From Ocean View, Va., captured September 26, 1922.

Family LXIII.—LOBOTIDÆ. The tripletails

Body oblong, compressed; back elevated; anterior profile more or less concave; head moderate; snout short; eye small, anteriorly placed; mouth moderate, oblique; lower jaw projecting; teeth in the jaws pointed, small, none on vomer and palatines; preopercle serrate; scales of moderate size, rather strongly ctenoid; dorsal fin long, continuous, with 12 strong spines, the soft part elevated; caudal fin rounded; anal fin with three graduated spines, the soft part similar to that of dorsal and opposite it. This family consists of a single genus.

108. Genus LOBOTES Cuvier. Tripletails

The characters of the genus are included in the family description. A single species of wide distribution occurs on the Atlantic coast of the Americas and is not uncommon in Chesapeake Bay.

139. Lobotes surinamensis (Bloch). Tripletail; Flasher; "Lumpfish"; "Strawberry bass."

Holocentrus surinamensis Bloch, Naturg. Ausl. Fische, IV, 1790, p. 98, Pl. CCLXIII; Surinam.
Lobotes surinamensis Uhler and Lugger, 1876, ed. I, p. 135; ed. II, p. 115; Jordan and Evermann, 1896–1900, p. 1235, Pl. CXCIV, fig. 510; Fowler, 1912, p. 58.

Head 2.9; depth 2.05; D. XII, 16; A. III, 12; scales 48. Body deep, compressed; back elevated; anterior profile concave over the eyes; head moderate; snout tapering, 4.15 in head; eye 5.05; interorbital 3.45; mouth moderate, oblique; lower jaw projecting; maxillary reaching middle of eye, 2.65 in head; teeth in jaws small, pointed; preopercular margin strongly serrate, the serræ at angle much enlarged, longer than pupil; scales moderate, strongly ctenoid, extending more or less on the base of all the soft fins; dorsal with strong spines, the soft part much higher than the spines; caudal fin round; anal fin with three strong, graduated spines, the soft part shorter but similar in shape to soft dorsal; ventral fins long, reaching vent, inserted slightly behind base of pectorals; pectoral fins short, 1.95 in head.

Color brownish black, with darker blotches below base of dorsal and anal; pectoral fins pale; other fins all brownish to blackish; caudal with a broad, pale margin. (The pale pectoral and pale margin of the caudal are probably characteristic of young fish only.) The color becomes darker after death. Large fish examined by us in Norfolk fish markets were black everywhere on body and fins. Young fish are sometimes marked with yellow and brown. A specimen observed at Key West had the yellow and brown colors of an autumn leaf.

A single specimen, 175 millimeters (6⅞ inches) long, was preserved, and it forms the basis for the foregoing description. This species is easily recognized by the deep, compressed body and the high, soft rays in the dorsal and anal fins, giving the fish the appearance of having three tails.

The lumpfish is taken in very limited numbers in the lower parts of the bay—that is, at Cape Charles and Lynnhaven Roads to Buckroe Beach, Va. No records of the yearly catch of this species are available. It is estimated, however, that the catch for 1922 did not exceed 1,000 pounds, all taken in pound nets.

A few fish are caught throughout the summer and fall, September and October yielding the largest number. Virtually the entire catch is marketed in Norfolk, where the species is known either as lumpfish or strawberry bass. The size of the fish observed in the market ranged from 5 to 25 pounds. A fish 738 millimeters (29 inches) long weighed 25 pounds.

Although widely distributed, this fish is not abundant anywhere. It is said to attain a maximum length of 3 feet.

Habitat.—Massachusetts to Uruguay.

Chesapeake localities.—(a) Previous records: "Occasionally caught in the lower part of the Chesapeake Bay" (Uhler and Lugger, 1876); observed in the Norfolk fish markets (Fowler, 1912). (b) Specimen in collection: Lynnhaven Roads, Va.; also observed at Cape Charles, Norfolk, Buckroe Beach, and Ocean View, Va.

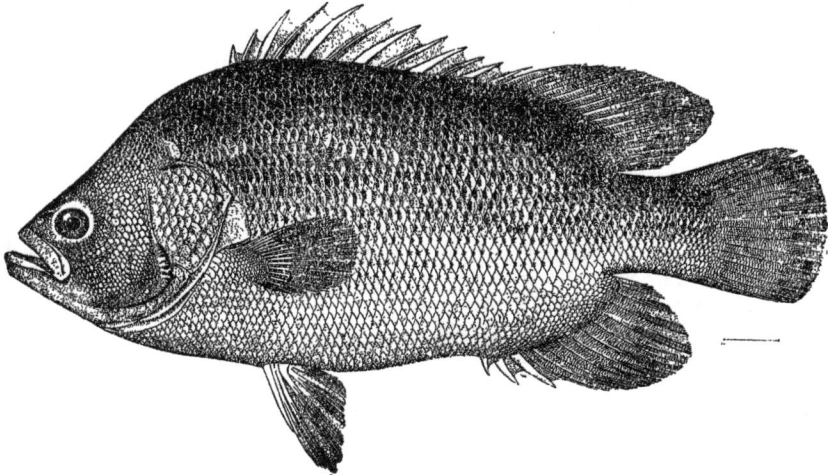

Fig. 146.—*Lobotes surinamensis*

Family LXIV.—LUTIANIDÆ. The snappers

Body elongate, compressed; head rather large; mouth usually large, terminal or with the lower jaw slightly projecting; teeth rather strong, present on jaws and usually on vomer, palatines, and tongue; premaxillaries protractile; maxillary long, without a supplemental bone; nostrils rather close together, neither with a tube; opercles without spines; gills 4; gill membranes free from the isthmus; pseudobranchiæ large; lateral line present; scales ctenoid, adherent; air bladder present; dorsal single or double, with 10 to 12 strong spines; caudal fin merely emarginate to deeply forked; ventral fins thoracic, with I, 5 rays.

109. Genus LUTIANUS Bloch. Snappers

Body elongate, compressed; back more or less elevated; head long; snout pointed; mouth large; jaws with bands of teeth, the outer ones usually enlarged, the upper jaw usually with two to four canines anteriorly; pointed teeth also present on vomer, palatines, and tongue; preopercular margin serrate; gill rakers rather few; scales ctenoid, wanting on head or present only at nape; soft dorsal and anal scaly at base; dorsal spines 10 or 11, not separated from the soft rays; caudal fin emarginate or slightly forked; anal fin with III, 7 to 9 rays.

140. Lutianus griseus (Linnæus).   Gray snapper; Mangrove snapper.

Labrus griseus Linnæus, Syst. Nat., ed. X, 1758, p. 283; Bahamas.

Lutjanus caxis Bean, 1891, p. 91.

Neomænis griseus Jordan and Evermann, 1896-1900, p. 1255; Evermann and Hildebrand, 1910, p. 161.

Head 2.6 to 2.65; depth 2.35 to 2.6; D. X, 14; A. III, 8; scales 48 to 51.   Body elongate; back moderately elevated; profile straight or slightly concave over snout; head moderate; snout rather pointed, 3.25 to 3.4 in head; eye 3.7 to 4.25; interorbital 5.4 to 5.8; mouth large, terminal; maxillary reaching to or a little beyond anterior margin of pupil, 2.5 to 2.55 in head; teeth present on jaws, vomer, palatines, and tongue; upper jaw with one or two pairs of canines; lower jaw with the outer series enlarged; vomerine teeth in an anchor-shaped patch, with a definite, median, backward projection; preopercular margin finely serrate; gill rakers rather few and short, eight or nine on lower limb of first arch; scales of moderate size, present on base of soft part of vertical fins, the rows above lateral line not parallel with it; dorsal fin continuous, without definite notch, the soft part higher than the spines; caudal fin concave, the upper lobe longest; anal fin with three spines, the second and third of about equal length, the soft part shorter but similar to that of dorsal; ventral fins moderate, inserted a little behind base of pectorals; pectoral fins rather short, 1.35 to 1.5 in head.

Color in alcohol dark brown above, becoming pale underneath; rows of scales on sides marked by definite dark longitudinal stripes; fins light brown to pale; margin of membranes of spinous dorsal black; soft dorsal and base of caudal with small brown spots.   The color in life, according to published accounts, is dark green above, coppery red below; spinous dorsal dusky, with red margin; soft dorsal dusky with white edge anteriorly; caudal olivaceous or reddish black; anal reddish with white margin; ventrals pale or whitish, sometimes with faint red markings; pectorals pale.

This fish is able to change its color to agree with its surroundings.   In southern Florida, where we have studied its habits, it was noted that at times fish became pale gray above with light red markings below, while again the color might be dark gray-green above and on sides, with bright red predominating on lower part of body.   The longitudinal stripes are most prominent when the fish is darkest.

Six small specimens, ranging in length from 105 to 111 millimeters (4⅛ to 4⅜ inches), were secured.   This is the only snapper taken in Chesapeake Bay.   It is characterized by the absence of both a black spot on the sides and crossbars and by the presence of dark longitudinal stripes along the rows of scales, which run obliquely upward and backward above the lateral line and are not parallel with it.

The gray snapper is a food fish of importance in southern Florida, the West Indies, and Panama. It occurs as a straggler in the southern part of Chesapeake Bay, where only the young are taken. One specimen was caught 2 miles from the mouth of a small creek, the east branch of the Carrotman River, a most unusual locality for a fish that enters Chesapeake Bay as a straggler.   The species ascends streams and not infrequently is taken in brackish water.   It is reported to reach a maximum weight of 18 pounds.

Habitat.—Massachusetts to Brazil; occurring north of Florida only as a straggler.

Chesapeake localities.—(*a*). Previous records: Cape Charles city, Old Point Comfort, and Ocean View, Va.   (*b*) Specimens in collection: Lower Rappahannock River, lower York River, Cape Charles, Buckroe Beach, and Lynnhaven Roads, Va.

Family LXV.—POMADASIDÆ.   The grunts

Body more or less elongate, compressed; back usually elevated; head rather large; snout pointed or blunt; mouth usually terminal, large or small, low and more or less horizontal; premaxillaries protractile; maxillary without a supplemental bone, slipping under preorbital; teeth in jaws only, pointed or conical, no canines; preopercle usually serrate; gills 4, a slit behind the fourth; lateral line concurrent with the back, usually not extending on caudal fin; scales moderate, firm, ctenoid; dorsal fin long, with 10 to 14 rather strong spines, depressible in a groove; caudal fin more or less concave; anal fin with three spines, the soft part similar to that of dorsal; ventral fins thoracic, with I, 5 rays. The fishes of this family are chiefly from tropical waters.   Only one species is common and of commercial importance in Chesapeake Bay.

110. Genus ORTHOPRISTIS Girard. Pigfishes

Body moderately elongate, compressed; back elevated; head rather deep; snout usually long;
mouth small, low; teeth in the jaws small, pointed, in bands; preopercle usually finely serrate; scales
small, series above lateral line not parallel with it; dorsal fin long, not deeply notched, the spines
rather slender, usual number of rays XII or XIII, 12 to 15; caudal fin not deeply forked, lunate;
anal spines 3, notably shorter than those of dorsal. A single species is included in the fauna of
Chesapeake Bay.

141. Orthopristis chrysopterus (Linnæus). Pigfish; Hogfish.

Perca chrysoptera Linnæus, Syst. Nat., ed. XII, 1766, p. 485; Charleston.
Orthopristis fulvomaculatus Uhler and Lugger, 1876, ed. I, p. 124; ed. II, p. 106.
Orthopristis chrysopterus Bean, 1891, p. 90; Jordan and Evermann, 1896–1900, p. 1338, Pl. CCX, fig. 541; Smith and Bean,
1899, p. 187; Evermann and Hildebrand, 1910, p. 161; Fowler, 1918, p. 18, and 1923, pp. 7 and 34.

Head 2.7 to 3.05; depth 2.3 to 2.65; D. XIII, 15 to 17; A. III, 12 or 13; scales 71 to 77. Body
elongate, compressed; back elevated; head moderate; snout long, tapering, 2.2 to 3 in head; eye
3.6 to 5; interorbital 3.85 to 4.7; mouth moderate, terminal, a little oblique; maxillary reaching
vertical from first nostril, 3 to 3.4 in head; teeth in the jaws small, pointed, in broad bands; gill
rakers short, 12 on lower limb of first arch; scales rather small, ctenoid, firm, in oblique rows above
lateral line and horizontal rows below it, extending on base of caudal, ventrals, and pectorals, also
forming a low sheath on base of dorsal and anal; dorsal fin continuous, rather low, the spines rather
slender, pungent, origin of fin over or slightly in advance of base of pectorals; caudal fin deeply
concave, the upper lobe longest; anal fin with three rather strong, graduated spines, the soft part
similar to that of dorsal; ventral fins moderate, inserted a little behind base of pectorals; pectoral
fins rather long, 1.2 to 1.55 in head.

Color of fresh specimen bluish with purplish reflections above, becoming paler to silvery below;
sides of head and back with golden or brassy markings, variable, forming more or less distinct
lines; dorsal clear, with bronze spots; caudal and pectorals plain translucent; anal whitish to dusky,
base and middle parts sometimes tinged with yellow; ventrals white to slightly dusky. Color in
alcohol largely brownish, the purplish reflections frequently remaining; small specimens paler in
color than large ones. Stripes are visible only in the young among the preserved specimens.

Many specimens of the pigfish, ranging from 60 to 285 millimeters (2⅜ to 11¼ inches) in
length, were preserved. This fish is recognized by the rather deep, compressed body, the long,
pointed snout, and the bluish-purplish ground color of the back, with more or less distinct lighter
to yellowish stripes. The young are proportionately deeper than the adults and the snout is less
strongly produced. The differences in color, due to age, has been mentioned in the description.

The food of this fish in Chesapeake Bay, according to the contents of 43 stomachs, consists
mainly of annelids, with crustaceans, mollusks, insect larvæ, fish, and vegetable débris entering in
minor quantities.

Spawning takes place in the spring. Fish examined late in May had their sexual organs well
developed, and during June spawning fish were observed. The extent of the spawning season is
not known, for adult fish were absent from our collections during the summer. By early fall, when
large fish were again caught, all were found to be spent.

During 1920, the hogfish ranked eighteenth among the various fishes from Chesapeake Bay,
both in quantity and value, the catch being 31,725 pounds, worth $2,348.

The entire catch is credited to Virginia, where the hogfish ranked seventeenth in both quantity
and value. Of the entire amount, 50 per cent were caught with hand lines, 26 per cent in pound

nets, 21 per cent in haul seines, and 3 per cent in gill nets. The counties with the largest catches were Norfolk, 12,010 pounds; Warwick, 5,310 pounds; and Princess Anne, 5,000 pounds.

The fishing season extends from April to October, the most productive months being May, June, September, and October. The hogfish is virtually unknown above Solomons, Md., and is seldom caught within the waters of that State. In the vicinity of Crisfield, Md., however, it may be taken with hook and line and gill nets, but the catch usually is very small. In the lower part of the bay it is most common in the vicinity of Cape Charles, Buckroe Beach, and Ocean View, Va. During the late fall small, unmarketable fish, 4 to 6 inches long, are exceedingly abundant, and a catch of 50 or more of these fish within a few hours by a hook-and-line fisherman fishing for "spots" is not unusual.

This fish is much esteemed in the Chesapeake region, and during 1921 and 1922 the retail price averaged about 20 cents a pound. The size of the fish observed in the markets usually ranges between one-third and 1 pound. The maximum recorded weight is 2 pounds. The names most

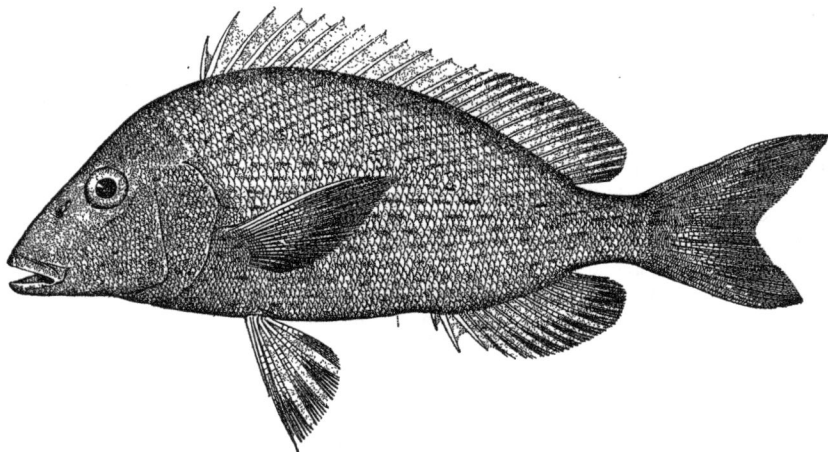

Fig. 147.—*Orthopristis chrysopterus*. From a specimen 9¾ inches long

generally used in the vicinity are "hogfish" and "pigfish." The pigfish is not taken in commercial quantities north of Chesapeake Bay. Southward, however, and particularly in North Carolina, it is a food fish of great importance.

Unlike many fish that are in prime condition just prior to their spawning period, the hogfish is notably thin when in full roe. In the fall, several months after spawning, it has gained considerably in weight. The weights of the following fish, taken in May, may be compared with those listed in the table for fish caught in October: 6½ inches, 2 ounces; 7 inches, 2.8 ounces; 8 inches, 3.7 ounces; 10½ inches, 7.2 ounces.

Habitat.—New York to Mexico.

Chesapeake localities.—(a) Previous records: From various sections of the southern part of Chesapeake Bay; one record from the Potomac River at Gunston, Va. (b) Specimens in collection: From many localities, from Love Point, Md., southward to Cape Charles and Cape Henry, Va.

Comparison of lengths and weights of hogfish caught during October

Number of fish weighed and measured	Length	Weight	Number of fish weighed and measured	Length	Weight
	Inches	*Ounces*		*Inches*	*Ounces*
1	3½	0.4	23	6¼	2.1
2	4¼	.6	15	6½	2.3
7	4½	.8	13	6¾	2.7
5	4¾	.9	10	7	3.0
9	5	1.0	2	8	4.9
14	5¼	1.3	1	8¼	5.6
23	5½	1.5	2	10	9.3
7	5¾	1.7	1	11¼	12.0
13	6	1.8			

111. Genus HÆMULON Cuvier.　Grunts

Body oblong, compressed; back more or less elevated; mouth large, horizontal; maxillary long, curved, extending to below eye; teeth in jaws in narrow, villiform bands; preopercle serrate; chin with a central groove behind symphysis; scales above lateral line in series, not parallel with it; soft parts of vertical fins densely scaled; dorsal fin more or less notched, usually with 12, rarely with 11, spines; caudal fin more or less forked; anal fin with three spines, the second enlarged; mouth red within.

142. Hæmulon plumieri (Lacépède).　Grunt; Black grunt.

Labrus plumieri Lacépède, Hist. Nat. Poiss., III, 1802, p. 480, Pl. II, fig. 2; Martinique.
Hæmulon formosum Uhler and Lugger, 1876, ed. I, p. 123; ed. II, p. 105.
Hæmulon plumieri Jordan and Evermann, 1896–1900, p. 1304, Pl. CCV, fig. 532.

Head 2.7 to 2.8; depth 2.4 to 2.5; D. XII, 15 or 16; A. III, 8 or 9; scales 49 to 53. Body elongate, compressed; back elevated; snout long, pointed, 1.9 to 2.3 in head; eye 3.9 to 5.2; mouth large, terminal, horizontal; maxillary reaching about middle of eye, 1.9 to 2.05 in head; gill rakers rather short, 14 or 15 on lower limb of first arch; scales of moderate size, those above lateral line enlarged, the series very oblique, four rows between origin of dorsal and lateral line; scales extending on base of pectoral fins, the soft parts of the other fins densely scaled; dorsal fin long, scarcely notched; caudal fin forked, the upper lobe longest; anal fin with three spines, the second one the strongest, reaching slightly past the tip of the third when deflexed; pectoral fins moderate, 1.3 to 1.6 in head.

Color of preserved specimens grayish brown; scales on sides each with a large silvery area; sides of head with about 12 more or less wavy, horizontal, blue stripes, not extending beyond head; vertical fins dusky; paired fins mostly pale. The color in life, according to published accounts, is bluish gray, the scales with greenish bronze spots forming oblique lines; the horizontal stripes on head are bright blue.

This species was not obtained during the present investigation. It is known from Chesapeake Bay only from a record by Uhler and Lugger (1876). From South Carolina southward it is a food fish of importance, being especially abundant at Key West, Fla. The maximum recorded weight for the species is 3 pounds.

Habitat.—Virginia to Brazil, occurring only as a straggler north of the coast of South Carolina.

Chesapeake localities.—(a) Previous record: "Also lives in the salt waters not remote from the ocean, mouth of the Potomac River, etc." (Uhler and Lugger, 1876.) (b) Specimens in collection: None.

112. Genus BATHYSTOMA (Scudder) Putnam.　Tom tates

This genus is very close to *Hæmulon.* Normally, however, it has 13 instead of 12 dorsal spines, the body is more elongate, and the back is lower.

143. Bathystoma rimator (Jordan and Swain).　Tom tate; Red-mouthed grunt.

Hæmulon chrysopteron Uhler and Lugger, 1876, ed. I, p. 124; ed. II, p. 105; not of Linnæus.
Hæmulon rimator Jordan and Swain, Proc., U. S. Nat. Mus., VII, 1884, p. 308; Charleston, Key West, and Pensacola.
Bathystoma rimator Jordan and Evermann, 1896–1900, p. 1308, Pl. CCVI, fig. 534.

Body 2.8 to 2.9; depth 2.85 to 3.1; D. XIII, 13 to 15; A. III, 8 or 9; scales 50 to 59. Body quite elongate, compressed; back little elevated; head rather long; snout tapering, 2.3 to 3 in head; eye

3.25 to 3.75; mouth large, terminal, slightly oblique; maxillary reaching about middle of eye, 1.8 to 2.1 in head; teeth in jaws in villiform bands, the outer ones enlarged; preopercle finely serrate; gill rakers rather short and slender, 13 to 15 on lower limb of first arch; scales moderate, ctenoid, six rows between origin of dorsal and lateral line; vertical fins densely scaled; dorsal fin long, low; caudal fin forked; pectoral fins rather short, 1.3 to 1.4 in head.

Color brownish or grayish above, silvery below; sides with two yellow stripes (most distinct in young); base of caudal with a large black spot; dorsal and caudal dusky, other fins mostly yellow.

This fish was not taken during the present investigation. It is known from the Chesapeake only from a record by Uhler and Lugger (1876). It is common at Charleston, S. C., and southward. A maximum weight of 1 pound is reported.

Habitat.—Virginia to Trinidad, occurring only as a straggler north of the coast of South Carolina.

Chesapeake localities.—(*a*) Previous records: "Occurs occasionally in the lower part of the Chesapeake Bay" (Uhler and Lugger, 1876). (*b*) Specimens in collection: None.

Family LXVI.—SPARIDÆ. The porgies

Body oblong or ovate, usually notably compressed; back more or less elevated; mouth rather small, nearly or quite horizontal; premaxillaries little protractile; maxillary slipping under preorbital for most of its length; supplemental bone present; preorbital usually broad; teeth strong, those on anterior part of jaws frequently incisorlike, lateral teeth blunt molars, none on vomer or palatines; gills 4, a slit behind the fourth; gill membranes separate, free from isthmus; opercle without spines; lateral line complete, not extending on caudal fin, concurrent with outline of back; scales moderate, firm, finely serrate; dorsal fin long, continuous or notched, with 10 to 12 spines, depressible in a groove; caudal fin usually forked; anal fin with three spines, the soft part similar to that of dorsal; ventral fins subthoracic, with I, 5 rays.

KEY TO THE GENERA

a. Front teeth very narrow, not notched; dorsal spines rather high, slender, the second one more than half the length of head_____Stenotomus, p. 261

aa. Front teeth broad, with or without a notch; dorsal spines shorter, the second less than half the length of head.

　b. Body with dark crossbars.

　　c. Incisor teeth deeply notched; size small_____Lagodon, p. 265

　　cc. Incisor teeth entire or only slightly notched; size large_____Archosargus, p. 267

　bb. Body without dark crossbars; black spot on caudal peduncle; incisor teeth broad, not notched _____Diplodus, p. 268

113. Genus STENOTOMUS Gill. The scups

Body rather deep; back elevated; head pointed; eye small, placed high; incisor teeth rather narrow, not notched; gill rakers short, about nine on lower limb of first arch; top of head, snout, and orbital region naked, the rest of body scaly; antrorse dorsal spine present, attached to interneural bone by a long process; dorsal with 12 spines, the first less than half the length of the second. Two apparently closely related species are known, and both were taken in Chesapeake Bay. A key to the species is omitted because it is difficult to show specific differences briefly. The two species are compared and contrasted under *Stenotomus aculeatus.*

144. Stenotomus chrysops (Linnæus). Scup; Porgy; "Maiden"; "Fair maid"; "Ironsides."

Sparus chrysops Linnæus, Syst. Nat., ed. XII, 1766, p. 471; Charleston, S. C.
Stenotomus argyrops Uhler and Lugger, 1876, ed. I, p. 122; ed. II, p. 104.
Stenotomus chrysops Bean, 1891, p. 90; Jordan and Evermann, 1896–1900, p. 1346, Pl. CCXI, fig. 544.

Head 2.95 to 3.45; depth 1.95 to 2.25; D. XII, 12; A. III, 11 or 12; scales 49 or 50. Body rather deep, compressed; back elevated, reaching its highest point under anterior dorsal spines; dorsal profile straight to slightly concave over eyes, convex elsewhere to caudal peduncle; depth of caudal peduncle 2.55 to 3.6 in head; head rather short, deep; snout more or less pointed, 2.3 to 2.55

in head; eye 2.5 to 3.9; interorbital 3.1 to 4; mouth rather small, terminal, oblique; maxillary scarcely reaching eye, 2.8 to 3.25 in head; incisor teeth on anterior part of jaws narrow and not notched, followed by smaller incisorlike teeth; two more or less definite rows of molar teeth laterally; gill rakers very short, nine more or less developed on lower limb of first arch; scales firm, finely ctenoid, extending on base of caudal, forming a very low sheath on dorsal and anal; dorsal fin long, continuous, the spinous portion higher than the soft part, the spines very slender, the second one frequently somewhat produced and reaching beyond the tip of the third when deflexed; caudal fin forked; anal fin with three spines, shorter but stronger than the dorsal spines, the soft part similar to that of the dorsal; ventral fins rather long and narrow, one or two of the outer rays with a slight filament, inserted a little behind base of pectorals; pectoral fins long, reaching beyond tips of ventrals in adults, proportionately shorter in young, 2.85 to 3.45 in length.

Color bluish silvery above, plain silvery below; young with about six dark crossbars; fins mostly plain translucent; soft part of dorsal and sometimes the anal with brownish spots, these most distinct in the smaller specimens; axil usually with a dusky spot.

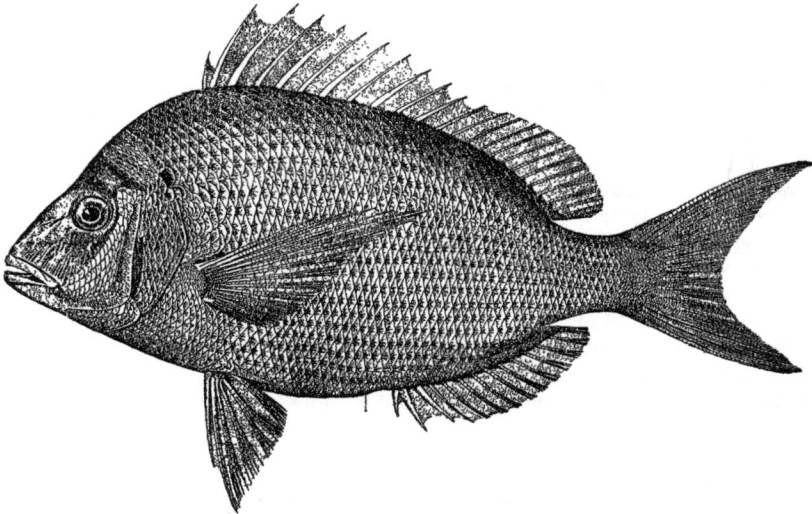

FIG. 148.—*Stenotomus chrysops*

Many specimens of this species, ranging from 70 to 260 millimeters (2¾ to 10¼ inches) in length, were preserved. The young differ from the adult in having distinct dark crossbars and proportionately shorter spines and rays in the fins. The narrow incisor teeth without notches, and the rather short to moderately long dorsal spines serve to separate the two closely related species of the genus from related forms.

The food of the scup, according to the contents of 24 stomachs, consists of crustaceans, mollusks, worms, insect larvæ, and small fish.

Spawning takes place in the spring, principally in May and June. The eggs are reported to be transparent, spherical, and about 0.85 to 0.9 millimeter in diameter, and to hatch in about 40 hours at a mean temperature of approximately 71° F.[21]

No ripe fish were observed in Chesapeake Bay, and it is not known whether spawning occurs there. The few large fish that appear in April and early in May are soon replaced by smaller ones (6 inches, or less, in length), which apparently are immature. Bean (1903, p. 559) states that off New York large spawning fish appear first in May. The chief spawning period in that vicinity occurs in June. Bigelow and Welsh (1925, p. 268) state that in southern New England spawning takes place chiefly in June, but that the period extends from May to August.

---

[21] For an account of the embryology and larval development of the scup, see Kuntz and Radcliffe, 1918, pp. 102 to 105, figs. 30 to 37.

The rate of growth, as given by Bean (1903, p. 560), is as follows: July 3, length ½ to 1½ inches; August 2, 1½ to 2 inches; September 6, 2 to 3 inches; September 29, 3 to 4 inches; November 1, 4 inches. In Chesapeake Bay we seined 30 scup, 63 to 92 millimeters (2½ to 3⅝ inches) in length, on September 23; 9 scups, 83 to 106 millimeters (3⅛ to 4⅛ inches), on October 6; and 5 fish, 115 to 123 millimeters (4½ to 4⅞ inches), on October 11. On May 23, at Cape Charles, 50 scups, 121 to 155 millimeters (4¾ to 6⅛ inches) in length, were taken with a 180-fathom seine, and on May 25, in Lynnhaven Roads, 25 fish ranging from 5 to 6 inches in length were caught in a pound net. No other young fish were observed. Taking Doctor Bean's records of the rate of growth as a guide, its seems fair to conclude that the fish taken in Chesapeake Bay in September and October were the result of the hatch of the preceding spring, and it seems probable then that the fish taken in May are approximately 1 year old.

FIG. 149.—Egg with embryo

During 1920 the scup ranked twenty-fourth in quantity and twenty-second in value in Chesapeake Bay, the catch being 7,165 pounds, worth $585. As the scup is taken only in the lower part of the bay, the total catch is credited to Virginia, where the fish takes the same rank as for the entire bay. Nearly the entire catch was taken in pound nets in Elizabeth City and Northampton Counties.

This fish is taken in the Chesapeake from April until late October. The small annual catch is caught in the bay below the York River. The first fish generally appear early in April and are large in size, weighing 1 to 3 pounds. Later the fish are slightly more plentiful and smaller. The greater part of the catch, as already stated, is taken with pound nets. A catch as great as 25 pounds in one day is seldom made by one set of nets, and frequently not more than this amount is taken during an entire month. Scups are caught occasionally with haul seines at Ocean View in September and October. In 54 hauls made by three 300-fathom haul seines at Ocean View from September 20 to October 27, 1922, the scup was taken on only one day (October 8), when 25 fish, each about 9 inches long, were present in a catch. The scup takes the hook freely, but it is seldom caught by that method in the Chesapeake. Small unmarketable fish are rather common near the mouth of the bay from May until October. At Cape Charles, on May 23, 1922, for example, 50 scups, measuring from 5 to 6 inches in length, were caught in one haul with a seine 180 fathoms long. Frequently moderate numbers of fish of this size also are caught with pound nets and discarded.

The local catch of scup is marketed in the Chesapeake region, but the fish does not meet with the same high regard that it does in New Jersey and northward. A closely related species, *S. aculeatus*, is occasionally caught and marketed along with *S. chrysops*, as the fishermen do not separate the species.

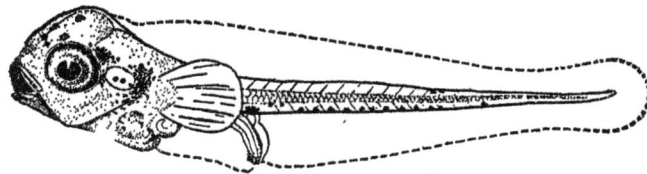

FIG. 150.—Larva, 3 days old, 2.8 millimeters long

Various names have been given to this species, but those most commonly used in the Chesapeake region are "maiden," "fair maid," and "ironsides"; the last name is in allusion to the hard, platelike scales.

This fish is an abundant and important food fish in the northern part of its range. South of Virginia it is not abundant and of no commercial importance. The scup is taken in commercial numbers along the Atlantic coast from Virginia to Cape Cod, Mass., the center of abundance being from New Jersey to Rhode Island. It is by far the most valuable food fish taken in Rhode Island, where in 1919[22] the catch amounted to 8,261,140 pounds, worth $817,846. In New York, in 1921,[23] it ranked third in quantity and fifth in value, the catch amounting to 1,297,375 pounds,

[22] Fishery Industries of the United States. Report of the Division of Statistics and Methods of the Fisheries for 1920. By Lewis Radcliffe. Appendix V, Report of the Commissioner of Fisheries for 1921 (1922). Bureau of Fisheries Document No. 908, p. 128. Washington.

[23] Fishery Industries of the United States. Report of the Division of Fishery Industries for 1922. By Harden F. Taylor. Appendix V, Report of the Commissioner of Fisheries for 1923 (1924). Bureau of Fisheries Document No. 954, p. 68. Washington.

worth $76,253. In New Jersey, during 1921, it ranked second in quantity and third in value, the catch amounting to 4,115,552 pounds, worth $200,046.

The usual size of market fish in the Chesapeake and along the Atlantic coast is from ½ to 2 pounds. The species attains a maximum weight of 4 pounds. The following weights were obtained from Chesapeake Bay fish: Four and three-fourths inches, 0.9 ounces; 5 inches, 1.1 ounces; 5¼ inches, 1.2 ounces; 5¾ inches, 1.4 ounces; 6 inches, 1.6 ounces; 6¼ inches, 2 ounces; 9 inches, 6.7 ounces; 13½ inches, 1 pound 2 ounces.

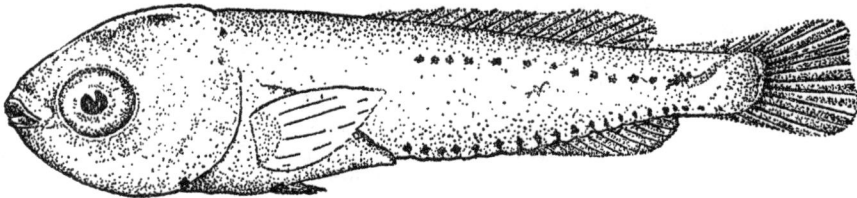

FIG. 151.—Young, 10.5 millimeters long

Habitat.—Maine to South Carolina; common from Virginia to Cape Cod.

Chesapeake localities.—(*a*) Previous records: Southern part of Chesapeake Bay (Uhler and Lugger, 1876) and Cape Charles city. (*b*) Specimens in collection: Lower York River, Cape Charles, Buckroe Beach, Ocean View, and Lynnhaven Roads, Va.

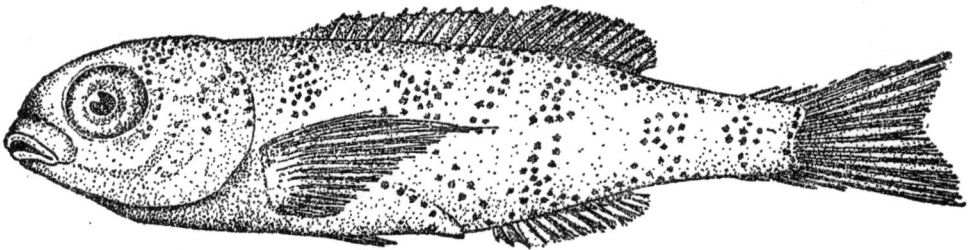

FIG. 152.—Young, 25 millimeters long

145. Stenotomus aculeatus (Cuvier and Valenciennes). Southern scup; Pinfish.

Chrysophrys aculeatus Cuvier and Valenciennes, Hist. Nat. Poiss., VI, 1830, p. 137; Charleston, S. C.
Stenotomus aculeatus Jordan and Evermann, 1896–1900, p. 1346, Pl. COXII, fig. 545.

Head 3.3 to 3.4; depth 2.2 to 2.45; eye 3 to 3.3 in head; snout 2.55 to 2.8; interorbital 3.1 to 3.5; maxillary 3.1 to 3.15; caudal peduncle 2.8 to 2.95; fourth dorsal spine, 1.55 to 1.9; pectoral fin 3.35 to 3.9 in length; D. XII, 12; A. III, 11; scales 50 or 51.

Three specimens in the Chesapeake Bay collection, respectively 115, 125, and 150 millimeters (4½, 5, and 6 inches) in length, appear to be referable to this species. Comparing specimens of like size, they differ from *S. chrysops* in having a more slender body and caudal peduncle, slightly larger eye, broader interorbital, and lower dorsal spines. The second spine does not reach the tip of the third when deflexed, whereas in *S. chrysops* the second spine frequently reaches beyond the tip of the third when deflexed. No difference in color is noticeable in preserved specimens. No notes on color in life were obtained, as the species was not recognized in the field.

The following series of proportions is based on three specimens of *S. chrysops* of the same length as those of *S. aculeatus* at hand and will serve to show the differences mentioned:

Head 3.3 to 3.45; depth 1.95 to 2; eye 3 to 3.3 in head; snout 2.75 to 3.05; interorbital 3.55 to 4; maxillary 3.05 to 3.1; caudal peduncle 2.5 to 2.6; fourth dorsal spine, 1.25 to 1.45; pectoral fins 3.4 to 3.65 in length.

A somewhat greater difference than is brought out by the foregoing customary proportions may be obtained by dividing the depth of the body by the diameter of the eye, because the proportions resulting serve to emphasize that the eye is larger and the depth smaller in *S. aculeatus*. Comparing the same series of specimens as in the proportions already given, the eye is contained in the depth in *S. aculeatus* 4.1, 4.32, and 5 times, whereas in *S. chrysops* it is contained 5.6, 5.6, and 5.75 times.

The difference in the dorsal contour of the body given in current descriptions, namely, that the outline of the body in *S. aculeatus* declines more rapidly from the first dorsal spine backward than in *S. chrysops*, is not evident from the small specimens at hand.

The present species is of southern distribution, replacing *S. chrysops*. It has previously not been reported as far north as Virginia. It is said to be rather common from Cape Hatteras southward. However, it nowhere reaches the large commercial importance attained by its congener from the Chesapeake Bay northward.

Habitat.—Virginia to Texas, very rare or wanting at Key West, Fla.

Chesapeake localities.—(*a*) Previous records: None. (*b*) Specimens in collection: Cape Charles, Va. (extreme point of cape), May 21, 1922, seine.

114. Genus LAGODON Holbrook. Pinfishes

In externally visible characters this genus is close to Probatocephalus, differing principally in the deeply notched teeth. The essential character of the genus is in the form of the skull, which is described in current works as follows: Supraoccipital and temporal crests nowhere coalescent; interorbital area not swollen; frontal bone in the interorbital area thin, concave in transverse section; temporal crest low, separated from supraoccipital crest by a flattish area, extending forward on each side of supraoccipital crest to the groove of premaxillary spines. The genus contains a single species.

146. Lagodon rhomboides (Linnæus). Pinfish.

Sparus rhomboides Linnæus, Syst. Nat., ed. XII, 1766, p. 470; Charleston, S. C.

Lagodon rhomboides Uhler and Lugger, 1876, ed. I, p. 122; ed. II, p. 104; Bean, 1891, p. 90; Jordan and Evermann, 1896–1900, p. 1358, Pl. CCXV, fig. 552.

Head 3.1 to 3.4; depth 2.15 to 2.35; D. XII, 11; A. III, 11; scales 62 to 66. Body oblong, variable in depth, compressed; back elevated; head moderate; snout rather pointed, 2.75 to 3.1 in head; eye 3.25 to 4.1; interorbital 2.9 to 3.55; mouth rather small, nearly horizontal, terminal; maxillary scarcely reaching eye, 2.95 to 3.3 in head; each jaw with eight broad, deeply notched incisors anteriorly on edge of jaws, followed by two rows of low, broad, blunt teeth; gill rakers short and slender, 12 on lower limb of first arch; scales rather small, firm, ctenoid, extending on base of caudal and forming a scaly sheath on soft part of dorsal and anal; dorsal fin long, continous, rather low, the spines rather slender, extremely sharply pointed, preceded by an antrorse spine, origin of fin a little in advance of base of pectorals; caudal fin forked; anal fin with three rather strong, sharply pointed spines, second and third of about equal length, the soft part of fin similar to that of dorsal; ventral fins moderate, inserted about half an eye's diameter behind base of pectorals; pectorals long, pointed, reaching well beyond tips of ventrals, 2.9 to 3.55 in body.

Color dark green above; silvery below; a dark spot at shoulder; 4 to 6 dark crossbars on sides, varying in distinctness among individuals; sides with several light-blue and yellow longitudinal stripes (fading and nearly disappearing in spirits); dorsal plain, with faint yellowish-brown spots and with yellowish brown on distal parts of spinous portion; caudal and pectorals pale yellow; anal plain translucent on basal half, the rest of fin yellowish brown; ventrals pale, with yellowish-brown streak at middle of fin.

This fish is represented in the present collection by many specimens, ranging in length from 20 to 185 millimeters (4/5 to 7¼ inches). The young are less brightly colored than the adults, the longitudinal stripes being absent and the dark crossbars quite distinct. The species is recognized chiefly by the deeply notched incisor teeth, the rather slender and very sharp fin spines, the antrorse spine preceding the dorsal fin, and by the coloration.

The food of this fish, according to Smith (1907, p. 300), is quite varied, consisting of fish, worms, crustaceans, mollusks, and seaweed. The contents of 13 stomachs taken from fish caught in Chesapeake Bay contained the following foods, named in the order of their apparent importance: Vegetable débris, crustaceans, mollusks, and annelids.

The pinfish apparently does not spawn in Chesapeake Bay. According to present evidence it is a winter spawner. Bean (1903, p. 562) states that spawning takes place in the Gulf of Mexico in winter or early spring. Smith (1907, p. 300) records that specimens examined at Beaufort, N. C., in June and July had no obvious reproductive organs, but that eggs were noted in a female on August 6, and a ripe male was taken on November 20. The capture of young fish in Chesapeake Bay in the spring is further evidence that spawning may occur durng the winter. The collection of numerous larval pinfish by one of us (Hildebrand) at Beaufort, N. C., during the winter of 1925–26 proves beyond a doubt that at Beaufort, at least, spawning takes place during late fall and winter. The following two catches of young pinfish were seined in Chesapeake Bay: May 22, 1921, creek tributary to Lynnhaven Bay, three fish, 23 to 29 millimeters (⅞ to 1⅛ inches) in length; May 22, 1922, Cape Charles, Va., many specimens, 20 to 27 millimeters (⅘ to 1 inch) in length. The only midsummer catch of pinfish consisted of seven fish, ranging in length from 96 to 105 millimeters (3¾ to 4⅛ inches), and was taken with a seine in the lower York River on July 10. During October the majority of the fish caught were from 136 to 160 millimeters (5⅓ to 6¼ inches) long. One catch of pinfish, taken on October 10 in a 300-fathom haul seine, consisted of 850 pin-

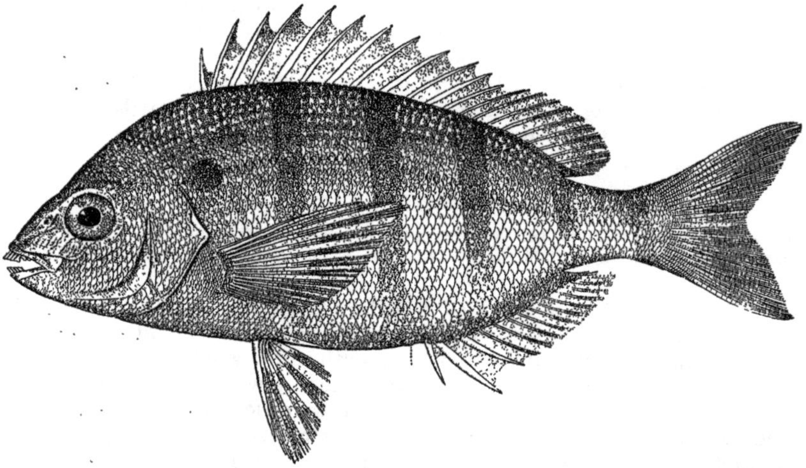

FIG. 153.—*Lagodon rhomboides.* From a specimen 6 inches long

fish, 170 to 185 millimeters (6¾ to 7¼ inches) in length. As the average length of the individuals of this lot was 1 inch or more greater than of other fish caught during October, it seems probable that they were older fish.

The pinfish in Chesapeake Bay inhabits only the southern section, where fish of marketable size are taken in small numbers during the summer and fall. Its commercial importance is very small, and the marketable catch of 1922 probably did not exceed 1,000 pounds, valued at about $40. Small fish, 4 to 7 inches long, are sometimes very common in the fall but are discarded by the fishermen. One day in October, 1922, about 5,000 fish of this small size were caught in a pound net at Lynnhaven Roads, Va.; and at Ocean View, Va., many were taken in haul seines throughout the month of October. Fishermen operating a set of two pound nets at Lynnhaven Roads caught and marketed 3,500 pounds of pinfish from May 7 to 11, 1918. These fish were somewhat larger than usual and represented an unusual run.

The fish caught in the Chesapeake are marketed chiefly in Norfolk, Va. The average size of fish observed in the markets is about one-third of a pound. The maximum size recorded for the species is 13 inches. (Schroeder, 1924, p. 26.) The following weights were obtained from 130 pinfish caught in Chesapeake Bay: Five inches, 0.9 ounce; 5¼ inches, 1.4 ounces; 5½ inches, 1.6 ounces; 5¾ inches, 1.9 ounces; 6 inches, 2.1 ounces; 6¼ inches, 2.4 ounces; 6½ inches, 2.8 ounces; 6¾ inches, 3.1 ounces; 7 inches, 3.4 ounces; 7¼ inches, 3.5 ounces.

Habitat.—Massachusetts to Texas; common from Virginia southward.

Chesapeake localities.—(*a*) Previous records: Lower part of Chesapeake Bay (Uhler and Lugger, 1876) and Cape Charles city, Va. (*b*) Specimens in collection: Lower York River, Cape Charles, Buckroe Beach, Ocean View, and Lynnhaven Roads, Va.

115. Genus ARCHOSARGUS Gill. Sheepshead

Body rather robust, deep, compressed; mouth moderate; jaws anteriorly with broad incisors, with entire or only slightly notched margins; jaws laterally with coarse molars; posterior nostril slitlike; gill rakers very short; dorsal fin long, continuous, preceded by an antrorse spine; spines strong, the soft part shorter than the spinous portion; caudal fin slightly forked; anal fin with three strong spines, the second enlarged; ventral fins subthoracic. The Virginia specimens with broad, black crossbars. A single species occurs in Chesapeake Bay.

147. Archosargus probatocephalus (Walbaum). Sheepshead.

Sparus probatocephalus Walbaum, Artedi Piscium, 1792, 295; New York.
Archosargus probatocephalus Uhler and Lugger, 1876, ed. I, p. 121; ed. II, p. 103; McDonald, 1882, p. 12; Bean, 1891, p. 90; Smith, 1892, p. 71; Jordan and Evermann, 1896-1900, p. 1361, Pl. CCXVI, fig. 554.

Head 3.05 to 3.25; depth 1.9 to 2.4; D. XI or XII, 11 to 13; A. III, 10 or 11; scales 44 to 49. Body deep, compressed; back elevated; head short, deep; snout short, 2.1 to 2.6 in head;

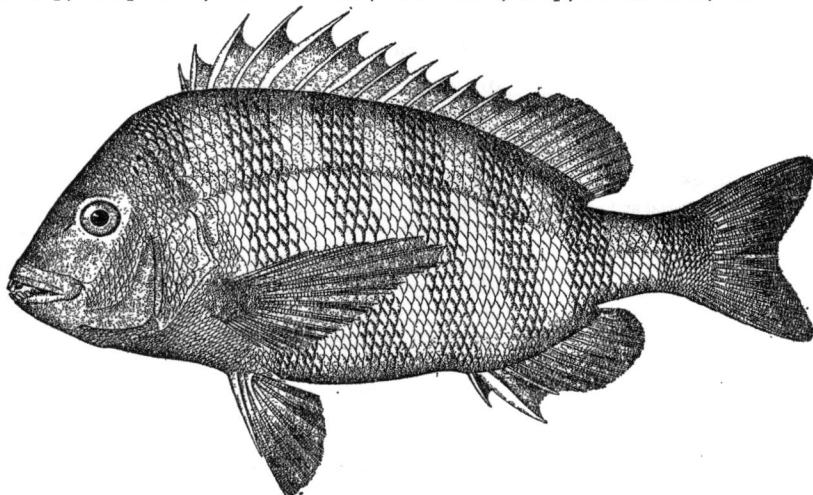

Fig. 154.—*Archosargus probatocephalus.* From a specimen 15 inches long

eye 2.75 to 4.55; mouth moderate, nearly horizontal; maxillary reaching about to vertical from anterior margin of eye, 2.7 to 3.3; teeth in the jaws strong, anterior teeth incisorlike, the posterior teeth broad, strong molars; gill rakers short, six or seven on lower limb of first arch; scales finely serrate; dorsal fin with very strong spines, the spinous portion longer than the soft part; caudal fin with a shallow fork; anal fin with three spines, the second much enlarged, the soft part of fin similar to that of dorsal; pectorals long, 2.5 to 3.7 in length.

Color greenish yellow; sides with seven black crossbars; dorsal, anal, and ventral fins mostly dusky or black; caudal and pectoral fins greenish.

The foregoing description is based upon specimens from Beaufort, N. C., ranging in length from 20 to 240 millimeters (⅘ to 9½ inches). Only a few very small specimens were preserved from Chesapeake Bay, as the large individuals seen were too bulky to preserve conveniently and no intermediate sizes were taken. This species, the only one of the genus occurring in Chesapeake Bay, usually is easily recognized by its color, entire incisor teeth, and large size.

The sheepshead feeds mainly on mollusks and crustaceans, for the crushing of which its teeth are well adapted. Relative to spawning, Smith (1907, p. 301) says: "At the spawning season, which is in spring, the sheepshead swim in schools and appear to prefer sandy shores. The eggs are about

0.03 inch in diameter, and more than 1,500,000 are in a fluid quart. They float at the surface and hatch rapidly, only 40 hours being required in water of 76° or 77° F."

The sheepshead is taken in the Chesapeake in very limited numbers, from the Rappahannock River southward. In 1920 the catch was only 863 pounds, worth $129, all taken with pound nets.

Some years ago this fish was an important commercial species in the bay, but the catch gradually has diminished, until at the present time the species has almost entirely disappeared. The small catch is readily absorbed by the Norfolk markets, as the sheepshead is a food fish of fine flavor. The Chesapeake fish are large, the size usually ranging from 5 to 15 pounds. The maximum recorded weight for the sheepshead is 30 pounds, but it rarely exceeds 20 pounds.

In some sections of its range this fish furnishes much sport for the angler, as it is said formerly to have done in Chesapeake Bay, for it is a very game fish, being among the gamest of salt-water fishes. It is often common along breakwaters, stone jetties, piles, and other objects in the water that are overgrown with barnacles, oysters, etc. It is in such places where the angler must seek the species, and the most commonly used bait consists of small crabs.

Habitat.—Cape Cod, Mass., to Texas, rarely to the Bay of Fundy.

Chesapeake localities.—(a) Previous records: "Frequents the oyster localities of all parts of Chesapeake Bay" (Uhler and Lugger, 1876); lower Potomac, Cape Charles city and Norfolk, Va. (b) Specimens in collection: Lynnhaven Roads, Va.; observed in the Norfolk market and at Ocean View, Va.

116. Genus DIPLODUS Rafinesque.   Spotted-tailed pinfish

Body ovate, compressed; back notably elevated; incisor teeth broad, not notched; molar teeth in several rows; gill rakers short; dorsal spines about 12; color silvery, with dark area on caudal peduncle.

148. Diplodus holbrookii (Bean).   Spot-tailed pinfish; Sailor's choice.

Sargus holbrookii Bean, Forest and Stream, June 13, 1878; Charleston, S. C.
Diplodus holbrookii Bean, 1891, p. 90; Jordan and Evermann, 1896–1900, p. 1362, Pl. CCXVII, figs. 555 and 555a.

Head 3.65; depth 2.1; D. XII, 14; A. III, 13; scales 55 to 57. Body more or less elliptical, compressed; dorsal profile regularly rounded; eye rather small, 4.35 in head; mouth large, almost horizontal; maxillary failing to reach front of eye, 3.35 in head; four incisor teeth in each jaw, directed obliquely forward, three series of molars in upper jaw, two in the lower; gill rakers very short, about 14 on lower limb of first arch; dorsal fin continuous, rather low, longest spine less than half the head; caudal fin forked; anal fin with three spines the second somewhat enlarged, the soft part of fin similar to that of dorsal; pectoral fins pointed, reaching origin of anal, about 3.35 in body.

Color dull blue above, lower part of sides and below silvery; a conspicuous black blotch or band on anterior part of caudal peduncle; opercular margin black; base of pectorals black. The young with about five narrow, vertical, dark stripes on back and sides, with an equal number of short intermediate stripes on back.

No specimens of this species were secured during the present investigation. It is known from Chesapeake Bay only from a record by B. A. Bean, based on seven specimens collected by W. P. Seal at Cape Charles city, Va., in 1890.

The species is not uncommon farther southward, and it is frequently seen along breakwaters and piers on the coast of North Carolina. One of us (Hildebrand) measured a specimen at Beaufort, N. C., 14 inches in length, which probably is the maximum size attained. This fish is nowhere taken in sufficient quantity to be of commercial importance.

Habitat.—Virginia to Cedar Keys, Fla.

Chesapeake localities.—(a) Previous record: Cape Charles city. (b) Specimens in collection: None.

This species, like the preceding, is of southern distribution, being common in tropical waters on the Atlantic coast of America. The maximum recorded size of this fish is 5 inches. It is nowhere of commercial importance.

Habitat.—Massachusetts to Brazil; occurring only as a straggler north of Beaufort, N. C.

Chesapeake localities.—(*a*) Previous records: Cape Charles city, Va. (*b*) Specimens in collection: None.

Family LXIX.—SCIÆNIDÆ. The croakers and drums

Body elongate, more or less compressed; head rather large, the bones more or less cavernous; mouth large or small; teeth in one or more series on jaws, none on vomer, palatines, pterygoids, or tongue; barbels sometimes present on chin; no supplemental maxillary bone; premaxillaries protractile; gill membranes not united, free from the isthmus; branchiostegals 7; lateral line continuous, extending on caudal fin; scales large or small, present on head; dorsal fins continuous or separate; anal fin short, with one or two spines; caudal fin usually square or emarginate; air bladder usually large (absent in Menticirrhus); vertebræ about 10+14.

KEY TO THE GENERA

a. No barbels on lower jaw.
 b. Teeth very small, those in lower jaw deciduous, wanting in adult; body comparatively short and deep; mouth small, horizontal; preopercular margin entire; a dark spot behind upper angle of gill opening _____ Leiostomus, p. 271
 bb. Teeth well developed, permanent in each jaw; no dark spot behind upper angle of gill opening.
  c. Mouth horizontal; gill rakers short and thick; one (sometimes several) black spot at base of caudal_____ Sciænops, p. 276
  cc. Mouth more or less oblique to nearly vertical; gill rakers rather long and slender; no black spot at base of caudal.
   d. Preopercle without bony serræ; snout shorter than eye; mouth very oblique
   _____ Larimus, p. 278
   dd. Preopercle serrate; snout not shorter than eye; mouth moderately oblique.
    e. Head not very broad; interorbital space not spongy to the touch_ Bairdiella, p. 279
    ee. Head broad; interorbital space very cavernous, more or less spongy to the touch _____ Stellifer, p. 282
aa. One to many barbels on lower jaw.
 f. Lower jaw with numerous barbels.
  g. Preopercular margin strongly serrate; body covered with rather small scales, 64 to 72 in lateral series_____ Micropogon, p. 283
  gg. Preopercular margin entire; body covered with rather large scales, 41 to 45 in a lateral series_____ Pogonias, p. 287
 ff. Lower jaw with a single, short, thickish barbel.
  h. Anal fin with two spines, the second one somewhat enlarged__ Umbrina, p. 289
  hh. Anal fin with a single weak spine_____ Menticirrhus, p. 290

119. Genus LEIOSTOMUS Lacépède. Spots

Body comparatively short, compressed; back elevated; head short, obtuse; snout blunt; mouth small, horizontal; teeth wanting in lower jaw (in adult); preopercle entire; gill rakers short; dorsal fins contiguous, the first rather high, consisting of 10 spines. A single species of this genus is known.

152. Leiostomus xanthurus Lacépède. Spot; "Croaker"; "Silver gudgeon"; Layfayette.

Liostomus xanthurus Lacépède, Hist. Nat. Poiss., IV, 1803, p. 439, Pl. X, fig. 1; Carolina. McDonald, 1882, p. 12; Bean, 1891, p. 89; Smith, 1892, p. 72; Jordan and Evermann, 1896–1900, p. 1458, Pl. CCXXIII, fig. 569; Smith and Bean, 1899, p. 187; Evermann and Hildebrand, 1910, p. 162; Fowler, 1912, p. 59, 1918, p. 18, and 1923, p. 7; Welsh and Breder, 1923, p. 177.
Liostomus obliquus Uhler and Lugger, 1876, ed. I, p. 118; ed. II, p. 100. Not *L. obliquus* (Mitchell).

Head 2.95 to 3.6; depth 2.55 to 3.6; D. X–I, 30 to 34; A. II, 12 or 13; scales 72 to 77. Body rather deep, compressed; back strongly elevated in adult; head moderate; snout blunt, 2.75 to 3.4

in head; eye 3 to 3.95; interorbital 3 to 3.85; mouth horizontal; lower jaw shorter than the upper, included; maxillary reaching nearly opposite middle of eye, 2.65 to 3.2 in head; teeth in the jaws minute, in villiform bands, wanting in lower jaw in adult; chin and snout with several pores; no barbels; gill rakers short, 22 to 23 on lower limb of first arch; scales rather small, ctenoid, extending on the caudal fin and covering most of it in adults, a few scales also present on the base of the other fins; dorsal fins contiguous, the first composed of slender spines, the middle ones longest, notably higher than any of the rays in the soft part; caudal fin truncate in very young, notably concave in adult, the upper rays the longest; anal fin with two stiff spines, origin of fin under middle of soft dorsal; ventral fins moderate, inserted a little behind base of pectorals; pectoral fins long, reaching well beyond tips of ventrals in adult, failing to reach this point in young, 0.95 to 1.45 in head.

Color bluish gray with golden reflections above; silvery underneath; sides with 12 to 15 oblique yellowish (dusky in preserved specimens) bars, in fish ranging upward of 50 millimeters, and again becoming indistinct in very large fish; a large, yellowish black shoulder spot present, except in very young; fins mostly pale yellow; dorsal and caudal fins more or less dusky; anal and ventrals also partly dusky in large examples. Young fish, 40 millimeters and less in length, mostly pale; sides of head silvery; sides of body and back each with a row of dark blotches composed of dusky punctulations, besides other irregularly placed dusky points.

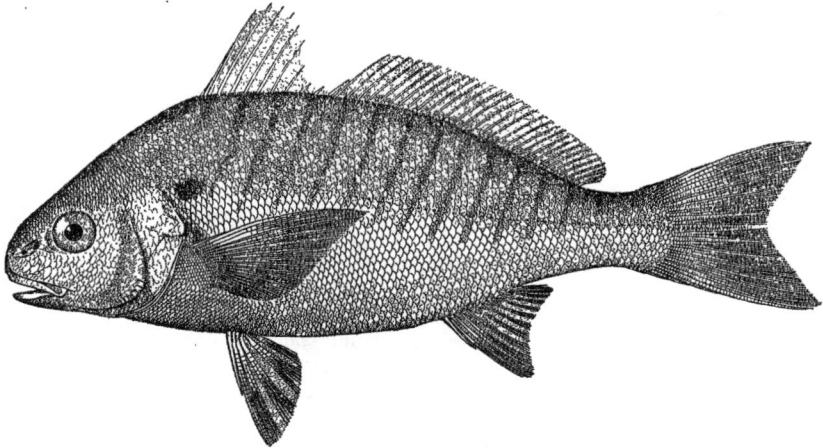

FIG. 155.—*Leiostomus xanthurus.* From a specimen 8 inches long

Many specimens of this species, ranging from 15 to 345 millimeters (⅝ to 13½ inches) in length, were preserved. The very young differ notably in color from the adult, as shown in the description. The young also are notably more slender, with the back proportionally much less strongly elevated. The comparatively short, compressed body, short, obtuse head, rather small horizontal mouth, and the oblique bars, and particularly the dark shoulder spot, distinguish this fish from related forms.

The male of this species makes a croaking or drumming sound, but it is not a loud one, owing probably to the thinness of the walls of the air bladder and the feeble development of the drumming muscles.

The food of the spot, as shown by the contents of 157 stomachs removed from specimens taken in Chesapeake Bay, consists mainly of small and minute crustaceans and annelids, together with smaller amounts of small mollusks, fish, and vegetable débris.

The spot grows rather rapidly during the first summer. The following sizes were collected during the spring and summer: April, 23 to 25 millimeters (about 1 inch); May, 26 to 72 millimeters (1 to 2⅞ inches); June, 21 to 86 millimeters (⅞ to 3⅜ inches); July, 32 to 82 millimeters (1¼ to 3¼ inches). The wide range in the size of young fish taken in the Chesapeake throughout most of the year makes it difficult to determine a correct average rate of growth. As the spawning season quite probably is a protracted one, a considerable variation in size among the young would be expected. The larger fish apparently come from those parent fish that spawned first (late fall), and the smaller fish result from the later (early winter) spawners. The monthly increase in length appears to be quite rapid during the summer and early fall. No spots less than 4 inches in length were collected from September to November. The predominating sizes trawled by the *Fish Hawk* in December, however, were only 85 to 95 millimeters ($3\frac{1}{3}$ to $3\frac{3}{5}$ inches) in length; while late in January, out of a total catch of 383 spots taken in various parts of the bay, 354 ranged in length from 80 to 110 millimeters ($3\frac{1}{4}$ to $4\frac{1}{3}$ inches). Whether these small fish are runts that remained in Chesapeake Bay over the winter while the larger fish migrated out, or are younger fish, is not known. If they were younger fish one would be obliged to conclude that the species has a double spawning period. In that event the smaller and younger fish would be expected to occur earlier in collections. The absence of these smaller fish in catches made during September, October, and November, when thousands of spots, almost all over 4 inches long, were taken with fine-meshed collecting seines alongshore and in beam trawls offshore, together with the fact that spots with large roe have been observed only during the fall and early winter, however, tends to discredit such a theory.

Length frequencies of 1,321 spots, Leiostomus xanthurus

[Measurements in millimeters, grouped in 5-millimeter intervals]

Total length, millimeters	Mar.		Apr.		May		June		July		Aug.		Sept.		Oct.		Nov.		Dec.		Jan.	
	1-15	16-31	1-15	16-30	1-15	16-31	1-15	16-30	1-15	16-31	1-15	16-31	1-15	16-30	1-15	16-31	1-15	16-30	1-15	16-31	1-15	16-31
15-19	1		11			8																
20-24					2	46		2														
25-29						76		2														
30-34						131		3														
35-39					3	80		12														
40-44					1	31		21		4												
45-49						6		22		5												
50-54					1	5		13		10												
55-59					1			10		13												
60-64								5		2												
65-69					2	2		7		7		1										
70-74					1	1		3		11		1										
75-79								1		3		2						1	3			10
80-84								2		6		8							6			46
85-89								2				5							16			75
90-94								1		2		14		2					13			75
95-99								1		1		8		3					2			68
100-104										2		19		5		3			3			46
105-109										1		4		3		2		2	4			28
110-114												6		3		19		3	3			7
115-119									2			6		7		17		4	2			4
120-124									1			2		4		11		3				6
125-129												1		2		13		3				7
130-134												1		5		15		2				
135-139									1			1		3		4						1
140-144												1		1		3		2				
145-149					1				1			2		1		11		1				1
150-154														5		4		1				
155-159							1							1		5		1				
160-164														1		1		1				
165-169									1					1		1		2				
170-174					1				2					1		6		1				
180-184																4		1				
185-189																						
190-194										2				1		2						
195-199										2						1						
200-204										1						1						
205-209										1				1		1						
210-214																2						
215-219																2						
250-254																1						
295-299																1						
Total	1		11		12	387	1	107	8	73		82		50		135		28	52			374

Collections of spots made at Beaufort in shallow water during the winter of 1926-27 consist almost exclusively of small fish, their average length being considerably below that of fish taken before cold weather came, which appears to indicate that the smallest young of the season have a tendency to remain in their early habitat, whereas the larger ones migrate to deeper and warmer water. We have measured a considerable number of spots collected throughout most of the year at Beaufort, N. C., and find that, allowing for a somewhat longer growing season, the growth corresponds fairly well with that of Chesapeake Bay. It appears from the data collected that the Chesapeake Bay spot attains a length of about 5 inches at one year of age.

Spawning takes place in late autumn and probably during the winter, and apparently at sea, for in the fall a general exodus of large fish with maturing roe takes place from the bay, the height of this migration occurring during late September and throughout October. The gonads of 104 spots caught at Ocean View, Va., on October 16, 1922, ranging in length from 114 to 268 millimeters (4½ to 10½ inches), were examined for the status of development. The smallest fish in this lot that had ripening roe was a female 214 millimeters (8½ inches) in length. The ovaries and testes of the larger fish were in various stages of development, suggesting that the spawning season is a protracted one. A female and a male, respectively 214 and 224 millimeters (8½ and 8⅞ inches) long, had very immature gonads, which probably would not have ripened by the coming winter. Fish with large roe, however, have been observed only during autumn and early in the winter. In the spring the spot is thin and poor, as shown elsewhere, indicating that it probably has spent much energy in the process of reproduction. Nearly all spots that remain in the deeper parts of the Chesapeake during the winter are immature fish, less than 6 inches in length. In fact, there is

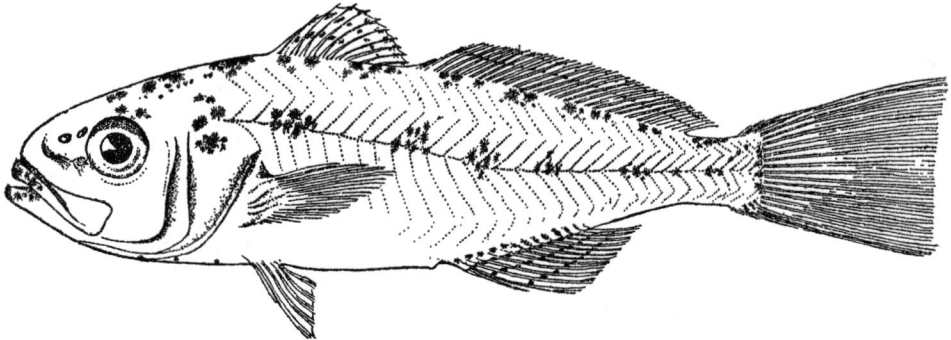

FIG. 156.—*Leiostomus xanthurus.* Young, 27 millimeters long

no evidence available (as already indicated) that spawning takes place within the bay. However, the young appear to enter when quite small.

The spot is one of the most important food fishes found in Chesapeake Bay. During 1920 it ranked seventh in quantity and sixth in value, the catch being 837,845 pounds, worth $63,138.

In Maryland it ranked eighth in quantity and ninth in value, the catch being 51,692 pounds, worth $3,138. Of this amount, 48 per cent was caught with pound nets, 46 per cent with haul seines, 3 per cent with purse seines, and 3 per cent with lines. Kent County is credited with the largest catch, namely 20,710 pounds, followed by Dorchester with 8,500 and Calvert with 5,010 pounds.

In Virginia the spot ranked sixth, both in quantity and value, the catch being 786,153 pounds, worth $60,000. Of this amount, 50 per cent was caught with pound nets, 33 per cent with haul seines, 12 per cent with gill nets, and 5 per cent with lines The bulk of the catch was taken in three counties, namely, Norfolk, 260,800; Princess Anne, 148,000; and Elizabeth City, 142,400 pounds.

The spot is caught in the Chesapeake from April until November, but the largest part of the catch is taken during September and October. Records obtained from a set of two pound nets at Lynnhaven Roads, Va., give the first catch of marketable spots in quantities of 10 pounds or more on the following dates: April 5, 1910, May 17, 1912, April 25, 1916, April 21, 1917, May 1, 1918, April 14, 1919, March 29, 1920, April 25, 1921, and April 4, 1922. The catch of marketable fish taken during April and May is usually very small. Large numbers of 6 and 7 inch spots often appear in April, but these are discarded by the fishermen as they are not in prime condition and have no marketable value. Large quantities of these small spots are destroyed annually by fishermen who do not take the time at the pound nets to return them to the water alive. This wastage is notably evident in the vicinity of Buckroe Beach and the lower York River, Va.

Much the greater part of the annual catch of spots is caught in the lower part of the bay, particularly in the fall, when they are caught with large haul seines. In the vicinity of Ocean View, Norfolk County, Va., two large haul seines were operated in 1920 by stationary shore equipment. About 164,000 pounds of spots were caught with these two seines, constituting a large part of the entire catch from Chesapeake Bay. From September 23 to October 27, 1922, in 40 separate hauls with the same two seines, 270,420 spots, weighing about 169,000 pounds, were caught, making an average of 4,225 pounds per haul. The largest single haul of spots ever taken in Chesapeake Bay, as far as known, was made by Lambert brothers (owners of one of the above-mentioned seines) on October 23, 1922, which consisted of 90,000 spots, weighing 50,000 pounds.

Haul-seine fishing for spot is begun in July, and sometimes large catches are made during the summer, but usually only small numbers are caught before September. The haul seines generally vary in length from 100 to 300 fathoms, and the largest seines are operated in the vicinity of Ocean View, Va. The Ocean View seine is 1,800 feet long, 25 feet deep, with 1-inch bar mesh in the center and 1¼-inch mesh at the ends. Both leads and corks are placed about 18 inches apart. To use such a large seine it is necessary to find a large area of water that has the proper depth and a comparatively smooth, clean bottom. Two of the largest seines at Ocean View are operated by crews of about 22 men each. The seine is set out a short while before low tide in order that most of the hauling may be done on slack water. The entire seine is carefully arranged in a seine boat, which is towed by a power boat while making a haul. The boats are run from the beach in a perpendicular line until a 200-fathom hauling line has been paid out. Then the 300 fathoms of seine are put out in the shape of a half ellipse. To the end of the seine put overboard last a hauling line, 350 fathoms in length, is attached, and this is paid out in an oblique line until shore is reached. By experience the fishermen are able to judge these distances very closely. After the seine has been paid out and the boats have returned to shore hauling is commenced by means of a winch operated by electric power. For about an hour no work, excepting that of two men who coil the lines as the seine is slowly being drawn to shore, is required of the fishermen. After the staffs or ends of the seine reach shore, the entire crew is needed to complete the haul. By means of a hook attached to loops made on the lead line, the power winch continues to do the heavy work of hauling, but now on only one end of the seine. When one end has been brought ashore a certain distance the other end is worked similarly. After about two-thirds of the seine has been beached the fishermen haul in the remainder of it without the aid of the winch. The closer the seine gets to the shore, the more care must be exercised in "footing" the lead line and apportioning the strain on the seine. If the catch of fish is small the seine can be drawn upon the beach by the crew. Very often, however, a large catch is made, which necessitates scooping the fish out with small hand nets, each of which is operated by two men. An average haul was timed as follows:

1. Started to prepare equipment for making haul at 7.20 p. m.
2. Beginning of haul, seine boat left beach at 8.11 p. m.
3. Seine boat returned to beach after paying out seine at 8.23 p. m.
4. Seine and catch of fish landed at 10.15 p. m.

When a catch of spots is landed the fish are scattered on the sandy beach and thereby become coated with sand. They are then packed carefully in "trays" and hauled to market by auto truck, or, if held over night at the fishery, they are placed in a cold-storage room. A tray is a shallow wooden box that holds about 120 ½-pound spots. The process of packing the fish in trays is called "setting up."

A large part of the catch of spots is shipped to various parts of Virginia and Maryland and a small part to outside markets. The spot is the favorite fish of Norfolk and many parts of Virginia, and for this reason the market is rarely glutted. During the heavy runs in the fall many consumers buy a tray of spots to salt down for use in the winter, and many are held in cold storage and disposed of during the winter, when fish are scarce locally. Ocean View, Va., is famous for its fine spots, and the trade name "Ocean View spots" is commonly used in the markets. During October the fish are in prime condition and large numbers are caught by anglers, who preserve their catch by salting.

From 1920 to 1922 the retail price of spots varied from 15 to 30 cents a pound. During the fall, however, an entire trayful of fish of about 60 pounds could be bought at a price varying from $4 to $10. The fish caught in the spring and summer usually weigh one-third to one-fourth of a pound, but the fish caught in the fall weigh from one-half to 1 pound each. The largest fish seen during the present investigation were taken at Ocean View, Va., on October 27, 1926. One of these individuals measured 13⅜ inches in length and weighed 19 ounces, and another had a length of 13⅛ inches and a weight of 22 ounces. These fish apparently represent the maximum length attained by the species.

The spot is generally common in coastal waters of the Middle Atlantic States and southward. It ascends brackish and fresh-water streams, and occasionally is taken in strictly fresh water. "Spot" is the most generally used common name in Chesapeake Bay. In the vicinity of the mouth of the Potomac it is sometimes called "croaker," and at Baltimore the name "silver gudgeon" was heard.

Habitat.—Massachusetts to Texas.

Chesapeake localities.—(a) Previous records: Various localities. (b) Specimens in collection: From every locality visited, from Havre de Grace, Md., to the entrance of the bay.

Following is a table of comparisons of length and weights of spots taken in the lower parts of Chesapeake Bay during April and May and during September and October. In a measure the table brings out the fact, well known to fishermen and dealers, that fish are not as "fat" in the spring of the year as they are in the fall.

Length, in inches	Spring		Fall		Length, in inches	Spring		Fall	
	Weight	Number of fish weighed and measured	Weight	Number of fish weighed and measured		Weight	Number of fish weighed and measured	Weight	Number of fish weighed and measured
	Ounces		*Ounces*			*Ounces*		*Ounces*	
4–4⅜			0.5	2	8–8⅜	4.0	16	4.0	37
4½–4⅞			.75	50	8½–8⅞	5.0	2	5.0	18
5–5⅜			1.00	117	9–9⅜	5.3	1	6.4	19
5½–5⅞	1.1	6	1.3	142	9½–9⅞	6.6	1	7.9	23
6–6⅜	1.5	24	1.75	107	10–10⅜			9.0	26
6½–6⅞	1.9	35	2.3	66	10½–10⅞			10.0	10
7–7⅜	2.7	32	2.9	70	12½–12⅞			18.0	1
7½–7⅞	3.2	14	3.4	48	13–13⅜	19.0	1	22.0	1

120. Genus SCIÆNOPS Gill. Red drums

Body elongate, compressed; back moderately arched; teeth in the jaws well developed; preopercle serrate in young, becoming entire with age; slits and pores about the mouth well developed; no barbels; no scales on soft dorsal; caudal fin pointed in very young, becoming square to slightly concave with age. A single species, reaching a large size, is known.

153. Sciænops ocellatus (Linnæus). "Drum"; "Red drum"; Redfish; Channel bass.

Perca ocellata Linnæus, Syst. Nat., ed. XII, 1766, p. 483; South Carolina.

Sciænops ocellatus Uhler and Lugger, 1876, ed. I, p. 119; ed. II, p. 100; Bean, 1891, p. 89; Jordan and Evermann, 1896–1900, p. 1453, Pl. CCXXII, fig. 567; Evermann and Hildebrand, 1910, p. 162; Welsh and Breder, 1923, p. 184.

Head 2.85 to 3.3; depth 3.35 to 3.95; D. X–I, 23 to 25; A. II, 8; scales 40 to 45. Body elongate; back somewhat elevated; ventral outline nearly straight; head rather long and low; snout conical, 3.3 to 3.8; eye 3.15 to 4.75; interorbital 3.7 to 4.6; mouth horizontal; lower jaw included, with large pores but no barbels; maxillary reaching nearly opposite posterior margin of eye, 2.1 to 2.45 in head; teeth in the jaws in villiform bands, the outer ones in the upper jaw enlarged; preopercular margin coarsely serrate; gill rakers short, 8 or 9 on lower limb of first arch; scales rather large, firm, strongly ctenoid, reduced in size on head; dorsal fins contiguous, the first composed of rather stiff, pungent spines, not much higher than the second; caudal fin pointed in very young, becoming straight to slightly concave in larger fish; second anal spine thick, much shorter than the longest soft rays;

ventral fins moderate, inserted a little behind base of pectorals; pectoral fins rather small, 1.55 to 1.85 in head.

Color of a 42½-inch specimen silvery, tinged with greenish bronze above; white below; scales on sides with dark centers, forming stripes; one irregular jet black spot at base of caudal above lateral line; dorsal and caudal fins dusky; anal and ventrals white; pectorals bright rusty on outer

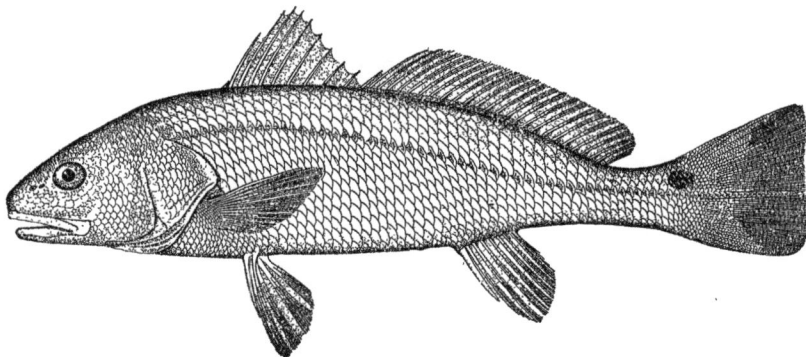

FIG. 157.—*Sciænops ocellatus*. From a specimen 12 inches long

part. Some fish turn red after death. The number of spots at base of caudal varies from one to several. One specimen, 15 inches long, for example, had four spots on one side and eight on the other, and they extended from the base of the caudal along the sides. Most frequently, however, only a single spot is present on each side at the base of the caudal. Young of about 100 millimeters (4 inches) and less in length have large black spots or blotches distributed over the entire side and back. These spots are present, being more diffuse, however, in our smallest specimens (20 millimeters in length). The young, up to 40 millimeters in length, have a dark vertical bar on base of caudal.

The collection contains four specimens, ranging in length from 165 to 225 millimeters (6½ to 9 inches), and many small ones ranging from 20 to 90 millimeters (⅘ to 3½ inches) in length. Many large examples, too large to preserve conveniently, were observed.

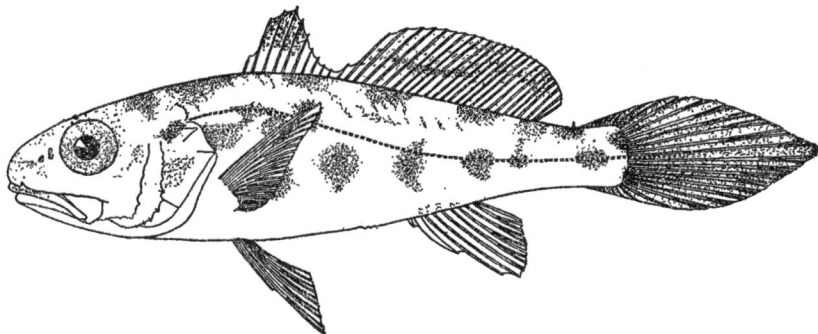

FIG. 158.—Young, 42 millimeters long

The young differ from the adults chiefly in color and in the shape of the caudal fin. These differences have been pointed out in the description. This drum is characterized by the elongate body, the absence of barbels about the mouth, and the presence of one or more black spots at base of caudal.

The food of this fish, according to the contents of 15 stomachs taken from fish ranging in length from 30 to 1,075 millimeters (1¼ to 42½ inches), consists of crustaceans, the smaller fish having fed principally on Gammarus and Mysis and the larger ones on shrimp.

Little is known about the spawning and life history of this fish. Welsh and Breder (1923, p. 184) offer the following: "Spawning occurs chiefly in the late fall or early winter, although from the size of some young fish taken in Florida waters in January it is probable that some spawning may take place as early as September. The eggs and larval stages have not been studied."

The smallest specimens (20 millimeters and upward) in the present collection appear to be the youngest of record. In view of the fact that records of young red drum are scarce, the following catches made in Chesapeake Bay are given:

Date	Locality	Number of specimens	Length	
			Millimeters	*Inches*
1921				
Sept. 19	Crisfield, Md.	6	24–34	1.0–1.3
Sept. 20	Cape Charles, Va.	7	20–42	.8–1.6
Oct. 7	Buckroe Beach, Va.	6	44–53	1.7–2.1
Oct. 11	York River (lower)	23	26–46	1 –1.8
Oct. 15	Rappahannock River (lower)	1	49	2
Oct. 26	Patuxent River (lower)	45	25–45	1 –1.8
Nov. 21	Crisfield, Md.	2	48–54	1.9–2.1
Nov. 23	Cape Charles, Va.	28	39–90	1.5–3.5
July 1	York River (lower)	4	165–225	6.5–8.8

The red drum has but small commercial importance in Chesapeake Bay. During 1920 the catch for the entire bay was 17,565 pounds, worth $280. Maryland is credited with 28 per cent of this amount, or 4,835 pounds, and Virginia with 72 per cent, or 12,730 pounds. Virtually the entire catch was taken with pound nets.

This fish is taken from May until October and is most abundant during the spring and fall. It is most common about the entrance of the bay, and is seldom taken above Chesapeake Beach, Md.

The size of market fish varies from ½ to about 40 pounds. Fish of more than 50 pounds are rare. During 1922, the smaller fish (10 pounds or less) sold at retail for 10 to 15 cents a pound, whereas the larger fish were cut into steaks, which sold for 5 to 10 cents per pound. The smaller fish are superior to the larger ones, which are coarse of flesh and lacking in flavor. In the lower Chesapeake markets the demand for this species is small. The names most used for the species in the Chesapeake region are "drum" and "red drum." Along the New Jersey coast, where this drum is a favorite with surf anglers, it is known as the "channel bass." Along the South Atlantic and Gulf coasts, it is an important food fish and is called "redfish."

Habitat.—Massachusetts to Texas; not taken in commercial numbers north of the coast of New Jersey.

Chesapeake localities.—(a) Previous records: Gloucester Point, Cape Charles city, "southern parts of the bay." (b) Specimens in collection (young): Solomons and Crisfield, Md., lower Rappahannock River, lower York River, Cape Charles, and Buckroe Beach, Va. Larger fish were observed at many points, from Solomons, Md., southward to the entrance of the bay.

121. Genus LARIMUS Cuvier and Valenciennes. Banded drums

Body rather short, compressed; skull firm, not greatly cavernous; upper jaw with slits and pores little developed; no barbels; teeth all small, no canines; snout very short; mouth large, very oblique to vertical; preopercle with a membranous edge, without bony serræ; pseudobranchiæ well developed; gill rakers long and slender; second dorsal fin long; anal fin short.

154. Larimus fasciatus Holbrook. Banded drum; "Bastard perch."

Larimus fasciatus Holbrook, Ichthyol., South Carolina, 1860, p. 153, Pl. XXII, fig. 1; Charleston, S. C. Bean, 1891, p. 88; Jordan and Evermann, 1896–1900, p. 1424.

Head 2.9 to 3.7; depth 2.4 to 2.85; D. X–I, 24 to 27; A. II, 6 to 8; scales 50 to 56. Body rather deep, compressed; head short, deep; snout very short, blunt, 3.75 to 4.5 in head; eye 3 to 3.8; interorbital 2.8 to 3.06; mouth very oblique; lower jaw protruding; maxillary reaching about to vertical from center of eye, 1.85 to 2.5 in head; teeth all small, pointed, in a single series in each jaw; gill rakers long and slender, 23 to 25 on lower limb of first arch; scales moderate, ctenoid, extending forward on head, cheeks, and opercles; small scales present at least on the base of the fins; dorsal fins contiguous, the first with slender, flexible spines, the third and fourth the longest, rather longer than the longest soft rays; caudal fin slightly rounded in adult, lanceolate in young; anal fin small, the second spine large and strong, 2.5 to 2.65 in head; ventral fins rather large, inserted slightly behind base of pectorals; pectoral fins 1 to 1.35 in head.

Color in alcohol grayish above, silvery underneath; sides with seven to nine vertical black bars; fins plain with dusky punctulations; mostly yellowish in life.

This species is represented in the present collection by three specimens, respectively 190 millimeters (7½ inches), 200 millimeters (7⅞ inches), and 215 millimeters (8½ inches) long. The foregoing description is based upon these and 12 specimens from Beaufort, N. C., ranging in length from 30 to 205 millimeters (1¼ to 8⅛ inches). This fish is the only one of this genus of tropical fishes that ranges northward as far as Chesapeake Bay. It is readily recognized by its short body, nearly vertical mouth, and dark bands on the sides.

This species is not a stranger to the fishermen of the lower sections of Chesapeake Bay, where it is known as the "bastard perch," some of the fishermen believing it to be a cross between the "sand perch" (*Bairdiella chrysura*) and the black drum (*Pogonias cromis*). However, it is not common enough to rank as a food fish of any importance in Chesapeake Bay. Welsh and Breder (1923, p. 170) offer the following relative to its distribution: "South of this point (Cape Hatteras) and on the shores of the Gulf of Mexico it is one of the most abundant fishes, being taken in large numbers in the trawls of the shrimp fishermen." These authors, however, say that the majority of the fish caught in shrimp trawls are small, individuals exceeding 8 inches in length being rare; and that, furthermore, the species is of little or no economic importance because of its small size.

Habitat.—Massachusetts to Texas, occurring only as a straggler north of Chesapeake Bay.

Chesapeake localities.—(*a*) Previous record: Cape Charles city, Va. (*b*) Specimens in collection: From Lynnhaven Roads, Va. The species was observed only in the southern part of the bay.

122. Genus BAIRDIELLA Gill. Mademoiselle

Body moderately elongate; compressed; back elevated; mouth oblique; gill rakers rather long; preopercle with serrate margin, the lower spine curved downward and forward; skull little cavernous; coloration plain. A single species of this genus of tropical fishes occurs in Chesapeake Bay.

155. Bairdiella chrysura (Lacépède). "White perch"; "Sand perch"; "White sand perch"; "Virginia perch"; "Yellow-tail"; "Tint."

Dipterodon chrysurus Lacépède, Hist. Nat. Pois., III, 1803, p. 64; South Carolina.
Liostomus xanthurus Uhler and Lugger, 1876, ed. I, p. 117; ed. II, p. 99. (Not *L. xanthurus* Lacépède.)
Bairdiella chrysura Bean, 1891, p. 88; Jordan and Evermann, 1896–1900, p. 1433, Pl. CCXXII, fig. 566; Evermann and Hildebrand, 1910, p. 163; Fowler, 1918, p. 18; Welsh and Breder, 1923, p. 171.

Head 2.85 to 3.4; depth 2.8 to 3.15; D. XI or XII, 19 to 21; A. II, 9 or 10; scales 55 to 59. Body oblong, compressed; back moderately elevated; head moderate; snout conical, 3.75 to 4 in head; eye 2.85 to 4.15; interorbital 3.75 to 4.15; mouth a little oblique, terminal; maxillary reaching nearly or quite below posterior margin of pupil, 1.95 to 2.35 in head; teeth small, those in upper jaw in a band, mostly in a single series in lower jaw; preopercle serrate, a few of the spines at angle somewhat enlarged; gill rakers rather long, slender, 14 to 16 on lower limb of first arch; scales moderate, rather firm, ctenoid, small scales covering most of soft dorsal, caudal, and anal fins, also present on base of ventrals and pectorals; dorsal fins contiguous, the first composed of slender spines, the third and fourth spines the longest, higher than any of the rays in the soft part of the fin; caudal fin very slightly double, truncate in adult, the middle rays longest, broadly rounded in young; anal fin with one very short and one rather long strong spine, not quite reaching the tips of the soft rays immediately behind it, origin of fin somewhat behind middle of base of soft dorsal; ventral fins inserted a little behind base of pectoral, equal to or a little longer than pectorals; pectoral fins short, not reaching tips of ventrals, 1.4 to 1.55 in head.

Color olivaceous, greenish, or bluish gray above; lower part of sides and abdomen bright silvery; fins mostly yellowish; dorsals and caudal and sometimes the anal partly dusky.

Many specimens of this species, ranging in length from 22 to 225 millimeters (⅞ to 8⅞ inches), were preserved. The young do not differ greatly from the adults. Very young of 22 millimeters in length are rather deeper and more strongly compressed anteriorly than the adults. Fish of this size are largely pale in color, with dusky punctulations and a dark spot on the opercle at the upper

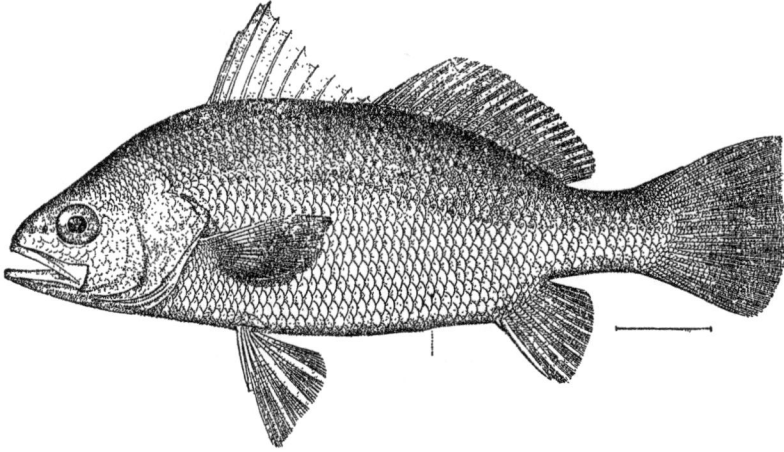

FIG. 159.—*Bairdiella chrysura*

posterior angle. Fish 40 millimeters and upward in length are colored essentially like the adult and are very similar in form and general appearance. This fish is readily recognized by its plain coloration, yellowish caudal fin, rather deep, compressed body, its oblique terminal mouth, and serrated preopercle.

The food of this fish in Chesapeake Bay, as shown by the stomach contents of 100 specimens examined, consists very largely of small and minute crustaceans. Foods of much less importance are annelids and fish, Only two individuals of the entire lot examined had fed on fish.

Spawning takes place in late spring and early summer. Ripe fish of both sexes were trawled in 12 fathoms off Crisfield as early as May 16. Many fish had already spawned by June 11. The eggs are small and are produced in large numbers. A single apparently nearly ripe female, 140 millimeters long, taken in Lynnhaven Roads on May 9, 1921, contained approximately 52,800 eggs. The embryology and larval development of this species has been described by Kuntz (1914, pp. 4–13) from material secured at Beaufort, N. C. The eggs are described as being spherical in form and 0.7 to 0.8 milli-

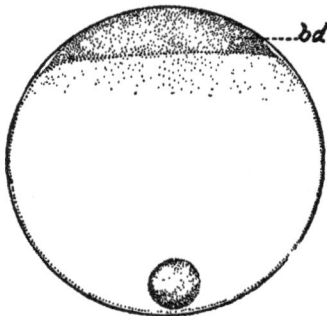

FIG. 160.—Egg recently fertilized, with fully developed blastodisc

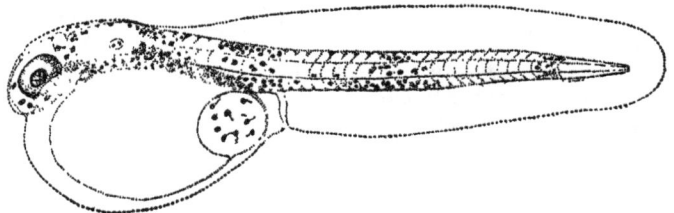

FIG. 161.—Newly hatched larva, 1.8 millimeters long

meter in diameter, pelagic, and hatched at laboratory temperature in about 18 hours. At the time of hatching the larvæ are 1.5 to 1.8 millimeters in length. By the time the young fish reaches a length of 30 millimeters the fins are fully differentiated, the body is covered with small, deeply embedded scales, and they show many of the diagnostic characters of the adult.

Welsh and Breder (1923, p. 174) contribute the following relative to the rate of growth:

By the first winter a length of from 6 to 14 centimeters (2½ to 5½ inches) is attained, depending on the time of hatching, the average length for May-hatched fish being about 12 centimeters (4¾ inches) and for June-hatched fish about 10 centimeters (4 inches). During the winter months growth practically stops. The average increment of growth the second season is about 6 centimeters (2⅜ inches), with a length for the second winter of from 12 to 20 centimeters (4¾ to 8 inches). The first spawning occurs in the third season, when the fish are 2 years old and between 15 and 21 centimeters in length (6 to 8¼ inches). After the first spawning the growth is slow, the largest fish of which scales were examined having reached a length of 23 centimeters (9 inches) at the age of 6 years.

The foregoing data by Welsh and Breder were derived from the measurement of specimens taken at Beaufort, N. C., and in Chesapeake Bay, supplemented by the study of scales.

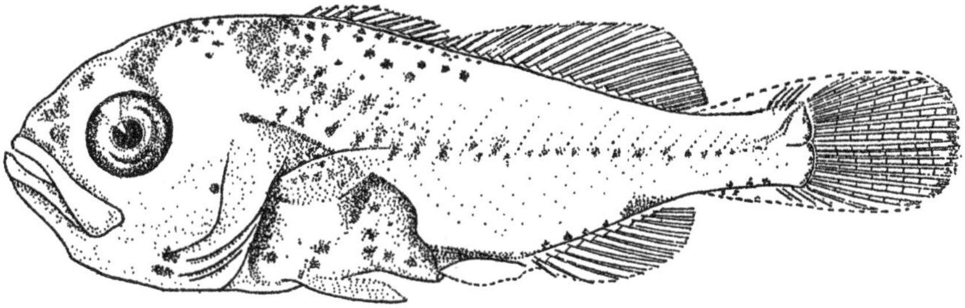

FIG. 162.—Young, 11 millimeters long

Data relative to the rate of growth derived from a large series of measurements made of young fish caught in Chesapeake Bay during 1921 and 1922 do not differ notably from those given by Welsh and Breder. The following variations (that is, extremes in length) were secured from certain collections apparently consisting of individuals in their first summer: July 8 to 12, 23 to 58 millimeters (⅞ to 2⅛ inches), about 500 fish; July 23 to 31, 20 to 85 millimeters (⅘ to 3⅓ inches), about 1,000 fish; August 4 to 8, 25 to 90 millimeters (1 to 3⅓ inches), about 400 fish; September

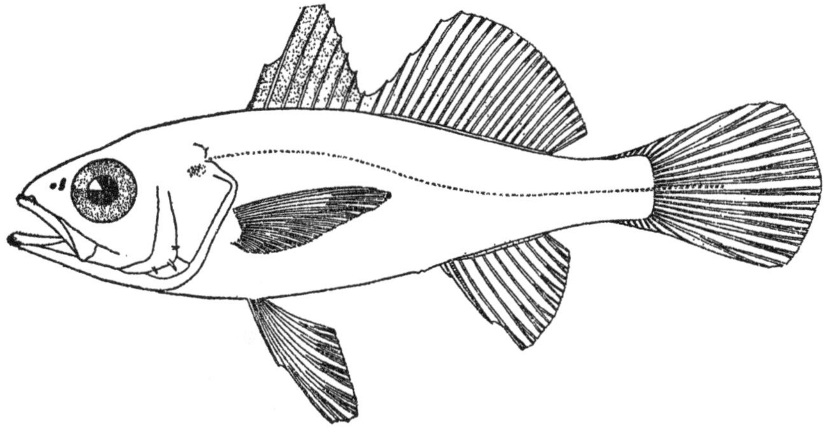

FIG. 163.—Young, 33 millimeters long

14 to 16, 40 to 109 millimeters (1⅗ to 4⅓ inches), 40 fish; October 5 to 19, 55 to 98 millimeters (2⅙ to 3⅞ inches), 70 fish; November 22 and 23, 76 to 117 millimeters (3 to 4½ inches), 40 fish. It was found from measurements of larger fish (that is, fish ranging from about 4½ to 9 inches), taken during September and October, that these fell into two groups, a decided break occurring in a frequency curve between the two. The predominating sizes of the individuals composing the group of smaller fishes ranged from 130 to 145 millimeters (5⅛ to 5¾ inches). These fish were believed to be in their second summer. The predominating sizes of the next group ranged from 165 to 195 millimeters (6½ to 7⅔ inches), and these probably were in their third summer.

Some of the young fish of this species stay in the bay throughout the winter and were not infrequently taken in deep water.

The sand perch is a very abundant fish in Chesapeake Bay, but because of its small size a comparatively small part of the catch is marketed, the remainder, in large part, being wasted. Thousands of discarded sand perch frequently are seen floating on the surface of the water or strewn on the beaches about the fishing camps. It is a food fish of good flavor, and a large percentage of the catch could be utilized if a demand for the fish were created. A large part of the commercial catch consists of fish 7½ to 8½ inches in length, weighing about one-fourth pound each.

The sand perch is caught chiefly with pound nets, haul seines, and hook and line. The season extends from April until November, October being the month when the fish are most abundant. It is most common in the lower part of the bay, decreasing in abundance northward, and north of Baltimore it is rarely taken.

Various names have been given to this fish in Chesapeake Bay. "Sand perch," "white sand perch," and "white perch" are the most commonly used. Confusion is caused by calling this species "white perch," as this name is well established for *Morone americana*. Names less commonly used are "Virginia perch," "yellow-tail," and "tint."

The largest fish observed in Chesapeake Bay was 9¼ inches in length, which is about the maximum size attained by this species, and it weighed 6 ounces. Fish of slightly less than 9 inches are common and of sufficient size to be utilized as food. Small quantities are sold in the Norfolk and Baltimore markets, and some are peddled inland among the country people. The retail price during 1922 was 10 to 15 cents per pound.

Habitat.—New York to Texas.

Chesapeake localities.—(*a*) Previous records: Various sections of Chesapeake Bay, from the mouth of the Potomac River southward. (*b*) Specimens in collection: From many localities, from Annapolis, Md., to Cape Charles and Cape Henry, Va. Apparently rare north of Annapolis.

Comparison of lengths and weights of sand perch

Number of fish weighed and measured	Length	Weight	Number of fish weighed and measured	Length	Weight
	Inches	*Ounces*		*Inches*	*Ounces*
1	3	0.3	5	6¼	1.6
3	3½	.4	18	6½	2.2
4	3¾	.4	20	6¾	2.3
7	4	.5	26	7	2.6
8	4¼	.6	34	7¼	3.0
11	4½	.7	29	7½	3.3
12	4¾	.8	33	7¾	3.6
18	5	1.0	37	8	4.0
40	5¼	1.1	28	8¼	4.2
51	5½	1.2	15	8½	4.5
18	5¾	1.3	5	9	5.0
15	6	1.5	5	9¼	5.3

NOTE.—Most of these fish were from total catches made with collecting seines and are therefore unselected. The scarcity of fish 6 to 6¾ inches long, already mentioned, is illustrated in a measure by this table. It is probable that most of the fish of less than 6½ inches are immature.

123. Genus STELLIFER (Cuvier) Oken

This genus is distinguished from related genera by the very cavernous construction of the bones of the skull, the septa being reduced to the thinness of the walls of honeycomb. The skull is rather broad, somewhat depressed between the eyes, and more or less spongy to the touch.

156. Stellifer lanceolatus (Holbrook).

Homoprion lanceolatus Holbrook, Ichth, South Carolina, ed. I, 1855, p. 168, Pl. XXIII; Beaufort, S. C.
Stellifer lanceolatus Jordan and Evermann, 1896–1900, p. 1443.

Head 3.45; depth 2.75; D. XII–I, 21; A. II, 8; scales 49. Body oblong, compressed; head rather low; snout blunt, 3.5 in head; eye 4.05; interorbital broad, 2.4; mouth moderate, oblique; lower jaw included; maxillary reaching about opposite posterior margin of pupil, 2.25 in head; teeth in the jaws in bands, the one on lower jaw very narrow; preopercular margin with enlarged spines; gill rakers slender, 21 on lower limb of first arch; scales firm, ctenoid, extending more or less on all of the fins; dorsal fins contiguous, the first with slender, flexible spines; the longest spines not much higher than the longest rays of the second dorsal; caudal fin lanceolate; second anal spine enlarged, not quite as long as the soft rays following it; ventral fins rather small, inserted under base of pectorals; pectoral fins rather long, 1.05 in head.

Color in alcohol uniform silvery, darker above than below; fins mostly plain translucent, the spinous dorsal with black margin.

A single specimen, 165 millimeters (6½ inches) in length, was secured, and it forms the basis for the foregoing description.

This small drum apparently has not been recorded from any locality north of Beaufort, N. C. It probably seldom exceeds the length of the fish in hand, and although abundant on the South Atlantic and Gulf coasts it has no commercial value. The young are said to resemble young croakers and spots, from which however, they may be distinguished by the larger head and strongly oblique mouth. Welsh and Breder (1923, p. 175) state that spawning occurs in late spring or early summer, May and June being the principal months during which spawning takes place on the Atlantic coast. The eggs and larvæ have not been studied, and the species as yet has no common name.

Habitat.—Chesapeake Bay to Texas; not common north of South Carolina.

Chesapeake localities.—(*a*) Previous records: None. (*b*) Specimen in collection: Taken in a pound net in Lynnhaven Roads, Va., July, 1921.

124. Genus MICROPOGON Cuvier and Valenciennes. Croakers

Body elongate, compressed; back somewhat elevated; preopercle strongly serrate; teeth in the jaws in villiform bands; chin with a row of short, slender barbels on each side; gill rakers short; spinous dorsal consisting of 10 to 11 spines; anal with two spines, the second strong and of moderate length. A single species of this genus is known from Chesapeake Bay.

157. Micropogon undulatus (Linnæus). Croaker; Crocus; "Hardhead"; "King Billy."

Perca undulatus Linnæus, Syst. Nat., ed. XII, 1766, p. 483; South Carolina.

Micropogon undulatus Uhler and Lugger, 1876, ed. I, p. 119; ed. II, p. 102; Bean, 1891, p. 89; Smith, 1892, p. 72; Jordan and Evermann, 1896–1900, p. 1461, Pl. CCXXIV, fig. 570; Evermann and Hildebrand, 1910, p. 162; Fowler, 1912, p. 56, and 1918, p. 18; Welsh and Breder, 1923, p. 180.

Head 2.95 to 3.4; depth 2.9 to 3.65; D. X–I, 28 or 29; A. II, 8; scales 64 to 72. Body elongate, compressed; back moderately elevated; head rather long; snout conical, projecting beyond the mouth in the adult and proportionately much longer than in the very young, 2.85 to 3.75 in head; eye 3.35 to 4.8; interorbital 3.35 to 3.8; mouth moderate, horizontal, inferior; maxillary reaching a little past front of eye to below middle of eye, 2.3 to 2.85 in head; teeth in the jaws all small, in broad villiform bands; chin with several pores and a row of short, slender barbels on each side; preopercle with strong, short spines on margin; gill rakers short, 14 to 16 on lower limb of first arch; scales moderate, reduced anteriorly above lateral line, strongly ctenoid, extending on the caudal but not on the other fins; dorsal fins contiguous, or more or less continuous in young, the first composed of slender spines, somewhat elevated, the third and fourth spines longest, higher than any of the rays in the soft part; caudal fin slightly double concave in adult, with the upper and middle rays longest, sharply pointed in very young; anal fin small, with two strong spines, the first very short, the second about two-thirds the length of the soft rays; ventral fins moderate, inserted under and slightly behind base of pectorals; pectorals rather long in adult, reaching well beyond tips of ventrals, scarcely reaching tips of ventrals in young, 1.15 to 1.5 in head.

Color greenish or grayish silvery above, silvery white below, highly irridescent in life; back and sides with numerous brassy or brownish spots arranged in oblique, wavy bars on sides, becoming less distinct in large individuals. Young of about 100 millimeters and less in length do not have the color pattern of the adult, for they are paler; the upper parts bear dark blotches, or mere points, in the very young, these spots becoming vertically elongated when the fish reaches a length of about 80 millimeters; dorsal fins with numerous dark spots; caudal and pectorals greenish dusky; base of pectorals dusky; anal and ventrals yellowish to orange.

Numerous specimens of this common fish, ranging in length from 10 to 355 millimeters (⅖ to 14 inches) in length, were preserved. The principal differences between the young and the adults have been pointed out in the description.   The croaker is most readily recognized by the inferior mouth, the series of short barbels on each side of the chin, and the very strongly serrated preopercle.

In this species both sexes are capable of making a croaking sound.   The croaking apparatus, as explained under *Cynoscion regalis*, consists of a pair of croaking muscles and the air bladder. The air bladder is peculiarly modified, for it has two hornlike appendages anteriorly and is slender posteriorly, ending in a sharply pointed, tail-like appendage.   The croaking sound may be heard for a considerable distance, and it may be emitted underneath the surface of the water and after the fish is removed from the water.

The croaker, with its inferior mouth and barbels on the chin, is adapted for bottom feeding. The food in Chesapeake Bay, as shown by 392 stomachs examined, consists of the following, named in the order of their apparent importance: Crustaceans, annelids, mollusks, ascidians, ophiurans, and fish.   Besides these, considerable sand and vegetable débris were present, which may have been taken incidentally in the capture of food.   The first three groups of food named are all important, the other three being represented merely as "traces."   It is noteworthy that only three of the croakers examined had fed, in part, on fish.   The crustaceans consumed were mostly small or minute, and the mollusks consisted of small bivalves and small gastropods.   These data appear to show that the croaker utilizes as food the lower and smaller forms of animal life, which have no

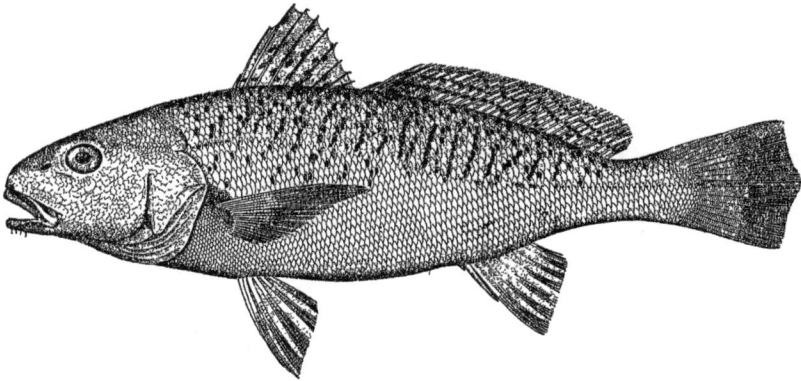

Fig. 164.—*Micropogon undulatus.*   From a specimen 12 inches long

direct commercial value and which in large part probably would not serve a useful purpose to man in any other way.

Welsh and Breder (1923, p. 180) state that the spawning season is a long one, extending from August to December, and possibly later in southern waters.   No ripe females appear to have been taken in August.   These authors, however, base their contention that spawning does take place as early as August upon the fact that on September 12, 1916, fry 32, 36, and 41 millimeters in length were taken in Chesapeake Bay and in New York Bay; fry 22.5, 28, and 29 millimeters in length were taken from September 7 to 21.   Fish with well-developed roe were common in Chesapeake Bay during October and the early part of November, which probably is the principal spawning period.   The eggs and larvæ of the croaker have not been studied, the smallest individuals known being those at hand, having a total length of about 10 millimeters.   In fish of this length most of the fins are already well formed, the mouth is large, and the body is largely unpigmented. The number of eggs produced by a single fish apparently is large.   The roe of a female 395 millimeters (15½ inches) long, caught in the mouth of the York River, on October 25, 1921, contained approximately 180,000 eggs of uniform development.

The statement made by Welsh and Breder (1923, p. 180) that spawning extends over a long period of time (August to December) is substantiated in a measure by the large variation in size of young fish taken during the fall and winter months, when many hundreds of specimens were collected, usually with the beam trawl, in the deeper waters of the bay.   The young fish were so numerous that occasionally as many as 5 and 6 quarts of fish were taken in a single haul.   Specimens taken in October ranged in length from 10 to 105 millimeters (⅖ to 4 inches); in November the range was from 15 to 116 millimeters (⅗ to 4½ inches); in December from 11 to 120 millimeters (⅖ to 4¾ inches); in January, 10 to 110 millimeters (⅖ to 4⅓ inches); and in March from 32 to 64 millimeters (1¼ to 2½ inches). It would appear from these data that spawning continues during at least a part of the winter.

It was found impossible to follow, through measurements, the rate of growth of young fish, even until the age of 1 year was reached. This, no doubt, is due largely to the long spawning period, as pointed out in the preceding paragraph. It is possible that an irregular rate of growth also is maintained, and that would add to the difficulty. Hundreds of fish were measured and length frequencies were plotted with the view of finding breaks, if they existed, upon which growth curves might be based. None were found, and scale studies were not attempted. We are unable, therefore, to give any information relative to the rate of growth.

The croaker is one of the most valuable and abundant food fishes caught in Chesapeake Bay. During 1920 it ranked second in quantity and third in value, the catch being 14,170,385 pounds and the value $393,162.

In Maryland it ranked third in quantity and sixth in value, the catch being 1,130,590 pounds, valued at $31,683. Of this amount, 54 per cent were caught with pound nets, 43 per cent with haul seines and drift nets, 2 per cent with lines, and 1 per cent with fyke nets. Dorchester County is credited with the largest catch, amounting to 380,945 pounds, followed by Calvert, with 279,400 pounds, and Talbot, with 134,800 pounds.

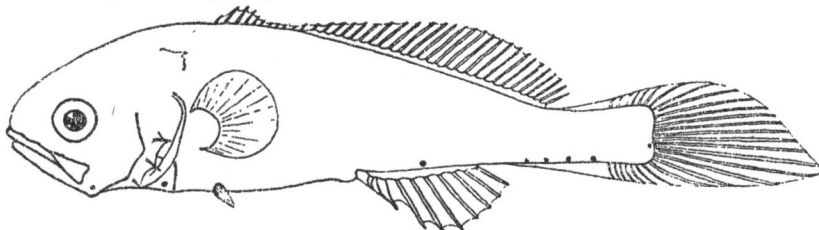

FIG. 165.—*Micropogon undulatus.* Young, 12.25 millimeters long

In Virginia it ranked second in both quantity and value, the catch being 13,039,795 pounds, valued at $361,479. Of this amount, 87 per cent were caught with pound nets, 6 per cent with lines, 3 per cent with drift nets, 2 per cent with haul seines, and 2 per cent with fyke nets. The five counties having the largest catches were Elizabeth City, 4,465,600 pounds; Gloucester, 1,981,300 pounds; Mathews, 1,803,955 pounds; York, 1,552,924 pounds; and Accomac, 1,067,529 pounds.

The croaker is caught in the bay from March until October, a few being taken in the lower part of the bay as late as the beginning of December. For a set of two pound nets at Lynnhaven Roads, Va., the first catch of croakers for each year mentioned is recorded as follows: March 23, 1910; March 29, 1912; April 12, 1916; April 4, 1917; March 21, 1918; March 17, 1919; March 24, 1920; March 15, 1921; March 21, 1922; and March 15, 1923. In the spring they suddenly appear in great numbers, and the bulk of the yearly catch is usually taken during March, April, and May. During 1922 no croakers were caught in the previously mentioned pound nets until March 21, when the first catch consisted of 8,675 pounds, with a catch of 10,400 pounds on the following day.

After about June 10 pound-net catches of croakers take a decided drop, but apparently this does not mean that the fish have left the bay. Although these fish are present during March and April in large numbers, they seldom bite on a hook. In the latter part of May and after June 1 they readily take the hook, however, and are caught in large quantities in that way throughout the summer and fall.

Drift nets are used to advantage in some localities, notably Cape Charles and the lower York and Rappahannock Rivers, Va. This apparatus is used where the bottom is smooth and free of débris. The so-called "drift net" is really nothing but a gill net ranging from about 200 to 400 fathoms in length and 20 feet in depth. It is weighted to such an extent that the tide carries it slowly over the bottom. Buoys placed at intervals indicate the progress of the net to the operators. When not fishing the net is wound on a large wooden reel, which is part of the equipment of the power boat used in operating the net.   Good catches are made by this method when fish are plentiful. Some of the advantages of drift-net fishing are as follows: The outfit is comparatively inexpensive; the boat, with its gear, can be moved from place to place without delay; a crew of two or three men may operate it successfully; and fishing is done directly from the power boat, therefore requiring no dories.

The early run of croakers sometimes is so great that markets are glutted for a time and the price falls to such an extent that it is not profitable to ship the fish.   At such times the croakers are either turned out of the nets or kept entrapped in the hope that the market may rise within a few days.   Such a glut fortunately does not occur often, and when it has occurred it has been of short duration.   The large catches are made during the height of the shad season and for that reason are not looked upon with much favor by the pound-net fishermen.

Losses sometimes are suffered by fishermen because the large catches can not be disposed of profitably when the prolific spring run occurs.   During this time the fish often are shipped in car-load lots to distant markets, which in turn become glutted.   The croaker is a good food fish, and because it is caught in abundance early in the spring when the weather is cool it would appear that the distribution could be extended to new and even more distant markets, where at present it is but little known and where cheap and wholesome fish are scarce.   During 1921 and 1922 the wholesale price ranged from $4 to $15 per barrel during most of the season, and the retail prices ranged from 5 to 20 cents a pound, the general average being about 12½ cents.

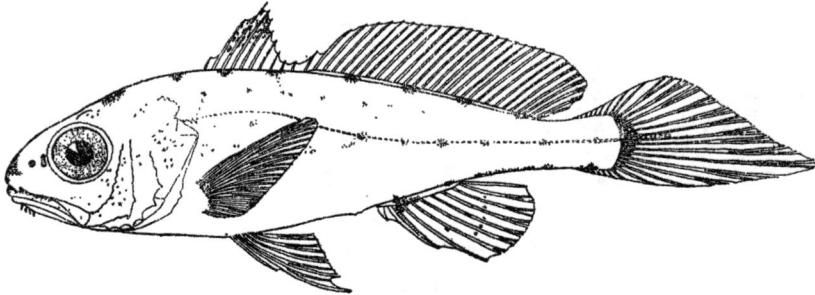

FIG. 166.—*Micropogon undulatus.*   Young, 34 millimeters long

The croaker, as already stated, usually appears suddenly in March or April in great abundance, and the first catches are made with pound nets operated near the entrance to the bay.   Thereafter the fish appear to migrate gradually up the bay, as the earliest catches at some distance from the entrance always are made somewhat later.   The croaker is common throughout the summer in the shallower shore waters.   As cool weather arrives, late in September and in October, the fish become scarce along the shores.   At this time, however, large fish are abundant in the deeper waters of the lower Potomac, Rappahannock, York, and James Rivers.   No adult croakers are taken during the winter by the commercial fishermen in their restricted operations, and, as indicated elsewhere, no large fish occur in our collections, which were made at that season.   It is quite certain, therefore, that this fish, like many others, leaves the bay upon the approach of winter.

The names "croaker," "crocus," and "hardhead" are used interchangeably throughout the Chesapeake region.   Small fish, less than 10 inches long, are called "pinhead croakers."   In the markets of the Chesapeake region the croaker is on sale from late March until late November, and because of its abundance and low price it is one of the most popular of all the food fishes.   In quality it is generally considered inferior to the spot, squeteague, and others.   Some persons, however, prefer it to almost any other species.   The croaker is most abundant in the southern part of the bay, decreasing gradually northward, and above Baltimore County, Md., it is taken only as a straggler.

The size of market fish usually ranges from one-half to 1½ pounds.   At times large quantities of "pinhead croakers" (fish 7 to 10 inches long) are caught, and these generally are most common in the spring.   On September 18, 1922, a set of two pound nets at Lynnhaven Roads, Va., caught 1,200 pounds of croakers, among which were many fish weighing 3 pounds or more, which is an extraordinary size. The five largest were 19 to 20 inches in length and ranged in weight from 3 pounds 6 ounces to 4 pounds 2 ounces. The last-mentioned fish is the largest recorded from Chesapeake Bay and about the maximum size attained by the species.

Habitat.—Massachusetts to Texas. Not common north of New Jersey.

Chesapeake localities.—(*a*) Previous records: Various parts of the southern sections of the bay. (*b*) Specimens in collection: From many localities, from Baltimore southward. Reported by fishermen as rare north of Baltimore, where it was not seen by us.

Comparison of lengths and weights of croakers

Number of fish weighed and measured	Length	Weight
	Inches	Ounces
4	3	0.2
1	4½	.5
1	5	.7
5	5½	1.0
12	6	1.3
19	6½	1.6
27	7	2.0
35	7½	2.5
78	8	3.1
58	8½	3.7
34	9	4.3
10	9½	5.3
18	10	6.2
19	10½	7.5
35	11	8.7

Number of fish weighed and measured	Length	Weight	
	Inches	Pounds	Ounces
37	11½		9.5
65	12		10.8
28	12½		12.2
36	13		14.0
16	13½	1	0
17	14	1	1.3
13	14½	1	3.0
16	15	1	5.0
5	15½	1	7.4
6	16	1	9.0
3	16½	1	12.8
1	17	2	4.0
2	19	3	14.0
2	19½	3	14.0
1	20	4	2.0

125. Genus POGONIAS Lacépède. Black drums

Body rather deep; back elevated, ventral outline nearly straight; mouth moderate; the jaws with bands of short teeth; lower pharyngeal bones united, armed with strong paved teeth; chin with numerous small barbels; preopercular margin entire; gill rakers short and blunt; air bladder large, thick, complicated in structure; dorsal fins contiguous, the first with long, slender spines; caudal fin nearly square; second anal spine greatly enlarged. A single species occurs in the North American fauna.

158. Pogonias cromis (Linnæus). Drum; Black drum.

Labrus cromis Linnæus, Syst. Nat., ed. XII, 1766, p. 479; Carolina.
Pogonias cromis Jordan and Evermann, 1896–1900, p. 1482, Pl. CCXXV, fig. 573.

Head 2.9 to 3.45; depth 2.65 to 2.8; D. X–I, 20 to 22; A. II, 6 or 7; scales 41 to 45. Body oblong, compressed; the back much elevated; ventral outline nearly straight; head moderately short; snout blunt, 2.85 to 3 in head; eye 2.85 to 3.95; interorbital 3 to 4; mouth horizontal; lower jaw included, with numerous small barbels, none of them exceeding half the length of eye; maxillary scarcely reaching middle of eye; 2.55 to 2.8 in head; teeth in jaws in broad bands, none of them especially enlarged; preopercular margin entire; gill rakers very short, 14 to 16 on lower limb of first arch; scales firm, ctenoid, reduced in size on head; dorsal fins contiguous, the first with stiff, slender spines, the third spine longest; notably higher than any of the rays in second fin; caudal fin subtruncate; anal fin short, the second spine much enlarged; ventral fins rather large, inserted slightly behind base of pectorals; pectoral fins long, pointed, 3.3 to 3.6 in length.

Color of a 37-inch male, silvery with brassy luster in life, becoming dark gray after death; grayish white below; all fins dusky or black. Color of specimens 7 inches in length, back and sides silvery; dusky white below; sides with four or five vertical black bars; all fins more or less dusky or black, except pectorals, which are plain. The caudal sometimes is plain translucent in young.

This species is represented in the present collection by 17 specimens, ranging in length from 75 to 235 millimeters (3 to 9¼ inches). Larger individuals, weighing upward of 40 pounds, were seen.

49826—28——19

The black drum is recognized by the numerous barbels on the chin, the entire preopercle, and the elevated back and straight ventral profile. The young are characterized by four to six broad, black bars on the sides.

The drum, with its subinferior mouth and barbels on the chin, is adapted for bottom feeding. It feeds largely on mollusks and crustaceans, which it is able to crush before swallowing. Schools of this fish are alleged to do great damage to oyster beds at times.

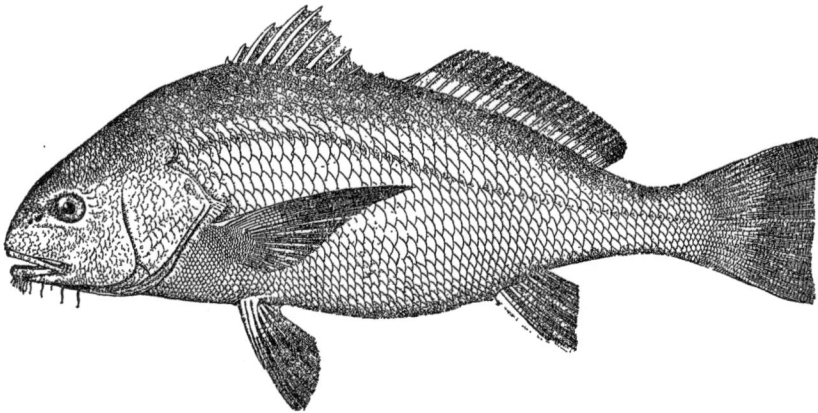

FIG. 167.—*Pogonias cromis.* Adult

Little is known of the life history of the black drum. The eggs and larvæ are undescribed. A fully ripe male, 37 inches long, was caught on May 22, 1922, at Cape Charles, Va., with hook and line, at a depth of 48 feet. Individuals less than 3 inches in length appear to be virtually unknown in collections.

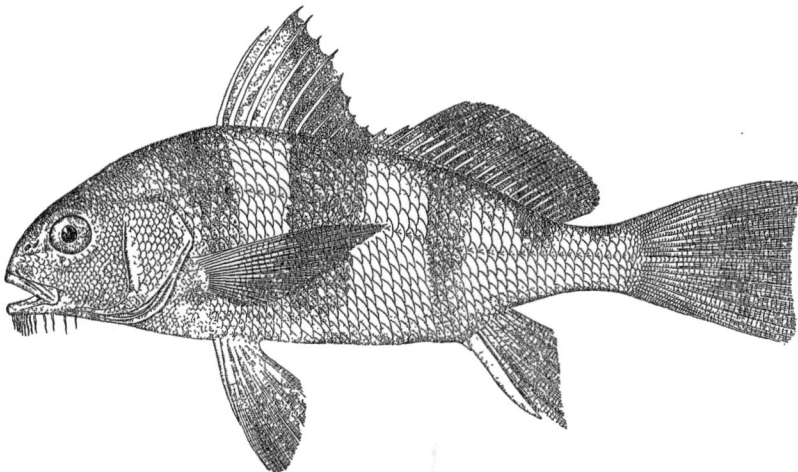

FIG. 168.—*Pogonias cromis.* Young, 10 inches long

Little is known of the rate of growth of the black drum. The smallest specimen, 75 millimeters (3 inches) in length, was taken with a small collecting seine at the mouth of Mill Creek, Solomons Island, Md., August 10, 1921. Other young fish all were taken in late September and during October. In the lower Rappahannock River three black drum, 8 inches long, were caught on October 18, and on October 24, 1921, a 9-inch specimen was taken in the lower Potomac River. During the fall of 1922, at Ocean View, Va., from September 27 to October 27, 20 black drum, ranging in length from 171 to 252 millimeters (6¾ to 10 inches), were caught on nine dates in 32 hauls of 1,800-foot seines.

This fish frequently is infested with parasites, and often is referred to by fishermen as a very "wormy" fish. None of the parasites of this fish, so far as known, are injurious to man, even though they might be eaten; and, furthermore, thorough cooking eliminates all possible chance of infection.

In Chesapeake Bay, during 1920, the black drum ranked twentieth in quantity and twenty-seventh in value, the catch being 23,700 pounds, worth $238.

In Maryland the total catch amounted to 700 pounds, worth $8—the least valuable of all the Maryland commercial fishes.

The catch in Virginia consisted of 23,000 pounds, the majority of which was taken with pound nets in Accomac County.

The black drum is caught from April until December and is most common in May and November. Most of the fish are caught in pound nets, a small number with hook and line, and a few with haul seines. Occasionally a school of fish is entrapped in a haul seine, but this happens less frequently in Chesapeake Bay than along the Atlantic coast.

During the season many individual fish are taken by anglers and fishermen of which no record can be obtained, and owing to their large size the aggregate catch probably is larger than indicated by the statistics collected in 1920.

The black drum is consumed in Baltimore, Crisfield, Norfolk, and other Chesapeake localities. Its value is not sufficient to make shipping to distant markets profitable. During 1921 and 1922 the price received by the fishermen ranged from 1 to 4 cents a pound. At retail, the fish is sold in steaks at 5 to 8 cents a pound. The flesh is coarse and not well flavored.

The names "drum" and "black drum" are used throughout the Chesapeake. The usual size of market fish ranges from 10 to 40 pounds; about 75 pounds is the maximum in the bay. The largest fish of which there is a record weighed 146 pounds and was taken in Florida. The following weights were obtained: Six and three-quarter inches, 2.8 ounces (1 fish); 7¼ inches, 2.9 ounces (2 fish); 7¾ inches, 3.8 ounces (1 fish); 8½ inches, 5.6 ounces (3 fish); 9½ inches, 7 ounces (1 fish); 10 inches, 9.5 ounces (1 fish); 37 inches, 34 pounds (1 fish).

Habitat.—Massachusetts to Argentina; common on the South Atlantic and Gulf coasts of the United States.

Chesapeake localities.—(a) Previous records: None. (b) Specimens in collection: Solomons, Md., mouth of the Potomac, mouth of the Rappahannock, and Ocean View, Va. Observed at various other localities in the bay from Solomons, Md., southward.

126. Genus UMBRINA Cuvier. Roncadores

Body moderately elongate; back more or less arched; head oblong; snout thick, extending beyond mouth; mouth horizontal, or nearly so; preopercle with a finely serrated bony margin; chin with a single short, thickish barbel; teeth in the jaws in villiform bands; first dorsal with 10 spines; anal fin with 2 spines, the second somewhat enlarged; caudal fin lunate or truncate; gill rakers present but short; air bladder well developed.

159. Umbrina coroides Cuvier and Valenciennes. Roncador.

Umbrina coroides Cuvier and Valenciennes, Hist. Nat. Poiss., V, 1830, p. 187, Pl. CXVII; Brazil. Jordan and Evermann, 1896–1900, p. 1466.

Head 3.55; depth 3.3; D. X-I, 29; A. II, 6; scales 58. Body elongate, compressed; back moderately elevated; head rather short; snout conical, projecting beyond the mouth, 3.3 in head; eye 2.9; interorbital 3.3; mouth moderate, inferior, horizontal; maxillary reaching under middle of eye, 2.55 in head; teeth in jaws small, in villiform bands; chin with a very short, thickish barbel; preopercular margin serrate; gill rakers about 11 on lower limb of first arch; scales rather small, ctenoid; dorsal fins continuous but deeply notched, the first with rather weak flexible spines; caudal fin injured, probably more or less rounded; anal fin very small, the second spine rather long and strong, 2.2 in head; ventral fins moderate, inserted under and a little posterior to pectoral fins; pectoral fins rather short, not reaching tips of ventrals, 1.45 in head.

Color of preserved specimen brownish above, lower parts silvery; sides with about nine dark crossbars; fins mostly plain, the dorsals and caudal with dusky punctulations.

A single specimen of this species, 50 millimeters (2 inches) in length, which is new to the fauna of Chesapeake Bay, was obtained.   This fish differs in several respects from specimens (4 examined) from the Atlantic coast of Panama.   The fish from Chesapeake Bay has 29 rays in the second dorsal, 58 transverse series of scales above the lateral line, and 11 gill rakers on the lower limb of the first arch.   The Panama fish have 24 or 25 rays in the soft dorsal, 46 to 50 scales (enumerated in the same way as for the local specimen), and 5 or 6 gill rakers on the lower limb of the first arch.   The smallest Panama fish is 160 millimeters (6½ inches) in length.   The size, therefore, is too unequal to show comparative differences in the shape and proportions of the body, etc., if, in fact, they exist.   *U. coroides* originally was described from Brazil, and although specimens from Florida and elsewhere have been considered identical with the Brazilian fish, the specimens in hand appear to show that two species may have been included.   The doubtful form, *U. broussonetii* Cuvier and Valenciennes, may possibly prove to be a valid species.   The material at hand, however, is too meager to show definitely the relationship.   The Chesapeake Bay specimen, therefore, is tentatively referred to *U. coroides*.

The species is not known to be of much commercial value anywhere within its range and probably does not attain a large size.   The largest individual seen by one of us (Hildeband) at Colon, Panama (where the species is occasionally taken), was only 8 inches long.

Habitat.—Florida to Brazil, now for the first time recorded northward of Florida.

Chesapeake localities.—(*a*) Previous records: None.   (*b*) Specimen in collection; Lynnhaven Roads, Va.; taken with seine on a sandy beach, September 27, 1921.

127. Genus MENTICIRRHUS Gill.   Whitings

Body elongate, little compressed; head rather long; snout conical, projecting beyond the mouth; mouth small, horizontal; teeth in jaws in bands; a single barbel at chin; gill rakers very short; first dorsal with 10 or 11 slender spines; anal with a single weak spine; air bladder absent.

KEY TO THE SPECIES

a. Scales on the chest not reduced and not notably smaller than on sides; pectoral fins reaching to or beyond tips of ventrals.

  b. Sides normally with dark, oblique bars, the last one on nape and the first one on body, meeting and forming a V; soft rays of anal usually 8, sometimes 9; longest dorsal spine produced, reaching far beyond origin of second dorsal in adult, 3.1 to 3.85 in length; scales 91 to 96, counting vertical series above lateral line_____*saxatilis*, p. 290

  bb. Sides plain or with obscure dark bars, not forming a V on sides; soft rays of anal usually 7, rarely 8; none of the dorsal spines produced in adult, the longest not reaching far beyond origin of soft dorsal 4.95 to 5.95 in length; scales 86 to 90, counting vertical series above lateral line_____*americanus*, p. 291

aa. Scales on chest much reduced, notably smaller than on sides; pectoral fins failing conspicuously to reach tips of ventrals; none of the dorsal spines produced, the longest 5.3 to 6.45 in length in adults; soft anal rays typically 7; scales 72 to 74, counting vertical series above lateral line; coloration plain silvery gray_____*littoralis*, p. 294

160. Menticirrhus saxatilis (Bloch and Schneider).   "Kingfish"; "Roundhead"; "Sea mullet"; "Sea mink"; Whiting; King whiting.

Johnius saxatilis Bloch and Schneider, Syst. Ichth., 1801, p. 75; New York.
Menticirrhus nebulosus Lugger, 1877, p. 78; Smith, 1892, p. 72.
Menticirrhus saxatilis Jordan and Evermann, 1896-1900, p. 1475; Smith and Bean, 1899, p. 187.

Head 3.05 to 4.1; depth 3.65 to 4.3; D. X–I, 24 to 26; A. I, 8 (sometimes I, 9); scales 91 to 96 (counting vertical series between enlarged scale at upper angle of opercle and base of caudal). Body elongate, compressed; back elevated; ventral outline nearly straight; head low; snout conical; projecting beyond the mouth, 2.9 to 3.5 in head; eye 2.8 to 4.6; interorbital 3.5 to 4.45; mouth horizontal, inferior; chin with a single short, thickish barbel; maxillary reaching opposite middle of eye, 1.35 to 2.85 in head; teeth in the jaws in bands, the outer ones in upper jaw somewhat enlarged; preopercle serrate; gill rakers very short, about six more or less developed on lower limb of first arch; scales small, firm, strongly ctenoid, not reduced in size on the breast; dorsal fins contiguous, the first with slender, flexible spines, the third one produced in the adult, reaching far beyond the anterior soft rays when deflexed, 3.1 to 3.85 in length in specimens 140 to 285 millimeters long; soft dorsal rather long and low; caudal fin with concave upper lobe and somewhat produced lower lobe, proportionately longer in young than in adult; anal fin moderate, with a single slender spine; ventral fins rather short, inserted about a half an eye's diameter behind base of pectorals; pectoral fins reaching to or a little beyond tips of ventrals, 1 to 1.45 in head.

Color dusky above, silvery underneath; some specimens much darker than others; sides with oblique bars running upward and backward; a horizontal stripe extending to end of lower lobe of caudal, often present on posterior part of body; two bars at nape running upward and forward, the second of these forming a V with the first bar on body; these markings are obscure on some individuals, often present on young as small as 30 millimeters in length; pectorals and spinous dorsal mostly black; other fins plain to dusky, varying among individuals.

Many specimens, ranging in length from 18 to 285 millimeters (¾ to 11¼ inches), were preserved. The fishermen do not distinguish this fish from others of the genus and they are not separated in the market. The color markings, forming a dark or black V on anterior part of sides, usually distinguishes this species from its relatives in Chesapeake Bay. In the adult the produced third dorsal spine, which reaches far beyond the beginning of the second dorsal, separates this species from its relatives. Other small but constant differences appear to exist in the size of the scales and number of fin rays. The young do not differ greatly from the adults, as indicated in the description.

Among 10 individuals, 9 had fed exclusively on crustaceans ranging in size from copepods to fairly large shrimp. One individual had fed on a squid.

The eggs of this whiting were obtained and hatched by Welsh and Breder (1923, p. 190). These authors, working at Atlantic City, N. J., state that spawning commences in June and continues until August. The eggs are reported to have an average diameter of 0.8 to 0.85 millimeter, being almost colorless, floating in sea water, and hatching in 46 to 50 hours in a water temperature of 68° to 70° F. The newly hatched fish were 2 to 2.5 millimeters in length.

Young *M. saxatilis* were collected throughout the summer and fall, often in company with young *M. americanus*. Fish 16 millimeters in length were taken late in June. By late September the size ranged from 35 to 154 millimeters (1⅖ to 6 inches). During October fish 50 to 185 millimeters (2 to 7¼ inches) were caught in the lower bay, and at the same time another size group, which ranged from 8½ to 11 inches in length, was present. All of the fish listed, with the exception of the last group, quite probably were the product of the previous spring and summer hatch. The group mentioned last no doubt consisted of older fish.

This species is included in the discussion of commercial importance of the whiting, under *M. americanus*. The exact proportion of each species comprising the total catch is not known. *M. saxatilis*, however, is said to attain its greatest abundance north of Chesapeake Bay.

The maximum size is about 3 pounds, but market fish usually range from ½ to 1½ pounds in weight.

Habitat.—Cape Cod, Mass., to Florida; rarely northward to Casco Bay, Me.

Chesapeake localities.—(a) Previous records: Southern part of Chesapeake Bay, mouth of the Potomac, Gunston Wharf, Va. (b) Specimens in collections: Crisfield, Md., Lower York River, Cape Charles, Buckroe Beach, Ocean View, and Lynnhaven Roads, Va.

161. Menticirrhus americanus (Linnæus). "Kingfish"; "Roundhead"; "Sea mullet"; "Sea mink"; Whiting; King whiting.

Cyprinus americanus Linnæus, Syst. Nat., ed. X, 1758, p. 321; Carolina.
Menticirrhus alburnus Uhler and Lugger, 1876, ed. II, p. 101; Bean, 1891, p. 89.
Menticirrhus americanus Jordan and Evermann, 1896–1900, p. 1474, Pl. CCXXV, fig. 572; Evermann and Hildebrand, 1910, p. 162; Fowler, 1918, p. 18.

Head 3.1 to 3.9; depth 3.56 to 4.1; D. X–I, 24 to 27; A. I, 7 (rarely I, 8); scales 86 to 90 (counting vertical series between enlarged scale at upper angle of opercle and base of caudal). Body elongate, compressed; back elevated; ventral outline nearly straight; head low; snout conical, projecting beyond the mouth, 3.1 to 4.15 in head; eye, 2.8 to 6.5; interorbital 3.1 to 4.4; mouth horizontal, inferior; chin with a single, short, thickish barbel at symphysis; maxillary reaching opposite middle of eye, 2.5 to 2.9 in head; teeth in bands in each jaw, the outer ones in upper jaw enlarged; preopercle serrate; gill rakers very short, only about six somewhat developed on lower limb of first arch; scales rather small, firm, strongly ctenoid, not reduced in size on the breast; dorsal fins contiguous, the first composed of slender, flexible spines, none of them produced, and not extending far beyond the anterior soft rays when deflexed, the longest one 4.95 to 5.95 in

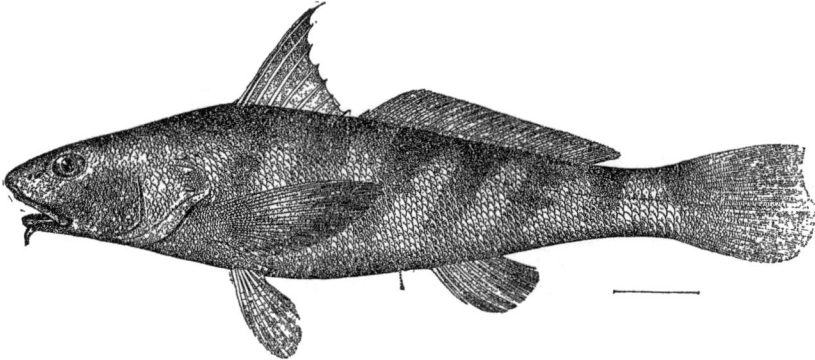

FIG. 169.—*Menticirrhus americanus*

length in specimens 130 to 293 millimeters long; soft dorsal rather long and low; caudal fin with concave upper lobe and produced, pointed lower lobe, this lobe proportionately longer in the young than in the adult; anal fin rather short, with a single rather weak spine; ventral fins rather small, inserted behind base of pectorals; pectoral fins rather large, reaching to or beyond tips of ventrals, 1.05 to 1.55 in head.

Color of two specimens, 11½ and 12½ inches long, silvery gray above; white below, with dusky markings; sides with seven to eight obscure dark bars, the one at nape and the one on caudal peduncle darkest and most persistent; spinous dorsal plain, membranes edged with black from first to sixth spine on one fish and from second to seventh spine on the other; soft dorsal plain, some of rays tipped with black; caudal dusky, with light brown on upper part; anal and ventrals white faintly marked with light brown; pectorals dusky or black. In some specimens the dark bars

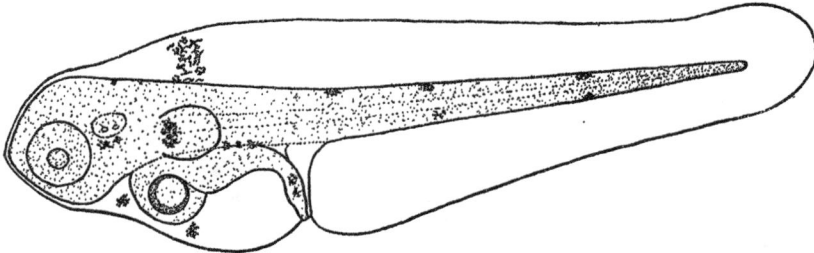

FIG. 170.—Recently hatched larva, 26.5 millimeters long

along back and sides are wanting and the belly may be plain white instead of with grayish or dusky markings. Young rather darker than the adults; the back and sides with dusky punctulations and sometimes with blackish blotches. The fin colors of 13 young, 80 to 135 millimeters (3⅛ to 5⅛ inches) in length, were as follows: Spinous dorsal slightly dusky; soft dorsal nearly plain, sometimes slightly dusky, in one specimen tinged with yellow brown; caudal slightly dusky, tinged with yellowish brown; anal, ventrals, and pectorals slightly dusky, sometimes white, tinged with yellowish brown. Two specimens of *M. saxatilis*, 77 and 128 millimeters (3 and 5 inches) in length, besides having dark V-shaped body markings, differed chiefly in the more dusky coloration of the two dorsals and the pectorals.

Many specimens ranging in length from 22 to 393 millimeters (⅞ to 15½ inches) were preserved. This fish usually is readily distinguished from *M. saxatilis* by the plainer coloration. Some specimens of *saxatilis*, however, are so obscurely marked that other characters must be relied upon for identification. The alleged difference in the number of soft rays in the second dorsal, mentioned in current descriptions, does not appear to exist. In 22 specimens of *M. americanus* examined for this character, five had 24 rays, sixteen had 25, and seven had 26. The number of anal rays is of some value, however, as *M. americanus* typically has 7, rarely 8, rays in the anal fin, and *M. saxatilis* typically has 8, rarely 9, rays. The longest dorsal spine in the adult in *americanus* is never as high as in *saxatilis*, and the scales are somewhat larger. *M. americanus* is most readily separated from *M. littoralis* by the size of the scales on the chest, which in *americanus* are scarcely smaller than the scales elsewhere on the body, whereas in *littoralis* they are notably reduced in size. The smallest specimens (22 millimeters) at hand do not differ notably from the adults except in color, as shown in the description, and in the proportionately longer and more pointed lower lobe of the caudal fin.

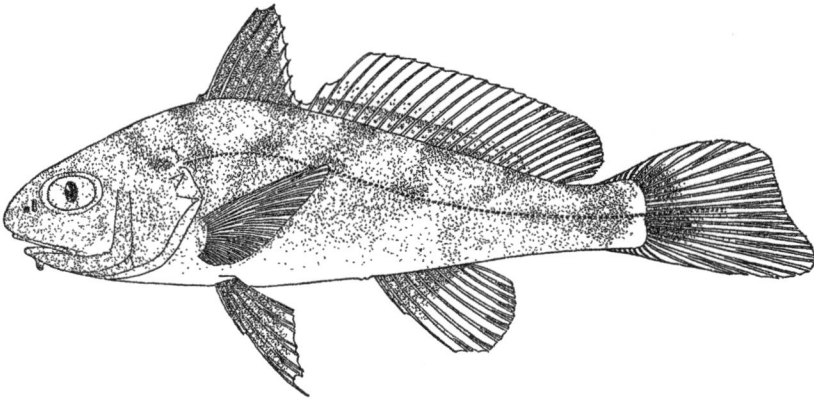

FIG. 171.—Young, 39 millimeters long

The food of this fish, as shown by the contents of 21 stomachs taken from specimens collected in Chesapeake Bay, consists of crustaceans and fish, the two foods being present in the proportion of about 85 per cent crustaceans (chiefly shrimp) and 15 per cent fish.

The spawning habits of this fish are not well known, and the eggs and larvæ have not been studied. Smith (1907, p. 332) states that this fish spawns in June at Beaufort, N. C. Welsh and Breder (1923, p. 186), however, found no ripe or spent fish on the coast of New Jersey as late as August, 1920, and these authors produce some evidence indicating that the fish probably spawns in the fall in the Gulf of Mexico. It is evident, therefore, that the information is far from complete.

On May 20, 1922, two ripe males, 286 and 310 millimeters (11¼ and 12¼ inches) in length, were trawled off Cape Charles, and on the following day a number of males and females, 11 to 12¼ inches long, with well-developed gonads, were seined along the beach at Cape Charles.

On June 11, 1921, many of the fish in the Norfolk market contained nearly ripe eggs, while some apparently had already spawned. The presence of ripe males in May and of nearly ripe males and females, together with spawned-out fish, in June, followed by the presence of young, three-fourths inch or more in length, early in the summer, indicates that this species spawns in Chesapeake Bay during the spring and probably early summer.

The commercial catch of kingfishes in the Chesapeake includes three species (*M. americanus*, *M. littoralis*, and *M. saxatilis*) which resemble each other so closely that they are not separated by the fishermen. However, one species (*M. americanus*) is more abundant than the others. The statistics that follow include all three species.

In Chesapeake Bay, during 1920, the kingfishes ranked twenty-first, both in quantity and value, the catch being 17,933 pounds, worth $1,606.

During 1920 the kingfishes were not taken in Maryland in commercial quantities. The catch, therefore, is credited entirely to Virginia. Considered among the fishes of Virginia, the kingfishes ranked twentieth in quantity and nineteenth in value. Of the total catch, 47 per cent were caught with lines, 30 per cent with pound nets, and 23 per cent with seines. The counties with the largest catches were Princess Anne, 6,600 pounds; Norfolk, 4,700 pounds; and Elizabeth City, 4,700 pounds.

The kingfishes are caught from April until November, the bulk of the catch being taken in the spring and late fall. Records obtained from a set of two pound nets at Lynnhaven Roads, Va., give the earliest catches as follows: April 21, 1916; April 17, 1917; April 16, 1918; April 22, 1919; April 16, 1920; April 27, 1921; and April 11, 1922. The table on page 32 gives the monthly catch for the years 1916 to 1922, taken in the same set of two pound nets situated in the same part of Lynnhaven Roads for the entire period.

According to this record, the catch of kingfish has been declining steadily since 1917, with a slight rise in 1922. The pound nets upon which this catch is based, because of their large size and peculiar location, probably yield more of this species than any other single set of apparatus in the bay. Because of this fact and the long period of time involved it is believed that these records indicate the general trend of abundance for the kingfishes in the Chesapeake.

The kingfishes are taken with haul seines in small quantities from time to time during the summer and fall, but they are not abundant enough to support a special fishery. A few are taken with gill nets of a type known as the drift net. The drift net is sunk to the bottom by weights and is employed chiefly to catch croakers and striped bass. The kingfish bites freely on hook and line during the summer and fall, and the aggregate catch taken by anglers is rather large.

This species is essentially a fish of the lower sections of the bay, but it sometimes is taken in small numbers as far north as Solomons, Md. However, it is rare above Solomons, and occurs only as a straggler at Love Point, Md.

Most of the catch is marketed in Baltimore and the vicinity of Norfolk, but occasionally when a good catch is made it is shipped to New York, where a good price usually is obtained. During 1920, 1921, and 1922 the price received by the fishermen was about 9 cents per pound, whereas the retail price ranged from 15 to 25 cents. During recent years these fish have not been caught in sufficient numbers to overstock the local markets, and the demand, although rather small, is constant.

Various names have been assigned to these fish in the Chesapeake, the most common being "kingfish" and "round head." They are sometimes called "sea mullet," "sea mink," or "whiting." To avoid confusion with the Florida kingfish (Scomberomorus) and the New England whiting (Merluccius), the name "king whiting" has been proposed.

The kingfishes are high-quality food fishes and are much esteemed by those who are well acquainted with them. For some reason they are not as well known generally as several other species that they surpass in quality.

The weight of market fish ranges from ½ to 1½ pounds and the maximum is 2½ pounds.

The following comparisons of length and weights of fresh fish may be of interest: Length 4 to 4⅞ inches (7 fish), average weight 0.4 ounce; 5 to 5⅞ inches (5 fish), 0.85 ounce; 6½ inches (11 fish), 1.8 ounces; 7½ inches (1 fish), 2.1 ounces; 9 inches (1 fish), 3.8 ounces; 11 to 11⅞ inches (9 fish), 7.45 ounces; 12 to 12¾ inches (6 fish), 10 ounces; 13 inches (1 fish), 11 ounces; 14½ inches (2 fish), 17.2 ounces; 15 inches (1 fish), 20.6 ounces; and 16½ inches (1 fish), 30 ounces.

Habitat.—New York to Texas; common from Chesapeake Bay southward.

Chesapeake localities.—(*a*) Previous records: St. George Island, Md., Cape Charles city and Ocean View, Va. (*b*) Specimens in collection: From various localities, from Solomons, Md., southward.

162. Menticirrhus littoralis (Holbrook). "Kingfish"; "Roundhead"; "Sea mullet"; Whiting; Surf whiting; Silver whiting.

Umbrina littoralis Holbrook, Ichth., South Carolina, ed. I, 1855, p. 142, Pl. XX, fig. 1; South Carolina.
Menticirrus littoralis Jordan and Evermann, 1896–1900, p. 1477.

Head 3.45 to 3.75; depth 3.6 to 4.35; D. X–I, 24 to 26; A. I, 7; scales 72 to 74 (counting vertical series between the enlarged scale at upper angle of gill opening and base of caudal). Body elongate, compressed; back elevated; ventral outline nearly straight; head low; snout conical, projecting beyond mouth, 2.9 to 3.25 in head; eye 3.1 to 4.1; interorbital 3.6 to 4.7; mouth horizontal, inferior; chin with a single, short, thickish barbel; maxillary reaching nearly or quite opposite middle of eye, 2.6 to 3.05 in head; teeth in the jaws in bands, none of them especially enlarged; preopercle serrate; gill rakers very short, seven or eight more or less developed on lower limb of first arch; scales rather large, firm, strongly ctenoid, notably reduced in size on the chest; dorsal fins contiguous; the first composed of slender, flexible spines, none of them produced, the longest scarcely reaching the origin of the second dorsal when deflexed, 5.3 to 6.45 in length in specimens 100 to 160 millimeters long; second dorsal rather long and low; caudal fin with concave upper lobe and pointed lower lobe, more produced in the young than in the adult; anal fin small, with a single weak spine; ventral fins moderate, inserted fully half an eye's diameter behind base of pectorals; pectoral fins short and broad, not nearly reaching tips of ventrals, 1.3 to 1.55 in head.

Color silvery gray above, paler below; sides without dark markings; fins mostly pale; the spinous dorsal and the lower lobe of the caudal usually with more or less dusky. Very young (60 millimeters and less in length) with dusky punctulations; the punctulations sometimes concentrated, forming dusky blotches on the back and two on base of caudal.

Many specimens, ranging from 22 to 160 millimeters ($\frac{7}{8}$ to $6\frac{1}{4}$ inches) in length, were preserved. All except the very young are readily separated from other species of Menticirrhus by the

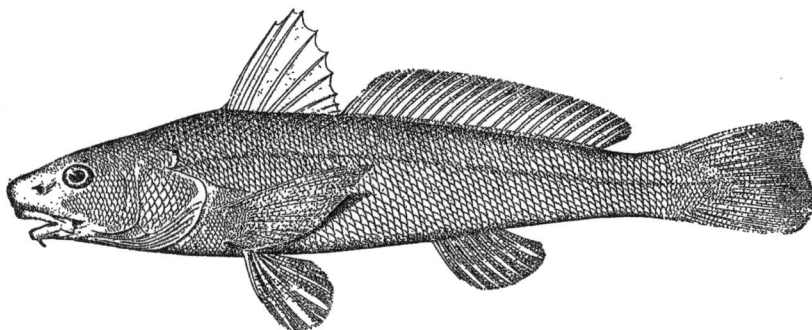

FIG. 172.—*Menticirrhus littoralis*

greatly reduced scales on the chest. The scales in this fish are somewhat larger, and the length of pectorals, as compared with the length of the ventrals, is less. All of these characters are rather difficult of application in the very young, which are separated from *M. americanus* not without trouble. The young differ from the adults chiefly in color and in the longer and more pointed lower lobe of the caudal.

The food of this species appears to be identical with that of *M. saxatilis*. Seven stomachs examined contained crustaceans only.

The eggs and also the young, smaller than the smallest ones at hand, have not been described. Smith (1907, p. 324) states that he took ripe fish in June at Cape Lookout; also that ripe eggs have been taken on several occasions between June 1 and June 10 at Beaufort, N. C.

The species apparently has not been recorded previously from Chesapeake Bay. It was not distinguished from *M. americanus* in the field. Judging from the number of specimens obtained, it probably is the least abundant of the three species occurring in Chesapeake Bay. Whether or not this southern species of whiting is taken in commercial numbers in Chesapeake Bay can not be stated here for the reason already stated—that the species was not recognized in the field.

Habitat.—"South Atlantic and Gulf coasts; rarely, if ever, straying north of North Carolina" (Smith, 1907). Now apparently for the first time recorded from Chesapeake Bay.

Chesapeake localities.—(a) Previous records: None. (b) Specimens in collections: Buckroe Beach, Ocean View, and Lynnhaven Roads, Va.

49826—28——20

Family LXX.—OTOLITHIDÆ. The weakfishes.

This family, according to Jordan (1923, p. 202), differs from the Sciænidæ principally in the different arrangement of the vertebræ, the Sciænidæ having typically 10+14 vertebræ, whereas the present family has 14+10. A single genus of the Chesapeake Bay fauna falls within the scope of the Otolithidæ.

128. Genus CYNOSCION Gill. Squeteagues; "Sea trouts"

Body elongate; head conical; mouth large, oblique; lower jaw protruding; teeth sharp, not protruding, two enlarged, recurved teeth at tip of upper jaw; no enlarged teeth on lower jaw; preopercle serrate; pseudobranchiæ present; dorsal spines slender, flexible; second dorsal long and low, more than twice the length of the anal; anal spines 2, very weak.

KEY TO THE SPECIES

a. Soft dorsal and anal scaleless; gill rakers comparatively short, eight on lower limb of first arch; 11 or 12 scales in a series between the origin of the anal and the lateral line; body with round black spots _____*nebulosus*, p. 296
aa. Soft dorsal and anal usually closely scaled; body not spotted with round black spots.
    b. Anal fin with nine soft rays; gill rakers few, nine on lower limb of first arch; eight scales in a series between origin of anal and lateral line; coloration nearly uniform _____*nothus*, p. 299
    bb. Anal fin with 11 or 12 soft rays; gill rakers more numerous, 11 to 13 on lower limb of first arch; 10 scales in a series between the origin of the anal and the lateral line; body usually with irregular dark blotches, sometimes forming wavy, oblique lines _____*regalis*, p. 300

163. Cynoscion nebulosus (Cuvier and Valenciennes). Spotted weakfish; Spotted squeteague; "Salmon trout"; "Simon trout"; "Spotted trout"; "Speckles"; "Speckled trout".

Otolithus nebulosus Cuvier and Valenciennes, Hist. Nat. Poiss., V, 1830, p. 79; locality unknown.
Cynoscion carolinensis McDonald, 1882, p. 12, fig. 2.
Cynoscion maculatus Bean, 1891, p. 88; Smith, 1892, p. 72.
Cynoscion nebulosus Jordan and Evermann, 1896–1900, p. 1409, Pl. CCXXI, fig. 563; Evermann and Hildebrand, 1910, p. 162; Fowler, 1912, p. 58; and 1918, p. 18; Welsh and Breder, 1923, p. 164.

Head 2.95 to 3.25; depth 3.4 to 4.35; D. X (rarely XI)–I, 24 to 26; A. II, 10 or 11; scales 90 to 102. Body elongate, somewhat compressed; back little elevated; head long and low; snout pointed, 3.75 to 4.2 in head; eye 4.45 to 5.35; interorbital 4.5 to 5.9; mouth large, oblique; lower jaw projecting; maxillary reaching nearly or quite opposite posterior margin of eye, 2.2 to 2.3 in head; teeth as in *C. regalis;* gill rakers rather short, 8 on lower limb of first arch; scales small, thin, ctenoid, extending forward on head, cheeks, and opercles, not present on fins, 11 or 12 between origin of anal and lateral line; dorsal fins contiguous or separate, spines of the first weak, flexible, the longest spines scarcely longer than the longest soft rays; caudal fin pointed in very young, becoming straight to somewhat emarginate in adults; anal fin small, the spines very weak, base of fin ending about an eye's diameter in advance of end of base of dorsal; ventral fins rather small, inserted a little behind base of pectorals, 1.85 to 2.25 in head.

Color dark gray above, with sky-blue reflections; pale, silvery below; upper part of sides marked with numerous round, black spots, the spots extending on dorsal and caudal fins. Very young with a broad, dark, lateral band; blotches of the same color on the back; base of caudal black. Fins pale to yellowish green; the dorsal and caudal spotted with black in the adult.

Many specimens of this common species, ranging in length from 25 to 245 millimeters, were preserved. The very young of this squeteague differ very markedly from the adult in color, as shown in the description, also in the more or less sharply pointed tail and other less striking characters. Individuals 5 inches and upward in length are readily distinguished from related species by the round, black spots situated on the upper parts of the body and on the dorsal and caudal fins. The scales, also, are smaller and are wanting on all the fins.

The food of the spotted squeteague appears to be identical with that of the gray squeteague. The contents of 20 stomachs consisted of fish and crustaceans, the large individuals having fed mainly on fish and the small ones on crustaceans.

Virtually nothing is known of the spawning habits of this fish, and fish with large roe seldom are seen. From the meager data at hand, it is believed to spawn in the spring, probably in May and June. Yarrow observed, at Beaufort, N. C., that females had quite large roe in April. (Smith, 1907, p. 312.) No ripe fish were seen in the Chesapeake region. A male 14 inches long, taken on July 22, and a female 19 inches long, taken on September 26, had gonads in such a state of development that they probably would have matured the following spring. It seems probable, because of the rather large number of young (small) fish taken, that this squeteague may spawn within Chesapeake Bay.

The rate of growth of the spotted squeteague is equally as little known as the spawning habits. Very young fish are rare in collections, and the smallest ones at hand, having a length of only 25 millimeters, are among the smallest that have been studied.

The following record of catches of young fish made in the Chesapeake are given in the hope that they may prove of value in future studies of the rate of growth of this species. It is perhaps noteworthy that all except the first specimen listed were taken north of the mouth of the Potomac River—Rappahannock River, one specimen, 75 millimeters long, taken on July 23; Solomons Island 52 fish, Chesapeake Beach 43 fish, Annapolis 2 fish, all taken from August 9 to 18, range in length 25 to 75 millimeters; Love Point and Oxford, 6 and 8 fish, respectively, taken from September 5 to 13, range in length from 51 to 89 millimeters.

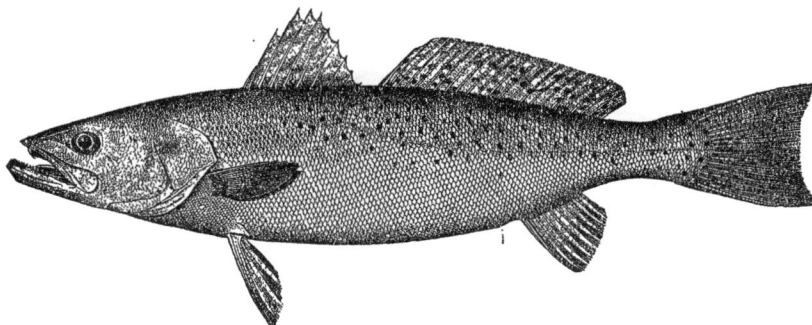

FIG. 173.—*Cynoscion nebulosus*

The spotted squeteague is one of the important food fishes of Chesapeake Bay. During 1920 it ranked ninth in quantity and eighth in value, the catch being 438,797 pounds, worth $43,879.

In Maryland it ranked thirteenth in quantity and eleventh in value, the catch being about 20,000 pounds, worth $2,000. The catch is about equally divided between the haul seines and the many pound nets found within the State.

In Virginia it ranked seventh in quantity and eighth in value, the catch being 418,797 pounds, worth $41,879. Of this amount, 53 per cent were caught with haul seines, 34 per cent with pound nets, 8 per cent with fyke nets, and 5 per cent with gill nets. Norfolk County was credited with the largest catch, 115,640 pounds, followed by Northampton with 94,077 and York with 67,845 pounds.

The spotted squeteague is caught in the Chesapeake from March until December, including two definite shorter periods of abundance—namely, from March until May and from September until November. This fish is not taken in large quantities in pound nets. Small numbers, however, are taken in nearly every pound, and the aggregate catch of the many traps forms a large portion of the total catch of this species. Most of the pound-net catch is taken in March, April, and May. A scattering few are taken during the summer, the catch increasing slightly in the fall. Quantities of this fish are caught by numerous fyke nets located in the lower parts of the larger rivers. It is also taken in small numbers with gill nets, principally about Tangier Island, Great Fox Island, and Pocomoke Sound. The largest part of the catch is taken with haul seines and set seines. A special fishery is carried on in the fall with the last-mentioned apparatus.

Spotted squeteague are caught incidentally with the haul seine along with spots, croakers, etc. The set seines, however, are employed chiefly for the spotted squeteague. During 1922, four set seines were operated in Chesapeake Bay, all located between Ocean View and Lynnhaven Inlet, Va. A set seine is similar to a haul seine, being more heavily leaded, however, and more closely corked. With one end lying on the beach, it is set out from the shore in nearly a perpendicular line, and at the outer end a 300-pound anchor is fastened. Between this anchor and the shore a number of 40-pound anchors are attached to the lead line, for the purpose of holding the seine in place. When the tide changes, these small anchors must be adjusted to avoid entanglement with the seine. Several fishermen remain in the immediate locality, while the net is set, and adjust anchors and otherwise observe the gear. The net is hauled ashore at slack, low tide, once in 24 hours. This is done by attaching a line to the outer end of the seine, and it is hauled to shore by making a semicircular sweep. A power winch, operated by a gasoline engine, is employed for this purpose. After the outer end has reached the shore, the procedure is similar to haul seining (described under *Leiostomus xanthurus*). The set seine is usually about 1,500 to 1,800 feet long, 14 to 25 feet deep, and of $1\frac{1}{8}$ to $1\frac{1}{2}$ inch bar mesh. A crew of 12 to 15 men is necessary to operate the net successfully. The fish do not gill themselves in the set seine but have a peculiar habit of lying close against the seine, near the lead line, and in this position they remain dormant for hours during the cool fall weather.

The relative efficiency of the set seine and the haul seine for catching this squeteague is illustrated by the following records, made from a series of hauls of both kinds of seines made near Ocean View, Va. These data are based upon the catches made by two haul seines and one set seine and represent the total of all the hauls observed. The catch taken with the set seine from Septem-

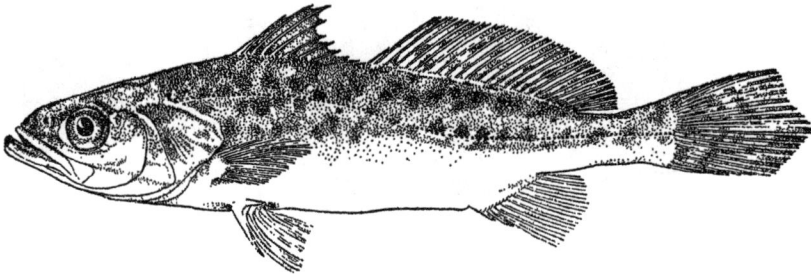

FIG. 174.—*Cynoscion nebulosus.* Young, 12 millimeters long

ber 26 to October 27, 1922, in a total of 15 hauls, consisted of 11,216 fish; average number of fish per haul, 747; average weight per fish, $2\frac{1}{4}$ pounds; weight of largest fish, $7\frac{1}{2}$ pounds. The catch taken with two haul seines from September 23 to October 27, 1922, in a total of 33 hauls, consisted of 749 fish; average number of fish per haul, 23; average weight per fish, $1\frac{1}{2}$ pounds; weight of largest fish, 6 pounds.

The first seine catches are made in the lower part of the bay about September 15, and continue until cold weather arrives. The spotted squeteague apparently leave the bay in the fall for a more southern habitat, and it is during this outward migration that they are intercepted by the fishermen's nets. This squeteague, like *C. regalis*, is rare in the bay north of Annapolis, and the fishing season is shorter in the upper stretches of the bay than near the entrance.

The spotted squeteague is always in demand in the fish markets, where it brings a good price at all seasons. During 1921 and 1922 the retail prices usually ranged between 25 and 30 cents per pound. In the fall it is shipped to Baltimore, Washington, New York, and other markets. Its fine appearance, firm flesh, and good flavor place this fish in high esteem.

Many names have been given to this species. In the Chesapeake region only a few are in common use, "speckles," "speckled trout," "Simon trout," and "salmon trout" being used interchangeably.

The largest fish observed by us weighed 16 pounds. It was taken in a pound net at Ocean View during the early part of April, 1922. This fish was said by dealers to be one of the largest ever caught in the bay, and in fact, it probably represents the maximum size attained by the species. Fish weighing 8 to 12 pounds are not uncommon in the spring, but during the fall individuals weighing over 8 pounds are rare.

Habitat.—New York to Texas.

Chesapeake localities.—(*a*) Previous records: Lower Potomac, Cape Charles city, Old Point Comfort, and Norfolk. (*b*) Specimens in collection: From numerous localities, from Annapolis, Md., to Cape Charles and Cape Henry, Va.

Comparison of lengths and weights of spotted squeteagues

Number of fish weighed and measured	Length	Weight	Number of fish weighed and measured	Length	Weight	
	Inches	*Ounces*		*Inches*	*Lbs.*	*Oz.*
1	5	0.6	1	12		9.5
2	7¼	2.0	1	15¾	1	5.6
1	7¾	2.3	1	17¼	2	1.0
1	8	2.6	1	18	2	10.0
1	8½	3.5	1	19	2	8.6
1	9	3.8	2	20½	3	5.7
2	10¼	6.0	2	22½	4	15.0
1	11¼	7.0	1	27	8	6.5
1	11½	9.0				

164. Cynoscion nothus (Holbrook). " Bastard trout "; Silver squeteague.

Otolithus nothus Holbrook, Ichth., South Carolina, 1860, p. 134, Pl. XIX, fig. 1; South Carolina. Lugger, 1878, p. 111.
Cynoscion nothus Jordan and Evermann, 1896–1900, p. 1406, Pl. CCXX, fig. 561.

Head 3.2; depth 3.3 to 3.5; D. X–I, 28 or 29; A. II, 9; scales 68 or 69. Body compressed, rather deep; back more strongly elevated than in related species; head long; snout moderate, 3.9 to 4.05 in head; eye 3.9 to 4.25; interorbital 4.55; mouth large, oblique; lower jaw projecting; maxillary reaching vertical from posterior margin of pupil, 2.25 in head; teeth as in *C. regalis*; gill rakers 9 on lower limb of first arch; scales rather large, thin, ctenoid, extending on head and fins as in *C. regalis,* about eight between origin of anal and lateral line; caudal fin round; anal fin very small; fins otherwise as in *C. regalis.*

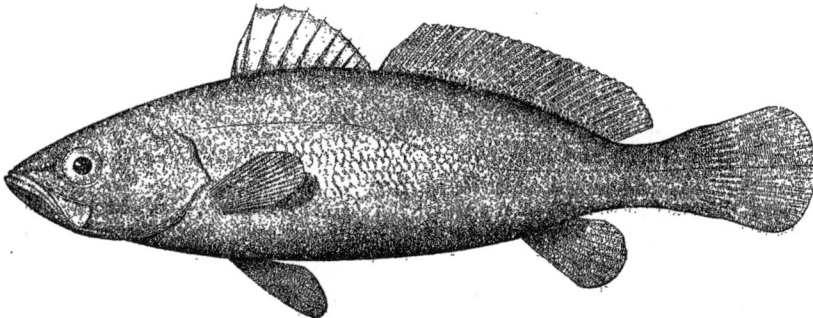

Fig. 175.—*Cynoscion nothus.* From a specimen 8½ inches long

Color plain greenish blue above; silvery below; no dark spots or reticulations; fins all plain; axil of pectoral dusky.

Two specimens, each 215 millimeters (8½ inches) in length, were preserved and form the basis for the foregoing description. This species, as already indicated, differs from *C. regalis,* its nearest relative, in the deeper and more compressed body. These differences are real and unmistakable when specimens of even size are compared. The scales are larger, the gill rakers are fewer, the anal fin is shorter, and the caudal fin is rounded in the specimens in hand, whereas in specimens of the same size of *C. regalis* the margin is nearly straight. The color is plainer than is usual for *C. regalis,* although this character is rather unreliable, as the writers have seen individuals of *C. regalis* that were equally as plain in color.

The figure published by Jordan and Evermann (1896–1900, Pl. CCXX, fig. 561) and republished by Smith (1907, p. 309, fig. 137) is misleading, as it appears to have been based on an unusually slender fish. Furthermore, the scales are represented as much smaller than they are in the specimens at hand and in the original color plate of Holbrook.

Welsh and Breder (1923, p. 169) found that *regalis* and *nothus* are closely related, and they state that further study may show them to completely intergrade.   Coles (1916, pp. 30 and 31) concluded that *C. nothus* is an abnormal *regalis*, saying that his specimens always were caught with *regalis*, the body was not more compressed, and the only obvious difference between the two was in the color.   One of us (Hildebrand), working at Beaufort, N. C. (i. e., in the same general vicinity where Doctor Coles obtained his specimens), also found specimens of Cynoscion that agreed in color with *nothus*, but no other tangible difference between them and specimens of *regalis* appeared to be present; he, too, arrived at the tentative conclusion that *nothus* was a plain-colored *regalis*.   The specimens examined by Coles and Hildebrand in the vicinity of Beaufort, N. C., quite probably were abnormally colored *regalis*, for the differences between a true *nothus*, such as we believe to have in hand now, and a *regalis* are so evident and so numerous that they scarcely would have been overlooked.

The fishermen of the lower part of Chesapeake Bay recognize this species and call it the "bastard trout."   It is taken only now and then, and it is not numerous enough to be of commercial importance.   On the coast of the Gulf States it is said to be a food fish of some importance.

Habitat.—Maryland to Texas.

Chesapeake localities.—(*a*) Previous record: Baltimore, Md.   (*b*) Specimens in collection: From Lynnhaven Roads, Va.; taken in pound nets, August 22, 1921.

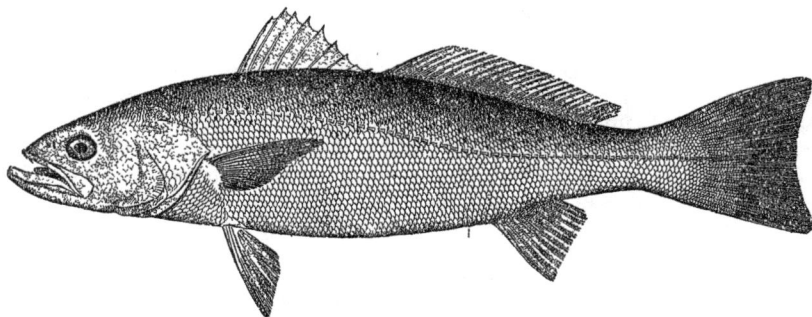

FIG. 176.—*Cynoscion regalis*

165. Cynoscion regalis (Bloch and Schneider).   Weakfish; Squeteague; "Trout;" "Gray trout."

Johnius regalis Bloch and Schneider, Syst. Ichth., 1801, p. 75; New York.
Cynoscion regalis, Uhler and Lugger, 1876, ed. I, p. 116; ed. II, p. 98; McDonald, 1882, p. 12, fig. 3; Bean, 1891, p. 88; Jordan and Evermann, 1896-1900, p. 1407, Pl. CCXX, fig. 562; Fowler, 1918, p. 18; Welsh and Breder, 1923, p. 150.

Head 2.9 to 3.3; depth 3.5 to 4.25; D. X–I, 25 to 28; A. II, 11 or 12; scales 76 to 86.   Body elongate, not much compressed; head long and low; snout pointed, 4.85 to 5.15 in head; eye 3.1 to 5.6; interorbital 4.1 to 4.75; mouth large, oblique; lower jaw projecting; maxillary reaching posterior margin of pupil or beyond, 2.1 to 2.4 in head; teeth in jaws pointed, in two series anteriorly, becoming single posteriorly, at least in lower jaw, two enlarged, recurved teeth usually present in anterior part of upper jaw; gill rakers long, 11 to 13 on lower limb of first arch; scales rather thin, finely ctenoid, extending on head, cheeks, and opercles; reduced scales also present on fins, about 10 rows between origin of anal and lateral line; dorsal fins contiguous in young, well separated in adult, the first composed of flexible spines, the third and fourth the longest, somewhat higher than the longest soft rays, soft part of dorsal long, with nearly straight margin; caudal fin with concave margin in adult, round in young; anal fin small, situated under posterior part of dorsal, its base ending a little in advance of that of dorsal; ventral fins rather small, inserted a little behind base of pectorals; pectoral fins short, failing to reach tips of ventrals, 1.65 to 1.95 in head.

Color largely greenish above and silvery underneath, upper parts with metallic reflections of purple and gold; upper parts of sides marked with black, dark green, and bronze blotches, mostly arranged in oblique wavy lines; dorsal and caudal dusky, with yellowish green tinge; anal and ventrals bright yellow; pectorals pale outside; axil dusky.

Many specimens of this common species, ranging in length from 30 to 460 millimeters (1¼ to 18 inches), were preserved. The young are proportionately deeper and more strongly compressed than the adult. The caudal fin is round or even pointed, and the very young are largely pale in color, with more or less definite dark bars on the back. This species is most readily distinguished from related species by the greenish color of the back, bearing darker markings, which form mostly oblique wavy lines.

The sexes in the adults of this species may be separated externally by feeling the abdominal wall. In the female the wall is uniformly thin. The male, however, has the walls thickened on each side along the lower ventral edge, due to the croaking or drumming muscles that are situated there. The female has no croaking muscles and can not make a croaking sound. The air bladder, which is used in making the sounds, is present in both sexes. Special muscles, however, as in the male, are required to produce croaking or drumming. The process is described in part by Smith (1907, p. 307) as follows: "The muscle, with the aponeurosis, is in close relation with the large air bladder and by its rapid contraction produces a drumming sound with the aid of the tense air bladder, which acts as a resonator." The value, or the purpose served, to the fish by croaking or drumming is not known. The females within a school usually average a somewhat larger size than the males.

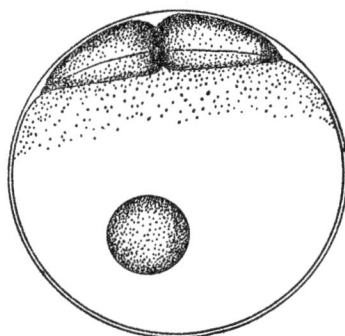

FIG. 177.—*Cynoscion regalis.* Egg in two-cell stage

The food of this fish, as indicated by the contents of 85 stomachs taken from specimens collected in Chesapeake Bay, consists of the following: Small fish (mostly anchovies and silversides) and small crustaceans (largely Mysis). Large individuals had fed almost exclusively on fish and the young mainly on small to minute crustaceans. The food of the squeteague is discussed at considerable length by Welsh and Breder (1923, pp. 159 to 164). These authors show that the food varies somewhat with the locality. Fish and crustaceans, however, are everywhere the principal foods, although mollusks and annelids also are eaten at times. It is noted also by Welsh and Breder that "small invertebrates" constitute the principal food of small-sized fish.

The reader is referred to the work of Welsh and Breder (1923, pp. 150–158) for an extended account of the spawning, embryology, and growth of the squeteague. These authors state that spawning takes place in the larger bays and possibly in the ocean. The season is a protracted one,

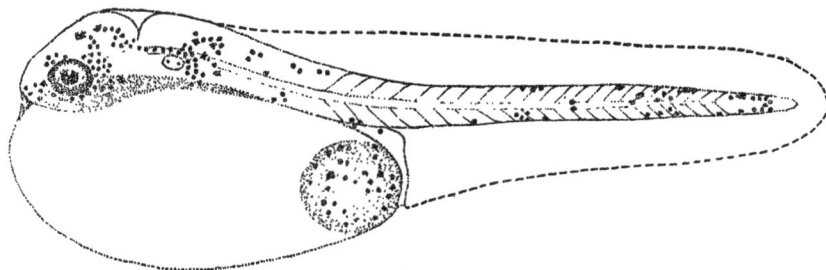

FIG. 178.—Newly hatched larva, 1.75 millimeters long

commencing in May and continuing until September. The great majority of the fish, however, are reported to spawn between the middle of May and the middle of June, the season appearing to be little affected by latitude, spawning occurring at approximately the same time from the Carolinas to Cape Cod.

The account of the development of the eggs and larvæ given by the same authors in their account of this weakfish was taken from the field notes of Lewis Radcliffe, who took the eggs upon which the account is based in the lower part of Chesapeake Bay. The work was done aboard the *Fish Hawk* and constitutes a part of the present investigation. The eggs are described as being pelagic, spherical, about 1 millimeter in diameter, buoyant, and transparent. Hatching occurred in from 12 to 14 hours after fertilization at temperatures of from 68° to 70° F. It seems probable, from the examination of the condition of the reproductive organs, that virtually all spawning takes place in Chesapeake Bay and vicinity during May, June, and July. The fish examined in May were nearly ripe, and no spent fish were seen. In June most of the fish examined still contained roe, but some spent fish also were seen, and as late as July 9 many of the fish examined had not yet spawned.

No ripe squeteagues were observed in Chesapeake Bay, but as spawning is known to occur in Delaware Bay (Welsh and Breder, 1923, p. 150), it is probable that it also occurs around the entrance of the Chesapeake. Most of the spawning in this region is accomplished during May,

FIG. 179.—Larva, 6.5 millimeters long

June, and July. On May 21 we caught (with hook and line off Cape Charles city) 14 females and 4 males, 14 to 18 inches in length, having nearly mature gonads. On May 24 many of the Norfolk market fish were distended with eggs or milt. On June 11 many fish had large roe, and some had already spawned. On June 27, at Buckroe Beach, 1,000 pounds of squeteagues just landed from near-by pound nets (mostly females, 10 to 13 inches in length) were full of spawn, which protruded through the thin wall of the abdomen. As late as July 9 many fish caught in the lower York River had not yet spawned. A large series of fish, both males and females, examined during October, 1922, at Ocean View, Va., showed that individuals of 200 to 286 millimeters (about 8 to 11¼ inches) would probably spawn for the first time the following spring.

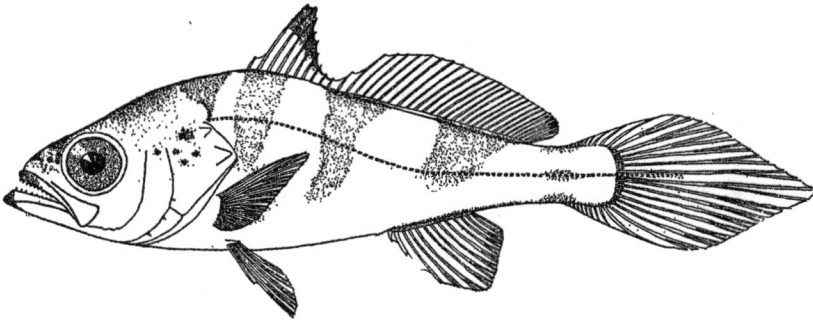

FIG. 180.—Young, 32 millimeters long

The young are reported to grow rapidly. However, observers do not agree with respect to the rapidity of their growth, and before it can be stated what the usual or average rate of growth is, further and more exhaustive studies must be made. Eigenmann (1902, p. 47) states that this fish reaches a "marketable size" in one year. However, it is not clear at which length this author considered the fish to have attained a marketable size. Welsh and Breder (1923) produce very limited data intended to show that a fish 30 millimeters long on July 1 may reach a length of 180 millimeters by November 1. One of us (Hildebrand), working at Beaufort, N. C., found (unpublished), for example, that on about August 1 two definite size groups of young squeteague could be taken. The fish composing the group of youngest fish ranged from 50 to 120 millimeters in length. These fish were thought to be the young of the same season. The fish of the other group, representing older fish yet sexually immature, averaged about 220 millimeters in length and were thought to be fish 1 or a little over 1 year old.

Only two catches of very young squeteagues were made in Chesapeake Bay. The earliest catch consisted of 10 fish, 30 to 38 millimeters (1¼ to 1½ inches) long, seined at Buckroe Beach, Va., on June 28, 1921; and the other one was taken on July 27, 1916, when several young squeteagues, 36 to 49 millimeters (1⅜ to 2 inches) in length, were trawled off Windmill Point, lower Rappahannock River. No young were caught during August, but many were taken during the fall. Although the difficulty encountered by other investigators in separating squeteagues into age groups was met, some success in this respect was obtained by measuring entire catches taken in various types of fishing gear. Most of these fish were caught during September, October, and November in the lower parts of the bay, particularly at Ocean View, but catches made in other parts of the bay (that is, as far north as Solomons) agreed with these in every respect. As a result of consolidating the total number of young squeteagues taken in various seine and beam-trawl hauls from September 25 to October 31, 1921 and 1922, the distribution of sizes given in Figure 182 was found. It will be seen from this graph that the predominating size of the young during October, 1921 and 1922, in the lower parts of Chesapeake Bay was from 3½ to 5 inches in length. Late in October, 1915, of 107 squeteagues trawled by the *Fish Hawk* in the lower Chesapeake, 103 were 90 to 155 millimeters (3½ to 6⅛ inches) in length, three were 185 to 190 millimeters (about 7½ inches), and one was 225 milli-

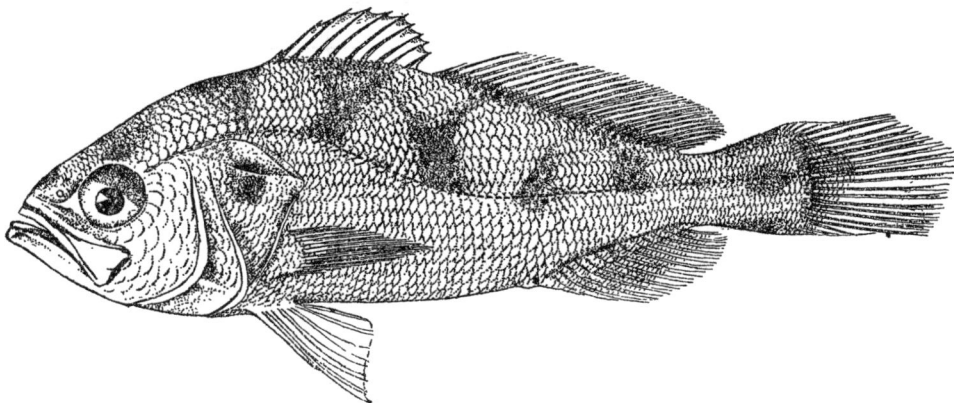

FIG. 181.—Young, probably about 45 millimeters long

meters (about 9 inches). The only squeteague taken (trawled) in December, 1915, was 90 millimeters (3½ inches) in length, and a single fish trawled in December, 1920, was 75 millimeters (3 inches) in length. The predominating sizes of the next group of fish, taken in the fall of 1921 and 1922, were 200 to 263 millimeters (about 8 to 10½ inches). A pronounced scarcity of fish between 160 and 190 millimeters (6¼ to 7½ inches) was found at this time. It is evident, therefore, that in the fall, at the age of about ½ year, the Chesapeake Bay squeteague has attained a length of from 3½ to 5 or even 6 inches, and that the usual size at 1½ years is 8 to 10½ inches. The size of older fish is difficult to determine unless scales are utilized for study.

The squeteague is one of the most valuable food fishes taken in Chesapeake Bay. During 1920 it ranked fourth, both in quantity and value, the catch being 7,918,713 pounds, worth $390,101.

In Maryland it ranked fifth in quantity and fourth in value, the catch being 678,470 pounds, worth $44,143. Of this amount 70 per cent were caught with pound nets, 16 per cent with haul seines, 8 per cent with purse nets, 3 per cent with lines, 2 per cent with gill nets, and 1 per cent with fyke nets. Kent County is credited with the greatest catch, namely, 255,600 pounds, followed by Somerset with 121,123 and Anne Arundel with 78,825 pounds.

In Virginia it ranked fourth in quantity and third in value, the catch being 7,240,243 pounds, worth $345,985. Of this amount, 96 per cent were caught in pound nets, 2 per cent with lines, and 2 per cent with other apparatus. Elizabeth City County is credited with the greatest catch namely, 2,034,840 pounds, followed by Mathews with 1,338,462 and Gloucester with 1,292,970 pounds.

The fishing season extends from about April 15 until November 15. The first appearance of the fish varies from year to year and is one or two weeks later in the upper part of the bay than near the entrance. Records obtained from a set of two pound nets at Lynnhaven Roads, Va., give the first appearance of the squeteague as follows: April 1, 1910; April 5, 1912; April 12, 1916; April 16, 1917; April 2, 1918; April 11, 1919; April 28, 1920; April 5, 1921; and April 1, 1922. The catch during April is usually small, but during May, June, July, and October large numbers are taken with pound nets. In the lower part of the bay some of the best pound-net catches of this species are made during November. It is taken with haul seines from May until October. Hook and line fishing is most productive from June until the end of October. In the vicinity of Norfolk, Va., good catches are sometimes made with pound nets as late as December 1, but in the upper sections of the bay fishing operations usually have ceased by November 1. The squeteague is caught in large numbers in all parts of the bay, from Baltimore southward. North of Baltimore the water freshens rapidly and the catch of squeteagues diminishes perceptibly, until at Turkey Point a scattering few are caught only in the fall, at which time the water usually is slightly brackish.

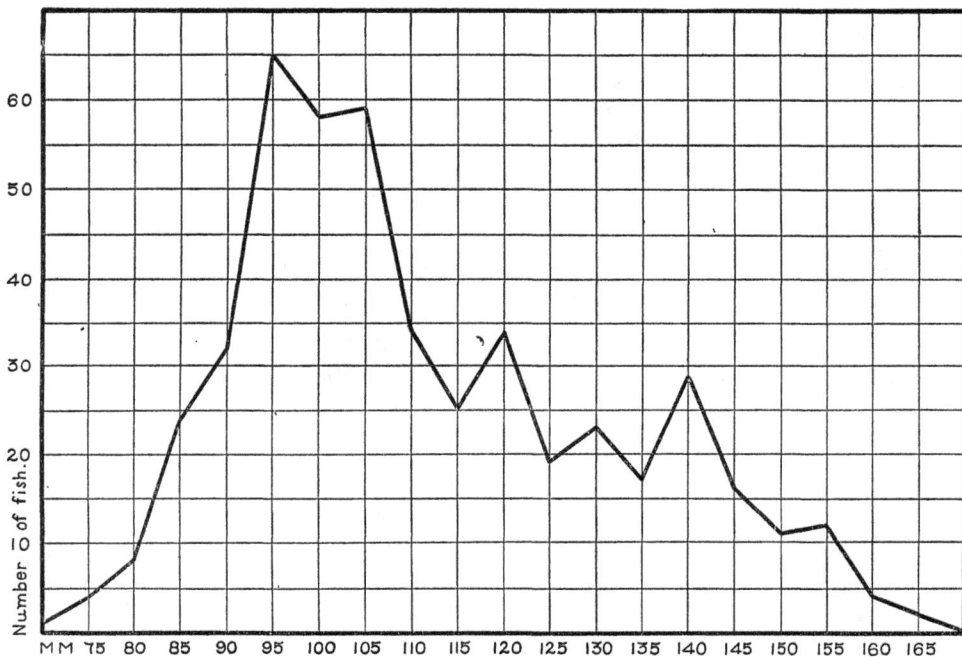

Fig. 182.—Graphic representation showing the frequency of lengths of 480 young squeteagues (*Cynoscion regalis*) taken from September 25 to October 31, 1921 and 1922, in the lower part of Chesapeake Bay

A large part of the Chesapeake catch of squeteague is shipped to outside markets. The best prices are obtained early in the spring and late in the fall, when the species is scarce along the upper Atlantic coast. Sometimes during May and June the price drops to the extent that it is scarcely profitable to ship the fish to market, and at such times the smaller sizes often are discarded by the fishermen. The markets are seldom glutted, for the fish is well known and it is shipped to all cities in the East. The highest price received by the fishermen during 1921 and 1922 was about 20 cents, the lowest 2 cents, and the average for the season about 5 cents per pound.

The squeteague is known throughout the Chesapeake region as "trout" or "gray trout," and is the weakfish of New York. The species bears no relationship to the true trouts and salmons, and the name "trout" tends to lead to confusion; but it is so firmly established in the Chesapeake and southward that a change of name would be very difficult.

In many markets the squeteague is the principal fish sold. Its fine appearance, good flavor, and the long season during which it may be caught contribute to its favor with both the consumer and the marketman. The majority of the Chesapeake catches consist of fish weighing from ½ to 3 pounds. Larger fish are often taken and are particularly common during the spring and late fall. For example, on May 26, 1921, 500 pounds of unusually large fish were taken from a pound net in Lynnhaven Roads. The 29 largest fish weighed 210 pounds, averaging about 7 pounds each, and the largest one weighed 11 pounds 14 ounces. The largest fish observed by us weighed 16 pounds and was taken in a pound net at Ocean View, Va., on May 27, 1921. Fish weighing 12 to 15 pounds are seen occasionally, and a weight of 6 to 10 pounds is not uncommon. The maximum size recorded is 30 pounds. (See Welsh and Breder, 1923, p. 158.)

Habitat.—Massachusetts to the east coast of Florida; especially abundant from North Carolina northward.

Chesapeake localities.—(*a*) Previous records: "Chesapeake Bay, near the ocean" (Uhler and Lugger, 1876); Cape Charles city and Norfolk, Va. (*b*) Specimens in collection: From many localities, from Annapolis, Md., southward to Cape Charles and Cape Henry.

Comparison of lengths and weights of squeteagues

Number of fish weighed and measured	Length	Weight	Number of fish weighed and measured	Length	Weight	
	Inches	*Ounces*		*Inches*	*Lbs.*	*Ozs.*
3	3	0.2	7	13		11.0
7	3½	.3	6	13½		11.9
6	4	.4	10	14		13.5
9	4½	.6	5	14½		15.7
16	5½	.8	2	15	1	.7
11	6	1.1	4	15½	1	2.3
12	6½	1.4	2	16	1	3.0
17	7½	2.3	3	16½	1	7.2
37	8	2.8	4	17	1	8.3
6	8½	3.2	4	17½	1	9.4
10	9	4.0	2	18	1	11.4
12	9½	4.6	4	18½	1	13.0
33	10	5.2	1	20	2	
6	10½	5.9	2	28	9	
10	11	6.7	1	30	9	10.0
16	11½	8.1	1	31	10	
10	12	9.0	1	33	11	14.0
6	12½	10.0				

Family LXXI.—BRANCHIOSTEGIDÆ. The tilefishes

Body elongate, more or less compressed; anterior profile strongly convex; preopercle denticulate; vomer and palatines toothless; scales rather small; nape with a large fleshy appendage; labial folds with somewhat similar appendages at sides; soft dorsal and anal fins rather short, composed of about 13 to 15 rays; ventral fins thoracic, composed of I, 5 rays.

129. Genus LOPHOLATILUS Goode and Bean. Tilefishes

Body stout, somewhat compressed; mouth moderate; maxillary reaching eye; each jaw with an outer series of canines and an inner band of villiform teeth; vomer and palatines toothless; nape with a high, fleshy appendage, resembling an adipose fin; lower jaw at sides with fleshy prolongations, extending backward beyond angle of mouth.

166. Lopholatilus chamæleonticeps Goode and Bean. Tilefish.

Lopholatilus chamæleonticeps Goode and Bean, Proc., U. S. Nat. Mus., II, 1879, p. 205; Nantucket Shoals. Jordan and Evermann, 1896–1900, p 2278.

Head, 3; depth 3.5; D. VII, 14 or 15; A. 14 or 15; scales about 93. Body robust, somewhat compressed; head rather large, its upper profile strongly convex; snout rather blunt; eye small, placed high, 6.5 in head; mouth moderate; maxillary reaching eye; upper jaw with an outer series of strong canines and an inner band of villiform teeth; lower jaw with a series of large canines; dorsal fin continuous, the spines scarcely shorter than the soft rays; caudal fin rather deeply concave; anal fin about half as long as the dorsal; ventral fins inserted below the pectorals; pectoral fins pointed, notably shorter than head; nape with a finlike fleshy flap in front of dorsal; a similar but smaller fleshy flap on side of lower jaw, near angle of mouth.

Color brilliant; the back and upper part of sides bluish or olive-green; this color changing to yellow or rose on lower part of sides; belly rosy with a median white line; head tinged with red on sides, white underneath; back and sides, above the level of pectorals, thickly dotted with small, irregular, yellowish spots, these spots most conspicuous at the nape; dorsal dusky, with yellow spots, the soft part with a pale margin; anal pale pinkish, clouded with purple and bluish iridescence; pectorals pale sooty brown, with purplish near bases; fleshy flap at nape greenish yellow.

This fish does not occur in the Chesapeake collection. The foregoing description is compiled from published accounts of the species. The brilliant colors and the fleshy flap at nape readily identify this fish.

The food (Bigelow and Welsh, 1925, p. 356) consists mainly of bottom-dwelling invertebrates including crabs, squid, shrimp, shelled mollusks, annelids, sea urchins, sea cucumbers, and anemones. Occasionally fish are included in the diet. The tilefish evidently is distinctly a ground fish, seldom, if ever, rising to the surface.

Spawning of the tilefish takes place in July. The eggs are reported to be buoyant and about 1.35 millimeters in diameter. Nothing is known about the larval stages nor the rate of growth.

The tilefish is reported to reach a weight of 50 pounds, but the usual size probably does not exceed 35 pounds. This fish is of no commercial importance as far south as off the mouth of Chesapeake Bay. Off New York and the southern New England coast it is of considerable importance.

Habitat.—Along the outer edge of the continental shelf, off the coast of New England, to opposite the mouth of Chesapeake Bay, usually in water varying from 50 to 200 fathoms in depth.

Chesapeake localities.—(a) Previous records: None. (b) Specimens in collection: None. The species is included because a specimen was observed in the market in Cape Charles city, Va., by Dr. Paul Bartsch, curator, division of mollusks, United States National Museum, in July, 1913. The exact locality where this fish was taken is not known to the writers. It is probable, however, that the fish was taken off the mouth of the bay and that the species does not properly belong to the fauna of the Chesapeake. Doctor Bartsch reports that the fish was the attraction of the entire fishing community.

Family LXXII.—EPHIPPIDÆ. The spadefishes

Body much compressed, very deep; back high, strongly arched; ventral outline less so; mouth small, terminal; jaws with bands of slender, sharp, movable teeth; premaxillary slightly protractile; nostrils double; gill rakers short; gill membranes broadly united to the isthmus; scales ctenoid, of small or moderate size; lateral line strongly arched; dorsal fins 2, the first with 8 to 11 spines, the second longer, with anterior rays produced; caudal fin broad, either square or concave; anal with 3 or 4 spines, the soft rays similar to those of dorsal; ventrals thoracic; pectoral fins short. A single genus and species of this family is found on the Atlantic coast of the Americas.

130. Genus CHÆTODIPTERUS Lacépède. The spadefishes

Body much compressed, nearly as deep as long; the back greatly elevated; snout short and blunt; vomer and palatines toothless; preopercle finely serrate; branchiostegals 6; lateral line concurrent with the back; scales small, about 60 to 70 in a lateral series; dorsal fins slightly disconnected, the first consisting of about eight spines, the third spine longest; anal fin with three spines, the second longest. A single species of this genus occurs on the Atlantic coast of the Americas.

167. Chætodipterus faber (Broussonet). Porgy; Spadefish; Moonfish; Angelfish.

Chætodon faber Broussonet, Ichth. Sistens, Pisc., 1782, p. 17, Pl. VI; Jamaica.
Parephippus faber Uhler and Lugger, 1876, ed. I, p. 107; ed. II, p. 89.
Chætodipterus faber Bean, 1891, p. 86; Jordan and Evermann, 1896-1900, p. 1668, pl. CCXLVII, fig. 619; Smith and Bean, 1899, p. 187; Evermann and Hildebrand, 1910, p. 162.

Head 2.9 to 3.2; depth 1.15 to 1.2; D. VIII–I, 23; A. III, 18 to 20; scales about 65 to 75. Body very deep, strongly compressed; the back greatly elevated; head short; snout very blunt, 2.3 to 2.58 in head; eye 2.7 to 3.25; interorbital 2.65 to 3; mouth small, nearly terminal; maxillary reaching eye in young, not as far back in adult, 3 to 3.35 in head; teeth in brushlike bands in each jaw, the outer series somewhat enlarged; gill rakers very short, 10 to 13 on lower limb of first arch; lateral line arched like the back; scales small, ctenoid; dorsal fins contiguous, the third spine somewhat produced, equal to or longer than head in adult, proportionately shorter in young; the soft part of fin anteriorly somewhat elevated; caudal fin with concave margin in adult, round in young; anal fin with three short spines, the soft part similar to that of dorsal; ventral fin inserted under base of pectorals, one or more of the outer rays sometimes produced, reaching to or beyond origin of anal; pectoral fins very short, 1.45 to 1.6 in head.

Color variable, from grayish to greenish and yellowish; sides with four to six black, vertical bands, the first on head and passing through the eye, the last on caudal peduncle; these bars sometimes becoming obscure in large examples; fins mostly grayish green to dusky, the membranes attached to the produced spine of dorsal and the ventrals often black; caudal fin with a blackish bar at base, the remainder of the fin being plain translucent in small specimens.

FIG. 183.—*Chætodipterus faber*. From a specimen 7½ inches long

Thirteen small specimens, ranging in length from 55 to 120 millimeters (2⅙ to 4¾ inches) were preserved, and adult fish frequently were observed in the Norfolk markets. This fish is readily recognized by the very deep and strongly compressed body, which is nearly as deep as long. The sides bear from four to six broad, black bands.

The food contained in four stomachs of small specimens (68 to 82 millimeters) consisted chiefly of vegetable débris with a few minute crustaceans intermixed. Smith (1907, p. 335) says: "It frequents rocky patches, wrecks, and piling in search of food, which consists of small crustaceans, worms, etc."

Spawning takes place during the summer. Fish with well-developed roe were taken at Crisfield, Md., on May 26, 1916. Smith (1907, p. 335) says: "At Beaufort (N. C.) ripe male and female fish have been found early in June. The eggs are quite small, being less than 1 millimeter in diameter."

The same author believes that fish about 3 inches long, which may be seined in August, are the young of the year. Young fish, probably the product of the same year's hatch, were seined in the Chesapeake at Ocean View, Va., during 1922, as follows: September 18, 1 fish, length 55 millimeters

(2⅛ inches); October 2, 8 fish, 69 to 83 millimeters (2⅝ to 3¼ inches); October 7, 1 fish, 78 milli-meters (3 inches); October 11, 1 fish, 100 millimeters (4 inches); and October 13, 1 fish, 80 milli-meters (3¼ inches). From October 23 to 25, 1915, the *Fish Hawk* trawled 12 spadefish, 65 to 85 millimeters (2½ to 3⅛ inches) long, in 7½ to 22 fathoms of water off the mouths of the Potomac and Rappahannock Rivers. Nothing definite is known beyond this concerning the rate of growth' or when maturity is reached.

The spadefish is of minor commercial importance in the lower part of Chesapeake Bay. During 1922 the catch was about 1,000 pounds, worth $80, all taken in pound nets.

The season extends from May to October, during which time fish occasionally are caught by the pound nets in the lower York River, Cape Charles, Buckroe Beach, Ocean View, and Lynn-haven Roads, Va.

The name "porgie" is given this fish in the lower Chesapeake, where it is well known to most of the fishermen because of its occasional yet persistent appearance and its comparatively large

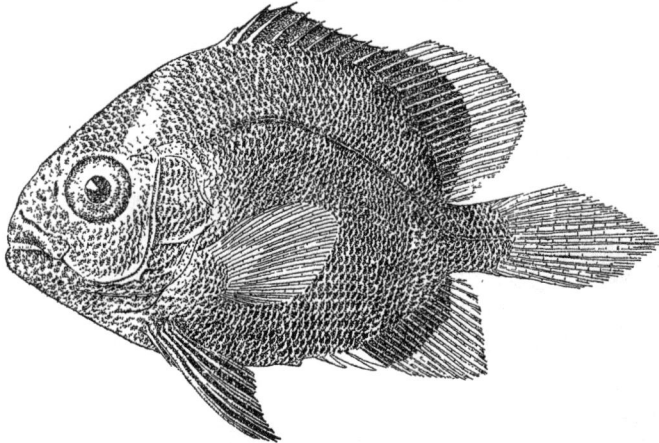

FIG. 184.—*Chætodipterus faber.* Young, 17.5 millimeters long

size. The usual size of market fish is between 3 and 5 pounds. However, fish of 10 or 12 pounds are not uncommon. The small catch is readily absorbed by the Norfolk markets. The porgy is a good food fish and held in high regard almost everywhere along its range. In North Carolina, during 1918, the catch amounted to about 9,000 pounds. In the vicinity of Key West, during the same year, about 1,000 pounds were caught.

This fish is reported to reach a length of 3 feet, but the average is probably less than 1 foot.

Habitat.—Cape Cod to Rio Janeiro, Brazil; rare north of Chesapeake Bay.

Chesapeake localities.—(a) Previous records: Gunston wharf, Cape Charles city, Hampton Roads, Norfolk, and Ocean View, Va. (b) Specimens in collection: Off Point No Point, off mouth of Potomac River, and Crisfield, Md.; mouth of Rappahannock River; Mobjack Bay, Lower York River, Buckroe Beach, Ocean View, Lynnhaven Roads, and Lynnhaven Inlet, Va.

Family LXXIII.—CHÆTODONTIDÆ. The butterfly fishes

Body short and deep, usually strongly compressed; head short; mouth small, with numerous bristlelike teeth on the jaws, none on vomer or palatines; gill membranes attached to the isthmus; pseudobranchiæ large; scales small to moderate, ctenoid; dorsal fin single, continuous, the soft part frequently elevated anteriorly, densely scaled; anal fin with three or four spines, the soft part similar to that of the dorsal; ventral fins thoracic, consisting of I, 5 rays. Color usually brilliant.

131. Genus CHÆTODON Linnæus. The butterfly fishes

Body short, deep, much compressed; head short; snout pointed; mouth small, terminal; teeth in the jaws in bands, numerous, slender, and flexible; scales firm, ctenoid; dorsal fin long, continuous, with about 12 or 13 spines; caudal fin straight or round; anal fin with three spines, the soft part similar to that of the dorsal; ventral fins thoracic, with a strong spine. A single species of this genus of tropical fishes comes within the scope of the present work.

168. Chætodon ocellatus Bloch. Butterfly fish.

Chætodon ocellatus Bloch, Ichthyol., 1787, p. 11, Pl. 210, fig. 2; Jordan and Evermann, 1896–1900, p. 1674, Pl. CCXLIX, fig. 621.

Head 2.55; depth 1.55; D. XII, 21; A. III, 17; scales 36. Body very deep, strongly compressed; anterior profile nearly straight (notably concave in adult); head short and deep; snout pointed, 2.95 in head; eye 2.3; interorbital 2.9; mouth very small, terminal; teeth in the jaws flexible, in bands; gill membranes attached to the isthmus; lateral line running high and ending under posterior part of dorsal fin; scales ctenoid, rather large on sides, reduced on head and caudal peduncle, the rows running obliquely upward and backward on upper portion of side; dorsal and anal both scaled; the dorsal long, its origin over upper angle of gill opening, the spines strong; caudal fin nearly straight; anal fin with three strong spines, the soft part similar to that of the dorsal; ventral fins inserted under base of pectorals; pectoral fins rather broad, 1.3 in head.

Color grayish to yellowish; a jet black bar, about two-thirds the width of eye, extending from the origin of the dorsal, through eye, to the lower margin of preopercle; a second indefinite bar running from middle of anal to middle of soft part of dorsal (this band is present only in the young); fins all more or less orange, middle of soft part of dorsal with a large black blotch.

A single specimen, 35 millimeters in length, was secured. Its short, deep form, pointed snout, and the black band extending from the nape, through the eye, to the lower margin of the head serve well to distinguish the butterfly fish from all other fish of Chesapeake Bay.

Nothing is known of the spawning and feeding habits of this showy little fish, which probably rarely exceeds a length of 6 inches. It has no commercial value.

Habitat.—Woods Hole, Mass., to the Isthmus of Panama; probably more common in the West Indies than elsewhere within its range.

Chesapeake localities.—(*a*) Previous records: None. (*b*) Specimen in collection: From end of Cape Charles, Va., caught September 20, 1921. Apparently very rare within the bay, as only this single specimen was seen.

Order CATAPHRACTI

Family LXXIV.—HEMITRIPTERIDÆ. The sea ravens

Body elongate, robust anteriorly; head large, bony; eyes large, placed high, in anterior half of head; interorbital space concave; mouth large; teeth in jaws in bands; preopercle with spinous processes; suborbital connected with preopercle by a bony stay; gill membranes united, free from the isthmus; gill arches 4, the slit behind the last obsolete; gill rakers rudimentary; body largely covered with prickles and dermal appendages; lateral line present; vent situated in anterior half of body; dorsal fins 2, separate, the first one with about 16 weak spines; anal fin with about 13 soft rays and no spines; caudal fin round; ventral fins thoracic, with one spine and three unbranched soft rays; pectoral fins moderate, with broad bases. Vertebræ 16+23; myodome much contracted behind.

132. Genus HEMITRIPTERUS Cuvier. Sea ravens

Body moderately elongate; head large, with numerous bony humps, ridges, and fleshy flaps; orbital rim much elevated; interorbital space deeply concave, followed by two blunt spines on each side; mouth large; teeth in broad bands on the jaws, vomer, and palatines; no slit behind the last gill; gill membranes broadly united, free from the isthmus; preopercle with stout, blunt spines; suborbital stay forming a sharp ridge; scales wanting; the skin covered with prickles and bony protuberances; spinous dorsal much longer than the soft part, with 16 to 18 spines; ventral fins with I, 3 rays; pectoral fins very broad.

169. Hemitripterus americanus (Gmelin). Sea raven; Red sculpin.

Scorpæna americana Gmelin, Syst. Nat., 1788, p. 1220; no definite type locality given.
Hemitripterus acadianus Uhler and Lugger, 1876, ed. I, p. 105; ed. II, p. 88.
Hemitripterus americanus Jordan and Evermann, 1896–1900, p. 2023; Evermann and Hildebrand, 1910, p. 163.

Head 2.65; depth 3.75; D. XVI–I, 12; A. 13; scales 40. Body rather stout; head very large and bony; mouth large; maxillary reaching beyond posterior margin of eye, about 2 in head; both jaws with several rows of sharply pointed teeth; body largely covered with prickles, these enlarged along the back and lateral line; nasal spines strong; supraocular ridge much elevated, with dermal flaps and two blunt spines; three pairs of fleshy flaps on nasal bones and two on supraocular ridges; smaller cirri on maxillary, on preorbital, and several on lower jaw; interorbital space deeply concave; two blunt, occipital spines on each side, and two or three others on the outside of these; opercle small, with a bony ridge; preopercle with two blunt spines and one or two more below these; the first two or three dorsal spines longest, the fourth and fifth spines shorter than those farther back; caudal fin rounded; anal fin somewhat similar to the second dorsal and about opposite it; ventral fins fleshy, with three rays; pectorals broad, nearly reaching origin of anal.

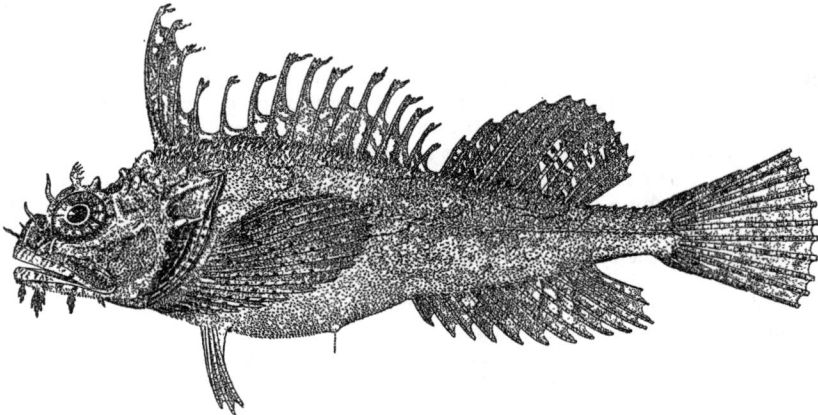

FIG. 185.—*Hemitripterus americanus.* From a specimen 4½ inches long

Color variable, reddish, reddish-purple, or yellowish-brown; always paler below; belly usually yellow; some individuals are variously marbled, others are of uniform coloration; fins variously barred with light and dark; anal and pectorals often with yellow rays.

This species was not taken during the present investigation. It is included here because of published records of its occurrence in Chesapeake Bay. The sea raven is the only sculpin known from Chesapeake Bay, and it is readily recognized by its rough, prickly skin and its bony head, with high ridges, numerous spines, and fleshy flaps.

The food of the sea raven is reported (Bigelow and Welsh, 1925, p. 332) to consist of invertebrates living on the bottom, such as mollusks, various crustaceans, sea urchins, and worms. It is said also to feed on fish. Bean (1903, p. 647) reports as follows relative to the eggs and spawning: "The sea raven spawns in November. Eggs observed on November 29, 1897, were in masses adhering tightly together. The egg at that date was five thirty-seconds of an inch (approximately 4 millimeters) in diameter and showed the form of the fish distinctly. Its color, when first deposited, is yellow, but soon changes to salmon and then to amber before hatching."

Maximum length, 25 inches; a fish 22½ inches long weighed 7 pounds 2 ounces. Fish 18 to 20 inches long and weighing 2 or 3 pounds are reported to be not uncommon in the Gulf of Maine. The species is nowhere of much commercial value, and of none whatsoever in the Chesapeake.

Habitat.—Labrador and Newfoundland, south to Chesapeake Bay; not abundant south of New Jersey.

Chesapeake localities.—(*a*) Previous records: Near the entrance of Chesapeake Bay (Uhler and Lugger, 1876), Cape Charles city, and Old Point Comfort, Va. (*b*) Specimens in collection: None. The species evidently is very rare in Chesapeake Bay.

Family LXXV.—CYCLOPTERIDÆ. The lump suckers

Body short and thick; back more or less elevated; head short and thick; suborbital stay present, thin and flattish; mouth small, terminal; teeth in the jaws simple, in bands; none on palatines or vomer; gill openings narrow, restricted to the sides; gill membranes broadly joined to the isthmus and to the shoulder girdle; gills 3½; pseudobranchiæ present; branchiostegals 6; skin smooth, tubercular, or spinous; dorsal fins 2, the anterior one sometimes hidden by the skin; soft dorsal and anal similar, without spines; caudal fin narrow, round, with few rays; ventrals thoracic, forming the bony center of a sucking disk; pectoral fins rather short, with very broad bases.

133. Genus CYCLOPTERUS Linnæus. Lumpfishes

Body more or less compressed toward the back, somewhat triangular in cross section; head short, thick, more or less quadrate in cross section; snout blunt, round; mouth terminal, turned slightly upward; skin covered with rough, bony tubercles; dorsal fins 2, the first visible only in the very young, completely hidden in the skin in the adult; ventral disk moderately large.

170. Cyclopterus lumpus Linnæus. Lumpfish; Lump sucker.

Cyclopterus lumpus Linnæus, Syst. Nat., ed. X, 1758, p. 260; Baltic and North Seas. Jordan and Evermann, 1896–1900, p. 2096, Pl. CCCXIII, fig, 757; Bean, 1907, p. 178; Kendall, 1914, p. [1].

Head 5 (in entire length); depth 2; D. VI to VIII–9 to 11 (the first dorsal visible only in very young); A. 9 to 11. Body massive, with the dorsal profile much more strongly arched than the ventral and concave over the head; body more or less triangular in cross section and with seven longitudinal ridges, one of these on the median line of the back as a cartilaginous flap inclosing the first dorsal in the adult and dividing into two ridges between the dorsal fins; another ridge on each side over the eye; another ridge paralleling it and extending from somewhat below posterior point of opercle to lower edge of caudal peduncle; and another marking the boundary from side to belly; each ridge with large pointed tubercles; the skin between the ridges thickly studded with small knobs; snout short; eye small, as long as snout, 4 in head; mouth broad, terminal; teeth small, in bands; gill opening moderately wide; second dorsal and anal similar and placed opposite each other; caudal fin square to slightly rounded posteriorly; ventral fins modified into six pairs of fleshy knobs in the center of the sucking disk, surrounded by a roughly circular flap of skin, the entire disk about as wide as head and situated close behind the throat; pectoral fins large, very broad at base, nearly meeting at throat.

Color variable, yellowish to greenish in young; adult males reddish; females bluish to brownish; spots, blotches, cloudings, and other marks not infrequent. The young often take on the color of their surroundings very closely.

This species was not taken during the present investigation. The foregoing description is compiled from published accounts. The peculiar shape, the body ridges with their bony tubercles, and the rough skin readily separate the lumpfish from all other fishes of Chesapeake Bay.

The food of the lumpfish appears to be quite varied, consisting of crustaceans, mollusks, worms, jellyfish, and various other invertebrates. Fish also are eaten.

The spawning period of this fish evidently is a protracted one, occuring in general during late winter and spring. The only two specimens recorded from Chesapeake Bay were ripe females; one was taken on April 14, 1907, and the other on April 29, 1914. An inshore migration is said to take place during the spawning season, and the eggs are deposited in comparatively shallow water. A large number of eggs are produced by a single female; they are about 2.2 to 2.6 millimeters in diameter and they stick together in masses and sink.

The pelagic habits of the young, described by Bigelow and Welsh (1925, pp. 336–337) were observed by one of us (Schroeder) in the Gulf of Maine during the summer of 1925, when many fry about 1 inch long were found among floating masses of rockweed.

312 BULLETIN OF THE BUREAU OF FISHERIES

The maximum length attained by the lumpfish appears to be 20 inches, but few exceed 14 to 16 inches. (Bigelow and Welsh, 1925, p. 336.) This fish is not eaten in the United States and is of no commercial value.

Habitat.—Both sides of the North Atlantic; on the American side from Western Greenland south to Chesapeake Bay; rare south of New Jersey.

Chesapeake localities.—(a) Previous records: Buckroe Beach and Wolf Trap Light, near Old Point Comfort, Va. (b) Specimens in collection: None. The species evidently is very rare in Chesapeake Bay.

Family LXXVI.—TRIGLIDÆ. The sea robins or gurnards

Body elongate, fusiform, deepest at nape; head large, completely inclosed in bony plates bearing spines, granules, and striations; mouth large, terminal; teeth small, in bands on jaws, vomer, and palatines; maxillary slipping under preorbital; premaxillaries protractile; no supplemental maxillary; gill arches 4; gill membranes not attached to isthmus; lateral line present; scales or bony plates on body; air bladder and pyloric cæca usually present; dorsal fins 2, the spines short; caudal fin rather long; anal fin without spines, similar to second dorsal; ventral fins thoracic, wide apart, I, 5 rays; pectorals long, winglike, the three lowermost rays detached, free from each other, developed as feelers.

134. Genus PRIONOTUS Lacépède. Sea robins

The characters of the genus are included in the family description. The species are mostly rather small, and as they are not eaten in this country they have no commercial value here.

KEY TO THE SPECIES

a. Mouth large, the maxillary reaching nearly opposite anterior margin of eye; no cross groove on top of head; gill rakers rather numerous, 15 to 20 on lower limb of first arch; lateral line in a black streak _____ *evolans*, p. 312

aa. Mouth smaller, the maxillary not nearly reaching eye; a shallow cross groove present on top of head; gill rakers less numerous, 10 to 13 developed on lower limb of first arch; lateral line not in a black streak.

  b. Body moderately robust, the depth 4.15 to 5.15 in the length; mouth rather large, the maxillary 1.2 to 1.3 in snout; cross groove on top of head arched backward___ *carolinus*, p. 314

  bb. Body less robust, the depth 5.55 to 5.65 in the length; mouth somewhat smaller, the maxillary 1.4 to 1.55 in snout; cross groove on head straight, not noticeably arched backward_____ *affinis* sp. nov., p. 315

171. Prionotus evolans (Linnæus). Sea robin; Flying fish.

Trigla evolans Linnæus, Syst. Nat., ed. XII, 1766, p. 498; Carolina.
Prionotus strigatus Cuvier and Valenciennes, Hist. Nat. Poiss., IV, 1829, p. 86; Bean, 1891, p. 86; Jordan and Evermann, 1896–1900, p. 2167.
Prionotus lineatus Uhler and Lugger, 1876, ed. I, p. 102; ed. II, p. 85.
Prionotus evolans Jordan and Evermann, 1896–1900, p. 2168, Pl. CCCXX, fig. 772.

Head 2.2 to 2.6; depth 3.65 to 5.1; D. X–11 or 12; A. I–9 or 10; scales 73 to 86. Body moderately robust, about as broad as deep at nape, compressed posteriorly; head large, depressed, broader than deep; snout broad, its length 2.05 to 2.4 in head; eye 4.15 to 5.8; interorbital concave, 5.1 to 6.45; mouth large, horizontal; lower jaw included; maxillary reaching nearly opposite anterior margin of eye, 2 to 2.35 in head; teeth in the jaws in broad villiform bands; gill rakers longer and more numerous in young than in adult, the stumps, however, usually remaining, 15 to 20 somewhat developed on lower limb of first arch; scales rather small, ctenoid; spines on head and about snout not very large; no spine at center of radiations on cheek; no cross groove on top of head; dorsal fins separate, the first consisting of rather short, stiff spines, the third spine the longest, 2.4 to 2.9 in head; the longest rays of the second dorsal scarcely as long as the longest spines; caudal fin with slightly concave margin, sometimes truncate; anal fin similar to second dorsal; ventral fins well developed, inserted under and somewhat behind the base of pectorals; pectoral fins large, the free rays tapering, greatest length of fin 1.79 to 2.15 in body.

Color grayish above; pale underneath; the largest specimen spotted with pale markings; back usually with three or four longitudinal stripes, not visible in large examples; lateral line in a black streak; dorsal and caudal fins brownish, the first dorsal with a black blotch between the fourth and sixth rays, more or less ocellated in young; anal and ventrals pale; pectorals largely bluish black, usually with the uppermost ray white, the fin being crossed in the largest individuals at hand by wavy black lines. Color in life (based on 12 specimens, 9 to 15 inches in length, observed in the aquarium at Woods Hole, Mass.) pale green or light brown above, white below; four or five prominent to obscure, regularly-placed, dark, saddlelike blotches on back (these generally more prominent than in *carolinus*), the first blotch under first dorsal, the second between dorsal fins, the third and fourth under second dorsal, the fifth on caudal peduncle; a prominent dark brown stripe, usually broken posteriorly, along lower part of side; brown markings on head, forming bars and concentric lines (no orange markings on head, no dusky markings on branchiostegals and throat, and no brownish streaks on dorsal, as in *carolinus*); first dorsal yellowish brown, with large black blotch usually extending from the fourth to the sixth spines, occasionally only from the fourth to the fifth spines; caudal and anal pale brown or yellowish brown, with pale edges; ventrals yellowish or pale; pectorals orange to brown, with pale edges, center of fin washed with dusky (this area not divided into two bars, as in *carolinus*), prominent dark brown or black wavy cross lines present; filaments of pectorals white, with pale brown or orange, marked with narrow brown bars (the bars absent in *carolinus*).

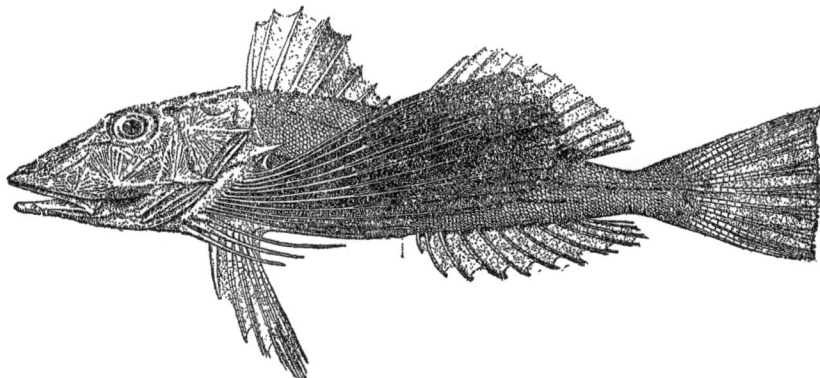

FIG. 186.—*Prionotus evolans*

This species, as here understood, is represented by 18 specimens, ranging from 60 to 340 millimeters (2⅜ to 13½ inches) in length. This species is distinguished from its relatives largely by the rather weak spines and weak striations on the head, by the rather numerous gill rakers, and by the color. The largest specimens in the collection appear to belong to the nominal species, *P. strigatus*. *P. strigatus*, as described, apparently differs from *P. evolans* chiefly in color. The specimens at hand indicate that the color pattern, namely the black-striped pectoral fins, develop with age. The largest specimen, 340 millimeters, has the pectoral fins very distinctly cross-striped; another, 300 millimeters, is somewhat less distinctly striped, and one of 245 millimeters has them obscurely cross-striped. The specimens at hand quite certainly are all of one species and here are referred to *P. evolans*.

Five stomachs of this sea robin were examined for food and found to contain only small crustaceans, principally Mysis. The spawning habits and the rate of growth are as yet unknown.

This species grows somewhat larger and is less common than *P. carolinus* in Chesapeake Bay. A maximum length of 18 inches has been recorded. It is not eaten and has no commercial value.

Habitat.—Massachusetts Bay to South Carolina.

Chesapeake localities.—(a) Previous records: "Lower part of Chesapeake Bay" (Uhler and Lugger, 1876) and Cape Charles city, Va. (b) Specimens in collection: Mobjack Bay, Cape Charles, off Fortress Monroe, Buckroe Beach, Lynnhaven Roads, and Ocean View, Va.

172. Prionotus carolinus (Linnæus). Sea robin; Gurnard; Flying fish.

Trigla carolina Linnæus, Mantissa Plantarum, Part II, 1771, p. 528; Carolina.
Prionotus carolinus Uhler and Lugger, 1876, ed. I, p. 103; ed. II, p. 86; Jordan and Evermann, 1896–1900, p. 2156, Pl. CCCXVIII, fig. 768; Smith and Bean, 1899, p. 187; Evermann and Hildebrand, 1910, p. 163.

Head 2.8 to 3.1; depth 4.15 to 5.15; D. X–13; A. I, 11; scales 100 to 107. Body rather robust, a little broader than deep under spinous dorsal, round or slightly compressed posteriorly; head moderately large, depressed; snout broad, its length 2 to 2.3 in head; eye 3 to 5; interorbital deeply concave, 7 to 9.3; mouth rather small, horizontal; lower jaw included; maxillary failing notably to reach eye, 1.75 to 2.6 in head, 1.2 to 1.3 in snout; teeth small, in villiform bands in each jaw; gill rakers slender, 10 to 13 on lower limb of first arch; scales small, ctenoid; striations and spines on head moderately developed; no spine at center of radiations on cheek; preopercular spine scarcely reaching base of humeral spine; serrations on margin of snout small; a shallow furrow with a backward curve across cranium, just posterior to the eyes; dorsal fins separate, the first composed of rather short slender spines, the third one the longest, 2 to 2.45 in head, scarcely longer than the longest rays of the second dorsal, not reaching past the tips of the posterior spines nor to origin of second dorsal when deflexed; caudal fin with rather deeply concave margin in adult, nearly straight in young; anal fin similar to second dorsal; ventral fins moderately developed, inserted under and posterior to bases of pectorals, reaching a little past origin of anal; pectoral fins rather short, reaching opposite base of sixth anal ray, 2.35 to 3.55 in body, the detached rays expanded at tips.

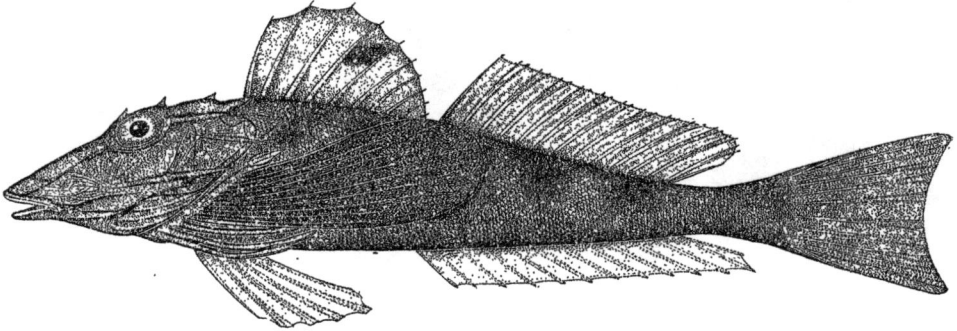

FIG. 187.—*Prionotus carolinus.* From a specimen 6 inches long

Color variable, usually grayish or reddish brown above, pale underneath; back with about five dark saddlelike blotches; young usually with small dark spots in addition to the large saddlelike blotches; dorsal fins more or less grayish, marked with pale spots and stripes, spinous dorsal with a black spot between the fourth and fifth spines; caudal fin uniform, grayish or brownish; anal and ventrals pale; pectorals brownish to blackish, sometimes spotted with darker markings. Color in life (based on eight fish, 8 to 12 inches long, observed in the aquarium at Woods Hole, Mass.) greenish to brownish above; white underneath; 3, 4, or 5 obscure, dark, irregularly-spaced, saddlelike blotches on back; head marked with orange; branchiostegals and throat with prominent dusky markings; dorsal fins pale to grayish, with irregular, brownish, longitudinal streaks; a prominent black spot on first dorsal on the membrane between the fourth and fifth spines; caudal brown, margin not pale; anal pale brown, without pale margin; ventrals pale yellow to brown; pectorals yellow or orange, with two broad dusky bars, one crossing middle of fin and the other on outer third, fin without brown wavy lines; pectoral filaments orange.

Many specimens of this species, ranging in length from 40 to 170 millimeters (1⅗ to 6¾ inches), were preserved. This species is recognized by its rather small mouth, robust body, short and blunt spines about the head, margin of caudal concave, and the dark saddlelike blotches on the back.

The food of this sea robin, as shown by 18 specimens taken in Chesapeake Bay, consists principally of crustaceans, which may be named in the order of their apparent importance, as follows: Mysis, amphipods, isopods, shrimp, and small crabs. One individual had fed on annelids and another on a small fish.

The spawning season at Beaufort, N. C., according to Smith (1907, p. 362), takes place in the spring. Welsh and Breder (field notes) found ripe fish at Atlantic City, N. J., from August 19 to 25, 1920. Kuntz and Radcliffe (1918, pp. 105 to 109), who give an account of the embryology and larval stages, state that the species spawns (presumably at Woods Hole, Mass., for that is where these authors worked) in June, July, and August. The egg is described as being 1 to 1.15 millimeters in diameter, lighter than sea water, slightly yellowish in color, but highly transparent. The incubation period at a temperature of 22° C. is given as approximately 60 hours. The newly hatched larvæ are stated to be about 2.8 millimeters long, and to have grown to a length of 3.1 to 3.4 millimeters in five days. In young fish, 8 to 10 millimeters long, the dorsal, anal, and caudal fins had become well differentiated, and the free rays of the pectoral fins, characteristic of the genus, were already present. Young fish 25 to 30 millimeters in length have all the fins well formed, the head shows the bony structure characteristic of the adult, and the fish are gradually acquiring the diagnostic characters of the species.

The following catches of young fish were made in the Chesapeake: December, 2 fish, 48 and 50 millimeters (2 inches); March, 2 fish, 46 and 53 millimeters (2 inches); April, 48 fish, 33 to 82 millimeters (1⅓ to 3¼ inches); May, 10 fish, 45 to 68 millimeters (1¾ to 2½ inches).

This species is reported to reach a maximum length of 15 or 16 inches; rarely, however, exceeding a length of 1 foot. No individuals among the numerous ones taken in Chesapeake Bay exceed a length of 11¾ inches. The species is very common in Chesapeake Bay, occupying both shallow and deep water, apparently being more numerous in deep than in shallow water. It was taken in considerable numbers throughout the year in the beam trawl, which usually was hauled in the deeper waters of the bay. The greatest depth recorded in the log for this species is 23 fathoms. This sea robin is not used for food and has no commercial value.

Habitat.—Bay of Fundy to South Carolina; common from Cape Cod to Cape Lookout.

Chesapeake localities.—(*a*) Previous records: Mouth of the Potomac River, Md., Gunston Wharf, Old Point Comfort, Hampton Roads, Norfolk, Ocean View, and off Cape Henry, Va. (*b*) Specimens in collection: Many localities, from Love Point, Md., to the mouth of the bay.

173. Prionotus affinis sp. nov. Sea robin.

Type No. 87654, U. S. National Museum; length 203 millimeters; off Kent Island, Md.

Head 2.95 to 3; depth 5.55 to 5.65; D. X–13; A. 12; scales about 108. Body quite slender, the depth under spinous dorsal nearly equal to the width, somewhat compressed posteriorly; head moderately large, quite strongly depressed; snout rather long, depressed, its length 1.95 to 2.05 in head; eye 5.2; interorbital deeply concave, 6.15 to 6.9; mouth rather small, horizontal; lower jaw included; maxillary not nearly reaching eye, 2.9 to 3 in head, 1.4 to 1.55 in snout; teeth as in *P. carolinus;* gill rakers rather short, 11 or 12 somewhat developed on lower limb of first arch; scales small, ctenoid, somewhat reduced in size on chest; spines and striations on head somewhat more prominent; otherwise as in *P. carolinus,* the preopercular spine reaching well beyond base of humeral spine; serrations on margin of snout quite prominent; a straight, shallow furrow across the cranium just posterior to the eyes; dorsal fins separate, the first composed of rather strong, stiff spines, the third and fourth of about equal length, 2 to 2.1 in head, somewhat longer than the longest soft rays, not reaching beyond the tips of the posterior spines when deflexed nor to origin of second dorsal; caudal fin with rather deeply concave margin, the rays of upper lobe slightly the longest; anal fin similar to second dorsal but not quite as high, its origin slightly behind that of the second dorsal; ventral fins moderately developed, inserted under and somewhat posterior to bases of pectorals, reaching somewhat past origin of anal; pectoral fins of moderate length, reaching about opposite base of the sixth anal ray, 2.4 to 2.45 in length of body, the free rays distally with membranous expansions.

Color in alcohol brownish above, with small darker brownish markings, but no large blotches; pale underneath; gill membranes dusky; dorsal fins brownish, with pale to white spots or longitudinal markings; a black spot between the fourth and fifth dorsal spines; caudal fin uniform brownish; anal fin pale, with a dark band across middle; ventral fins pale; pectorals dusky, the lower free rays pale.

Two specimens, 203 and 325 millimeters (8 and 12¾ inches) in length, occur in the collection, which are closely related to *P. carolinus,* differing, however, in the more slender and more elongate body, slightly smaller mouth, somewhat larger and more prominent sculpture of the head, and in color. The furrow on head just posterior to the eyes is straight and not bent backward as in *P.*

carolinus. No indications of dark saddlelike blotches are present on the back. These specimens differ from *P. scitulus,* a related species, known from North Carolina southward, principally in having a broader interorbital, shorter dorsal spines, none of them reaching beyond the tips of the posterior spines when deflexed and not to origin of second dorsal, whereas in *P. scitulus* the longest spines reach well beyond the tips of the posterior spines and to origin of second dorsal. The caudal fin in the present species is rather deeply concave posteriorly. In *P. scitulus* the margin is nearly straight. We are unable to identify the specimens in hand with any known species and, although very closely related to *P. carolinus,* they differ slightly in so many characters that we are convinced they represent a distinct species. We therefore are obliged to propose a new name. The genus, however, is in need of a critical study, and until such a study is made the true relationships of the species will not be known. We have reviewed the literature pertaining to the Atlantic coast species and had specimens of other species for comparison and for study only from Chesapeake Bay and Beaufort, N. C.

The species probably is rare in Chesapeake Bay. It was taken only in the beam trawl; one specimen was obtained at a depth of 14½ fathoms and the other at 10 fathoms.

Chesapeake localities.—Off Kent Island, Md., and off Old Point Comfort, Va.

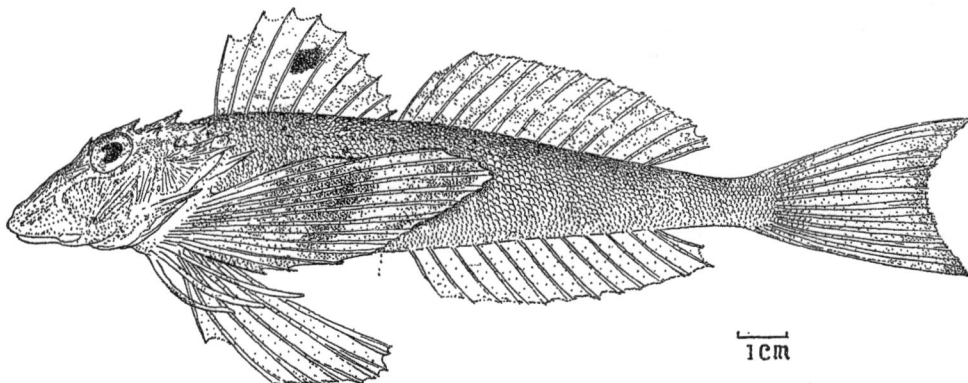

FIG. 188.—*Prionotus affinis* sp. nov. From the type, 8 inches long

Family LXXVII.—CEPHALACANTHIDÆ. The flying gurnards

Body elongate, rather broad, sides rather vertical; head blunt, quadrangular, nearly the entire surface bony, the bones about the eye united into a shield; a long, bony process ending in a sharp spine, extending backward from nape to beyond origin of dorsal; preorbital projecting beyond the jaws; preopercle extending backward as a long round spine, reaching beyond the base of ventrals; mouth small, inferior; teeth granular, present only on the jaws; gill openings restricted to the sides; pseudobranchiæ large; scales small, keeled; dorsal fin consisting of slender spines and rays; pectoral fins greatly enlarged, divided into two sections, the inner section much the longer. This family contains one genus, with a single representative in American waters.

135. Genus CEPHALACANTHUS Lacépède

The characters of the genus are included in the description of the family. The flying gurnards possess the power of flight, but to a much less degree than the true flying fishes.

174. Cephalacanthus volitans (Linnæus). Flying fish; Flying robin.

Trigla volitans Linnæus, Syst. Nat., ed. X, 1758, p. 302; Mediterranean Sea and oceans within the Tropics.
Dactylopterus volitans Uhler and Lugger, 1876, ed. I, p. 101; ed. II, p. 85.
Cephalacanthus volitans Jordan and Evermann, 1896–1900, p. 2183, Pl. CCCXXIII, fig. 778.

Head 3.95; depth 5.5; D. VI–I, 8; A. 6; scales 59. Body elongate, depressed, somewhat broader than deep; head depressed; snout short, blunt, its length 2.65 in head; eye 3; interorbital broad, concave, 1.85; mouth moderate; upper jaw projecting; maxillary reaching nearly to pupil, 2.6 in head; teeth in the jaws blunt, in bands; preopercle with a very large long spine, its free portion exceeding somewhat the width of interorbital, not quite reaching the tip of the nuchal spine; scales on back and sides with prominent keels; first two spines of the dorsal separate; second dorsal and anal similar in shape; caudal fin deeply concave; ventral fins inserted under posterior part of pectorals, rather narrow, pointed; pectoral fins in two sections, the upper short and with six rays, the lower section long and broad, reaching base of caudal, 1.35 in body.

Color in life more or less variegated; the back and sides brownish green, with shades of red; white underneath; spinous dorsal barred and spotted with purple, brown, and yellow; membranes of soft dorsal plain, the rays alternately spotted with yellow and red; caudal fin with two or three irregular, vertical, wine-colored bars, with yellowish interspaces; anal with three pale reddish bars, its outer edge yellowish; ventrals deep orange; pectorals mostly black, the outer third of fin with bright blue bars and spots and a margin of the same color, inner third of fin with five or six bright blue streaks, fin everywhere with obscure reddish blotches. In preserved specimens the bright colors fade, but most of the markings remain either as light or dark spots and bars.

A single specimen 170 millimeters (6¾ inches) in length was seen and preserved. This specimen forms the basis for the foregoing description. The flying fish is readily recognized by its broad, depressed head and body, very large preopercular and nuchal spines, long pectorals, and the absence of "feelers."

This flying fish is reported to feed on various small crustaceans. Nothing has come to our notice in the literature concerning the spawning habits of this fish, and it is probable that these are largely unknown. The species is said to be able to maintain itself for considerable distances in the air by means of its flat body and large, winglike pectoral fins. A maximum length of 1 foot is reported.

Habitat.—Both coasts of the Atlantic Ocean; on the American coast from Cape Cod, Mass., to Brazil; common southward; rare north of North Carolina, only occasionally straying northward to Cape Cod.

Chesapeake localities.—(*a*) Previous records: Lower part of Chesapeake Bay (Uhler and Lugger, 1876). (*b*) Specimen in collection: From Ocean View, Va., taken in an 1,800-foot haul seine on October 2, 1922. Very rare in Chesapeake Bay.

Order PHARYNGOGNATHI

Family LXXVIII.—LABRIDÆ. The labrid or lipped fishes

Body moderately elongate, greatly compressed in some species; mouth terminal, usually of small or moderate size; lips usually thick, with longitudinal folds; premaxillaries protractile; maxillary without a supplemental bone; teeth in the jaws strong, prominent, separate or more or less fused at the base; lower pharyngeal bones united and bearing strong conical or tubercular teeth; nostrils double, without flaps; branchiostegals 5 or 6; gill arches 3½; dorsal fin continuous, usually long, the spines varying from 3 to 20; anal with two to six spines; ventrals thoracic, with one weak spine and five soft rays. This family contains about 20 American genera, only two of which are known to occur in Chesapeake Bay.

KEY TO THE GENERA

a. Anterior profile high and strongly convex; preopercular margin entire; about 70 scales in a lateral series; cheeks and opercles largely naked_____Tautoga, p. 317
aa. Anterior profile not greatly elevated and only gently convex; preopercular margin serrate; about 40 scales in a lateral series; cheeks and opercles mostly covered with scales
_____ Tautogolabrus, p. 320

136. Genus TAUTOGA Mitchill. Tautogs

Body elongate, moderately deep and compressed; anterior profile rather strongly arched head nearly as deep as long; eye small, placed high; mouth rather small; lips quite broad and thick; teeth in the jaws strong, the anterior ones more or less incisorlike; scales rather small, about 70 in a lateral series; cheeks and opercles largely naked; dorsal fin long, continuous, the soft part short; caudal fin short, round to slightly truncate; anal fin with three stout spines, the soft part similar to that of the dorsal. This genus consists of a single species.

175. Tautoga onitis (Linnæus). Tautog; Blackfish; Black porgy; Chub; Salt-water chub.

Labrus onitis Linnæus, Syst. Nat., ed. X, 1758, p. 286. (No type locality given.)

Tautoga onitis Uhler and Lugger, 1876, ed. I, p. 106; ed. II, p. 89; Jordan and Evermann, 1896–1900, p. 1578, Pl. CCXXXVII, fig. 596; Evermann and Hildebrand, 1910, p. 162; Fowler, 1912, p. 56.

Hiatula onitis Bean, 1891, p. 86.

Head 3.25 to 3.56; depth 2.55 to 2.95; D. XVI or XVII, 10; A. III, 7 or 8; scales 69 to 73. Body rather deep, compressed; the back rather strongly elevated; caudal peduncle short and deep, its depth equal to or a little greater than distance from tip of snout to margin of preopercle; head rather short; snout blunt, 2.25 to 3.25 in head; eye small, 3.05 to 6; interorbital 3.85 to 5.1; preopercular margin smooth; mouth moderate, slightly subinferior; lips rather broad and thick; maxillary reaching nearly opposite anterior margin of eye in young, not nearly as far back in adult, 3 to 3.55 in head; teeth in the jaws strong, the anterior ones in young more or less compressed, incisorlike becoming more rounded and caninelike in large examples; gill membranes broadly united, free from the isthmus; gill rakers short, about nine on lower limb of first arch; lateral line complete and continuous; scales rather small, thin, with smooth edges, reduced in size on belly and chest, largely absent on cheeks and opercles; dorsal fin continuous, the spines stiff, not quite as high as the soft part of fin; caudal fin broadly rounded; anal fin with three rather short strong spines, the soft part similar to that of the dorsal; ventral fins moderate, inserted about an eye's diameter behind base of pectorals; pectoral fins broad, their length 1.15 to 1.5 in head.

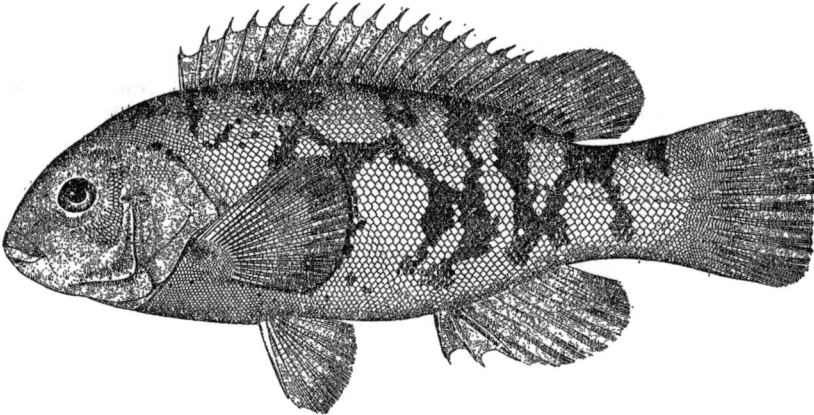

FIG. 189.—*Tautoga onitis.* From a specimen 6¾ inches long

Color dull black to greenish black or brownish above, with more or less distinct, irregular, blackish bars or blotches; the bars and blotches most distinct in young; fins plain, mostly like the ground color of body.

In regions off the Atlantic coast where this species is abundant two color patterns are recognized, one being plain blackish and the other having irregular blackish or brownish bars on a pale ground. Both varieties are found even among the largest fish, but very small fish nearly always are marked with blotches and irregular bars.

Many specimens of this species, ranging in length from 55 to 320 millimeters, were preserved. It is most readily recognized by its strongly convex anterior profile; thick lips; large incisorlike teeth, and very deep caudal peduncle.

The food of the tautog is varied, consisting largely of various small mollusks and crustaceans of suitable size. It is most numerous among rocks, old wrecks, piling, etc., were it feeds on barnacles, crabs, and other animals common in such places. The food of the tautog and the sheepshead are practically identical, and they occupy the same feeding grounds within the region where their ranges overlap. The tautog takes the hook readily and is of some importance as a "game" fish. Large numbers are caught with hook and line by sportsmen as well as commercial fishermen, particularly off the coasts of Long Island and New Jersey.

The tautog spawns principally during June at Woods Hole and probably somewhat earlier in Chesapeake Bay. The eggs are buoyant, have no oil globule, and are from 0.9 to 1 millimeter in diameter.[24] They hatch in 42 to 45 hours at a temperature of 68° to 72° F. The young at hatching are 2.2 millimeters long. The yolk sac is absorbed when the fish reaches a length of 3.3 millimeters. At this time the mouth is fully formed. At a length of 10 millimeters many of the characters of the adult fish are already evident. It has been suggested that fish seined along shores, measuring from 3 to 6 inches in length, are a year old. Nothing definite is known concerning the rate of growth, however, nor as to the age at which they mature.

Not one specimen had been collected in Chesapeake Bay during 1921 until September 23, when 208 were seined in six hauls of a 250-foot bag seine along Cape Charles beach. Excepting three adult fish, all these ranged from 55 to 115 millimeters (2¼ to 4½ inches) in length. Seining in the same locality on November 23, 1921, only three tautogs were taken, 3½ to 4 inches in length, indicating that the young had left the immediate shores. One fish, 150 millimeters (6 inches) in length, was seined on October 12, and on December 4 one specimen of the same length was taken

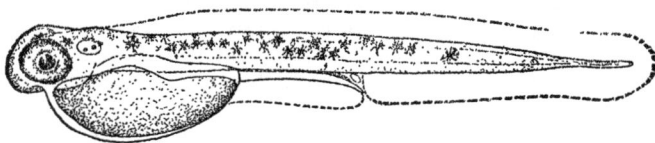

FIG. 191.—Recently hatched larva, 2.9 millimeters long

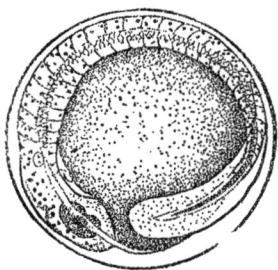

FIG. 192.—Larva, 5 millimeters long

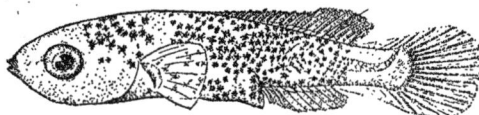

FIG. 190.—Egg with large embryo

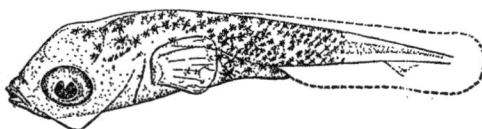

FIG. 193.—Young fish, 10 millimeters long

in the beam trawl. Whether the latter are a year older than the 2¼ to 4½ inch fish caught in October and November can not be determined from the meager data at hand. On April 13 one tautog 158 millimeters (6¼ inches) in length, was seined.

During 1922 the total catch of tautog in Chesapeake Bay is estimated to have been 2,000 pounds, worth about $80 to the fishermen. Almost all of this catch was taken in Virginia with hook and line and with pound nets.

The tautog may be caught throughout the year but is most common during the spring and fall. This fish usually frequents the vicinity of rock piles and old wrecks and is not considered a pound-net fish. However, stragglers are caught throughout the fishing season by the many pound nets in the Chesapeake, and the combined catch made with the apparatus is worthy of note.

The tautog is caught in the bay principally from the York River to the capes. It is rare above the Rappahannock. The combined daily catch of 120 pound nets in the vicinity of Buckroe beach, during April, 1922, was 10 to 40 tautog. Buckroe Beach, Ocean View, and Cape Charles are the chief fishing grounds. During September, 1922, as many as 60 rowboats, each with one to three persons, were fishing for spots off Ocean View; the average catch of tautogs was 1 to every 4 boats per day. However, the bottom was sandy and therefore not favorable to tautog fishing. It is

[24] For details relative to the embryology and development of the young refer to Kuntz and Radcliffe, 1918, pp. 92 to 99, figs. 1 to 17.

said that during the late fall tautogs are commonly caught off the large stone breakwater at Cape Charles city, and that fish of 8 or 9 pounds are sometimes taken. One fish about 13 inches in length was taken from a pound net at Chesapeake Beach, Md., on April 29, 1922, establishing the most northerly record for the bay.

The small annual catch is utilized in the Chesapeake markets; principally in Norfolk and Baltimore. The fishermen include this species among the mixed or miscellaneous fish, as it is rarely taken in sufficient numbers at one time to pack in separate boxes. During 1921 and 1922 the retail price ranged from 10 to 15 cents a pound. The tautog is a food fish of good flavor and is much esteemed along the North Atlantic coast, where it assumes considerable importance.

In Chesapeake Bay the tautog is usually called "salt-water chub" or "chub"; at Oxford, Md., fishermen call it "black porgy"; and at Solomons, where it is rare, it is called "blackfish." The last name is in common use at New York.

The maximum weight reported for the tautog is 22½ pounds. This fish was caught off New York in 1876 and was 36½ inches long. Fishermen reported individuals from Cape Charles city weighing 8 or 9 pounds. The largest from Chesapeake Bay seen during the present investigation weighed 4 pounds. Larger ones evidently are comparatively rare in the vicinity. The weight of adult fish varies widely, some being much deeper, and therefore heavier, than others. One Chesapeake specimen, 13 inches long, weighed 1 pound and 7 ounces, while a fish 14 inches long weighed only 13½ ounces.

Habitat.—Bay of Fundy to South Carolina; not taken in commercial numbers south of Chesapeake Bay.

Chesapeake localities.—(a) Previous records: Southern part of the bay, Cape Charles city, Old Point Comfort, and Norfolk. (b) Specimens in collection: Chesapeake Beach and Solomons, Md.; Tangier Island, Lewisetta, Lower York River, Cape Charles, Old Point Comfort, Buckroe Beach, and Ocean View, Va. Rather rare north of Cape Charles city and the mouth of the York River, and most common in the vicinity of Cape Charles.

137. Genus TAUTOGOLABRUS Günther. Cunners

This genus differs from Tautoga principally in the more elongate body; lower head, with a more gently convex upper profile; serrate preopercular margin; larger scales, about 40 in a lateral series; and in having the cheeks and opercles more nearly fully scaled. This genus, like Tautoga, consists of a single species.

176. Tautogolabrus adspersus (Walbaum). Cunner.

Labrus adspersus Walbaum, Artedi Piscium, 1792, p. 254; New York.
Tautogolabrus adspersus Lugger, 1877, p. 71; Jordan and Evermann, 1896–1900, p. 1577, Pl. CCXXXVI, fig. 595; Bigelow and Welsh, 1925, p. 281, fig. 131.

Head 3.45; depth 3.45; D. XVIII, 9; A. III, 8; scales 41. Body moderately deep, compressed; caudal peduncle deep, its depth about equal to postorbital part of head; head rather low, moderately long; snout pointed, 3.5 in head; eye 2.8; interorbital 4.7; mouth moderate, terminal; lips thin; maxillary scarcely reaching anterior margin of eye, 3.5 in head; teeth in bands on anterior part of jaws, becoming uniserial laterally, the outer ones anteriorly enlarged, caninelike; gill membranes united but free from the isthmus; preopercular margin serrate; lateral line complete and continuous, running high anteriorly but becoming median on caudal peduncle; scales moderate, thin, with smooth membranous edges, reduced on chest, present on cheeks and opercles; dorsal fin long, continuous, the spines stiff, pungent, not quite as high as the soft rays, caudal fin round; anal fin with three rather strong spines, the soft part similar to that of dorsal; ventral fins moderate, inserted slightly behind base of pectorals; pectoral fins moderately broad, round, their length 1.45 in head.

Color in alcohol uniform brownish yellow; dorsal fin with a black spot on base of anterior soft rays. Bigelow and Welsh (1925, p. 282) make the following observation concerning the color of this fish: "To describe the color of the cunner is to list all the colors of the bottoms on which it lives, it being one of the most variable of fishes."

A single small specimen of this species, 60 millimeters in length, was obtained. This cunner is most readily distinguished from its nearest relative, the tautog, by the scaly gill covers, the pointed snout, and the gently convex profile.

The cunner is reported to be virtually omnivorous in its feeding habits, feeding on almost all animals of suitable size that occur in the waters that it inhabits; often eel grass also is found in the stomach. Furthermore, it is reported to be a scavenger.

Spawning takes place in June, July, and August. The eggs are buoyant, transparent, and only 0.75 to 0.85 millimeter in diameter and without an oil globule. In temperatures of 70° to 72° they hatch in about 40 hours. At hatching the larvæ are about 2 to 2.22 millimeters in length.[25] According to Bigelow and Welsh (1925, p. 285), the young in the Gulf of Maine may reach a length of 2½ to 3½ inches by the autumn of the season during which they were hatched.

The cunner is found in abundance from New Jersey to Maine. Along the New England coast it lives chiefly close to shore, particularly in bays and sounds, and is one of the chief fish caught by youthful anglers. It prefers rocky bottom covered with marine growths. Along the northern New Jersey coast the cunner is present in large numbers throughout the year. In this region it is found chiefly offshore, on the rocky ledges frequented by tautog, sea bass, and other species. The usual size of New Jersey fish taken within several miles of shore is 5 to 8 inches, but about 8 miles offshore, on the 17-fathom bank, large cunners, 10 to 12 inches or more in length, are commonly taken. The cunner's habit of stealing bait from the hooks of fishermen is well known, and New York bank fishermen know that more bait is required to feed the cunner (or bergall, as it is called there) than all other species for which he may be fishing combined.

The cunner reaches a maximum length of about 15 inches. It is a food fish of good flavor and of some commercial importance.

Habitat.—Labrador to Virginia. Rare south of New Jersey. This fish was recorded from the Atlantic coast of Worcester and adjoining counties, Maryland, by Lugger (1877, p. 71). This record appears to have been overlooked, as the southernmost range given by recent authors is New Jersey. The capture of the small specimen in hand extends the known range to the mouth of Chesapeake Bay.

Chesapeake localities.—(a) Previous records: None. (b) Specimen in collection: Sixty millimeters long, taken by seining, Cape Charles, Va., September 20, 1921.

Family LXXIX.—SCARIDÆ.   The parrot fishes

Body oblong, moderately compressed; mouth moderate, terminal; teeth in the jaws coalesced, at least at base, often forming continuous plates; frequently with one or more canines above the cutting edge; no teeth on vomer or palatines; scales large, cycloid, 23 to 26 in a lateral series; dorsal fin continuous, its rays constantly IX, 10, the spines weak and flexible or stiff and pungent; anal constantly III, 9. A single genus and species of this family of tropical fishes comes within the scope of the present work.

138. Genus SCARUS Forskål.   Parrot fishes

Body rather robust, compressed; head moderately short and deep; snout blunt; upper lip laterally double, the inner fold becoming very narrow or disappearing anteriorly; teeth in the jaws fully coalesced, forming continuous plates, with a single median suture; gill membranes scarcely united to the isthmus; scales large, 22 to 26 in a lateral series; dorsal fin constantly with 9 flexible spines and 10 soft rays; anal fin with three flexible spines (the first one very small and often hidden in the skin) and 9 soft rays. This is a rather large genus, the members being of the warm and tropical seas.

---

[25] For details concerning the embryology and the larval development of the cunner, see Agassiz (1882, p. 290, Pls. XIII to XV); Agassiz and Whitman (1885, p. 18, Pls. VII to XIX); and Kuntz and Radcliffe (1918, p. 99, figs. 18 to 29).

177. Scarus cæruleus (Bloch). Blue parrot fish.

Coryphæna cæruleus, Bloch, Auslandische Fish., II, 1786, p. 120, pl. 176; Bahamas.
Scarus cæruleus Jordan and Evermann, 1896–1900, p. 1652, Pl. CCXLIV, fig. 613; Smith and Kendall, 1898, p. 170.

Head 3.15 to 3.6; depth 2.85 to 3.2; D. IX, 10; A. III, 9; scales 24 to 26. Body elongate, moderately compressed; head not much longer than deep; snout very blunt, with a well-developed fleshy pad on its upper surface in adults, its length 2.35 to 2.7 in head; eye 5.15 to 6.4; mouth small, reaching about halfway to eye; lower jaw included; teeth fully coalesced, forming continuous plates; each plate with an evident median suture; no free canines; gill membranes slightly connected; scales large, not much reduced on chest, five in advance of ventrals; most of head scaly; two rows of six scales each and a third row consisting of two scales on cheek, scales of upper row much larger than those of second row; lateral line interrupted under posterior rays of dorsal, beginning again lower down on caudal peduncle, the pores more or less branched; dorsal with nine flexible spines, each one with a fleshy tip; caudal fin notably concave, with the angles produced in adult fish; anal with three flexible spines, the first one very small; ventrals a little shorter than the pectorals; pectorals 1.3 to 1.55 in head.

Color dark green to slightly grayish green above, becoming a lighter shade on sides and underneath; no stripes or bars present on preserved specimens; lips deep blue-green; dorsal and anal deep blue-green, almost black, each with a bright green margin; caudal slightly paler than the dorsal and anal, the outer rays bright green; ventrals and pectorals mostly greenish; axillary spot absent; teeth white.

This fish was not seen during the present investigation. The record is taken from Jordan and Evermann (1896–1900, p. 1652) and Smith and Kendall (1898, p. 170). These records are both based on an identification made from the jaws of one specimen taken in 1894 in a pound net set in the Potomac River off St. George Island. After the receipt of this jaw, an illustration was sent by Smith and Kendall to a J. E. N. Sterling, at Cape Charles city, Va., with an inquiry whether this fish had been caught in the vicinity. The reply was that from 6 to 10 fish resembling the figure and corresponding to the description that had been supplied were obtained in pound nets between Cape Charles and Hunger Creek.

This fish, being the only one of the family known from Chesapeake Bay, may be recognized readily by the coalesced teeth, which have a continuous cutting edge and resemble somewhat the beak of a bird.

Nothing definite is known concerning its food, breeding habits, or rate of growth. It is reported to reach a length of 2 to 3 feet.

Habitat.—Maryland to Panama; apparently rare on the coast of the United States; probably most common in the West Indies.

Chesapeake localities.—(a) Previous record: St. George Island, Md. (b) Specimens in collection: None.

Order GOBIOIDEA

Family LXXX.—GOBIIDÆ. The gobies

Body oblong or elongate, compressed or not; scales present or wanting; skin of head continuous with the covering of the eyes; premaxillaries protractile; opercle unarmed; preopercle unarmed or with a short spine; gill openings largely restricted to the sides, the membranes united to the isthmus; gills 4, a slit behind the fourth; no lateral line; teeth various, usually small; dorsal fins separate or connected, the spinous dorsal with two to eight flexible spines, rarely wanting; anal usually with a single weak spine, similar to soft dorsal; ventral fins close together, separate or united, each composed of I, 5 rays (rarely I, 4); the ventral fins, when united, forming a sucking disk, a cross fold between their bases completing the cup; caudal fin convex or pointed. Most of the members of this family are of small size. Some of them live in fresh water, others in salt water, and many of them occupy brackish water or live indiscriminately in salt or fresh water.

KEY TO THE GENERA

a. Body entirely naked; mouth nearly horizontal; second dorsal and anal short, each with 10 to 14 rays-- Gobiosoma, p. 323

aa. Scales present on at least most of body; mouth oblique to nearly vertical.

    b. Teeth in the jaws in bands, immovable; lower jaw rather strong, round anteriorly; second dorsal and anal rather long, each with 15 to 18 rays---------------- Microgobius, p. 325

    bb. Teeth in the jaws in a single series, movable; lower jaw very thin and angular anteriorly; second dorsal and anal short, each with about 11 or 12 rays-- Mugilostoma gen. nov., p. 327

139. Genus GOBIOSOMA Girard. Naked gobies

Body rather slender; mouth moderate, nearly horizontal; teeth pointed, in several series or in a band, the outer ones enlarged; no canines; skin entirely naked; no barbels or dermal flaps; spinous dorsal normally with 7 spines, rarely 5 or 6; second dorsal and anal short, with 10 to 14 rays; ventral fins united, forming a sucking disk.

KEY TO THE SPECIES

a. Body moderately robust, its depth 3.95 to 4.8 in its length; second dorsal normally with 13 rays, infrequently with 12 or 14; ventral disk short, reaching about half the distance from its base to the vent--- *bosci*, p. 323

aa. Body slender, its depth 6 to 7.15 in its length; second dorsal normally with 12 rays, infrequently with 11 or 13; ventral disk long, reaching two-thirds the distance from its base to the vent-- *ginsburgi* sp. nov., p. 324

178. Gobiosoma bosci (Lacépède). Clinging goby; Variegated goby; Naked goby.

Gobius bosci Lacépède, Hist. Nat. Poiss., II, 1798, p. 555, Pl. XVI, fig. 1; Charleston, S. C.

Gobiosoma bosci Bean, 1891, p. 86; Jordan and Evermann, 1896–1900, p. 2259; Smith and Bean, 1899, p. 187; Evermann and Hildebrand, 1910, p. 163.

Head 3.15 to 3.5; depth 3.95 to 4.8; D. VII or VIII–13 (infrequently 12 or 14); A. 11 (rarely 10). Body robust; head depressed, broader than deep; snout short, tapering, 3.45 to 4.25 in head; eye 3.25 to 4.67; interorbital bone about the width of pupil; mouth moderate, terminal; maxillary reaching about opposite middle of eye, 2.25 to 2.4 in head; teeth in the jaws pointed, in bands in each jaw, with some of the outer teeth enlarged; gill openings mostly lateral, the membranes joined to the isthmus; pores usually present on the cheeks; dorsal fins close together, the first consisting of very slender spines, the margin convex; second dorsal and anal similar and opposite each other, the rays of the dorsal reaching to or beyond the base of the upper short rays of the caudal; caudal fins short, round, shorter than head; ventral fins short, forming a disk, the disk scarcely reaching more than half the distance from its base to the vent, its length 1.65 to 2.45 in head; pectoral fins broad, a little shorter than head, 3.25 to 4.7 in the length of body.

Color in life greenish to dusky above; pale underneath; nape and sides with very narrow pale crossbars; pectoral fins mostly greenish; other fins mostly blackish; the caudal of a somewhat lighter shade than the dorsals. Considerable variation in color among individuals has been noted, some specimens being notably darker than others, and the males in general appear to be darker than females, with the pale crossbars showing less distinctly.

Many specimens, ranging from 25 to 60 millimeters (1 to 2⅜ inches) in length, were preserved. This goby is entirely naked. Its body is quite robust, the head is depressed, and the ventral disk is very short, scarcely reaching more than halfway to vent.

In a series of 37 specimens examined for the food contents of the stomachs, 12 had fed on small crustaceans (mainly Gammarus), 14 had eaten annelids (Chætopods), 2 had fed on fish, and 2 on ova of unknown origin. Two individuals had fed on both Gammarus and annelids, and 9 stomachs were empty.

Spawning apparently takes place from June to October. The gonads in many of the specimens taken in July are in an advanced stage of development, and it seems probable that most of the spawning takes place during this month. However, a few specimens taken during the early part of October were not entirely spent.   Kuntz (1916, pp. 423–426), who took the eggs and described their development, found that at Beaufort, N. C., ripe fish were comparatively scarce during August, and he states that the height of the spawning season evidently was past.   This author states that the mature, unfertilized eggs are approximately spherical in form and about 0.5 millimeter in diameter, yellow in color, and opaque.   They are heavier than sea water, and when stripped from the female they aggregate in a compact clump.   As soon as fertilized, the egg begins to expand and becomes elliptical.   The incubation period at ordinary summer temperature in the laboratory was approximately five days.   The newly hatched larvæ are approximately 2 millimeters in length and almost transparent.   Young fish 10 millimeters in length show many of the diagonostic characters of the adult.   The head already has the characteristic shape of the species, the fins are well differentiated, and the sucking disk formed by the ventral fins is well developed.

This goby is an abundant species in Chesapeake Bay, inhabiting shallow grassy flats off the immediate shores.   During April and May it was found in comparatively small numbers along the shores, but from June to October it was present in abundance.   The largest catch made during our collecting consisted of 511 fish taken in 8 hauls of a 30-foot seine in the lower York River on October 11.   In November it was found to be comparatively scarce in localities where it had been

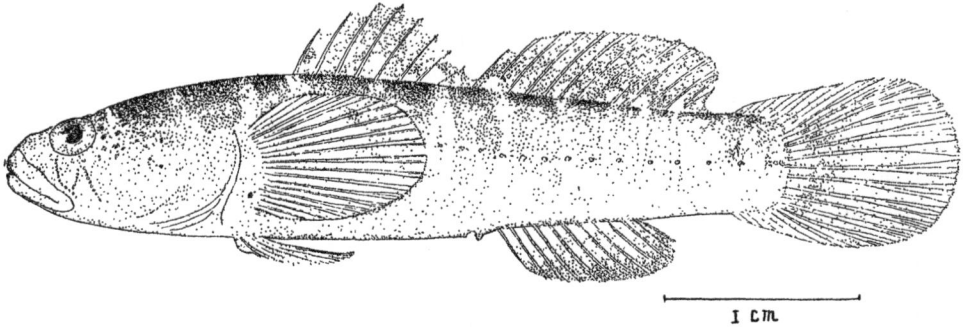

FIG. 194.—*Gobiosoma bosci*.   Adult 50 millimeters long

abundant the previous month.   This goby was found chiefly from the York River to Annapolis and Love Point.   Only a few specimens were taken in the Patapsco River below Baltimore and in the vicinity of Havre de Grace.   From the capes to Buckroe Beach it was not as common as from the York River northward.   It was not infrequently taken in water only slightly brackish, and several specimens seined at Havre de Grace were from fresh water.   The maximum size is about 2½ inches.

Habitat.—Massachusetts to Florida.

Chesapeake localities.—(a) Previous records: Gunston Wharf, Cape Charles city, Gloucester Point, and Hampton, Va.   (b) Specimens in collection: From numerous localities, taken in fresh, brackish, and salt water from Havre de Grace, Md., to the mouth of the bay.

179. Gobiosoma ginsburgi sp. nov.

Type No. 87655, U. S. National Museum; length 45 millimeters; Cape Charles, Va.

Head 3.45 to 3.8; depth 6 to 7.15; D. VII–12 (infrequently VII–11 or 13); A. 11.   Body rather slender; head somewhat depressed, broader than deep; snout short, tapering, 3.8 to 4.2 in head; eye 3.8 to 4.6; interorbital bone about the width of pupil; mouth terminal, slightly oblique; maxillary reaching somewhat beyond middle of eye, 2.05 to 2.35 in head; teeth in the jaws pointed, in bands, some of the outer teeth enlarged; gill openings mostly lateral, the membranes joined to the isthmus; pores evident and in series on cheeks; dorsal fins separate, the first with rather short, very slender spines and convex margin; second dorsal and anal similar and opposite each other, the rays of the dorsal scarcely reaching the base of upper rays of caudal; caudal fin moderate, round, about as long as head; ventral disk long, reaching about two-thirds the distance from its base to the vent, its length 1.25 to 1.45 in head; pectoral fins moderately broad, equal to or very slightly shorter than head, 3.65 to 4.2 in length of body.

Color of preserved specimens, brownish; body with about six or seven rather ill-defined, whitish crossbars; lateral line usually with longitudinally elongated dark spots; a few similar spots on median line of back in advance of dorsal; lower surface of head spotted with black; a black bar on mandible; ventral disk dusky, at least at base; other fins pale to slightly dusky; the dorsal fins and the caudal sometimes with indications of dark spots or bars; anal fin with a dark margin.

This species is represented by 26 specimens ranging in length from 37 to 52 millimeters (1½ to 2 inches). This goby is named for our colleague, Isaac Ginsburg, who made many of the preliminary identifications of the present collection and first called our attention to the fact that apparently two species were included in the genus Gobiosoma. This led to a detailed study and finally to the description of this new species. This naked goby differs from its relative, *G. bosci*, the only other naked goby recognized from the Atlantic coast of the United States, principally in having a more slender body, generally higher fins, somewhat shorter second dorsal, and in color. The difference in the height of the fins is most noticeable in the ventrals composing the sucking disk. In *G. bosci* the disk extends only about half the distance from its base to the vent, whereas in the present species it reaches fully two-thirds the distance to the vent. In counting the rays of the second dorsal in *G. bosci*, in 49 specimens, 4 had 12 rays, 41 had 13, and 4 had 14. In 26 specimens of the present species, 1 had 11 rays, 21 had 12, and 4 had 13. In making these enumerations the first simple ray was included and the last two, which apparently are united at the base, were counted as one.

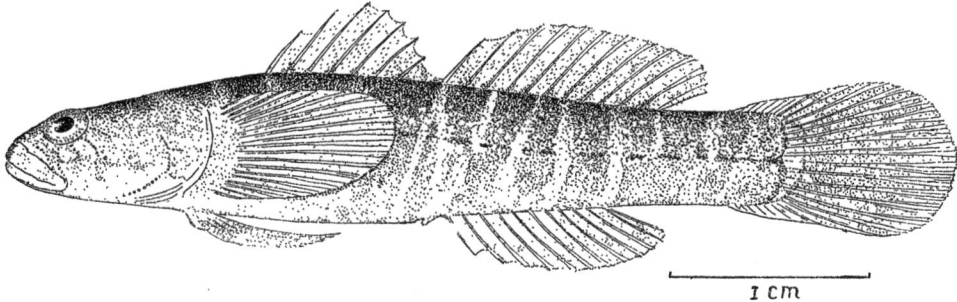

FIG. 195.—*Gobiosoma ginsburgi* sp. nov. From the type, 45 millimeters long

Three specimens were examined for food. In one the stomach was empty; the other two had fed on small crustaceans, chiefly Gammarus. A ripe or nearly ripe female occurs in a lot of specimens taken from May 21 to 23, 1922.

The size attained, judging from the specimens in hand, may be somewhat smaller than in *G. bosci*, and it evidently is much less numerous. *G. bosci* was taken only in shallow water, whereas the present species, although found in shallow water, was taken principally in deeper water, reaching upward of 25 fathoms. Most of the deep-water catches were made with the beam trawl from October to March.

Chesapeake localities.—Solomons, off Barren Island, and Crisfield, Md.; Cape Charles, lower York River, off Old Point Comfort, off Thimble Shoals Light, and Buckroe Beach, Va.

140. Genus MICROGOBIUS Poey. Gobies

Body elongate, more or less compressed; mouth quite large and very oblique to nearly vertical; outer teeth in the jaws enlarged, rather strong; scales cycloid or weakly ctenoid, present on most of the body; first dorsal with 7 or 8 spines; second dorsal and anal with 15 to 18 rays. The species are small and confined chiefly to the shores.

KEY TO THE SPECIES

a. Body rather deep, quite strongly compressed, the depth 4.7 to 5.4 in the length of body; mouth only moderately oblique; ventral disk long, reaching to or a little beyond origin of anal --- *holmesi*, p. 326

aa. Body more elongate and less strongly compressed, the depth 4.75 to 5 in the length; mouth nearly vertical; ventral disk short, reaching only about two-thirds the distance from its origin to the origin of anal_____ *eulepis,* p. 327

180. Microgobius holmesi Smith. Holmes goby.

Microgobius holmesi Smith, North Carolina Geol. and Econ. Surv., Vol. II, 1907, p. 366, fig. 168; Beaufort, N. C.

Head 3.1 to 3.8; depth 4.7 to 5.4; D. VII or VIII–16 or 17; A. 17; scales about 45. Body elongate, quite strongly compressed; head rather large, compressed; snout short, 3.5 to 4.1 in head; eye 2.95 to 3.3; interorbital very narrow 11.3 to 14; mouth large, very oblique; maxillary reaching under middle of eye, 2 to 2.4 in head; teeth in the jaws simple, those of the upper jaw with an outer enlarged series well separated from an irregular inner series of smaller ones; lower jaw with an outer series of enlarged teeth, well separated from a narrow band of small, inner teeth; gill openings large, the membranes narrowly attached to the isthmus; lateral line indistinct; scales cycloid, present posteriorly, wanting on head and body from about middle of base of first dorsal forward, largest and most distinct on caudal peduncle; dorsal fins well separated, the first consisting of very slender, short spines; second dorsal and anal similar and opposite each other; caudal fin long, pointed, notably longer than head; ventral disk long, reaching to or a little beyond origin of anal; distance from origin of ventrals to origin of anal somewhat shorter than head; pectoral fins rather large, about as long as head, 3.05 to 3.6 in length of body.

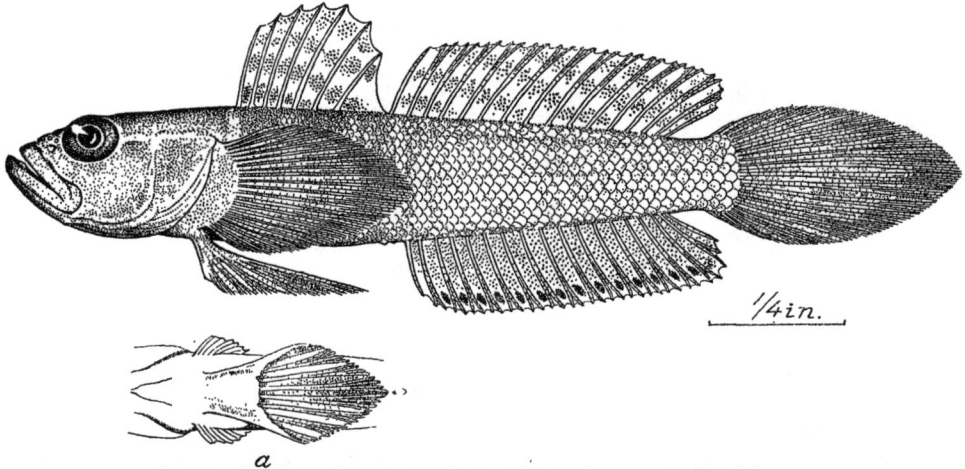

Fig. 196.—*Microgobius holmesi.* Adult, about 2 inches long. *a.* Ventral disk

Color of a fresh specimen plain light blue; opercle edged with yellowish green; several bluish vertical bars above abdomen; specimens faded to a pale straw color in alcohol, without noticeable markings.

This little goby is represented by 10 specimens, ranging in length from 22 to 47 millimeters (⅞ to 1⅞ inches). It is most readily distinguished from its nearest relative, *M. eulepis,* by its deeper and more compressed body and a longer ventral disk, which reaches to or beyond the origin of the anal; whereas in *M. eulepis* it reaches only about two-thirds the distance from its base to the origin of the anal.

Two stomachs were examined for food. One was empty and the other contained fragments of a small crustacean. The spawning habits of this fish are largely unknown as yet. The senior author, however, took some ripe or nearly ripe fish on July 10, 1914, at Beaufort, N. C. The largest of the ripe fish was 2 inches in length, which probably is about the maximum size attained.

Habitat.—Chesapeake Bay to Beaufort; to date taken only at Lewisetta, Va., and Beaufort, N. C.

Chesapeake localities.—(*a*) Previous records: None. (*b*) Specimens in collection: From Lewisetta, Va., taken in a brackish pond on mud and sand bottom on August 6 and 8, 1921. Not seen elsewhere.

181. Microgobius eulepis Eigenmann and Eigenmann. Scaled goby.

Microgobius eulepis Eigenmann and Eigenmann, Proc., Calif. Ac. Sci., 1888, p. 69; Fortress Monroe, Va. Jordan and Evermann, 1896-1900, p. 2244; Evermann and Hildebrand, 1910, p. 163.

Head 3.9 to 4; depth 4.75 to 5; D. VII, 16; A. 17; scales probably about 50. Body elongate, slender, moderately compressed; head moderately large; snout short; 2.2 to 2.5 in head; eye 3.1 to 3.4; interorbital very narrow, 10.6 to 12.1; mouth large, nearly vertical; maxillary reaching nearly opposite anterior margin of pupil, 2.1 to 2.25 in head; teeth in the jaws simple, in narrow bands, with the outer ones in each jaw somewhat enlarged; gill openings large, the membranes narrowly attached to the isthmus; lateral line not visible; scales cycloid, wanting on head and body in advance of first dorsal; dorsal fins well separated, the first consisting of low flexible spines; second dorsal and anal similar and opposite each other; caudal fin moderately long and pointed, a little longer than the head; ventral disk rather short, reaching only about two-thirds the distance from its base to origin of anal; distance from origin of ventrals to origin of anal slightly longer than head; pectoral fins large, 3.65 in length of body.

Color of a fresh specimen pale bluish; head below eye bright greenish; a bright bluish blotch on abdomen behind pectoral fin; first dorsal pale, edged with black and yellow; second dorsal red at base, yellow in middle, and with a pale outer edge; other fins plain. The alcoholic specimens at hand have faded to a nearly uniform pale color.

Only two specimens of this goby of equal length—namely, 50 millimeters (2 inches)—are at hand. The most noticeable differences between this species and *M. holmesi*, as pointed out in the discussion of the last-mentioned species, are the differences in the shape and depth of the body and the length of the ventral disk. When specimens are compared, it is evident, also, that the mouth in *M. eulepis* is much more vertical, but this character is difficult to describe and to use unless both species are at hand.

Smith (1907, p. 368) reports that a female distended with nearly ripe eggs was taken on May 18, 1905, at Beaufort, N. C. This is all that is known about the spawning of this goby. The specimens at hand, which are 2 inches in length, appear to be the largest known. The species appears to be a rare one.

Habitat.—Chesapeake Bay to Beaufort, N. C.

Chesapeake localities.—(a) Previous records: Fortress Monroe and the mouth of Hampton Creek, Va. (b) Specimens in collection: From the mouth on the Patuxent River, Md.; both specimens at hand were taken on April 28, 1922, one with a beam trawl operated from the *Fish Hawk*, and the other along the beach with a 30-foot collecting seine. A third specimen, which has not been seen by us, is recorded in field notes by Lewis Radcliffe and was taken in a tow net operated from the *Fish Hawk* near Cape Charles, Va.

141. Genus MUGILOSTOMA gen. nov.

Type *Mugilostoma gobio* sp. nov.

Body elongate, more or less compressed; head compressed; gill openings quite large, the membranes rather narrowly connected with the isthmus; mouth moderate, with weak jaws; the lower jaw thin and angular anteriorly, shaped as in the mullets (Mugil); teeth movable, in a single series, set on the edge of the thin jaws; ventral fins probably united, forming a sucking disk. (The membrane connecting the fins in the specimen at hand apparently is broken, being still present, however, at the base of the fins. A membranous cross fold, forming a part of the sucking disk, is quite evidently present.)

182. Mugilostoma gobio sp. nov.

Type No. 87656, U. S. National Museum; length 27 millimeters; Lynnhaven Bay, Va.

Head 4; depth 5.25; D. VI–11; A. 12; scales about 30. Body very elongate, compressed, tapering gradually toward the tail; caudal peduncle 2.4 in head; head moderate, somewhat deeper than broad; snout very short, 6 in head; eye 2.9; interorbital very narrow, 10.6; mouth rather large, oblique; lower jaw slightly in advance of the upper, very thin, without evident lips, and angular anteriorly as in the mullets (Mugil); maxillary very narrow, reaching about middle of eye, 2.65 in head; teeth simple, movable, in a single series on edge of each jaw; gill openings mod-

49826—28——22

erately large, the membranes rather narrowly attached to the isthmus; scales quite large, ctenoid, lost on anterior part of body in the specimen in hand, therefore making the number counted for a lateral series uncertain; dorsal fins well separated, the first consisting of very slender, flexible spines; second dorsal and anal similar and opposite each other and separated from the caudal by a distance equal to the greatest depth of the body; caudal fin somewhat pointed, 3.5 in length of body; ventral fins probably united (the membrane apparently is broken in the specimen in hand, the fins however, being still connected at base; a cross fold of skin, forming a part of the sucking disk, is quite evident), reaching about two-thirds the distance from their bases to origin of anal, 1.25 in head; pectoral fins moderately large, the middle rays longest, about as long as head.

Color in alcohol brownish above, somewhat paler underneath; sides with irregular, large, brown blotches, darker than the ground color; base of caudal with two quadrate black spots, one on the upper half of the base and the other on the lower half; the fins otherwise slightly dusky to colorless.

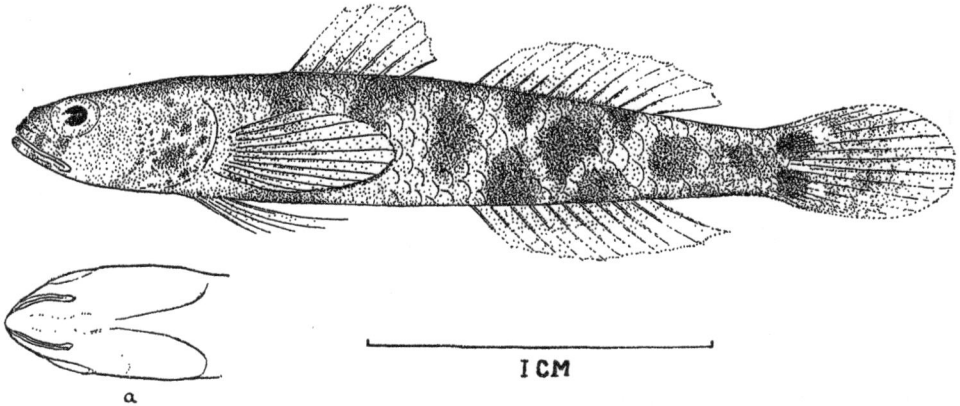

FIG. 197.—*Mugilostoma gobio* gen. et. sp. nov. From the type, 27 millimeters long. *a.* Ventral surface of head, showing mulletlike mouth

A single specimen of this singular fish, 27 millimeters ($1\frac{1}{16}$ inches) in length, is at hand. It seems to differ from all other gobies known from American waters in the thin, triangular lower jaw, which is shaped and formed as in the mullets (Mugil). The teeth differ from all other gobies of Chesapeake Bay in being set in a single series on the outer edge of the lower jaw and in being movable.

Chesapeake localities—Brackish marsh, connected with Lynnhaven Bay, Va., by a creek; taken in a small collecting seine in company with *Fundulus ocellaris*, September 26, 1921.

Order DISCOCEPHALI

Family LXXXI.—ECHENEIDIDÆ. The remoras

Body elongate or slender; head depressed above, with a large oval disk consisting of crosswise partitions or laminæ and a single lengthwise septum; lower jaw projecting beyond upper; mouth wide; teeth villiform, present on jaws, vomer, palatines, and usually on tongue; gill arches 4; gill membranes free from the isthmus; branchiostegals 7; scales minute, cycloid; air bladder wanting; dorsal and anal fins long and low; ventral fins thoracic; pectoral fins placed high.

142. Genus ECHENEIS Linnæus. Remoras; Shark suckers

Body slender, fusiform, disk long, with 20 to 28 laminæ; soft dorsal with numerous short rays; anal similar, the anterior rays somewhat elongated; caudal slightly concave behind; ventrals long, the inner rays narrowly adnate to abdomen; pectorals pointed, the rays soft and flexible.

183. Echeneis naucrates Linnæus.   Pilot fish; Shark's pilot; Shark sucker; Remora.

Echeneis naucrates Linnæus, Syst. Nat., ed. X, 1758, p. 261; Indian Ocean. Jordan and Evermann, 1896–1900, p. 2269, Pl. CCCXXIX, fig. 796.

Leptecheneis naucrates Uhler and Lugger, 1876, ed. I, p. 138; ed. II, p. 117.

Echeneis naucratoides Uhler and Lugger, 1876, ed. I, p. 138; ed. II, p. 118.

Head 5.25 to 5.45; depth about 8 to 10; D. 28 to 34; A. 30 to 32.   Body elongate, more or less cylindrical; head depressed above; sucking disk large, with 20 or 21 laminæ; snout broad, flat, 1.95 to 2 in head; eye 5.05 to 5.15; interorbital 1.65 to 1.8; mouth broad, superior; lower jaw strongly projecting, angulate at tip; maxillary reaching anterior nostril, 2.65 to 2.8 in head; teeth in jaws in broad villiform bands, those in lower jaw mostly exposed; dorsal fin long, somewhat elevated anteriorly; caudal fin posteriorly nearly straight; anal fin similar to dorsal and opposite it; ventral fins rather long and narrow, the inner ray of each fin connected by membrane at base; this membrane on median line attached to abdomen by another membrane; pectoral fins moderate, pointed, the upper rays being longest, 1.1 in head.

Color in alcohol dark brown on back; sides lighter brown; belly still lighter; sides with a dark, longitudinal band; dorsal fin black, anteriorly at least with a pale margin; caudal fin black, the tips of outer rays pale; anal fin dusky brown with a broad pale margin; ventrals dusky brown; pectorals black.

Four specimens of this species, ranging in length from 390 to 485 millimeters (15½ to 19⅛ inches) are at hand.   This is the only remora known from Chesapeake Bay.   It is most readily recognized by the very slender body, the long sucking disk on top of the head, which has 20 or more pairs of transverse plates, and by the long, pointed, projecting lower lip.

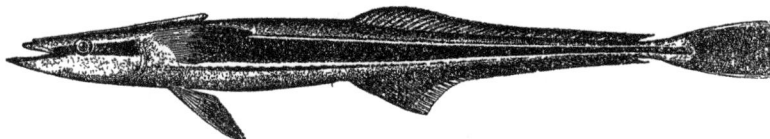

FIG. 198.—*Echeneis naucrates*

This remora is reported to reach a length of nearly 3 feet.   It is usually found attached to sharks and occasionally to turtles and other large aquatic animals, but it also swims independently, and sometimes it is taken with hook and line.   The remoras are not parasitic on the animals to which they attach themselves by means of the large sucking disk over the head, for this is only their method of "stealing a ride."   *E. naucrates* apparently is rare in Chesapeake Bay, possibly because sharks, too, are rather uncommon.   Only four specimens were seen during the present investigation, and these were all found in two pound nets on the same date.

Habitat.—All warm seas; northward to Massachusetts Bay on the Atlantic coast of the United States.

Chesapeake localities.—(*a*) Previous records: "Chesapeake Bay" and "southern parts of Chesapeake Bay" (Uhler and Lugger, 1876).   (*b*) Specimens in collection: Taken on June 15, 1921, all from two pound nets operated in Lynnhaven Roads, Va.

Order JUGULARES

Family LXXXII.—URANOSCOPIDÆ.   The stargazers

Body elongate, conic, more or less compressed; widest and usually deepest at occiput; head large, broad, partly covered with bony plates; eyes small, superior, placed anteriorly; mouth vertical; teeth moderate, present on jaws, vomer, and palatines; premaxillary protractile; maxillary broad, without a supplemental bone; gill openings wide; gill membranes nearly separate, free from the isthmus; gills 3½, a slit behind the last; pseudobranchiæ present; branchiostegals 6; scales, if present, small; spinous dorsal small or wanting, the soft dorsal long; caudal fin not forked; anal fin large; ventral fins jugular, close together, with I, 5 rays; pectoral fins large, broad, with oblique bases, the lower rays rapidly decreasing in length.

143. Genus ASTROSCOPUS Brevoort. Electric stargazers

Body robust; upper surface of head not entirely covered with bone, the occipital plate ceasing far behind eyes; a bony Y-shaped process on head, the forks of the Y reaching forward to inter-orbital, the vertical limb extending backward to occipital plate; the area between the forks of the Y and extending forward to upper lip covered by naked skin; a somewhat quadrangular naked area on each side of the Y covering the electric organs; head smoother in the adult, in young individuals largely covered with bone and with spines; anterior nostril round, situated in front of eye, fringed; posterior nostril represented externally as a crescent-shaped groove, terminating behind eye, fringed; lips fringed; back and sides covered with close-set scales in adult; first dorsal with four or five short, pungent spines.

184. Astroscopus guttatus (Abbott). Electric toad; Stargazer.

Uranoscopus guttatus Abbott, Proc., Ac. Nat. Sci., Phila., 1860 (1861), p. 365, Pl. VII; Cape May, N. J.
Astroscopus anoplus Uhler and Lugger, 1876, ed. I, p. 99; ed. II, p. 83.
Astroscopus guttatus Jordan and Evermann, 1896–1900, p. 2310; Evermann and Hildebrand, 1910, p. 163.

Head 2.4 to 2.7; depth 2.7 to 3.55; D. IV or V–13 or 14; A. I, 12. Body robust, anteriorly quite as wide as deep, posteriorly compressed; head broad, flat above; snout very broad and short, 4.5 to 5.3 in head; eye superior, very small, 5.75 to 13; interorbital very broad, 3 to 3.95; mouth broad, vertical; lower jaw forming anterior margin of head; both lips provided with fringes; nostrils provided with shorter fringes; an elliptical area between and behind eyes with a double row of fringes; maxillary broad posteriorly, reaching under or beyond the eye, 1.95 to 2.15 in head; teeth small, in bands on jaws, also present on vomer and palatines; scales very small, not evident on head, chest, and abdomen; upper part of head largely rough and bony, this sculpture forming a Y on anterior part of head, the limbs of this Y a little longer than the straight part, the straight portion of the Y very wide, as broad as eye; total length of the Y shorter than interorbital, 4.73 to 5.2 in head; a somewhat quadrate naked area present on each side of the Y, these areas being the seat of electric organs; two very short, blunt spines on edge of snout in front of eye; dorsal fins separate, the first composed of short, sharp spines; second dorsal much higher; caudal fin with somewhat convex margin; anal fin more or less enveloped in skin, especially in the adult; ventral fins inserted at the throat, under the anterior part of gill opening, some of the rays distally more or less free; pectoral fins large, the lower rays short, the longest ones 1.05 to 1.3 in head.

Color of a fresh specimen, 280 millimeters in length, largely dusky above; dirty white underneath; upper half with many small, irregular, white spots, increasing somewhat in size posteriorly, extending backward on upper part of side to end of second dorsal (in most specimens at hand the spots do not extend as far back on body as the end of second dorsal). Upper part of caudal peduncle with five irregular whitish blotches; caudal peduncle laterally with an irregular, dark, longitudinal band; lower half of sides with obscure dark blotches; each side of chin with a large black blotch; membranes of spinous dorsal black, with a few pale stripes; second dorsal anteriorly at base with pale spots similar to those on body, the fin with alternating black and white bars; caudal with similar alternating bars; anal fin pale, with a single black bar; ventral fins mostly pale, with dusky points on distal parts; pectorals brownish, base of upper rays with white spots like body, distal parts black. The color varies considerably as to shade among specimens, but the general pattern in adults is about as described. The smallest specimen (60 millimeters) at hand bears no pale spots, and in spirits the upper parts are uniform brownish.

The stargazer is represented by 29 specimens, ranging in length from 60 to 310 millimeters (2⅜ to 12¼ inches). They evidently are all of one species, regardless of the fact that much variation in color exists among individuals of even size, and more pronounced differences among those of uneven size. The specimens here described have been compared with others from Beaufort, N. C., and we find a distinct difference in the shape of the Y on the head. In the Chesapeake specimens it is notably shorter and broader than in Beaufort specimens, and the straight part is slightly shorter than one of the limbs, while the reverse is true in the Beaufort specimens. The total length of the Y in Chesapeake specimens is shorter than the width of the interorbital space. In Beaufort fish the total length of the Y is equal to or slightly greater than the width of the inter-orbital space. The following table of measurements illustrates some of these differences. Three specimens from each locality were measured.

	Chesapeake Bay			Beaufort		
	I	II	III	I	II	III
Total length of Y in head	4.73	4.94	5.20	3.24	3.44	3.78
Total length of Y in interorbital	1.21	1.24	1.35	.82	.88	.90
Length of straight part of Y in head	9.50	9.60	10.80	5.25	5.90	6.15
Length of limb of Y in head	6.80	7.75	8.25	7.38	8.35	9.20
Width of straight part of Y in head	10.00	11.15	13.20	18.50	20.50	22.00

The other alleged differences pointed out by Jordan and Evermann (1896–1900, pp. 2307–2310) between the northern stargazer, *A. guttatus*, and the southern one, *A. y-græcum*, namely, that the two spines in front of the eye are longer in the northern species and that the pale spots are smaller, can not be substantiated. All Chesapeake Bay records quite certainly are referable to *A. guttatus*.

The stargazer is readily separated from all other fish of Chesapeake Bay by the broad head, with the very small eyes situated on top of it (from which the name "stargazer" originated), and by the vertical mouth and fringed lips. The rather rough and bony upper surface of the head, with sculpturing forming a Y, also is very characteristic.

The electric toad receives its name from the shock it is able to give to the one who handles it. This shock is produced by electric organs situated in the smooth, naked areas lying posterior to the eyes. A fish 6 inches in length is able to give a very perceptible shock, and larger fish give a proportionately stronger one. It is not yet known

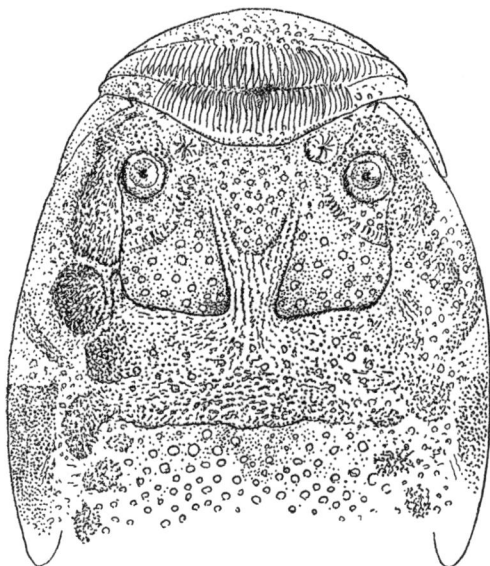

FIG. 199.—*Astroscopus y-græcum.* Dorsal surface of head; from a specimen 7⅛ inches long

whether use is made of the electric organs in the capture of food or in self-defense, or both.

Six fish, ranging in length from 4⅛ to 5⅛ inches, were examined for food. Four of these had fed on fish and two on isopods. This fish lives well in the aquarium, and when sand is provided buries itself, leaving only the eyes and lips exposed. When it is thus buried it is said to be lying in wait for prey. It can live for a long time out of water; one specimen, placed on ice, remained alive for 15 hours.

Very little is known of the spawning habits and rate of growth of this stargazer. Very young

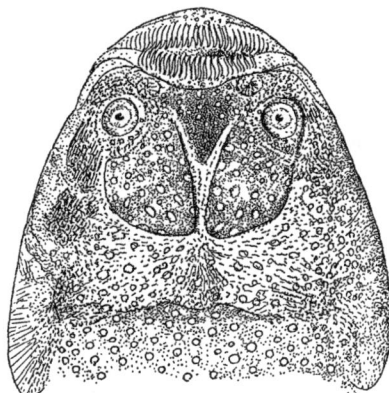

FIG. 200.—*Astroscopus guttatus.* Dorsal surface of head; from a specimen 9⅜ inches long

fish are rare in collections. Bean (1903, p. 659) records a specimen 1 inch long, taken on August 1 at Ocean City, N. J., and another, 2½ inches long, caught on August 26 at Longport, N. J. The smallest specimen in the Chesapeake collection—60 millimeters (2⅜ inches) long—together with another 93 millimeters (3¾ inches) in length, were seined at Buckroe Beach, Va., on October 5, 1921. At Ocean View, Va., 23 small stargazers were caught in collecting and commercial haul seines from September 25 to October 18, 1922. One of these fish was 151 millimeters (6 inches) long, and the others ranged from 91 to 138 millimeters (3⅗ to 5½ inches) in length.   The foregoing catches were made on 11 dates, the greatest number of fish in one haul being five.   Adult fish were observed by us in pound-net catches on May 25 and throughout November at Lynnhaven Roads, and again on October 22 at Lewisetta, Va.   In the lower parts of the bay more fish were caught during November than at any other period, the latest catch observed having been taken on November 29; but it is taken occasionally throughout the summer.   The species is nowhere abundant. It is eaten by the colored fishermen in at least one locality.

The largest specimen at hand, having a length of 12¼ inches, may represent about the maximum size attained.

Habitat.—New York to Virginia.

Chesapeake localities.—(*a*) Previous records "occurs occasionally in the southern part of Chesapeake Bay" (Uhler and Lugger, 1876), Norfolk and Gloucester Point, Va.   (*b*) Specimens in the collection: Lewisetta, Buckroe Beach, Lynnhaven Roads, and Ocean View, Va.

Family LXXXIII.—BLENNIIDÆ.   The blennies

Body moderately or greatly elongated, more or less compressed; mouth usually small, sometimes large, never vertical; teeth various; no spines on head; skin naked or covered with small or moderate-sized cycloid or ctenoid scales; lateral line single, double, or absent; dorsal fin very long, the anterior part and sometimes the entire fin with spines; caudal fin sometimes connected with dorsal and anal, sometimes free; usually round; anal fin similar to posterior part of dorsal; ventral fins small or wanting; placed far forward (jugular) if present, composed of one spine and one to three soft rays; pectorals varying from large to rudimentary.   Three genera of this family of small to medium-sized fishes are reported from Chesapeake Bay.

KEY TO THE GENERA

a. No fanglike or canine teeth in posterior part of either jaw; gill openings small, restricted to the sides, the membranes broadly united with the isthmus.
   b. Head rather pointed; snout well in advance of the forehead; upper margin of eye with or without a small tentacle_____Chasmodes, p. 332
   bb. Head very blunt; snout only a little in advance of forehead; upper margin of eye with a well developed tentacle_____Hypsoblennius, p. 334
aa. A fanglike tooth on each side of lower jaw; gill openings wide, the membranes free from the isthmus, or at least forming a fold across it_____Blennius, p. 335

144. Genus CHASMODES Cuvier and Valenciennes.   Blennies

Body oblong, compressed; head rather pointed; mouth large; maxillary extending to or beyond posterior margin of the eye; premaxillaries not protractile; teeth rather long, slender, in a single series, present only on front part of jaws, no canines; gill openings very short; scales wanting; a small tentacle on upper part of eye present or wanting; dorsal fin anteriorly, with very slender spines; caudal fin round, either united to or free from the dorsal; anal fin similar to soft part of dorsal; ventral fins jugular, with I, 3 rays; pectoral fins large.   A single species is known from Chesapeake Bay.

185. Chasmodes bosquianus (Lacépède).   Banded blenny.

Blennius bosquianus Lacépède, Hist. Nat. Poiss., II, 1800, p. 493; South Carolina.
Chasmodes quadrifasciatus Uhler and Lugger, 1876, ed. I, p. 98; ed. II, p. 81.
Chasmodes bosquianus Lugger, 1877, p. 70; Bean, 1891, p. 85; Jordan and Evermann, 1896–1900, p. 2394; Evermann and Hildebrand, 1910, p. 163.

Head 3.25 to 3.7; depth 2.9 to 3.8; D. XI or XII, 18 to 20; A. 18 to 20.   Body compressed, deepest slightly behind base of pectorals; head compressed, with moderately convex upper profile; snout not very blunt, 3.15 to 3.65 in head; eye moderate, lateral, 2.55 to 4.6; interorbital convex the bone 9.8 to 11; mouth moderately large, a little oblique, terminal; maxillary reaching nearly or quite to posterior margin of eye, 1.6 to 2.2 in head; teeth present only on anterior part of jaws, in a single series, pointed and curved inward at tips; gill opening restricted to the side, not much longer than the eye; a minute tentacle sometimes present on upper margin of the eye; dorsal fin very long, slightly joined to the caudal, its origin over the concealed margin of the preopercle; caudal fin round; anal fin long and low, distinct from the caudal, its origin under the first soft rays of the dorsal; ventral fins very narrow, inserted in advance of pectorals; pectoral fins broad, 1 to 1.35 in head.

Color of female in alcohol brownish; some individuals much darker than others; sides with pale crossbars or sometimes simply with whitish blotches; frequently also with irregular dark bars or blotches; sometimes with more or less wavy longitudinal lines; head above with very small dark dots; the fins dark spotted or barred; base of caudal with an obscure dark spot. Color of adult males usually darker and more uniform; sides with pale, wavy lines, sometimes with roundish pale spots, somewhat broader than the lines, situated in the course of the lines; upper part of head with small dark dots; fins mostly dark brown, with pale dots; spinous dorsal with a black spot between the first and second spines and frequently with smaller dark spots and dots, usually with a pale longitudinal stripe; no spot at base of caudal. The young resemble the adult female, being somewhat lighter in color, however, and having larger pale spots and bars.

Many specimens of this species, ranging in length from 25 to 90 millimeters (1 to 3½ inches), were preserved and were before the writers when the foregoing description was prepared. The comparatively great variation in color between the sexes and among examples of the same sex is indicated in the description. The sexes, because of the dissimilarity in color, have several times been described as different species. The banded blenny is most readily distinguished from other blennies of Chesapeake Bay by the comparatively pointed snout, the very small and frequently absent tentacle on the upper margin of the eye, and the convex interorbital.

The food, as shown by the contents of 18 stomachs, consists of small crustaceans, small mollusks, and insect larvæ. The small crustaceans, which constituted by far the greater portion of the food, were principally isopods and amphipods.

Spawning takes place from April to August. The eggs apparently are deposited in shells and probably on other objects, to which they adhere. On May 22, 1922, at Cherrystone Island, Va., a "nest," consisting of both valves of an oyster shell with almost the entire inside of the shell covered with eyed eggs, was picked up by hand. It so happened that the parent fish (the male) was inclosed in the shell as it was taken from the water. The nest and the fish were placed together in a bucket containing water. Even under these conditions the male fish remained close to the nest and could scarcely be driven away. Later the nest and the fish were preserved and are now before us.

In this connection the following quotation from Lugger (1877, p. 70) is of interest:

This peculiar little fish seems to be rather common in many parts of the Chesapeake Bay in localities where oysters are found. All the specimens were obtained from different oyster bars, and invariably inhabiting the dead and empty oyster shells called "snuffboxes." When taken out of their retreat they move about very slowly in the water but show great activity when disturbed. They drop from the shell very promptly as soon as the oyster is taken out of the water.

This blenny is reported to reach a maximum length of 4 inches. The largest individual seen during the present investigation, however, did not exceed 3½ inches. The banded blenny is not rare in Chesapeake Bay. On the other hand, it was never taken in large numbers. It was found to be most common in the lower York, Rappahannock, Potomac, and Patuxent Rivers, the largest catches being made from July to October. The latest catch made alongshore was on November 23 at Cape Charles. It was taken on clay, mud, and sand bottom. A few specimens were taken by the *Fish Hawk* during the winter months in water ranging from 9 to 15 fathoms in depth. The species, of course, is too small to be of commercial value.

Habitat.—New York to Florida; rare north of Maryland.

Chesapeake localities.—(a) Previous records: St. Marys River and near the mouth of the Potomac River (Uhler and Lugger, 1876); many parts of Chesapeake Bay, where oysters are found (Lugger, 1877), Blackistone Island, Md., and Cape Charles city, Va. (b) Specimens in collection: Many localities from Annapolis, Md., to Cape Charles and Lynnhaven Bay, Va.

145. Genus HYPSOBLENNIUS Gill. Blennies

Body elongate, compressed; head short, its profile steep; snout blunt; mouth small, horizontal; maxillary extending about to middle of eye; teeth slender, in a single series in each jaw, no canines; gill openings reduced, restricted to the sides; skin naked; a tentacle on upper margin of eye; dorsal fin long, the anterior part with slender, pungent spines; caudal round; anal somewhat similar to soft part of dorsal; ventrals inserted under the throat, with I, 3 rays; pectorals rather large. One species is known from Chesapeake Bay.

186. Hypsoblennius hentz (LeSueur). Blenny; Spotted seaweed fish.

Blennius hentz LeSueur, Journ., Ac. Nat. Sci., Phila., IV, 1825, p. 363; Charleston, S. C.
Hypsoblennius punctatus Bean, 1891, p. 85.
Hypsoblennius hentz Jordan and Evermann, 1896–1900, p. 2390, Pl. CCCXXXIX, fig. 823; Evermann and Hildebrand, 1910, p. 163.

Head 3.35 to 3.75; depth 2.95 to 3.3; D. XII, 13 to 15 (usually 15); A. 17 to 19 (usually 18). Body compressed, deepest over pectorals, tapering gradually from there to the tail; head short and deep, its anterior profile very steep; snout short, not much in advance of forehead, 2.8 to 3.25 in head; eye placed high, lateral, 3.15 to 3.8; interorbital deeply concave, the bone 7.8 to 10.4; mouth broad, terminal, horizontal; maxillary reaching below middle of eye, 2.5 to 3.05 in head; teeth in jaws only, in a single close-set series, slightly flattened and curved inward; each nostril

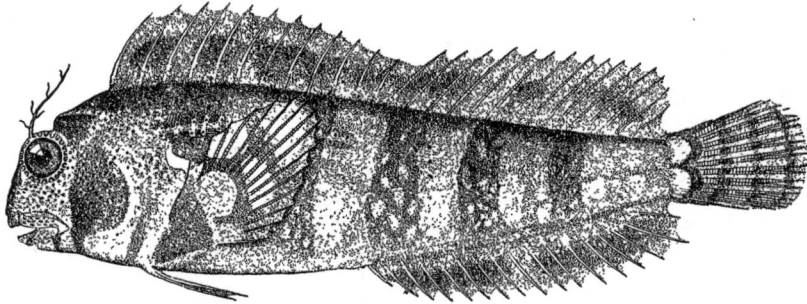

FIG. 201.—*Hypsoblennius hentz.* Adult, 4⅛ inches long

with a simple, short, fleshy tentacle; a branched tentacle on upper margin of eye, this tentacle sometimes shorter than eye and sometimes more than twice the length of eye (it seems probable that this tentacle is longer in males than in females, yet a very large variation in its length exists among individuals of the same sex); gill openings restricted to the sides, about twice as long as eye; dorsal fin long, continuous, the spines slender but pungent, not quite as high as the soft part of fin, slightly attached to base of caudal; caudal fin round; anal fin lower than soft dorsal, free from caudal, its origin under posterior spines of dorsal; ventral fins well developed, inserted in advance of pectorals; pectorals rather large, about as long as head.

Color in alcohol brownish, some specimens darker than others (this difference in color apparently bearing no relationship to sex); sides and head with dark spots, these largest on lower part of sides and smallest dorsally on head; some individuals less profusely spotted but with indefinite cross bars extending on dorsal fin; chin nearly always with two and occasionally with three dark bars; ventral fins nearly black; the other fins paler than the body, variously spotted; the caudal usually cross-barred; anal fin with the free tips of rays pale.

This blenny is represented in the present collection by 15 specimens, ranging in length from 43 to 93 millimeters (1¾ to 3¾ inches). This species is readily distinguished from other blennies of Chesapeake Bay by the very steep, almost vertical forehead, and by the deeply concave interorbital. The branched tentacle over the eye also helps to identify it.

The food of this blenny, as shown by the contents of five stomachs, consists of small crustaceans, mollusks, and ascidians. The comparatively large quantity of plant fragments present suggests that it may also feed on vegetable matter.

FISHES OF CHESAPEAKE BAY

335

Ripe or nearly ripe fish were taken in May, July, and September, indicating that spawning may take place throughout the summer.

The maximum length given for this fish in published accounts is 4 inches, which slightly exceeds the length of the largest specimen at hand. It was infrequently taken in nets in shallow water, the latest date along shore being November 23, at Cape Charles, and somewhat more frequently during the winter with the beam trawl operated from the *Fish Hawk* in water ranging from 13 to 25 fathoms in depth. This blenny may not be as rare as indicated by the collection, for it may live, like some of the other blennies, where it is inaccessible to nets.

Habitat.—Chesapeake Bay to Florida.

Chesapeake localities.—(a) Previous records: Cape Charles city, Old Point Comfort, and Ocean View, Va. (b) Specimens in collection: From off Bloody Point, off Cove Point, off Barren Island, off Cedar Point, off Hooper Island, off Point No Point, and Crisfield, Md., and off Smith Point, Cape Charles, lower York River, off Old Point Comfort, Ocean View, and Lynnhaven Roads, Va.

146. Genus BLENNIUS Linnæus. Blennies

Body oblong, compressed; head short, its upper profile usually bluntly rounded; mouth small, horizontal; teeth in a single series in each jaw, long and slender, curved inward; lower jaw in addition with a stout fanglike tooth on each side; premaxillaries not protractile; gill openings wide, the membranes free from the isthmus, or at least forming a broad fold across it; scales wanting; dorsal fin entire or more or less notched, the spines slender; ventrals well developed, I, 3 rays; pectorals moderately developed.

187. Blennius fucorum Cuvier and Valenciennes. Seaweed blenny.

Blennius fucorum Cuvier and Valenciennes, Hist. Nat. Poiss., XI, 1836, p. 263; open seas south of the Azores. Uhler and Lugger, 1876, ed. I, p. 97; ed. II, p. 81; Jordan and Evermann, 1896–1900, p. 2379.

This blenny was recorded from Chesapeake Bay by Uhler and Lugger (1876), whose description we quote in full. It has not been seen there by other investigators.

"Body small, cylindrical, and scaleless; head large, deeper than long; the large and very prominent eyes project beyond the face; a thread-shaped cirrus, bifid at tip, and nearly as long as the head, projects from the upper part of each orbit. Soiled greenish, brownish above, with numerous brown spots on the cheeks and sides of the body; throat and belly faintly rosaceous. Length, 1 to 2 inches. Fin rays, D. 11, 17; P. 14; V. 3; A. 18; C. 14."

It is rather remarkable that this blenny, usually found among seaweed in the open seas, should have been taken well up Chesapeake Bay.

Habitat.—Atlantic Ocean, in floating seaweed, and from Chesapeake Bay.

Chesapeake localities.—(a) Previous record: Oyster region south of Tangier Sound (Uhler and Lugger, 1876). (b) Specimens in collection: None.

Family LXXXIV.—OPHIDIIDÆ. The cusk eels

Body elongate, compressed, more or less eel-shaped; head large; lower jaw included; both jaws and usually vomer and palatines with villiform or blunt teeth; premaxillaries protractile; gill openings wide, the membranes separate, anteriorly narrowly joined to the isthmus behind ventrals; pseudobranchiæ small; gills 4, a slit behind the fourth; scales small, covering body and occasionally the head; air bladder and pyloric cæca present; vertical fins low, without spines, confluent around the tail; tail isocercal; ventral fins at the throat, each developed as a long, forked barbel. A single genus and species comes within the scope of the present work.

147. Genus RISSOLA Jordan and Evermann. Cusk eels

Body moderately elongate; lower jaws included; teeth in jaw pointed, those on vomer and palatines blunt; vent posterior, at origin of anal; no spines on opercles; scales present on body, wanting on head, elongate in shape and somewhat imbedded; air bladder placed just back of the cranium, very short, described as having a foramen posteriorly. This, however, appears to be a mistake, for the posterior end of the bladder really is closed by a small, solid, conical, nearly transparent body connected with the wall of the bladder by a very thin membrane. Two fresh specimens examined by us had this body in position as explained; a third alcoholic specimen had the little solid body pushed into the cavity of the bladder and this gave the bladder the appearance of having a foramen.

188. Rissola marginata (DeKay).   Cusk eel.

Ophidium marginatum DeKay, New York Fauna, Fishes, 1842, 315; New York Harbor.   Bean, 1891, p. 85.
Rissola marginata Jordan and Evermann, 1896–1900, p. 2480, Pl. CCCLIII, fig. 868.

Head 5.95 to 6.15; depth 7.3 to 8.2.   Body quite elongate, compressed, of about uniform depth from nape to middle of anal base, then tapering to tail; head compressed; snout moderately pointed, 3.5 to 4.05 in head; eye 3.05 to 3.3; interorbital (bone) 4.75 to 6.85; mouth moderately large, horizontal; lower jaw included; maxillary reaching nearly or quite to posterior margin of eye, 2 to 2.15 in head; teeth in jaws pointed, in bands; vomer and palatines with bands of blunt teeth; gill rakers short, four to five on lower limb of first arch; lateral line usually not quite complete; scales small, elongate, imbedded, wanting on head; dorsal and anal fins long, low, continuous with the round caudal; origin of dorsal over or a little behind middle of length of pectorals; origin of anal a little behind the beginning of the second one-third of body; ventral fins inserted below vertical from middle of eye, consisting of two filaments, the longest one about an eye's diameter shorter than head; pectoral fins moderately large, 1.2 to 1.3 in head.

Color in life grayish green; sides golden; belly snow white; ventral surface of head mostly golden; sides of head punctulated with brown; lateral line in a dark band; dorsal fin pale green with black margin, this color continued on caudal and posterior half of anal fin; margin on anterior half of anal white; ventrals white; pectorals golden with distal and lower margins white.   The color in alcohol fades to a light brown; the fins pale, with margins as in the live fish.

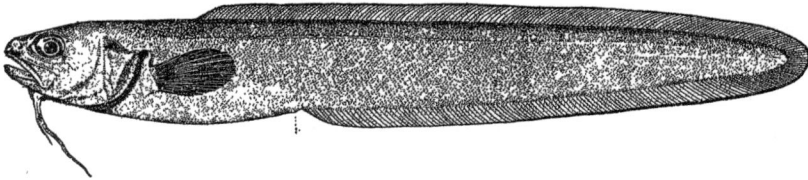

FIG. 202.—*Rissola marginata*.   Adult, 8⅕ inches long

The cusk eel is represented in the present collection by five specimens, ranging in length from 145 to 230 millimeters (5¾ to 9 inches) in length.   This fish is readily distinguished from all others of Chesapeake Bay by its long, somewhat eel-like body and by the white, filamentous ventral fins (each fin with two slender filaments) situated behind the chin and below the eyes.

Little is known about the life history of this fish.   It appears to be nocturnal in its habits.   An individual kept in the aquarium by one of us (Hildebrand), in the laboratory at Beaufort, N. C., remained concealed during the day in the sand placed in the aquarium.   Frequently only a current of water could be seen where its snout was near the surface.   At night, however, it frequently was seen swimming about in the aquarium, presumably searching for food.   Its tail is supported by heavy cartilage radiating from the last vertebra and is used in burrowing in the sand, for it always descends into the sand with the tail down.

Three examples were examined for food.   All had fed on small crustaceans and one had also eaten a small fish—a goby.   The ovary is single, showing a median fold, however.   It is not known when spawning takes place.   Three specimens examined were females, one taken on May 21, another on July 16, and the third on September 12; the ovaries of all appeared to be in an early stage of development.

It is difficult to know, of course, how common a species like the present one (which apparently remains burrowed in the sand throughout the day) really is, without doing considerable collecting at night.   Two of the specimens at hand were taken along a sandy beach with a seine at 3 o'clock in the morning, another was found dead on the beach, and two more were taken by the *Fish Hawk* in a beam trawl at a depth of 10 fathoms.   It is not known to the writers whether this haul by the *Fish Hawk* was made during the day or at night. The cusk eel, however, is not known to be abundant anywhere, and it probably is rather uncommon in Cheaspeake Bay.

Habitat.—New York to Texas; along sandy shores.

Chesapeake localities.—(a) Previous record: Cape Charles city, Va. (b) Specimens in collection: Cape Charles, off Cape Charles Light, and Lynnhaven Roads, Va.

Family LXXXV.—BATRACHOIDIDÆ. The toadfishes

Body robust, depressed anteriorly, compressed posteriorly; mouth large; teeth strong; gill openings chiefly lateral, the membranes united to the isthmus; scales present or wanting; air bladder present; dorsal fins 2, the first with two or three low spines; second dorsal and anal long, similar; caudal fin round and free from the dorsal and anal; ventral fins large, jugular; pectoral fins broad. A single genus and species comes within the scope of the present work.

148. Genus OPSANUS Rafinesque. Toadfishes

Body robust, notably depressed anteriorly, compressed posteriorly; head large, with numerous fleshy flaps; mouth very broad; teeth very strong, blunt, mostly in a single series, present on jaws, vomer, and palatines; opercle with two partly concealed spines; skin scaleless, wrinkled; lateral line obscure; three dorsal spines; axil of pectoral with a large foramen.

189. Opsanus tau (Linnæus). Toadfish.

Gadus tau Linnæus, Syst. Nat., ed. XII, 1766, p. 440; Carolina.

Batrachus tau Uhler and Lugger, 1876, ed. I, p. 98; ed. II, p. 82; Bean, 1891, p. 86; Smith, 1892, p. 72.

Opsanus tau Jordan and Evermann, 1896–1900, p. 2315; Smith and Bean, 1899, p. 187; Evermann and Hildebrand, 1910, p. 163; Fowler, 1912, p. 59.

Head 2.65 to 3.05; depth 3.55 to 4.7; D. III–26 or 27; A. 21 or 22. Body robust, anteriorly very broad, depressed (especially in adult), posteriorly compressed; head very large and broad; snout short and broad, 3.9 to 5 in head; eye 3.45 to 4.95; interorbital (bone) 6.55 to 12.85; mouth large and broad; lower jaw projecting; maxillary reaching well beyond eye, 1.55 to 2.15 in head; teeth strong, blunt, in a single series laterally in the jaws, anteriorly forming more or less of a band; a row of somewhat stronger teeth on vomer and palatines; opercle with three strong spines; skin smooth; barbels or tentacles present on head, about the mouth, and sometimes on sides of body; these enlarged over the eyes and on lower jaw; first dorsal with three short spines, enveloped in skin; second dorsal long and of about uniform height; caudal fin round; anal fin long, the rays distally more or less free; ventral fins rather small, the anterior rays enveloped in heavy skin; pectoral fins broad, fan-shaped, 1.4 to 1.65 in head.

Color in alcohol variable, grayish to brownish above, pale underneath, with profuse markings of darker and lighter color on the sides; fins all with dark and pale bars. Several adult toadfish observed in an aquarium at Woods Hole, Mass., were yellowish brown, mottled with darker brown on body and fins. The dark markings on dorsal and anal were in the form of irregular, oblique bars, and on the caudal in transverse bars. The outer half of pectorals was marked with concentric bars, the basal part being mottled like the body. This color pattern is typical (at least, of all those that we have seen from various localities between Maine and Florida) of the toadfish throughout its range.

Numerous specimens, ranging in length from 30 to 320 millimeters (1⅛ to 12⅝ inches), were preserved. Its scaleless skin, broad head with numerous fleshy fringes, its long, soft dorsal fin, and its fleshy ventral fins, placed under the throat, serve well to distinguish this fish from all others occurring in Chesapeake Bay.

The toadfish is omnivorous. The principal food, however, appears to consist of crustaceans. Among a lot of 31 individuals, 25 had either fed on crustaceans, in combination with mollusks, or on fish. Small crabs among the crustaceans appeared most frequently in the food, although shrimp (and in the smaller individuals, amphipods) and isopods also were present. Almost any kind of offal is eaten, and in places where garbage is thrown overboard toadfish are almost always present in comparatively large numbers.

In Chesapeake Bay spawning apparently takes place throughout the summer, as females with large eggs were taken from April 13 to October 25, 1922. The eggs are very large, being about 5 millimeters in diameter, and are laid under stones, in large shells, tin cans, old shoes, boiler tubes, etc. They adhere in a single layer to the surface upon which they are deposited. The nest is guarded by the male during the period of incubation, which is reported to cover a period of about three weeks. (For a comprehensive account of the spawning habits of the toadfish, see Gudger, 1910, pp. 1095 to 1106.) The larval toadfish remains attached to the yolk sac and the "nest" for several days after it breaks the egg case. When it finally becomes a free-swimming fish it is about 15 or 16 millimeters in length.

Owing to the protracted spawning season, it was difficult to follow the growth of the young in Chesapeake Bay. The smallest specimen, 30 millimeters (1⅛ inches) long, was caught on July 8 along with 15 fish ranging up to 137 millimeters (5⅜ inches); the larger fish obviously belonged to a different year group. A fish 63 millimeters (2½ inches) long, taken on June 23, may have been hatched the previous fall. Small fish were seined throughout the summer. A specimen 48 millimeters long, taken on April 20, and one 57 millimeters (2 to 2¼ inches) in length, taken April 26, undoubtedly were hatched the previous year, probably in the late summer or fall.

The toadfish is sluggish in its habits. It is very ugly in appearance, always being densely coated with slime. It often makes a croaking sound when removed from the water, and it erects its spines and snaps with its mouth at anything that comes near. Its sharp spines on the opercle

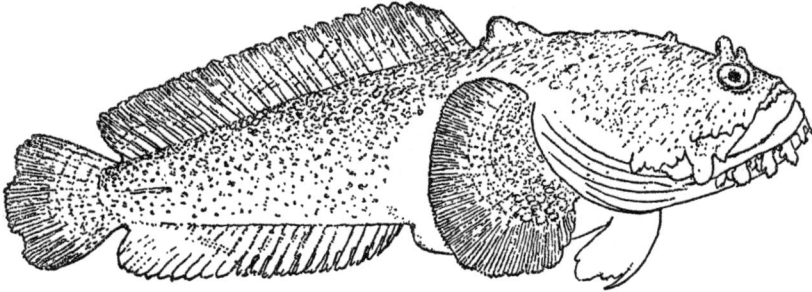

FIG. 203.—*Opsanus tau*

and in the dorsal fin, and its large mouth, provided with strong teeth and powerful jaws, are weapons to be feared, for with these it can inflict painful wounds. It is very tenacious of life, living for a comparatively long period of time out of water.

The maximum size attained by this fish, as given in published accounts, is 15 inches. Such a size, however, must be regarded as exceptional, as apparently few exceed a length of 12 inches. The largest individual seen in Chesapeake Bay during the present investigation was 12⅝ inches in length and weighed 1 pound 2 ounces. The toadfish is comparatively abundant in Chesapeake Bay and is taken throughout the summer in seines, with hooks and lines, and rather rarely in pound nets. It was taken by the *Fish Hawk* with the beam trawl, during the winter months, in water ranging from 5 to 27 fathoms in depth. This fish is much disliked by the fishermen because of its ugliness and its fighting habits and because it has no commercial value. The flesh is said to be of good appearance and fine flavor, but the fish apparently is not utilized on account of its repulsive appearance. Possibly a demand could be created if the fish were dressed by removing the head, skin, and internal organs before placing it on the market, as is sometimes done with the catfishes.

Habitat.—Maine to the West Indies; rare north of Cape Cod.

Chesapeake localities.—(a) Previous records: "Lives in the mud of the oyster regions of Chesapeake Bay, around the mouth of the Potomac River, and elsewhere in salt water" (Uhler and Lugger, 1876); Gunston Wharf, Cape Charles city, and Hampton Roads, Va. (b) Specimens in the collection: From many localities from Annapolis, Md., to the entrance of the bay.

Order XENOPTERYGII

Family LXXXVI.—GOBIESOCIDÆ. The clingfishes

Body rather elongate, broad, and depressed anteriorly; mouth moderate; upper jaw protractile; teeth usually rather strong, the anterior ones conical or incisorlike; no bony stay across cheek; opercle reduced to a spinelike projection, concealed in the skin and sometimes obsolete; pseudobranchiæ small or wanting; gills 2½ or 3; gill membranes broadly united, free, or united with the isthmus; scales entirely wanting; dorsal and anal similar, on posterior part of body and nearly or quite opposite each other, consisting of soft rays only; ventral fins far apart, each with one concealed spine and four or five soft rays; a large sucking disk present between the ventrals, the fins usually forming a part of it. This family is composed of small fishes that live chiefly in warm seas, clinging to stones and other objects by means of the sucking disk.

149. Genus GOBIESOX Lacépède. Clingfishes

Body anteriorly very broad and depressed, posteriorly slender; head large, rounded; mouth terminal; lower jaw with a series of strong incisors in front, their edges rounded or truncate; upper jaw with a series of strong teeth, sometimes with smaller teeth behind; no teeth on vomer or palatines; gills 3; gill membranes broadly united, free from the isthmus; sucking disk large.

190. Gobiesox strumosus Cope. Clingfish.

Gobiesox strumosus Cope, Proc., Ac. Nat. Sci., Phila., 1870, p. 121; Hilton Head, S. C. Uhler and Lugger, 1876, ed. II, p. 84; Jordan and Evermann, 1896–1900, p. 2333; Evermann and Hildebrand, 1910, p. 163.

Gobiesox virgatulus Jordan and Evermann, 1896–1900, p. 2333.

Head 2.45 to 2.7; depth 4.1 to 5; D. 10 to 12; A. 8 to 10. Body anteriorly broad, depressed; posteriorly compressed; caudal peduncle strongly compressed; vertebræ 12 to 14; head large, depressed, very broad, quite variable in width, its width 2.5 to 3.35 in length of body; snout very broad, forming with the rest of the head anteriorly an arc of a circle, its length 2.8 to 3.65 in the head; eyes small, partly superior, 3.65 to 4.75; interorbital broad, flat, 3.2 to 3.8; mouth wide, inferior, horizontal; maxillary concealed by the preorbital, reaching nearly or quite opposite middle of eye, 6 to 7.7 in head; teeth in the jaws with an irregular, enlarged outer series, these teeth anteriorly in the lower jaw incisorlike, the cutting edge entire, those of the upper jaw less strongly compressed and less incisorlike; both jaws anteriorly with smaller teeth behind the enlarged ones; cheeks full, bulging; opercle ending in a sharp spine; gill opening restricted, mostly lateral; dorsal fin placed far back, its origin nearer the end of the caudal than tip of snout; caudal fin round; anal fin shorter but otherwise similar to the dorsal, its origin a little nearer tip of caudal than anterior margin of ventral disk; ventral disk a little shorter than head.

Color in alcohol grayish to dusky, variable, some specimens much lighter than others; some specimens with distinct pale crossbars, others with indefinite blotches, still others without indications of pale bars or blotches, but with longitudinally elongate dark markings; the vertical fins usually dusky, with pale crossbars or blotches; the ventrals and pectorals pale.

Numerous specimens of this species, ranging in length from 19 to 60 millimeters (¾ to 2⅜ inches), were preserved. Much variation among individuals with respect to the width of the head and color appears to exist. These differences have been pointed out in the description. This fish is readily distinguished from all others of the Chesapeake by the very broad, depressed head and the large sucking disk between and behind the widely separated ventral fins. The species of this genus are not well defined, making identification difficult. It seems probable that some of the nominal species are not actually distinct. The chief diagnostic characters given in current works are the number of dorsal and anal rays. In the specimens at hand, which undoubtedly are all of one species, the range in the number of fin rays covers the nominal species, *G. strumosus* and *G. virgatulus*. In a lot of 30 specimens the rays in the dorsal fin varied from 10 to 13; that is, 4 had 10 rays, 16 had 11 rays, 9 had 12 rays, and 1 had 13 rays. In the same lot the number of anal rays varied from 8 to 10, as follows: 12 had 8 rays, 15 had 9 rays, and 3 had 10 rays.

We provisionally use the name *strumosus* for the specimens in hand, as it has priority over *virgatulus*. The true relationship of the species will necessarily remain in doubt until a more thorough study of the genus can be made.

The food of this little fish, as shown by the contents of 26 stomachs, consists mainly of isopods and amphipods and an occasional annelid.

Spawning evidently takes place in the spring, as most of the specimens taken in April and May contained well-developed gonads. Those taken later in the season appeared to be spent. The smallest ripe female seen was only 1⅜ inches in length.

It is difficult to determine the rate of growth of this fish because of the small size attained and because of the gradation of all sizes taken throughout the spring, summer, and fall. The smallest fish secured, a specimen 19 millimeters (four-fifths inch) in length, was taken on October 10.

The maximum size of *G. virgatulus*, which is herein considered identical with *strumosus*, as given by Jordan and Evermann (1896–1900, p. 2333), is 4 inches. It seems doubtful that the clingfish ever grows that large in Chesapeake Bay, as the largest of hundreds taken during the present investigation was only 2⅜ inches in length. This small fish, although previously only twice recorded from the Chesapeake, is quite common. It was rather scarce from Buckroe Beach to Cape Henry and at Cape Charles, but common to abundant from the York River to Annapolis. It was found in large numbers in the lower Rappahannock and Patuxent Rivers, where it was taken in company with pipefishes and sticklebacks on grassy bottom in 1 to 3 feet of water. Although rather common at Annapolis, none were seined along the shores at Love Point. One specimen, however, was trawled off Love Point on May 13 in 110 feet of water; none were found in the vicinity of Baltimore or Havre de Grace.

The clingfish, by means of its large sucking disk, is able to attach itself to shells, rocks, piling, etc., and it is not infrequently found adhering to such objects. For this reason it is sometimes not taken in nets in localities where it is comparatively common. A few individuals were taken by the *Fish Hawk* during the winter months in water ranging in depth from 7 to 17 fathoms.

Habitat.—Chesapeake Bay; probably to Florida or beyond.

Chesapeake localities.—(*a*) Previous records: Magothy River and St. Georges Island, Md. (*b*) Specimens in collection: From many localities between Annapolis, Md., and the mouth of the bay.

Order PLECTOGNATHI

Family LXXXVII.—BALISTIDÆ. The trigger fishes

Body usually rather deep, considerably compressed; snout long; eye small, placed high; mouth small, usually terminal; teeth in the jaws in a single series, frequently incisorlike; gill openings represented by oblique slits; preopercular bones externally not evident; scales more or less plate-like, bearing spines or bony tubercles; dorsal fins 2, the first spine high and strong; ventral fins represented by a single stout spine attached to the enlarged pubic bone. A single genus of this family of tropical fishes comes within the scope of the present work.

150. Genus BALISTES Linnæus. Trigger fishes

Body compressed, rather deep; snout long; eye small, placed very high; mouth small, terminal; gill opening an oblique slit with enlarged bony scutes behind it; teeth in the jaws irregular, usually very strong; scales platelike, usually bearing spinules; first dorsal with three spines, the first one much enlarged, erect and fixed when the second one is erect, readily laid back upon deflexing the second spine, hence the name "trigger fishes"; second dorsal and anal long, usually similar; ventrals represented by a single median spine. A single species has been taken in Chesapeake Bay. A second one is recorded from both north and south of the bay and may be expected within the bay. Accordingly, the following key, showing the chief differences of the two, is introduced:

KEY TO THE SPECIES

a. Head without prominent bands or stripes; outer caudal rays never greatly produced; D. III–27 to 29; A. 23 to 26; scales 54 to 62 _____ *carolinensis,* p. 341

aa. Head with prominent dark blue stripes and bars on side; outer rays of caudal greatly produced in adult, filamentous; D. III–29 or 30; A. 26 to 28; scales 60 to 62 _____ *vetula,* p. 341

191. Balistes carolinensis Gmelin.   Trigger fish; Leatherjacket; Turbot.

Balistes carolinensis Gmelin, Sys. Nat. I, 1788, p. 1468; Carolina. Jordan and Evermann, 1896–1900, p. 1701, Pl. CCLVIII, fig. 632.

Head 3; depth 1.8; D. III–27; A. 25; scales (counted from gill opening to base of caudal) 58. Body deep, rather strongly compressed; dorsal and ventral outlines about evenly convex; head deep; snout long, tapering, 1.32 in head; eye small, 4.4; interorbital 3.15; mouth small, terminal; teeth in jaws large and strong, the anterior ones more or less caninelike; gill opening reduced to an oblique slit; lateral line feebly developed, most distinct posteriorly; scales of moderate size, implanted in a thick leathery skin, the edges of scales not free, each scale covered with bony barbs; dorsal fins separate, the first consisting of three short, strong spines, the first one the longest, 1.45

FIG. 204.—*Balistes carolinensis.*   Adult, 12¼ inches long

in head; second dorsal and anal similar, highest anteriorly; caudal fin with concave margin; ventral fins represented by a single blunt spine, thickly covered with coarse barbs; pectoral fins short, round, 2.55 in head.

Color in alcohol grayish green; interorbital area blackish; indications of a blackish bar under base of spinous dorsal, extending to pectoral; another and more distinct one at origin of second dorsal; two dark blotches under base of second dorsal; another on caudal peduncle; fins more or less greenish to dusky; spinous dorsal with pale spots on membranes; caudal fin plain; all other fins with dark spots, streaks, or bars.

A single specimen, 250 millimeters (9¾ inches) in length is at hand and upon it the foregoing description is based. The trigger fish is recognized by the deep, compressed body; small eye, which is placed very high; small mouth, with large teeth; short gill opening; protruding pubic bone; and the very large first dorsal spine. The chief differences between this fish and its near West Indian relative, *B. vetula,* which to date has not been taken in Chesapeake Bay but which is not infrequently found at Woods Hole, are set forth in the accompanying key.

Nothing definite seems to be known about the food and feeding habits of the trigger fish. It is taken occasionally on the blackfish grounds off Beaufort, N. C., with hook and line, baited with pieces of fish; and at Key West, Fla., where it is known as "turbot," it is similarly taken in company with grunts.

Its spawning habits and rate of growth, too, are not known. R. L. Barney (field notes) took a female on the sea beach of Bogue Banks, Beaufort, N. C., on July 1, 1920, which contained "very ripe" spawn. The trigger fish is reported to reach a weight of 4 pounds, but the average is about 1 pound.

This fish evidently is rare in Chesapeake Bay, for it was not seen by the field men during the present investigation, and we have not seen a record or heard of its previous capture within the bay. The specimen at hand was caught with hook and line at the mouth of the Potomac River on September 21, 1922, by D. E. Knight, who presented the specimen to the Bureau of Fisheries.

Habitat.—Both coasts of the tropical Atlantic; on the American coast from Nova Scotia to the Dutch West Indies (St. Eustatius, Windward Islands); not common north of Florida.

Chesapeake records.—(*a*) Previous records: None. (*b*) Specimen in collection: From the mouth of the Potomac River, Md.

Family LXXXVIII.—MONACANTHIDÆ. The filefishes

Body much compressed, usually quite deep; mouth small, terminal or more or less superior; jaws with incisorlike teeth; those of the upper jaw in a double series, in a single series in lower jaw; gill openings mere slits; lateral line absent; scales small, bearing spines; first dorsal consisting of a single spine (rarely with a second rudimentary spine), smooth or barbed; second dorsal remote from the first and similar to the anal; ventrals either absent or represented by a long spine surmounting the pelvic bone. Two genera and two species of this family of warm-water fishes occur in Chesapeake Bay.

KEY TO THE GENERA

a. Dorsal spine long and comparatively strong, with two rows of retrorse hooks or barbs posteriorly; pelvic bones surmounted by a spine projecting through the skin of the abdomen; gill slit short, not longer than eye_____ Monacanthus, p. 342

aa. Dorsal spine rather weak and comparatively short, without barbs; pelvic bones without a spine; gill slits notably longer than eye_____ Ceratacanthus, p. 343

151. Genus MONACANTHUS (Cuvier) Oken. Foolfishes; Filefishes

Body short and deep, much compressed; mouth very small, terminal; gill slit oblique, shorter than eye; ventral flap and sometimes caudal peduncle spinous; dorsal spine large, posteriorly with two series of barbs; second dorsal remote from the spine and similar to the anal, each consisting of 25 or more rays; caudal fin broad, round; pectoral fins short and broad; a blunt, movable pelvic spine present. A single species comes within the scope of the present work.

192. Monacanthus hispidus (Linnæus). Foolfish; Filefish.

Balistes hispidus Linnæus, Syst. Nat., ed. XII, 1766, p. 405; Carolina.
Stephanolepis massachusettensis Uhler and Lugger, 1876, ed. I, p. 90; ed. II, p. 75.
Monacanthus hispidus Bean, 1891, p. 84; Jordan and Evermann, 1896–1900, p. 1715, Pl. CCLIX, fig. 635; Evermann and Hildebrand, 1910, p. 162.

Head 2.7 to 3.1 (measured to upper angle of gill opening); depth 1.5 to 1.65; D. I–32 to 34; A. 31 to 34. Body short and deep, strongly compressed; profile from snout to dorsal slightly concave; snout long, 1.4 to 1.75 in head; eye 2.2 to 3.2; interorbital 3.15 to 3.5; mouth very small, terminal; teeth in the jaws broad, with sharp cutting edges; gill opening an oblique slit, situated between the eye and the base of dorsal; scales minute, beset with short, rough bristles; first dorsal consisting of a single barbed spine inserted over posterior part of eye, remote from second dorsal; second dorsal and anal similar and opposite each other, first ray of dorsal sometimes filamentous; caudal fin with convex margin; ventral fins represented by a single median spine, extending beyond flap of skin attaching the spine to the abdomen; pectoral fins short, 1.95 to 2.2 in head.

Color variable, grayish or greenish; sides with irregular blackish blotches or with more or less definitely horizontally elongated black spots; caudal fin often dusky; other fins plain translucent.

This foolfish is represented in the present collection by 110 specimens, ranging from 31 to 80 millimeters in length. This fish is recognized by the short, deep body, rough skin, the prominent, barbed dorsal spine, and the rough ventral spine.

Numerous specimens examined at Beaufort, N. C., by Linton (1905, p. 401) had fed on bryozoans, small crustaceans and mollusks, gastropod eggs, annelids, small sea urchins, and algæ. Seven specimens examined from Chesapeake Bay, ranging in length from 60 to 80 millimeters, had fed mainly on annelids. One specimen contained fragments of shells of a mollusk, and three contained vegetable fragments.

Nothing is known concerning the spawning and breeding habits of this fish, and so far as we are aware no fish with gonads in an advanced state of development has been observed.

The foolfish reaches a maximum length of 10 inches. It is not eaten and has no commercial value.

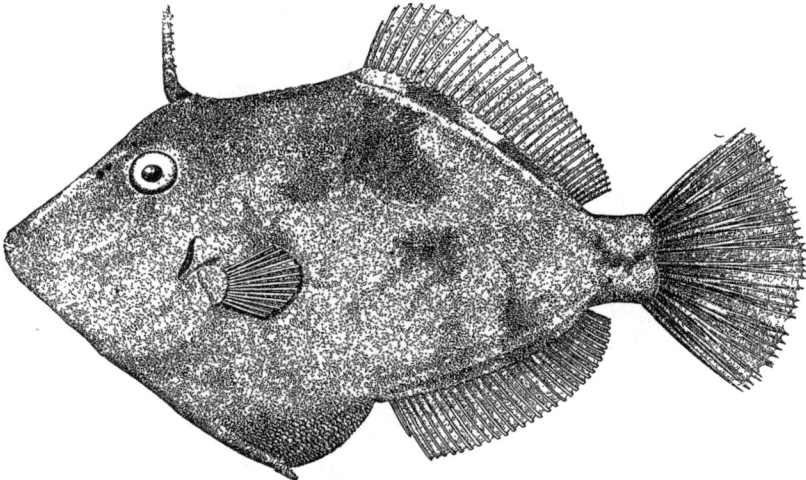

FIG. 205.—*Monacanthus hispidus*

Habitat.—Nova Scotia, south to Brazil; also recorded from the Canaries and Maderia in the eastern Atlantic. It is uncommon on the American coast north of Woods Hole, Mass., and it apparently does not occur regularly within Chesapeake Bay.

Chesapeake localities.—(a) Previous records: Cape Charles city and Ocean View, Va. (b) Specimens in collection: Cape Charles, Va., 110 specimens, all taken on September 23, 1921, in 6 hauls of a 250-foot bag seine. The species was not seen during other visits to Cape Charles, nor was it seen elsewhere in Chesapeake Bay during the present investigation.

152. Genus CERATACANTHUS Gill. Foolfishes; Filefishes

Body elongate, strongly compressed; mouth more or less superior; lower jaw projecting; gill opening consisting of a very oblique slit, much longer than eye; first dorsal consisting of a single, barbless spine; second dorsal remote from the first, its rays about 35 to 50; anal fin similar to soft dorsal; caudal fin more or less elongate, round, or somewhat pointed; pectoral fins very small; no external pelvic spine. A single species of this genus of warm-water fishes comes within the scope of the present work.

193. Ceratacanthus schœpfi (Walbaum). Filefish; Foolfish; Devilfish.

Balistes schœpfi Walbaum, Art. Gen. Pisc., 1792, p. 461; Long Island, N. Y.
Alutera cuspicauda Uhler and Lugger, 1876, ed. I, p. 89; ed. II, p. 74.
Ceratacanthus aurantiacus Lugger, 1877, p. 61.
Alutera schœpffi Bean, 1891, p. 84; Jordan and Evermann, 1898, p. 1718, Pl. CCLX, fig. 636; Fowler, 1912, p. 59.

Head 3.3 to 3.7 (measured to upper angle of gill opening); depth 1.95 to 2.65; D. I–34 to 37; A. 36 to 41. Body elongate, very strongly compressed, proportionately deeper in adult than in young; profile from snout to dorsal spine nearly straight to notably concave in small to moderate-sized specimens, concave over snout and convex over eyes in large individuals, measuring upward of 400 millimeters in length; snout long, 1.1 to 1.2 in head; eye 3.8 to 5.4; interorbital 3.9 to 5.2; mouth very small, superior, nearly vertical; teeth in the jaws broad, those of the lower jaw usually deeply notched, all with sharp cutting edges; gill opening consisting of an oblique slit, situated partly between the eye and base of pectoral; scales minute, not very evident in young, rough, being covered with short spines; first dorsal consisting of a single, rough, barbless spine situated over eye, remote from the second dorsal; second dorsal and anal similar and opposite each other; caudal fin long in small individuals, about half the length of body in specimens 145 millimeters long, shorter than snout in large examples (400 millimeters and upward); pectoral fins short, 2.45 to 3.1 in head.

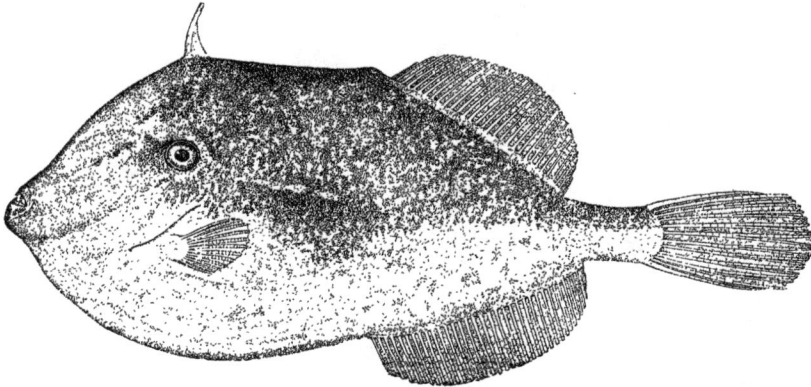

FIG. 206.—*Ceratacanthus schœpfi*

Color variable; specimens 6 to 8 inches long largely black or grayish; irregular dusky blotches, chiefly near bases of dorsal and anal fins; sides with well-defined, small, round, dusky or brownish spots, spaced irregularly, often forming a straight line along middle of sides, beginning below origin of soft dorsal and extending to base of caudal; nape plain or dusky; dorsal plain, tinged with yellow brown; caudal black, light brown at base; anal plain, outer edge tinged with brown; pectoral plain. A fish 18 inches in length was brownish on back and sides, spotted with yellow and orange; lower parts yellow and white; dorsal and anal dusky; caudal grayish, yellow at base; pectorals plain. Adult fish vary considerably in color pattern, the usual colors being brown, gray, yellow, orange, and white.

This fish is represented in the present collection by 13 specimens, ranging in length from 145 to 460 millimeters (5¾ to 18 inches). This foolfish differs from its relative, *M. hispidus*, principally in the more elongate body, a proportionately shorter dorsal spine, which has no barbs, and in the absence of a ventral spine. It also attains a much larger size. The young differ prominently from the adults in having the body much more slender and in having a proportionately much longer caudal fin.

Small examples examined by Linton (1905, p. 401) had fed on bryozoans, shrimp, amphipods, and sea lettuce. A specimen 460 millimeters long, from Chesapeake Bay, had fed exclusively on an unidentified plant resembling Naias. This plant filled the whole alimentary tract, which is of about uniform diameter throughout and about three times the total length of the fish.

Virtually nothing is known about the spawning and breeding habits of this fish. A large specimen taken in Lynnhaven Roads on May 17, 1921, had the ovaries somewhat developed, and they contained numerous eggs just distinguishable with the unaided eye. Eight fish caught at Ocean View, Va., on October 4 to 22, 1922, were 151 to 227 millimeters (6 to 9 inches) in length.

This foolfish is reported to attain a length of 2 feet. Although common within most of its range, it is not taken in large numbers and has no commercial value. It is called foolfish probably in part because of its awkward appearance and actions, and also because of its stupidity in escaping from a net. It often remains in a net when avenues for escape are plainly evident and when all other fish have left it. When removed from a net and placed in open water it usually remains quiet for some time before apparently knowing that it can swim away at will. Its rough skin, no doubt, has given rise to the name "filefish."

Habitat.—Portland, Me., to Brazil; uncommon north of Cape Cod.

Chesapeake localities.—(a) Previous records: St. Marys River, Md.; "southern part of Chesapeake Bay" (Lugger, 1877); Cape Charles and Hampton Roads, Va. (b) Specimens in collection: Cape Charles, Ocean View, and Lynnhaven Roads, Va.

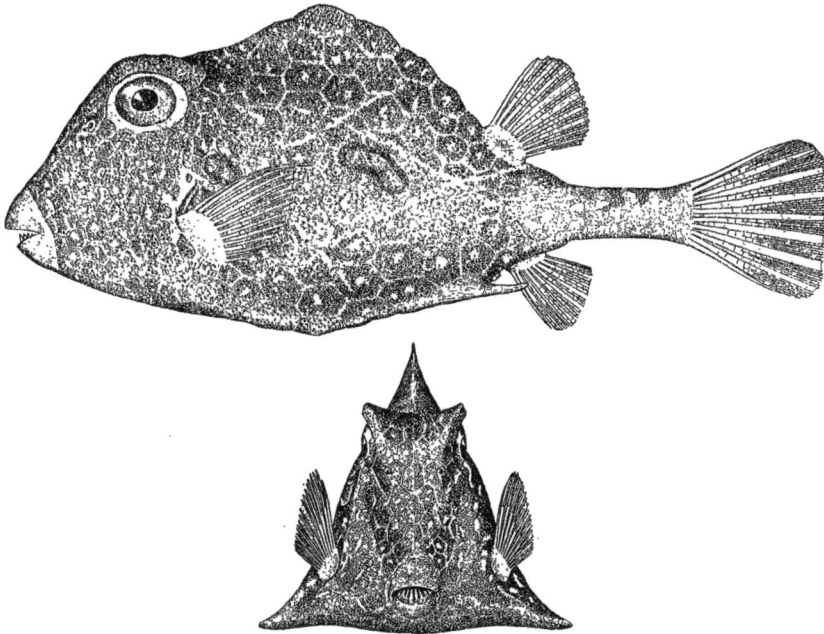

Fig. 207.—*Lactophrys trigonus*

Family LXXXIX.—OSTRACIIDÆ. The trunkfishes

Body short, cuboid, three, four, or five angled, covered by a hard, boxlike shell composed of polygonal plates, these wanting only on caudal peduncle, about the mouth and the bases of the fins; mouth small, terminal; a single row of pointed teeth in each jaw; gill opening consisting of a more or less vertical slit, situated below and posterior to eye; dorsal fin small, composed of soft rays only, placed far backward; caudal fin square or round; anal fin similar to the dorsal and nearly opposite it; ventral fins wanting; pectoral fins short.

153. Genus LACTOPHRYS Swainson. Three-angled trunkfishes

Body, in adult at least, three angled; ventral surface flat or concave; carapace closed behind anal fin; frontal and lateral spines present or wanting; dorsal fin with 9 or 10 rays. A single species of this genus is known to occur rarely in Chesapeake Bay.

194. Lactophrys trigonus (Linnæus).   Trunkfish; Shellfish; Boxfish.

Ostracion trigonus Linnæus, Syst. Nat., ed. X, 1758, p. 330; "India."
Lactophrys trigonus Uhler and Lugger, ed. I, p. 88; ed. II, p. 74; Jordan and Evermann, 1898, p. 1723, Pl. CCLXIII, fig. 641.

A single very small individual, described as follows, is at hand.   In this species the young differ greatly from the adult.

Head 2; depth 1.3; D. 9; A. 9.   Body more or less four-angled, with a prominent median ridge on back; the carapace open behind dorsal and without spines; head very deep; snout not much in advance of forehead, 3.2 in head; eye 4; interorbital 2.3; mouth very small, terminal; dorsal and anal fins similar; caudal fin round; pectoral fins broad, the upper rays longest, 3.5 in head.

Color in alcohol uniform brownish; fins plain translucent.

A single small specimen 20 millimeters in length was secured.   This is the only trunkfish known from Chesapeake Bay, and it is readily recognized by the hard, boxlike shell that covers the body.

The young are very different from the adult.   In large individuals the body is sharply three-angled, and the ventral ridge of the shell, somewhat in advance of the vent, bears a large, flat spine. This spine is undeveloped until the fish reaches a length of about 35 millimeters.   The body in the adult is more elongate and not as deep, the depth being contained in the length from 2.6 to 2.8 times.   The head in the adult is contained 3.8 times in the body.

This fish reaches a maximum length of about 1 foot.   Its flesh is of excellent flavor, and at Key West, Fla., and Colon, Panama, at least, it is used as food.

Habitat.—Woods Hole, Mass., south to Bahia, Brazil; apparently uncommon north of Florida. Only the young appear to have been taken north of Florida.

Chesapeake localities.—(*a*) Previous records: "Occurring very rarely in the salt waters of the southern part of Chesapeake Bay and around the extremity of St. Marys County" (Uhler and Lugger, 1876).   (*b*) Specimen in collection: Cape Charles, Va.; evidently very rare in Chesapeake Bay.

Family XC.—TETRAODONTIDÆ.   The swellfishes

Body oblong or elongate, usually about as broad as deep; belly usually capable of great inflation with water or air, or both; mouth small, terminal; teeth in the jaws fused, forming a continuous cutting edge, except for a median suture; gill slits small, situated in front of pectorals; scales present or absent; the skin often bearing prickles; lateral line conspicuous or not; dorsal fin inserted posteriorly, composed of soft rays only; caudal fin various in shape; anal fin similar to the dorsal and usually opposite it; ventral fins wanting; pectoral fins short and broad.

The members of this family mostly inhabit warm shore waters.   They are sluggish swimmers but find a measure of protection in their tough and often prickly skin and by greatly increasing their size, through inflation, which also causes the prickles to stand erect and to appear much more prominent than before inflation.   Some of the members of the family attain a considerable size, but none are of commercial importance, as their flesh is said to be rank and sometimes poisonous. Two genera and three species are known from Chesapeake Bay.

KEY TO THE GENERA

a. Body comparatively elongate; dorsal and anal fins rather long, each with 12 to 15 rays; skin largely smooth_____Lagocephalus, p. 347
aa. Body oblong, plump; dorsal and anal fins smaller, each with six to eight rays; skin largely prickly_____Tetraodon, p. 347

154. Genus LAGOCEPHALUS Swainson. Rabbit fishes or Puffers

Large puffers with smooth skin, except on the abdomen, where prickles are present; lower edge of caudal peduncle with a fold; dorsal and anal fins long, each with 12 to 15 rays; caudal fin concave behind. Two American species are known; only one of these comes within the scope of the present work.

195. Lagocephalus lævigatus (Linnæus). Puffer; Rabbit fish; Swellfish.

Tetraodon lævigatus Linnæus, Syst. Nat., ed. XII, 1766, p. 416; Charleston, S. C. Uhler and Lugger, 1876, ed. I, p. 87; ed. II, p. 73.

Lagocephalus lævigatus Jordan and Evermann, 1896–1900, p. 1728, Pl. CCLXII, fig. 642.

Head 3 to 3.75; depth about 3 to 4; D. 14; A. 13. Body elongate, somewhat deeper than broad; head rather long; snout conical, its length 1.7 to 2.75 in head; eye 4.15 to 4.8; interorbital 2.05 to 2.75; mouth small, nearly terminal; teeth in continuous plates, with a median suture, more or less beaklike; lateral line present, branched anteriorly; lower edge of body with a longitudinal fold or keel; abdomen with short spines; skin elsewhere smooth; dorsal and anal similar, the origin of dorsal somewhat in advance of that of anal; caudal fin deeply concave; pectoral fins short and broad, 1.55 to 1.95 in head.

Color dark greenish to dusky above; sides bright silvery; white underneath; gill opening black within; dorsal and caudal mostly dusky; anal and pectorals greenish to slightly dusky.

Only two specimens, 185 and 570 millimeters (7⅜ and 22½ inches) long, were preserved. Another large individual, 625 millimeters (24¾ inches) in length, was examined in the field. The

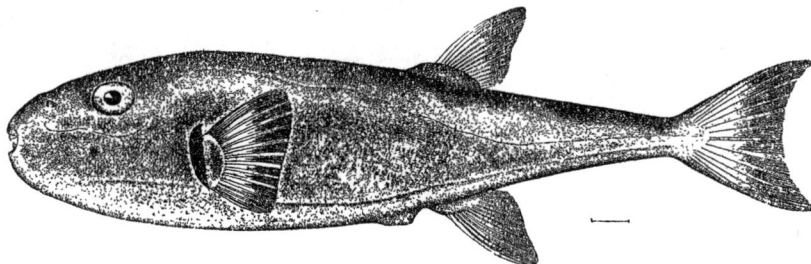

FIG. 208.—*Lagocephalus lævigatus*

foregoing description is based upon these three specimens, the only ones seen in Chesapeake Bay during the present investigation. This puffer is readily distinguished from the others known from the Chesapeake by the smooth, shining skin on the sides, the large dorsal and anal fins, and the concave tail fin.

This is the largest of the American puffers. It reaches a length of fully 2 feet. Its feeding and breeding habits are still virtually unknown. One small specimen, 7⅜ inches long, was taken on September 26. This puffer is rare in Chesapeake Bay, where it is known as "rabbit fish," because of the rabbitlike eyes. It nowhere has economic value.

Habitat.—Massachusetts to Brazil, uncommon north of Cape Hatteras.

Chesapeake localities.—(*a*) Previous record: Southern part of Chesapeake Bay (Uhler and Lugger, 1876). (*b*) Specimens in collection: Lynnhaven Roads, Va., taken in pound nets June 9 and September 26, 1921. Another specimen observed at Ocean View, Va., was taken in a haul seine on October 20, 1922.

155. Genus TETRAODON Linnæus. Puffers; Swellfishes

Body oblong, plump, capable of considerable inflation; nasal canal single, with two openings near tip; skin often smooth, sometimes with more or less distinct scalelike dermal development, often also with prickles, at least on back and abdomen, and not infrequently with dermal cirri; dorsal and anal fins similar, small, each consisting of six to eight rays; caudal fin usually straight or convex, rarely slightly concave. Two species of this genus have been recorded from Chesapeake Bay.

KEY TO THE SPECIES

a. Head rather broad; interorbital 2.2 to 2.85 in head; no white lines or reticulations on back and sides; a lateral series of vertically elongate dark spots_____*maculatus*, p. 348

aa. Head narrower; interorbital 4.6 to 8 in head; back and sides with white lines, forming reticulations; no definite series of lateral black spots_____*testudineus*, p. 349

196. Tetraodon maculatus Bloch and Schneider. Puffer; Swellfish; Swell toad; Balloonfish.

Tetrodon hispidus var. *maculatus* Bloch and Schneider, Syst. Ichth., 1801, p. 504; Long Island, N. Y.
Chilichthys turgidus Uhler and Lugger, 1876, ed. I, p. 88; ed. II, p. 73.
Tetrodon turgidus Bean, 1891, p. 84.
Spheroides maculatus Jordan and Evermann, 1896–1900, p. 1733, Pl. CCLXIV, fig. 645; Evermann and Hildebrand, 1910, p. 162.

Head 2.05 to 2.85; depth about 2 to 3; D. 8; A. 7. Body robust, about as broad as deep; head long; snout conical, 1.7 to 2 in head; eye small, 3.6 to 7.75; interorbital space 2.2 to 2.85; mouth small, terminal; teeth in plates, with a median suture, more or less beaklike; lateral line very feebly developed; no evident fold along lower edge of side; skin everywhere prickly, except on caudal peduncle; dorsal and anal fins similar, the dorsal somewhat in advance of anal; caudal fin with round margin; pectoral fins rather short and broad, 1.95 to 2.35 in head.

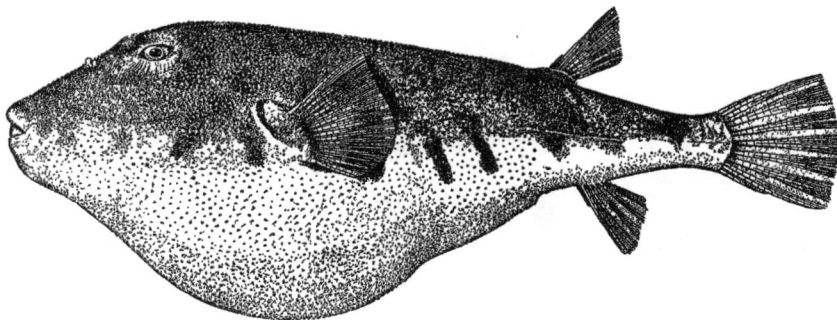

FIG. 209.— *Tetraodon maculatus*

Color in alcohol grayish to dark brown above; plain white underneath; most of the lighter colored specimens with spots darker than the ground color on the back; sides with a series of seven or eight vertically elongated dark spots from base of pectoral to base of caudal; frequently with one or two similar spots in advance of pectoral; many specimens with dark dots on side of head, these frequently extending back on side of body; fins all plain translucent. Color notes, taken in the field, of a living specimen 5 inches long, brownish above, with small green and black spots; white below; seven or eight irregular black bars along sides posterior to pectoral, and a black spot in front of pectoral; a dark brownish area at base of dorsal and another between dorsal and caudal; dorsal and anal plain; caudal yellow brown; pectoral pale yellowish brown.

Many specimens of this species, ranging in length from 25 to 260 millimeters (1 to 10¼ inches) were preserved. The young do not differ greatly from the adults. This puffer differs conspicuously from its relative, *T. testudineus*, in the absence of reticulating light lines on the back and sides and in the presence of a series of vertically elongate dark spots on the sides.

The food, according to 22 specimens examined, consists principally of small crustaceans, including crabs, shrimp, isopods, and amphipods. A few small mollusks, annelids, and traces of algæ also were present. Examinations for food made by Linton (1905, p. 402) at Beaufort, N. C., and Welsh and Breder (1922, p. 273) at Atlantic City, N. J., indicate that although other invertebrates enter into the food, small crustaceans predominate.

Ripe or nearly ripe fish were taken in Chesapeake Bay during May. Ripe males as small as 140 millimeters (5½ inches) were taken May 17 at Crisfield. Eggs apparently are produced in very large numbers. It is estimated that the ovaries of one specimen 265 millimeters (10½ inches) in length contained 176,000 eggs, all of uniform size. Welsh and Breder,[26] working at Atlantic City, N. J., took ripe females from July 30 to August 27. The eggs are described as transparent, spherical, demersal, and adhesive. In diameter they vary from 0.85 to 0.91 millimeter. Incubation occupies about 3 days and 10 hours at an average temperature of 67° F. The newly hatched larvæ are about 2.41 millimeters in length. By the time the young fish reach a length of 7.35 millimeters they already have many of the characters of the adult.

The two smallest puffers (each 1 inch in length) in the collection were seined on August 9 at Point Patience, lower Patuxent River. Sizes ranging from over 1 to nearly 4 inches are absent from our collection. In the fall large numbers of puffers from 4 to 10 inches long are caught in the lower parts of the bay, and these fish apparently are of various ages. One haul, made on October 2, consisted of 175 puffers, 117 to 163 millimeters (4½ to 6½ inches) long, the predominating lengths being 140 to 150 millimeters (5½ to 6 inches).

This puffer is taken in the Chesapeake from April to November, being most abundant in the catches during the spawning season in May and again in September and October. It is caught chiefly with hook and line, pound nets, and haul seines.

The species is said to attain a length of 14 inches but seldom exceeds 10 inches. It is quite abundant in the southern part of Chesapeake Bay, as well as elsewhere on the Atlantic coast from Cape Cod, Mass., southward, but it has no commercial value.

Habitat.—Portland, Me., to Florida; the only species of puffer abundant outside of the Tropics.

Chesapeake localities.—(a) Previous records: St. Marys River and St. Georges Island, Md., and Cape Charles city, Va. (b) Specimens in collection: From many localities from Love Point, Md., southward to the capes.

197. Tetraodon testudineus Linnæus. Puffer; Swellfish.

Tetraodon testudineus Linnæus, Syst. Nat., ed. X, 1758, p. 332. Type locality missing.
Chilichthys testudineus Lugger, 1877, p. 59.
Spheroides testudineus Jordan and Evermann, 1896-1900, p. 1734, Pl. CCLXV, figs. 646 and 646a.

Head 2.4 to 2.8; depth 3.15 to 4; D. 7 or 8; A. 6 or 7. Body robust; head rather broad; snout moderately long, 2 to 2.8 in head; eye 3.85 to 6.1; interorbital 4.6 to 8, more or less concave; skin with small prickles on back and somewhat larger ones on chest and abdomen; snout and tail smooth; no cirri; lateral line evident; dorsal and anal fins similar, the anal, however, somewhat smaller, its origin under posterior part of dorsal; caudal fin slightly convex; pectoral fins short and broad, 2 to 2.5 in head.

Color of back dark brown, broken on sides and becoming lighter; sides with black spots; the back and sides with narrow light lines forming reticulations; belly pale; caudal fin sometimes dusky; other fins pale greenish.

No specimens of this puffer were secured. The foregoing description is based on published accounts. The species is included because of a record by Lugger dating back to 1877. It has not been seen in the Chesapeake by other collectors. This fish is readily distinguished from its relative, *T. maculatus*, by the presence of light lines, forming reticulations on the back and sides, which *T. maculatus* does not possess.

This fish reaches a maximum length of about 10 inches. It has no commercial value.

Habitat.—Woods Hole, Mass., to Natal, Brazil; very common in the West Indies and on the Atlantic coast of Panama; very rare in the northern part of its range.

Chesapeake localities.—(a) Previous record: "A single fine specimen was obtained in the southern part of the Chesapeake Bay" (Lugger, 1877). (b) Specimens in collection: None; evidently extremely rare in Chesapeake Bay.

Family XCI.—DIODONTIDÆ. The porcupine fishes

Body short, depressed above; belly moderately inflatable; mouth moderate; terminal; jaws with continuous bony plates, having no median suture; nostril usually with a short tube with two lateral openings; gill-openings consisting of more or less vertical slits, placed in front of pectorals; body almost everywhere with bony spines; dorsal and anal fins similar, placed posteriorly and mostly opposite each other; caudal fin round; ventral fins wanting; pectoral fins short and broad.

[26] For a detailed account of the embryology and larval development of this puffer see Welsh and Breder, Zoologica, Vol .II, No. 12, 1922, pp. 261 to 276, figs. 80 to 96. New York.

KEY TO THE GENERA

a. Body with slender, sharply pointed, movable spines_____ Diodon, p. 350

aa. Body with shorter, less sharply pointed, immovable spines_____ Chilomycterus, p. 350

156. Genus DIODON Linnæus. Porcupine fishes

Body robust, inflatable; head short, broad; teeth in both jaws with a continuous cutting edge, no median suture; nasal tube simple, with two lateral openings; body everywhere with strong, pungent spines; dorsal and anal similar, posteriorly inserted; caudal fin round; ventral fins wanting; pectoral fins broad, the posterior margin truncate, the upper lobe longest. A single widely distributed species was once recorded from Chesapeake Bay.

198. Diodon hystrix Linnæus. Porcupine fish.

Diodon hystrix Linnæus, Syst. Nat., ed. X, 1758, p. 335; India. Uhler and Lugger, 1876, ed. I, p. 85; ed. II, p. 71; Jordan and Evermann, 1896–1900, p. 1745, Pl. CCLXVI, fig. 648.

? Trichodiodon pilosus Uhler and Lugger, 1876, ed. I, p. 87, ed. II, p. 72.

Head 2.2 to 2.4; depth 2.45 to 3; D. 13 or 14; A. 12 or 13. Body robust, rather broader than deep when not inflated; head depressed, notably broader than deep; snout very short, 2.8 to 3.1 in head; eye 2.85 to 3.45; interorbital very broad concave, 1.4 to 1.5; mouth rather broad; gill slit not longer than eye; skin thickly beset with strong sharp spines, longest on top of head, back, and sides; dorsal and anal fins similar, placed far back, the dorsal beginning somewhat in advance of the anal; caudal fin round; pectoral fins broad, the upper rays longest, 1.65 to 2.33 in head.

Color dusky above, white below, entire body and fins marked by small round spots.

This porcupine fish was not seen during the present investigation. It is included because of a record dating back to 1876 (Uhler and Lugger, 1876). The fish has not been seen in Chesapeake Bay by more recent investigators. The foregoing description is based upon published accounts of the species. The porcupine fish is readily recognized by the long, sharply pointed, movable spines, which cover nearly the entire body.

Although this fish has been known to science for several centuries and from many parts of the world, we are unable to find information relative to its feeding and breeding habits. The species is said to reach a length of 3 feet. It is regarded as a curiosity by many and is rather extensively sought by travelers. Mounted specimens are common in nearly all museums. The species apparently has no value as food.

Habitat.—All warm seas; northward on the American coast to Massachusetts; uncommon north of Florida.

Chesapeake localities.—(*a*) Previous record: In Chesapeake Bay off the southern extremity of St. Mary's County, Md. (Uhler and Lugger, 1876). (*b*) Specimens in collection: None; apparently very unusual in Chesapeake Bay.

157. Genus CHILOMYCTERUS Bibron. Bur fishes

Body broad, depressed, more or less inflatable; nasal tube simple, with two lateral openings; teeth in the jaws without median suture; dermal spines with three roots, immovable, triangular; caudal peduncle short. A single species is known from Chesapeake Bay.

199. Chilomycterus schœpfi (Walbaum). Bur fish; Spiny toadfish; Thorny toad.

Diodon schœpfi Walbaum, Artedi Pisc., 1792, p. 601; Long Island, N. Y.

Chilomycterus geometricus Uhler and Lugger, 1876, ed. I, p. 86; ed. II, p. 72.

Chilomycterus sp. Bean, 1891, p. 83.

Chilomycterus schœpfi Jordan and Evermann, 1896–1900, p. 1748, Pl. CCLXVI, fig. 649; Evermann and Hildebrand, 1910, p. 162.

Head 2.1 to 2.75; depth about 3 or 4 (when not inflated); D. 11 or 12; A. 10. Body robust, somewhat broader than deep; head short, broad; snout very short and broad, 2.15 to 2.35 in head; eye 3.8 to 4.5; interorbital broad, concave, 1.45 to 1.75; mouth small, terminal; teeth in a continuous plate in each jaw; scales wanting; skin everywhere, except on caudal peduncle, with more or less compressed three-rooted spines; two spines over orbit; one on middle of forehead;

dorsal and anal similar, placed posteriorly and opposite each other; caudal fin with round margin; pectoral fins short and broad, 1.85 to 2.35 in head.

Color pale to yellowish green; orange to white underneath (sometimes brownish to blackish n young); upper parts with 12 to 16 dark brown or blackish longitudinal stripes; a large, black ocellus above pectoral and another behind it and still another on side of base of dorsal; side below dorsal sometimes with a black blotch; a small specimen (90 millimeters) with roundish black spots on lower part of side; fins pale green to yellowish or pale orange.

This species is represented by 11 specimens, ranging in length from 90 to 225 millimeters (3½ to 9 inches). This spiny puffer differs from the only other spiny puffer (the porpupine fish) recorded from Chesapeake Bay, in having proportionately shorter, blunter spines, which are immovable. The spines of its relative are long, sharply pointed, and movable.

The food in the stomachs of six specimens of this spiny puffer, which were examined, consisted wholly of hermit crabs. Small hermit crabs were swallowed inclosed in the shells that they occupied, and in case of larger ones the shells apparently were first broken. In one specimen 30 undigested hermit crabs, with their shells, were present. Smith (1907, p. 351) says: "The strong, bony beak enables the fish to crush and eat mollusks and crustaceans, which are its principal food."

Nothing definite is known about the spawning of this fish, nor of the rate of growth. Some of the specimens taken at Ocean View, Va., during October, 1922, had nearly mature gonads. A maximum length of about 10 inches is attained.

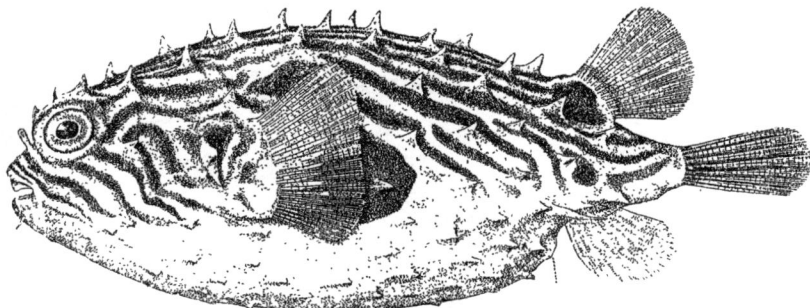

FIG. 210.— *Chilomycterus schœpfi*

The species was found during the present investigation only in the lower part of the bay, where it is rather uncommon. At Ocean View, Va., in 1,800-foot haul seines, 21 fish were caught from October 2 to October 23, 1922. Fishermen in this region said that occasionally as many as a dozen are caught in one day. All the fish seen were caught in May and October. The species has no commercial value.

Habitat.—Massachusetts Bay to Florida; rather common from Chesapeake Bay southward.

Chesapeake localities.—(a) Previous records: "Not uncommon along the coast (Maryland), entering bays" (Uhler and Lugger, 1876), and Cape Charles city, Va. (b) Specimens in collection: Mobjack Bay, Cape Charles, Lynnhaven Roads, and Ocean View, Va. Uncommon. Specimens at hand were taken with a beam trawl by the *Fish Hawk* and in pound nets and haul seines.

Order PEDICULATI

Family XCII.—LOPHIIDÆ. The anglers

Body broad anteriorly, diminishing rapidly in size from the shoulders backward; head large, very broad, depressed; mouth excessively large and broad; jaws, vomer, and palatines with bands of sharp teeth of uneven size; gills 3, the opening large, placed in lower axil of pectoral; gill rakers wanting; pseudobranchiæ present; scales wanting; head and sides with prominent dermal flaps; dorsal fins 2, widely separated, the spinous portion consisting of three separate tentaclelike spines on the head and three smaller ones connected by membrane; soft dorsal forming a normal fin; anal similar to soft dorsal; pectorals large and fleshy; ventrals jugular, far apart, with I, 5 rays.

49826—28——23

158. Genus LOPHIUS Linnæus. Anglers

Body anteriorly very broad; mouth exceedingly wide, superior; lower jaw much in advance of the upper; upper jaw protractile; first dorsal spine expanded at tip, overhanging the mouth, and forming a lure or bait for prey; gill openings below and behind the pectorals; size large; vertebræ 27 to 32.

200. Lophius piscatorius Linnæus. All-mouth; Angler; Goosefish.

Lophius piscatorius Linnæus, Syst. Nat., ed. X, 1758, p. 236; seas of Europe. Jordan and Evermann, 1896–1900, p. 2713, Pl. CCCLXXXVIII, fig. 952.

"Head as wide as long, and longer than body; eyes small, separated by a space about equal to snout; head very spinous in young, becoming less so with age; skin smooth; the head surrounded by a fringe of short dermal flaps, similar flaps on sides of body; a three-pointed humeral spine; dorsal rays III+III+10, the anterior spine with an expanded tip; anal rays 9, caudal margin straight; pectorals rounded, their bases constricted.

"Color above mottled brown, below white; caudal and pectorals black-edged." (Smith, 1907, p. 399.)

Only larvæ with yolk sac attached (probably just hatched) and a few others, taken at the same time and place but kept alive until the yolk was absorbed, are at hand. A single adult, 1,060

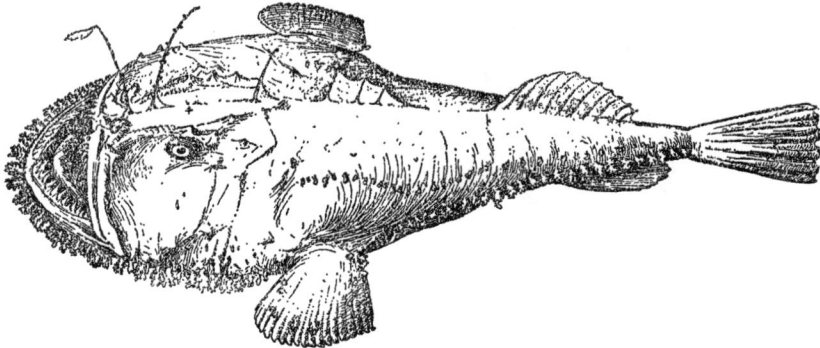

FIG. 211.—*Lophius piscatorius*

millimeters (41¾ inches) in length, was actually observed. This fish was regarded as too large for preservation. The following measurements, given in millimeters, were taken: Total length, 1,060; standard length, 835; distance between bases of pectoral fins, 780; width of mouth, 297; interorbital space, 100; distance from tip of snout to eye, 106; length of the first dorsal spine with "bait," 88; length of the second spine, 210. The color of upper parts was variegated, principally light and dark brown, lower parts were white to dusky. The all-mouth is easily recognized by its large size, smooth skin, broad, flat head, and the enormous mouth, to which the common name has reference.

The food consists of fish, crustaceans, water birds, and, in fact, any animal of suitable size. It is of record that a goosefish may contain food at one time half as heavy as the fish itself. One of us (Schroeder) observed a large goosefish swimming at the surface on Nantucket Shoals, Mass., August 23, 1925. This fish was easily captured and in its stomach was found a haddock 31 inches in length, weighing about 12 pounds. This meal was so large that the fish apparently was unable to leave the surface of the water. It has also been observed that this species uses its tag or "bait" (at the tip of the first dorsal spine) to lure fish close to its large mouth, which are then easily engulfed. For a complete account of the food eaten and of the insatiable appetite of this fish, see Bigelow and Welsh (1925, pp. 526 to 528).

The spawning period occurs during the summer and is reported to last a long time. Recently hatched young, with yolk sacs attached, were taken in the mouth of the bay by the *Fish Hawk* on June 10, 1916. The eggs float near the surface and are inclosed in a gelatinous substance that forms a sheet or veil often 20 to 30 feet long and 2 or 3 feet broad. It is supposed that each sheet is the product of a single female, and it has been estimated that one ovary may contain considerably more than 1,000,000 eggs. The eggs are spherical or slightly oval and 2.13 to 2.5 millimeters in diameter. The length of the period of incubation is not known. The recently hatched larvæ that are at hand (preserved in alcohol) are only about 3 millimeters long. Six days later larvæ of this same collection and age had dissolved the yolk sac and were 5 millimeters long (after preservation). These larvæ show a dorsal fin ray or spine in the finfold at the nape; the ventral fins, too, are present as a long membranous fold. Considerable dark pigment is present on the head, and three dark areas, about evenly spaced, appear on the axis of the body. For an extended account and illustrations of the embryology and larval development the reader is referred to Bigelow and Welsh (1925, pp. 528 to 532).

This fish is reported to reach a maximum length of 4 feet and a weight as great as 70 pounds. The single adult observed in Chesapeake Bay during the present investigation, as already stated, was nearly 42 inches long. The goosefish is regularly marketed in northern Europe, and its meat is reported to be white and of good flavor. It probably finds no sale in this country because of its repulsive appearance. Although previously not recorded from Chesapeake Bay, the species is not especially rare there. According to the fishermen of the southern sections of the bay, a few are taken each year in pound nets in late fall and early spring. It is probable that it occurs in this part of the bay in limited numbers throughout the winter. However, as no fishing is done there in the winter this can not be stated definitely.

Habitat.—Both coasts of the northern Atlantic; in shallow water on the American coast, from Newfoundland Banks and the Gulf of St. Lawrence to North Carolina, and in deep water as far south as the Barbadoes.

Chesapeake localities.—(a) Previous records: None. (b) Specimens in collection: From off Cape Charles, Va.; specimen observed in Lynnhaven Roads, Va.; also reliably reported by pound-net fishermen at Buckroe Beach, Va.

Family XCIII.—ANTENNARIIDÆ. The frogfishes

Body and head compressed; mouth large, vertical or very oblique; premaxillaries protractile; lower jaw projecting; teeth in the jaws in villiform bands; gill arches 2.5 or 3; gill openings very small, near the lower axil of pectorals; pseudobranchiæ wanting; spinous dorsal consisting of one to three detached, tentaclelike spines; soft dorsal long and high; anal similar but smaller; pectorals large; ventrals jugular, close together. This family is composed of small pelagic fishes, chiefly of the tropics, usually living among floating seaweed and becoming widely scattered by winds and currents.

159. Genus HISTRIO Fischer. Sargassum fishes; Mouse fishes

Body short, somewhat compressed; mouth small, oblique; palatine teeth present; skin smooth or with minute tubercules and with dermal tentacles; soft dorsal preceded by three spines, the first spine slender and expanded at tip, forming a lure or bait; ventral fins well developed, rather long; wrist and pectoral fins slender. This is a group of oddly shaped fishes that live in dense vegetation, to which the individuals attach themselves by means of their handlike pectorals. Two American species have been recognized; one of these occasionally drifts northward along the coast.

201. Histrio histrio (Linnæus). Sargassum fish; Mouse fish.

Lophius histrio Linnæus, Syst. Nat., ed. X, 1758, p. 237; open sea.
Pterophryne lævigata Uhler and Lugger, 1876, ed. I, p. 93; ed. II, p. 77.
Pterophryne histrio Jordan and Evermann, 1896–1900, p. 2716.

"Head 2¼; depth 1⅘; D. III–14; A. 7; V. 5. Skin of head and body as well as dorsal fins with fleshy tags, which are most numerous on the dorsal spines and abdomen. Wrist slender; ventrals large, nearly one-half as long as head. Dorsal and anal with the posterior rays not adnate to caudal peduncle; first dorsal spine bifurcate at tip. Yellowish, marbled with brown; three dark bands radiating from eye; vertical fins barred with brown; belly and sides with small white spots." (Jordan and Evermann, 1896–1900.)

No specimens of this species are at hand, nor was it observed by collectors or reported by fishermen during the present investigation. This account is included because of a record by Uhler and Lugger (1876), who mention a mouse fish, under the name *Pterophryne lævigata*, as occurring in the "oyster regions of Chesapeake Bay." It is difficult to understand what connection this pelagic fish could have with oyster regions. It probably was taken in such areas quite by accident. The mouse fish is recognized by its naked body, small oblique mouth, projecting lower jaw, hand-like pectoral fins, and its yellowish color, which is marbled or blotched with brown.

The habits of the sargassum fish are not well known. It is usually found among floating seaweed, to which it attaches itself by means of its handlike pectorals. In this way it is drifted far and wide by winds and currents. In the aquarium it is cannibalistic, attacking its fellows, biting off their fleshy appendages and swallowing its smaller companions. It is probably safe to conclude from such a display of voracity that it is naturally carnivorous in its habits.

The following account relative to the spawning of this fish is quoted from Smith (1907, p. 400). Nothing, so far as we are aware, has been added to our knowledge of this fish since this was written.

Our knowledge of the spawning habits and eggs of the species depends almost entirely on observations at the Government laboratories at Woods Hole and Beaufort. The spawning season is from July to October, and a number of captive specimens have laid their curious egg rafts while in aquaria. The eggs are deposited in a bandlike or ribbonlike mass, from 1.5 to more than 3 feet long, about 3 inches wide, and 0.25 inch thick; they are only one-fortieth inch in diameter and very numerous, and are held together by a transparent jelly, which is buoyant. Nothing is known about the embryology, as eggs have not been fertilized. On July 25, 1903, a fish 3.5 inches long, which had been at the Beaufort laboratory for seven weeks, laid a mass of eggs three times as large as the fish.

The maximum size attained by this fish is reported to be 6 inches. It is not taken in large numbers anywhere along the coast, and because of its peculiar shape, handlike pectoral fins, and bright color it is usually regarded as a curiosity. It has no commercial value. In Chesapeake Bay it evidently is extremely rare, and it does not occur there except as it may rarely be drifted into this water by favorable winds and currents.

Habitat.—Tropical Atlantic; occasionally drifted northward, probably in the Gulf Stream, on the coast of America as far as Woods Hole, Mass.

Chesapeake localities.—(*a*) Previous record: "Occurs in the oyster regions of Chesapeake Bay, but is perhaps quite uncommon" (Uhler and Lugger, 1876). (*b*) Specimens in collection: None.

Family XCIV.—OGCOCEPHALIDÆ   The batfishes

Body depressed, the trunk short and slender; head very broad, much depressed; snout more or less elevated, usually projecting; mouth not large, usually inferior, the lower jaw included; teeth pointed; gill openings small, above and behind axils of pectoral fins; skin covered with bony tubercles or spines; a rostral tentacle, retractile into a cavity under the rostral process, usually present; dorsal and anal small; ventrals present; pectorals well developed, with strongly angled base. Peculiarly shaped fishes, most of them apparently sluggish in their movements. Some of the species live along the shores in very shallow water and others inhabit the deep sea.

160. Genus OGCOCEPHALUS Fischer.   Batfishes

Body depressed; head broad, triangular or more or less disklike in form, not broader than long; snout provided with rostral projection, varying greatly in length; eyes rather large, lateral; mouth moderate, inferior; teeth in villiform bands on jaws, vomer, and palatines; gill opening small near inner axil of pectoral; gills 2½; skin rough, with bony tubercles; a dermal tentacle present under the rostral process, retractile into a well-developed cavity; ventrals I, 5, well separated; pectorals large, placed horizontally.

202. Ogcocephalus vespertilio (Linnæus). Batfishes.

Lophius vespertilio Linnæus, Syst. Nat., ed. X, 1758, p. 236; American seas.
Malthe vespertilio Uhler and Lugger, 1876, ed. I, p. 92; ed. II, p. 77.
Ogcocephalus vespertilio Jordan and Evermann, 1896–1900, p. 2737, Pl. CCCXCII, figs. 958, 958a, 958b.

"Head to gill opening 1.93; depth 5; D. 4; A. 4.

"Body rather robust, tapering backward, the caudal peduncle broader than deep; head broad, depressed, triangular, the greatest width equal to distance from tip of rostral process to inner angle of wrist of pectoral; snout acute, with long pointed process, 5.7 in head, 11 in body; eye wholly lateral, 7.4 in head; interorbital 9.15; mouth rather broad, mostly transverse; maxillary 6.7 in head; teeth small, villiform, in bands on jaws, vomer, and palatines; gill opening small, situated at inner angle of base of pectoral; skin with bony protuberances, varying in size, smallest on belly where the skin is shagreenlike, largest on back of tail, ventral surface of tail with bony plates; a large depression, longer than broad, under rostrum; this depression provided with a dermal tentacle, which has a stocklike base and a more or less definite triangular expansion at tip; dorsal fin small, its origin at vertical from posterior margin of wrist of pectoral; caudal fin round; anal fin small, its origin about equidistant from vertical of origin of dorsal and base of caudal; ventral fins rather long and narrow, inserted about midway between mouth and vent; pectoral fins with distinct wrist, inserted on posterior margin of the disklike head, the fin without wrist, 2.2 in head.

"Color very dark brown above, somewhat lighter brown below. There is a black area on snout below rostral process and two black areas over disk at shoulders; the distal part of the spiny processes on body paler than the ground color; a series of short dermal flaps on upper jaw and on lower margin of disk pale. Dorsal, caudal, anal, and pectorals very dark brown to nearly black, the caudal with greenish-yellow crossbar on middle of fin; ventral fins greenish." (Meek and Hildebrand, 1923–1927, p. 1018.)

This fish was not seen or taken during the present investigation. It is included in the present work on the general record of Uhler and Lugger (1876), who state that it is "rare in the southern part of Chesapeake Bay." The batfish is very readily recognized by its broad body, which has broad, winglike expansions anteriorly, to which the pectoral fins are attached. These winglike expansions give it an appearance somewhat resembling a bat, and this similarity gives origin to the name "batfish." Other peculiarities of this fish are the long-pointed snout and the rough skin beset with bony protuberances of various sizes.

Little is known of the habits of this fish. Its shape, of course, suggests very strongly that it lives and feeds on the bottom. Among the Florida Keys we have observed it rather frequently lying on the bottom, usually among marine growths, in very shallow water along the immediate shores. It swims sluggishly and frequently can be pursued and captured with a dip net.

The maximum size attained by the batfish is given in current works as 12 inches. The species is nowhere of commercial value.

Habitat.—Shallow waters of the Atlantic coast of tropical America, ranging from North Carolina (probably to Chesapeake Bay) to Brazil; rare northward of the Florida Keys.

Chesapeake localities.—(a) Previous record: "Rare in the southern part of Chesapeake Bay" (Uhler and Lugger, 1876). (b) Specimens in collection: None; not known north of Beaufort, N. C., except from the record of Uhler and Lugger.

GLOSSARY OF TECHNICAL TERMS

Abdomen.—The belly; the cavity containing the digestive and reproductive organs.
Adipose fin.—A peculiar fleshy fin, without rays but occasionally with a spine, occurring on the back behind the dorsal fin of most catfishes, salmons, etc.
Air bladder.—A sac filled with gas, lying beneath the backbone and in or behind the abdominal cavity; also known as swim bladder.
Anal.—Pertaining to the anus or vent.
Anal fin.—The fin on the median line behind the vent.
Antrorse.—Turned forward.
Anus.—The external opening of the intestine; the vent.
Articulate.—Jointed; said of soft fin rays.

Barbel.—An elongate, fleshy projection, usually about the head; also called whiskers; present in most catfishes.

Bicuspid.—Having two points.

Branchiæ.—The gills.

Branchiostegals.—Slender bones forming the support for the branchiostegal membranes, lying under the head and below the opercular bones.

Canines.—Long, conical teeth.

Cardiform.—Coarse, sharp teeth in the jaws of fishes.

Carinate.—Keeled; having a single ridge along median line.

Catadromous.—Running down; said of fish that descend to the sea to spawn.

Caudal.—Pertaining to the tail.

Caudal fin.—The fin on the tail.

Caudal peduncle.—The region between the anal and caudal fins.

Cephalic.—Pertaining to the head; as cephalic fins, meaning fins on the head, as in some of the rays.

Cirri.—Fringes.

Claspers.—Organs attached to the ventral fins of male sharks and skates.

Cœcum (plural cœca).—An appendage in the form of a blind sac, connected with the posterior end of the stomach or pylorus.

Compressed.—Flattened from side to side.

Ctenoid.—Rough-edged; said of scales when the posterior margin is spinous or pectinate.

Cycloid.—Smooth-edged; said of scales when the posterior margin is not rough; scales showing concentric lines or striations.

Deciduous.—Falling away or out.

Decurved.—Curved downward.

Depressed.—Flattened vertically.

Distal.—Remote from the point of attachment.

Dorsal.—Pertaining to the back.

Dorsal fin.—The fin on the median line of the back.

Emarginate.—Slightly notched at the end.

Falcate.—Scythe-shaped; long, narrow, and curved.

Filament.—Any slender, threadlike structure.

Filiform.—Thread form.

Fontanel.—An opening between the bones of the skull.

Foramen.—A hole or opening.

Frontal bone.—Anterior bone on top of head, usually paired.

Fusiform.—Spindle-shaped; tapering toward both ends.

Ganoid.—A group of fishes characterized by having the body more or less completely covered with bony platelike scales.

Gape.—Opening of the mouth.

Gill arches.—The bony arches to which the gills are attached.

Gill openings.—Openings reaching to or from the gills.

Gill rakers.—A series of bony projections placed along the inner edge of the gill arch.

Gills.—Organs for breathing the air contained in water.

Hæmal spine.—The lowermost projection of a caudal vertebra.

Heterocercal.—Term applied to the tails of fishes when vertically unequal, the backbone being deflected upward, as in the sharks.

Homocercal.—Term applied to the tails of fishes when equal, the backbone extending to the middle of base of caudal, as in most common fishes.

Imbricate.—Overlapping; said of scales that overlap like shingles in a roof.

Incisors.—Cutting teeth, usually in front of jaws.

Interorbital.—Space between the orbits or eyes.

Isthmus.—The region between the lower part of the gill openings.

Jugular.—Pertaining to the throat; said of ventral fins when attached in advance of the pectorals.

Keeled.—Having a ridge along the median line.

Lamellæ.—Thin plates or layers.

Lateral.—Referring to the side.

Lateral line.—A series of mucus pores along the side of the fish and containing sense organs; often appearing either as a colored or white stripe.

Mandible.—The lower jaw.

Marbled.—Variegated; clouded.

Maxillaries.—The outermost bones of the upper jaw, joined to the premaxillaries in front, and usually extending farther back than the latter.

Molar.—A broad grinding tooth.

Nape.—The back of the neck.

Nares.—Nostrils.

Nasal.—Pertaining to the nose.

Nuchal.—Referring to the nape.

Obtuse.—Blunt.

Occipital.—Relating to the occiput.

Occiput.—The back of the head.

Ocellated.—Having an ocellus or ocelli.

Ocellus.—An eyelike spot; a dark spot with a lighter border.

Opercle.—The thin, flat bone, one on each side of head, covering the gills; also called gill covers.

Orbit.—Eye socket.

Osseous.—Bony.

Oviparous.—Reproducing by means of eggs laid and hatched outside of the body.

Ovum (plural ova).—Egg.

Palate.—The roof of the mouth.

Palatines.—Bones of the roof of the mouth, one on each side of the vomer, often provided with teeth.

Papilla.—A small fleshy projection.

Papillose.—Covered with papillæ.

Parietal.—Bone of the side of the head.

Pectinate.—Having teeth like a comb.

Pectoral.—Pertaining to the breast.

Pectoral fins.—The anterior or uppermost paired fins, corresponding to the anterior limbs of the higher vertebrates.

Peritoneum.—The membrane lining the abdominal cavity.

Pharyngeal bones.—Bones behind the gills and at the beginning of the œsophagus, usually provided with teeth.

Plicate.—Folded; showing folds or wrinkles.

Plumbeous.—Lead colored; dull bluish gray.

Postorbital.—Behind the eye.

Premaxillaries.—The bones, one on each side, forming the front of the upper jaw, usually bearing most of the upper teeth.

Preopercle.—A thin bone lying just in front of the opercle.

Preorbital.—The bone lying just in front of the eyes.

Protractile.—Capable of being drawn forward.

Pseudobranchiæ.—Small gills developed on the inner side of the opercle.

Punctate.—Dotted with fine points.

Ray.—One of the bony or cartilaginous supports of a fin. Rays are either spiny or soft, the latter are either simple or branched.

Recurved.—Turned backward or toward the point of origin.

Reticulate.—Marked with a network of lines.

Retrorse.—Turned backward.

Rugose.—Rough, wrinkled.

Scute.—An external bony or horny plate.

Serrate.—Notched like the edge of a saw.

Setiform.—Having the form of a bristle.

Snout.—The portion of the head which projects beyond the eyes.

Spiracles.—Respiratory opening in the sharks and rays, corresponding to the nostrils in ordinary fishes.

Spinous.—Stiff or composed of spines.

Striate.—Striped or streaked.

Suborbital.—The bone immediately below the eye.

Supplemental maxillary.—A small bone, placed superficially on the upper part of the maxillary in many fishes.

Suture.—The line of union of two bones, as in the skull.

Symphysis.—The tip of chin; point of juncture of the two bones of lower jaw.

Synonymy.—A list of technical names applied to a certain genus or species.

Tail.—In ichthyology, the part posterior to the anal fin.

Temporal.—Referring to the region of the temples.

Thoracic.—Pertaining to the thorax; said of the ventral fins when attached beneath the pectorals.

Trenchant.—Compressed to a sharp edge.

Truncate.—With a square or straight margin.

Tubercle.—A small projection, like a pimple.

Type.—The particular specimen upon which the original description of the species was based or the species upon which was based the genus to which it belongs.

Type locality.—The particular place or locality at which the type was collected.

Vent.—The posterior opening of the alimentary canal.

Ventral.—Relating to the abdomen.

Ventral fins.—The paired fins behind, in front of or below, the pectoral fins, corresponding to the hind limbs in the higher vertebrates.

Ventral plates.—The plates lying on the belly.

Vertical fins.—The fins on the median line of the body; the dorsal, caudal, and anal fins.

Villiform.—Slender, minute teeth crowded into compact patches or bands.

Viviparous.—Bringing forth living young.

Vomer.—A bone in the center of the roof of the mouth, just behind the premaxillaries, often bearing teeth.

BIBLIOGRAPHY

AGASSIZ, ALEXANDER.
1882. On the young stages of some osseous fishes. Part III. Proceedings, American Academy of Arts and Sciences, new series, Vol. IX, whole series, Vol. XVII, June, 1881, to June, 1882 (1882), pp. 271–303, Pls. I–XX. Boston.

AGASSIZ, ALEXANDER, and C. O. WHITMAN.
1885. The development of osseous fishes. I. The pelagic stages of young fishes. Memoirs, Museum of Comparative Zoology, Vol. XIV, No. I, pt. 1, 1885, pp. 1–56, Pls. I–XIX. Cambridge.

ANDREWS, E. A.
1893. An undescribed acraniate, *Asymmetron lucayanum.* Studies from the Biological Laboratory, Johns Hopkins University, Vol. V, 1893, pp. 213–247, Pl. XIV, fig. 25. Baltimore.

ATKINS, C. G.
1887. The river fisheries of Maine. *In* The Fisheries and Fishery Industries of the United States, by George Brown Goode and associates, Section V, Vol. I, 1887, pp. 673–728. Washington.

BAIRD, SPENCER F.
1855. Report on fishes observed on the coasts of New Jersey and Long Island during the summer of 1854. Ninth Annual Report of the Smithsonian Institution, 1854 (1855), pp. 317–337. Washington.
1874. Report, Commissioner of Fish and Fisheries, 1872–73 (1874), pp. i–cii. [The shad, pp. xlviii–lix.] Washington.
1879. The carp (*Cyprinus carpio*). *In* Propagation of food fishes. Report, United States Commissioner of Fish and Fisheries, 1877 (1879), pp. *40–*44. Washington.

BEAN, BARTON A.
1891. Fishes collected by William P. Seal in Chesapeake Bay, at Cape Charles City, Virginia, September 16 to October 3, 1890. Proceedings, United States National Museum, Vol. XIV, 1891, pp. 83–94. Washington.
1907. A lump-fish from Chesapeake Bay. Forest and Stream, Vol. LXIX, July–December, 1907, pp. 178–179, 1 fig.

BEAN, TARLETON H.
1883. Notes on fishes observed at the head of Chesapeake Bay in the spring of 1882; and upon other species of the same region. Proceedings, United States National Museum, Vol. VI, 1883, pp. 365–367. Washington.
1888. Report on the fishes observed in Great Egg Harbor Bay, New Jersey, during the summer of 1887. Bulletin, United States Fish Commission, Vol. VII, 1887 (1889), pp. 129–154, Pls. I–III. Washington.
1901. The fishes of Long Island, with notes upon their distribution, common names, habits, and rate of growth. Sixth annual report of the Forest, Fish and Game Commissioner of the State of New York, 1900 (1901), pp. 375–478, 6 col. pls. Albany.
1903. Catalogue of the fishes of New York. New York State Museum, Bulletin 60, Zoology 9, 1903, 784 pp. Albany.

BIGELOW, HENRY B. and WILLIAM W. WELSH.
1925. Fishes of the Gulf of Maine. Bulletin, United States Bureau of Fisheries, Vol. XL, Pt. I, 1924 (1925), 567 pp., 278 figs. Washington.

CHIDESTER, F. E.
1920. The behavior of *Fundulus heteroclitus* on the salt marshes of New Jersey. The American Naturalist, Vol. LIV, No. 635, 1920, pp. 551–557. New York.

COLE, LEON J.
1905. The German carp in the United States. Report, United States Bureau of Fisheries, 1904 (1905), pp. 523–641, Pls. I–III, 4 figs. Washington.

COLES, RUSSELL J.
1910. Observations on the habits and distribution of certain fishes taken on the coast of North Carolina. Bulletin, American Museum of Natural History, Vol. XXVIII, 1910, pp. 337–348. New York.
1916. Is *Cynoscion nothus* an abnormal *regalis*? Copeia, No. 30, April 24, 1916, pp. 30–31. New York.

COPE, EDWARD D.
1889. Supplement on some new species of American and African fishes. Transactions, American Philosophical Society, Vol. XIII, Pt. III (1869), pp. 400–407. Philadelphia.

CRAWFORD, DONALD R.
1920. Notes on *Fundulus luciæ*. Aquatic Life, Vol. V, No. 7, July, 1920, pp. 75 and 76. Philadelphia.

DAY, FRANCIS.
1880–1884. The fishes of Great Britain and Ireland, 2 vols., 1880–1884, Pls. I–CLXXIX. London and Edinburgh.

EIGENMANN, CARL H.
1902. The eggs and development of the conger eel. Bulletin, United States Fish Commission, Vol. XXI, 1901 (1902), pp. 37–44, 15 figs. Washington.
1902a. Investigations into the history of the young squeteague. *Ibid.*, Vol. XXI, 1901 (1902), pp. 45–51, 9 figs. Washington.

EVERMANN, BARTON WARREN, and SAMUEL FREDERICK HILDEBRAND.
1910. On a collection of fishes from the lower Potomac, the entrance of Chesapeake Bay, and from streams flowing into these waters. Proceedings, Biological Society of Washington, Vol. XXIII, 1910, pp. 157–164. Washington.

EVERMANN, BARTON WARREN, and MILLARD CALEB MARSH.
1902. The fishes of Porto Rico. Bulletin, United States Fish Commission, Vol. XX, Pt. I, 1900 (1902), pp. 51–350, 49 pls., 112 figs., 3 fold. maps. Washington.

FIELD, IRVING A.
  1907. Unutilized fishes and their relation to the fishing industries. Report, United States Commissioner of Fisheries, 1906, 50 pp., 1 pl. Bureau of Fisheries Document No. 622. Washington.
FORBES, STEPHEN ALFRED, and ROBERT EARL RICHARDSON.
  1908. The fishes of Illinois. Natural History Survey of Illinois. Vol. III, Ichthyology, 1908, cxxxi, 357 pp., 55 pls., 76 figs., and Atlas of 103 maps. Danville.
FORD, E.
  1921. A contribution to our knowledge of the life histories of the dogfishes landed at Plymouth. Journal, Marine Biological Association of the United Kingdom, new series, Vol. XII, No. 3, September, 1921, pp. 468–505, 19 figs., Tables A–B. Plymouth, England.
FOWLER, HENRY W.
  1906. The fishes of New Jersey. Annual Report, New Jersey State Museum, Pt. II, 1905 (1906), pp. 35–477, 103 pls., text figs. Trenton.
  1912. Records of fishes for the Middle Atlantic States and Virginia. Proceedings, Academy of Natural Sciences of Philadelphia, third series, Vol. LXIV, 1912–13, pp. 34–59, 2 figs. Philadelphia.
  1918. Fishes from the Middle Atlantic States and Virginia. Occasional Papers of the Museum of Zoology, University of Michigan, No. 56, May 6, 1918, 19 pp., Pls. I–II. Ann Arbor.
  1923. Records of fishes for the Southern States. Proceedings, Biological Society of Washington, vol. 36, March 28, 1923, pp. 7–34. Washington.
GARMAN, SAMUEL.
  1913. The Plagiostomia (sharks, skates and rays). Memoirs Museum of Comparative Zoology, Vol. XXXVI, 1913, xiii, 515 pp., and atlas of 77 pls. Cambridge.
GARSTANG, WALTER.
  1897. On the variation, races, and migrations of the mackerel (*Scomber scomber*). Journal, Marine Biological Association, Vol. V (new series), 1897–1899, pp. 235–295, Tables I–XI, A–H. Plymouth, England.
GEISER, SAMUEL W.
  1923. Notes relative to the species of Gambusia in the United States. The American Midland Naturalist, Vol. VIII, Nos. 8–9, March–May, 1923, pp. 175–188, text figs. A–B, 1–18. Notre Dame, Ind.
GILL, THEODORE.
  1861. Synopsis of the Polynematoids. Proceedings, Academy of Natural Sciences of Philadelphia, vol. 13, second series, 1861 (1862), pp. 271–282. Philadelphia.
  1910. The story of the devil-fish. Smithsonian Miscellaneous Collections, Vol. LII, 1910, pp. 155–180, 55 figs. Washington.
GIRARD, CHARLES.
  1860. Ichthyological notices. Proceedings, Academy of Natural Sciences of Philadelphia, vol. 2, second series, 1859 (1860), pp. 157–161. Philadelphia.
GOODE, G. BROWN.
  1879. Copy of letter to Colonel Marshall McDonald. *In* Annual Report, Commissioner of Fisheries of Virginia for 1879, p. 14. Richmond.
GOODE, G. BROWN, and others.
  1884. The food fishes of the United States. *In* The Fisheries and Fishery Industries of the United States, by George Brown Goode and associates, Section I (text), Pt. III, 1884, pp. 163–682, and Section I (plates), 1884, pls. 35–252. Washington.
GOODE, GEORGE BROWN, and TARLETON BEAN.
  1879. A list of fishes of Essex County, including those of Massachusetts Bay, according to the latest results of the work of the United States Fish Commission. Bulletin, Essex Institute, Vol. XI, pp. 1–38.
  1896. Oceanic ichthyology. Memoirs, Museum of Comparative Zoology, Vol. XXII, 1896, pp. 1–553, 1 pl. and 28 figs., and atlas of CXXIII pls. Cambridge. *Also* Special Bulletin No. 2, United States National Museum, 1895, 553 pp., CXXIII pls. Washington.

GUDGER, EUGENE WILLIS.
1905. The breeding habits and the segmentation of the egg of the pipefish, *Siphostoma Floridæ*. Proceedings, United States National Museum, Vol. XXIX, No. 1431, 1906, pp. 447–500, Pls. V–XI, 2 figs. Washington.

GUDGER, E. W.
1907. A note on the hammerhead shark (Sphyrna zygæna) and its food. Science, new series, Vol. XXV, June 28, 1907, pp. 1005–6. New York.
1910. Habits and life history of the toadfish (*Opsanus tau*). Bulletin, United States Bureau of Fisheries, Vol. XXVIII, Pt. II, 1908 (1910), pp. 1073–1109, Pls. CVII–CXIII. Washington.
1912. Natural history notes on some Beaufort, N. C., fishes, 1910–11. No. I. Elasmobranchii, with special reference to utero-gestation. Proceedings, Biological Society of Washington, Vol. XXV, 1912, pp. 141–156. Washington.
1914. History of the spotted eagle ray, *Aëtobatus narinari*, together with a study of its external structures. Papers from the Tortugas Laboratory, Vol. VI, 1914. Carnegie Institution of Washington Publication No. 183, pp. 241–323, Pls. I–X, 19 text figs. Washington.
1918. Oral gestation in the gaff-topsail catfish *Felichthys felis*. Papers from the Department of Marine Biology of the Carnegie Institution of Washington, Vol. XII, 1918, Publication No. 252, pp. 25–52, Pls. I–IV. Washington.
1921. Notes on the morphology and habits of the nurse shark, *Ginglymostoma cirratum*. Copeia, No. 98, 1921, pp. 57–59. Philadelphia.

HESSEL, RUDOLPH.
1878. The carp and its culture in rivers and lakes and its introduction in America. Report, United States Commissioner of Fisheries for 1875–1876, Appendix C, VII, 1878, pp. 865–899, 5 figs. Washington.

HILDEBRAND, SAMUEL F.
1916. The U. S. Fisheries Biological Station at Beaufort, N. C., during 1914 and 1915. Science, new series, Vol. XLIII, No. 1105, 1916, pp. 303–307. New York.
1917. Notes on the life history of the minnows, *Gambusia affinis* and *Cyprinodon variegatus*. Appendix VI, Report, United States Commissoner of Fisheries, 1917 (1919), 15 pp., 4 figs. Bureau of Fisheries Document No. 857. Washington.
1919. Fishes in relation to mosquito control in ponds. United States Public Health Reports, May 23, 1919, pp. 1113–1128, 3 double pls., 3 figs. (Reprint No. 527.) Washington. *Also* Appendix IX, Report, United States Commissioner of Fisheries, 1918 (1920), 16 pp., Pls. I–VI, 3 figs. Bureau of Fisheries Document No. 874. Washington.
1921. Top minnows in relation to malaria control, with notes on their habits and distribution. United States Public Health Service Bulletin No. 114 (1921), 34 pp., 32 figs. Washington.
1922. Notes on habits and development of eggs and larvæ of the silversides, *Menidia menidia;* and *Menidia beryllina*. Bulletin, United States Bureau of Fisheries, Vol. XXXVIII, 1921–1922 (1923), pp. 113–120, figs. 85–98. Bureau of Fisheries Document No. 918. Washington.
1923. Annotated list of fishes collected in vicinity of Augusta, Ga., with description of a new darter. Bulletin, United States Bureau of Fisheries, Vol. XXXIX, 1923–1924 (1924), pp. 1–8, 1 pl. Bureau of Fisheries Document No. 940. Washington.
1925. A study of the top minnow, *Gambusia holbrooki*, in its relation to mosquito control. United States Public Health Service Bulletin No. 153 (1925), 136 pp., 15 figs., 74 tables, LII graphs. Washington.

HOWARD, H. H.
1920. Report on use of top minnow (*Gambusia affinis*) as an agent in mosquito control. International Health Board, Report No. 7486, 1920, 59 pp., 18 figs. New York. [Mimeograph copy.]

HUBBS, CARL L.
　　1922.　A list of the lancelets of the world with diagnoses of five new species of Branchiostoma. Occasional Pápers of the Museum of Zoology, University of Michigan, No. 105, January 2, 1922, 16 pp.　Ann Arbor.

HUYLER, A. J.
　　1876.　Shad as shrimp eaters.　Forest and stream, vol. 6, No. 15, May 18, 1876, p. 233. New York.

JACOT, ARTHUR PAUL.
　　1920.　Age, growth, and scale characters of the mullets, *Mugil cephalus* and *Mugil curema.* Transactions, American Microscopical Society, Vol. XXXIX, No. 3, July, 1920, pp. 199–229, Pls. XX–XXVI, 7 figs.　Menasha, Wis.

JORDAN, DAVID STARR, assisted by BARTON WARREN EVERMANN.
　　1917–1920.　The genera of fishes.　Leland Stanford Junior University Publications, University Series, 1917.　Pt. I (1917), pp. 1–161; Pt. II (1919), pp. i–ix, 162–284, i–xiii; Pt. III (1919), pp. 283–410, i–xv; Pt. IV (1920), pp. 411–576, i–xviii.　Stanford University.
　　1923.　A classification of fishes.　Stanford University Publications, University Series, Biological Sciences, Vol. III, No. 2, 1923, pp. 79–243, x.　Stanford University.

JORDAN, DAVID STARR, and BARTON WARREN EVERMANN.
　　1896–1900.　The fishes of North and Middle America.　Bulletin, United States National Museum, No. 47, Pt. I, 1896, pp. lx, 1–1240; Pt. II, 1898, pp. xxx, 1241–2183; Pt. III, 1898, pp. xxiv, 2183a–3136; Pt. IV, 1900, pp. ci, 3137–3313, Pls. I–CCCXCII.　Washington.

JORDAN, DAVID STARR, and CARL LEAVITT HUBBS.
　　1919.　A monographic review of the family Atherinidæ or silversides.　Leland Stanford Junior University Publications, University Series, 1919, 87 pp., Pls. I–XII.

JORDAN, DAVID STARR, and ALVIN SEALE.
　　1926.　Review of the Engraulidæ, with descriptions of new and rare species.　Bulletin, Museum of Comparative Zoology at Harvard College, Vol. LXVII, No. 11, pp. 355–418.　Cambridge.

KENDALL, W. C.
　　1902.　Notes on the silversides of the genus Menidia of the east coast of the United States, with descriptions of two new subspecies.　Report, United States Commissioner of Fish and Fisheries, 1901 (1902), pp. 241–267, 6 figs.　Washington.

KENDALL, WILLIAM C.
　　1914.　A new record for the lumpfish in Chesapeake Bay.　Copeia, No. 13, December 15, 1914.　New York.

KENDALL, WILLIAM CONVERSE.
　　1917.　The pikes: Their geographical distribution, habits, culture, and commercial importance. Appendix V, Report, United States Commissioner of Fisheries, 1917 (1919), 45 pp., 6 figs.　Bureau of Fisheries Document No. 853.　Washington.

KUNTZ, ALBERT.
　　1914.　The embryology and larval development of *Bairdiella chrysura* and *Anchovia mitchelli.* Bulletin, United States Bureau of Fisheries, Vol. XXXIII, 1913 (1915), pp. 3–19, 46 figs.　Bureau of Fisheries Document No. 795.　Washington.
　　1916.　Notes on the embryology and larval development of five species of teleostean fishes. Bulletin, United States Bureau of Fisheries, Vol. XXXIX, 1914 (1916), pp. 409–429, 68 figs.　Bureau of Fisheries Document No. 831.　Washington.

KUNTZ, ALBERT, and LEWIS RADCLIFFE.
　　1918.　Notes on the embryology and larval development of twelve teleostean fishes.　Bulletin, United States Bureau of Fisheries, Vol. XXXV, 1915–16 (1918), pp. 87–134, 126 figs. Bureau of Fisheries Documents No. 849.　Washington.

LEACH, GLEN C.
　　1925.　Artificial propagation of shad.　Appendix VIII, Report, United States Commissioner of Fisheries, 1924 (1925), pp. 459–486, 8 figs.　Bureau of Fisheries Document No. 981.　Washington.

LEIDY, JOSEPH.
1862. [Remarks on the food of the shad.] Proceedings, Academy of Natural Sciences of Philadelphia, vol. 13, second series, 1861 (1862), p. 2. Philadelphia.
1868. [Remarks on the food of the shad in the sea.] *Ibid.*, second series, 1868 (1868), p. 228. Philadelphia.

LEIM, A. H.
1924. The life history of the shad, *Alosa sapidissima* (Wilson), with special reference to the factors limiting its abundance. Contributions to Canadian Biology, new series, Vol. II, No. 11, 1925, pp. 163–284, 45 figs. Toronto.

LeSueur, C. A.
1817. Description of several species of Chondropterigious fishes of North America, with their varieties. Transactions, American Philosophical Society, new series, Vol. I, 1818, pp. 383–394, Pl. XII.

LINTON, EDWIN.
1905. Parasites of fishes of Beaufort, N. C. Bulletin, United States Bureau of Fisheries, Vol. XXIV, 1904 (1905), pp. 321–428, Pls. I–XXXIV. Washington.

LUGGER, OTTO.
1877. Additions to list of fishes of Maryland, published in report, January 1, 1876. *In* Report, Commissioner of Fisheries of Maryland, 1877, pp. 57–94. Baltimore.
1878. Additions to list of fishes of Maryland, previously published. *See* Reports, January 1, 1876, and January, 1877. *In* Report, Commissioner of Fisheries of Maryland, 1878, pp. 107–125. Baltimore.

MAST, S. O.
1916. Changes in shade, color, and pattern in fishes, and their bearing on the problems of adaptation and behavior, with especial reference to the flounders Paralichthys and Ancylopsetta. Bulletin, United States Bureau of Fisheries, Vol. XXXIV, 1914 (1916), pp. 173–238, Pls. XIX–XXXVII. Bureau of Fisheries Document No. 821. Washington.

McDONALD, MARSHALL.
1879. Annual Report, Commissioner of Fisheries of the State of Virginia, 1879, 23 pp., 3 figs. Richmond.
1880. Our fisheries. *In* Report upon the Fisheries and Oyster Industries of Tidewater Virginia, etc. (1880), pp. 5–18. Richmond.
1882. Annual Report, Commissioner of Fisheries [of Virginia], 1882, 17 pp., 4 figs. Richmond.
1884. The shad and the alewives. *In* The Fisheries and Fishery Industries of the United States, by George Brown Goode and associates, Section I, Pt. III, 1884, pp. 579–609. Washington.

MEEK, SETH E., and SAMUEL F. HILDEBRAND.
1923–1926. The marine fishes of Panama. Field Museum of Natural History, Publication No. 215, Zoological Series, Vol. XV, Pt. I, 1923, pp. III–XI, 1–330, Pls. I–XXIV; Publication No. 226, Vol. XV, Pt. II, 1925, pp. XIII–XIX, 331–707, Pls. XXV–LXXI; Publication No. —, Vol. XV, Pt. III, 1927, pp. 308–1020, Pls. LXXII–CII. Chicago.

MOORE, H. F.
1898. Observations on the herring and herring fisheries of the northeast coast, with special reference to the vicinity of Passamaquoddy Bay. Report, United States Commissioner of Fish and Fisheries, 1896 (1898), pp. 387–442, pl. 61, 1 fold. map. Washington.

MORDECAI, E. R.
1860. Food of the shad of the Atlantic coast of the United States (*Alosa præstabilis* DeKay) and the functions of the pyloric coeca. (1860.) King & Baird, 607 Sansom Street, Philadelphia. [An abstract of the above paper printed in Proceedings, Biologica Department. *In* Proceedings, Academy of Natural Sciences of Philadelphia, 1860, p. 9, Philadelphia. Reprinted in Bulletin, United States Fish Commission, 1881 (1882), Vol. I, pp. 277–282. Washington.]

Moseley, A.
> 1877. Annual Report, Fish Commissioner of the State of Virginia for 1877, 60. pp. Richmond.

Nichols, John Treadwell, and Robert Cushman Murphy.
> 1916. Long Island fauna—IV. The sharks. Brooklyn Museum, Science Bulletin, vol. 3, No. 1, 1916, 34 pp., 3 pls., 19 text figs. Brooklyn.

Peck, James I.
> 1894. On the food of the menhaden. Bulletin, United States Fish Commission, Vol. XIII, 1893 (1894), pp. 113–126, pls. 1–8. Washington.

Perley, M. H.
> 1851. Catalogue (in part) of the fishes of New Brunswick and Nova Scotia. *In* Report upon the fisheries of the Bay of Fundy, pp. 118–159. Fredericton.

Prime, W. C.
> 1876. Fly fishing for shad. Forest and Stream, Vol. VI, 1876, pp. 138–139.

Radcliffe, Lewis.
> 1916. The sharks and rays of Beaufort, N. C. Bulletin, United States Bureau of Fisheries, Vol. XXXIV, 1914 (1916), pp. 239–284, 26 text figs., Pls. XXXVIII–XLIX. Bureau of Fisheries Document No. 822. Washington.

Regan, C. Tate.
> 1910. The origin and evolution of the teleostean fishes of the order Heterosomata. The Annals and Magazine of Natural History, series 8, Vol. VI, 1910, pp. 484–496, 3 figs. London.

Rice, H. J.
> 1878. Notes on the development of the shad, *Alosa sapidissima* (?). *In* Report, Commissioner of Fisheries of Maryland, 1878, pp. 95–106. Baltimore.
> 1878. The *Amphioxus lanceolatus* off Old Point Comfort. Forest and Stream, vol. 10, No. 26, August 1, 1878, p. 503. New York.

Rice, Henry J.
> 1880. Observations upon the habits, structure, and development of *Amphioxus lanceolatus*. The American Naturalist, Vol. XIV, No. 1, 1880, pp. 1–19, Pls. I–II. Philadelphia.
> 1884. Experiments upon retarding the development of eggs of the shad, made in 1879, at the United States shad-hatching station at Havre de Grace, Md. Appendix B, Report, United States Fish Commission for 1881 (1884), pp. 787–794, 3 figs. Washington.

Ryder, John A.
> 1882. A contribution to the development and morphology of the lophobranchiates; (*Hippocampus antiquorum*, the sea horse). Bulletin, United States Fish Commission, Vol. I, 1881 (1882), pp. 191–199, Pl. XVII. Washington.
> 1884. On the retardation of the development of the ova of the shad (*Alosa sapidissima*), with observations on the egg, fungus and bacteria. Appendix B, Report, United States Fish Commission for 1881 (1884), pp. 795–811. Washington.
> 1887. On the development of osseous fishes, including marine and fresh-water forms. Appendix D, Report, United States Fish Commissioner for 1885 (1887), pp. 489–604, Pls. I–XXX, 2 figs. Washington.
> 1890. The sturgeons and sturgeon industries of the eastern coast of the United States, with an account of experiments bearing upon sturgeon culture. Bulletin, United States Fish Commission, Vol. VIII, 1888 (1890), pp. 231–328, Pls. XXXVII–LIX. Washington.

Schmidt, Johs.
> 1912. Ueber die Fortpflanzung des Aals und seine Laichplätze, eine zusammenfassende Uebersicht. Der Fischerbote, Jahrgang IV, Nr. 7, Juli, 1912, pp. 201–209, 2 figs. Hamburg.

Schroeder, William C.
> 1924. Fisheries of Key West and the clam industry of southern Florida. Appendix XII, Report, United States Commissioner of Fisheries, 1923 (1924), 74 pp., 29 figs. Bureau of Fisheries Document No. 962. Washington.

SMITH, HUGH M.
1892. Notes on a collection of fishes from the lower Potomac River, Maryland. Bulletin, United States Fish Commission, Vol. X, 1890 (1892), pp. 63–72, Pls. XVIII–XX. Washington.
1896. A review of the history and results of the attempts to acclimatize fish and other water animals in the Pacific States. Bulletin, United States Fish Commission, 1895 (1896), Vol. XV. [The shad, pp. 404–427, pl. 76.] Washington.
1907. The Fishes of North Carolina. North Carolina Geological and Economic Survey, Vol. II, 1907, xi, 453 pp., 21 pls., 187 figs. Raleigh.
SMITH, HUGH M., and BARTON A. BEAN.
1899. List of fishes known to inhabit the waters of the District of Columbia and vicinity. Bulletin, United States Fish Commission, Vol. XVIII, 1898 (1899), pp. 179–187. Washington.
SMITH, HUGH M., and WILLIAM C. KENDALL.
1898. Notes on the extension of the recorded range of certain fishes of the United States coast. Report, United States Commission of Fish and Fisheries, 1896 (1898), pp. 169–176. Washington.
SMITT, F. A.
1892–1895. A History of Scandinavian Fishes, by B. Fries, C. U. Ekström, and C. Sundevall. Second edition, revised and completed by F. A. Smitt. Pts. I and II, 1,240 pp., 380 figs., and vol. of LIII Pls. Stockholm.
SNYDER, J. P.
1917. Report of fisheries and fish-cultural conditions on the Eastern Shore of Maryland, 1917. 31 pp. Baltimore.
1919. Report of fish hatcheries, 1919. In Fourth annual report of the Conservation Commission of Maryland, 1919, pp. 50–66, 2 figs. Baltimore.
STEVENSON, CHARLES R.
1899. The shad fisheries of the Atlantic coast of the United States. Report, Commissioner of Fish and Fisheries, Vol. XXIV, 1898 (1899), pp. 101–269. Washington.
SUDLER, P. R., and RICHARD T. BROWNING.
1893. Eels. Report, Commissioners of Fisheries of Maryland for 1892 and 1893 (1894), p. 27. Annapolis.
SUMNER, FRANCIS B., RAYMOND C. OSBORN, and LEON J. COLE.
1913. A biological survey of the waters of Woods Hole and vicinity. Bulletin, United States Bureau of Fisheries, Vol. XXXI, Pt. II, 1911 (1913), pp. 549–794. Washington.
SURFACE, H. A.
1898. The lampreys of central New York. Bulletin, United States Fish Commission, Vol. XVII, 1897 (1898), pp. 209–215, pls. 10–11. Washington.
TRACY, HENRY C.
1910. Annotated list of fishes known to inhabit the waters of Rhode Island. Fortieth annual report, Commissioner of Inland Fisheries, Rhode Island, 1910, pp. 35–176. Providence.
UHLER, P. R., and OTTO LUGGER.
1876. List of fish of Maryland. In Report, Commissioners of Fisheries of Maryland, January 1, 1876 (1876) [first edition], pp. 81–208. Also Report, Commissioners of Fisheries of Maryland, January, 1876 (1876) [second edition, slightly revised], pp. 67–176. Annapolis.
WELSH, W. W., and C. M. BREDER, JR.
1922. A contribution to the life history of the puffer, Spheroides maculatus (Schneider) Zoologica, Vol. II, No. 12, 1922, pp. 261–276, figs. 80–96. New York.
1923. Contributions to the life histories of Sciænidæ of the eastern United States coast. Bulletin, United States Bureau of Fisheries, Vol. XXXIX, 1923–24 (1924), pp. 141–201, 60 figs. Bureau of Fisheries Document No. 945. Washington.

WILLEY, A.
  1923. Notes on the distribution of free-living Copepoda in Canadian waters. Contributions
         to Canadian biology, being studies from the Biological Station of Canada, new series,
         Vol. I, No. 16, 1923, pp. 305–334, 23 figs. Toronto. [Food of shad, pp. 313–320.]
WILLIAMS, STEPHEN R.
  1902. Changes accompanying the migration of the eye and observations on the *Tractus
         opticus* and *Tectum opticum* in *Pseudopleuronectes americanus*. Bulletin, Museum of
         Comparative Zoology, Vol. XL (1902–3), No. 1, 1902, pp. 1–57, 5 pls., figs. A–F.
         Cambridge.
WILSON, HENRY V.
  1891. The embryology of the sea bass (*Serranus atrarius*). Bulletin, United States Fish
         Commission, Vol. IX, 1889 (1891), pp. 209–277, Pls. LXXXIII–CVII, 12 figs.
         Washington.

GENERAL INDEX

to

FISHES OF CHESAPEAKE BAY

By

Samuel F. Hildebrand and William C. Schroeder

*

BUREAU OF FISHERIES DOCUMENT NO. 1024

Part I, Vol. XLIII, 1927, BULLETIN

U. S. BUREAU OF FISHERIES

UNITED STATES

GOVERNMENT PRINTING OFFICE

WASHINGTON

1928

367

www.ingramcontent.com/pod-product-compliance
Lightning Source LLC
Chambersburg PA
CBHW080759300326
41914CB00055B/953